Modern
Business Math

The McGraw Hill Series in Operations and Decision Sciences

Modern Business Math

1st edition

JEFFREY SLATER
North Shore Community College
Danvers, Massachusetts

SHARON M. WITTRY
Pikes Peak Community College
Colorado Springs, Colorado

MODERN BUSINESS MATH, FIRST EDITION

Published by McGraw Hill LLC, 1325 Avenue of the Americas, New York, NY 10019. Copyright © 2024 by McGraw Hill LLC. All rights reserved. Printed in the United States of America. 1st edition © 2024. No part of this publication may be reproduced or distributed in any form or by any means, or stored in a database or retrieval system, without the prior written consent of McGraw Hill LLC, including, but not limited to, in any network or other electronic storage or transmission, or broadcast for distance learning.

Some ancillaries, including electronic and print components, may not be available to customers outside the United States.

This book is printed on acid-free paper.

1 2 3 4 5 6 7 8 9 LWI 27 26 25 24 23

ISBN 978-1-26-629947-6
MHID 1-26-629947-5

Cover Image: PopTika/Shutterstock

All credits appearing on page or at the end of the book are considered to be an extension of the copyright page.

Library of Congress Cataloging-in-Publication Data

Cataloging-in-Publication Data has been requested from the Library of Congress

Acknowledgments

Academic Experts, Contributors

Sarah Alamilla

Marie Bok

Sheila Boysen

Derrick Cameron

Susan Courtney

Amy van de Graff

Joe Hanson

Kathy Johnson

Stephanie Klie

Cassie Koefod

Patty Kolarik

Michelle Laumb

Joseph Reihing

Tracy Smith

Ron Trucks

Company/Applications

Chapter 1

T-Mobile, Walmart—*Introduction*

Walt Disney—*Introduction; Multiplying and dividing whole numbers; Reading, writing and rounding whole numbers*

Chapter 2

Amazon—*Introduction; Types of fractions and conversion procedures*

M&Ms/Mars—*Fractions and multiplication*

Chapter 3

Lyft—*Introduction*

Netflix, Hulu—*Adding, subtracting, multiplying and dividing decimals*

Toyota, Sears—*Multiplication and division shortcuts for decimals*

Chapter 4

Ipswich Bank—*Checking account*

Chapter 5

Amazon—*Introduction*

Dunkin' Donuts—*Solving word problems for the unknown*

Chapter 6

Clorox—*Introduction*

Hershey—*Application of percents-portion formula*

Hasbro, PepsiCo—*Rounding percents*

Proctor & Gamble—*Calculating percent increases and decreases*

Chapter 7

UPS, Wal-Mart, Amazon—*Introduction*

Michael's—*Discounts*

Chapter 8

Gap—*Introduction*

Lululemon—*Markdowns and perishables*

Chapter 9

Hilton, Facebook—*Introduction*

IRS—*Computing payroll deductions*

Chapter 10

Consumer Federation of America—*Personal Finance: A Kiplinger Approach*

Chapter 11

JPMorgan, Wells Fargo—*Introduction*

The Gap—*Discounting an Interest-Bearing Note before Maturity*

Chapter 12

T. Rowe Price—*Personal Finance: A Kiplinger Approach*

Chapter 13

Boston Globe—*Introduction*

Fidelity—*Personal Finance: A Kiplinger Approach*

Chapter 14

Carvana—*Introduction*

Ford—*Amount financed, finance charge, and deferred payment*

Edmunds—*Truth in lending: APR defined and calculated*

Citibank—*Calculate finance charge on previous month's balance*

Chapter 15

Federal Reserve—*Introduction*

Chapter 16

Kraft Heinz—*Introduction*

Marriott, Macy's, Delta Airlines—*Ratio analysis*

Chapter 17

General Motors—*Introduction*

Chapter 18

Channel Capital Advisor—*Introduction*

Chapter 19

BDO U.S.A.—*Introduction*

Chapter 20

Zebra Insurance—*Personal Finance: A Kiplinger Approach*

Chapter 21

Tesla—*Introduction*

Hershey—*How to read stock quotations*

Franklin Templeton, Fidelity Investments—*How to read a mutual fund quotation*

GameStop—*End of chapter*

Chapter 22

National Small Business Association—*Introduction*

McKinsey—*Personal Finance: A Kiplinger Approach*

Contents

Modern
Business Math

Chapter 1

Whole Numbers: How to Dissect and Solve Word Problems

Learning Unit Objectives

LU 1–1: Reading, Writing, and Rounding Whole Numbers

1. Use place values to read and write numeric and verbal whole numbers.
2. Round whole numbers to the indicated position.
3. Use blueprint aid for dissecting and solving a word problem.

LU 1–2: Adding and Subtracting Whole Numbers

1. Add whole numbers; check and estimate addition computations.
2. Subtract whole numbers; check and estimate subtraction computations.

LU 1–3: Multiplying and Dividing Whole Numbers

1. Multiply whole numbers; check and estimate multiplication computations.
2. Divide whole numbers; check and estimate division computations.

Corporations outline pandemic's impact, from pay increases to cleaning supplies

By Inti Pacheco

A food distributer has paid $20 million for testing and plexiglass. **T-Mobile US** Inc. has spent $50 million on extra cleaning and safety gear. **Walmart** Inc. and three other big retail chains have put more than $3 billion into higher salaries, benefits and other Covid-19 measures.

Staying open during the pandemic wasn't cheap. Big companies say they spent anywhere from hundreds of thousands to almost a billion dollars in Covid-19-related costs. Some say they expect the costs to keep rising in coming quarters, even as they face uncertain demand from consumers.

Pacheco, Inti. "Staying Open." *The Wall Street Journal* (June 24, 2020).

℮ Essential Question

How can I use whole numbers to understand the business world and make my life easier?

🌐 Math Around the World

The Wall Street Journal chapter opener discusses how expensive Covid-19 has been for retailers. WalMart and three other large retail chains have spent $3 billion so far because of it. The *Wall Street Journal* clip below shows how the pandemic has affected Disney.

People of all ages make personal business decisions based on the answers to number questions. Numbers also determine most of the business decisions of companies. For example, go to the website of a company such as Disney and note the importance of numbers in the company's business decision-making process.

Disney has to use numbers to see

1. The effect of closing parks.
2. Profits and losses.
3. The expenditures necessary for new-product development.
4. Ways to improve customer satisfaction.

Your study of numbers begins with a review of basic computation skills that focuses on speed and accuracy. You may think, "But I can use my calculator." Even if your instructor allows you to use a calculator, you still must know the basic computation skills. You need these skills to know what to calculate, how to interpret your calculations, how to make estimates to recognize errors you made in using your calculator, and how to make calculations when you do not have a calculator.

The United States' numbering system is the **decimal system** or *base 10 system.* Your calculator gives the 10 single-digit numbers of the decimal system—0, 1, 2, 3, 4, 5, 6, 7, 8, and 9. The center of the decimal system is the **decimal point.** When you have a number with a decimal point, the numbers to the left of the decimal point are **whole numbers** and the numbers to the right of the decimal point are decimal numbers (discussed in Chapter 3). When you have a number *without* a decimal, the number is a whole number and the decimal is assumed to be after the number.

This chapter discusses reading, writing, and rounding whole numbers; adding and subtracting whole numbers; and multiplying and dividing whole numbers.

By R.T. Watson

Walt Disney Co. posted its first quarterly loss since 2001–nearly $5 billion–as the majority of its business segments reeled from global efforts to curb the spread of the corona-virus by shutting down public spaces around the world.

Watson, R.T. "Disney Posts First Loss Since 2001." The Wall Street Journal (August 05, 2020).

Learning Unit 1–1:
Reading, Writing, and Rounding Whole Numbers

Let's begin our study of whole numbers.

Learn: Reading and Writing Numeric and Verbal Whole Numbers

The decimal system is a *place-value* system based on the powers of 10. Any whole number can be written with the 10 digits of the decimal system because the position, or placement, of the digits in a number gives the value of the digits.

To determine the value of each digit in a number, we use a place-value chart (Figure 1.1) that divides numbers into named groups of three digits, with each group separated by a comma. To separate a number into groups, you begin with the last digit in the number and insert commas every three digits, moving from right to left. This divides the number into the named groups (units, thousands, millions, billions, trillions) shown in the place-value chart. Within each group, you have a ones, tens, and hundreds place. Keep in mind that the leftmost group may have fewer than three digits.

In Figure 1.1, the numeric number 1,605,743,891,412 illustrates place values. When you study the place-value chart, you can see that the value of each place in the chart is 10 times the value of the place to the right. We can illustrate this by analyzing the last four digits in the number 1,605,743,891,412:

$$1,412 = (1 \times 1,000) + (4 \times 100) + (1 \times 10) + (2 \times 1)$$

So we can also say, for example, that in the number 745, the "7" means seven hundred (700); in the number 75, the "7" means 7 tens (70), and the "5" means 5 ones (5).

To read and write a numeric number in verbal form, you begin at the left and read each group of three digits as if it were alone, adding the group name at the end (except the last units group and groups of all zeros). Using the place-value chart in Figure 1.1, the number 1,605,743,891,412 is read as one trillion, six hundred five billion, seven hundred forty-three million, eight hundred ninety-one thousand, four hundred twelve. You do not read zeros. They fill vacant spaces as placeholders so that you can correctly state the number values. Also, the numbers twenty-one to ninety-nine must have a hyphen. And most important, when you read or write whole numbers in verbal form, do not use the word *and*. In the decimal system, and indicates the decimal, which we discuss in Chapter 3.

FIGURE 1.1 Whole number place-value chart

Whole Number Groups

Trillions				Billions				Millions				Thousands				Units			
Hundred trillions	Ten trillions	Trillions	Comma	Hundred billions	Ten billions	Billions	Comma	Hundred millions	Ten millions	Millions	Comma	Hundred thousands	Ten thousands	Thousands	Comma	Hundreds	Tens	Ones (units)	Decimal Point
	1	,	6	0	5	,	7	4	3	,	8	9	1	,	4	1	2	.	

By reversing this process of changing a numeric number to a verbal number, you can use the place-value chart to change a verbal number to a numeric number. Remember that you must keep track of the place value of each digit. The place values of the digits in a number determine its total value.

Before we look at how to round whole numbers, we should look at how to convert a number indicating parts of a whole number to a whole number. We will use the *Wall Street Journal* clip "'Avengers' Posts Record $1.2 Billion Opening" as an example. This amount is 1 billion plus 200 million of an additional billion. The following steps explain how to convert decimal numbers into whole numbers.

Converting Parts of a Million, Billion, Trillion, Etc., to a Regular Whole Number

Step 1 Drop the decimal point and insert a comma.

Step 2 Add zeros so the leftmost digit ends in the word name of the amount you want to convert. Be sure to add commas as needed.

Example: Convert 1.2 million to a regular whole number.

Step 1   1.2 million

1,2   Change the decimal point to a comma.

Step 2   1,200,000   Add zeros and commas so the whole number indicates million.

Learn: Rounding Whole Numbers

Many of the whole numbers you read and hear are rounded numbers. Government statistics are usually rounded numbers. The financial reports of companies also use rounded numbers. All rounded numbers are *approximate* numbers. The more rounding you do, the more you approximate the number.

Rounded whole numbers are used for many reasons. With rounded whole numbers you can quickly estimate arithmetic results, check actual computations, report numbers that change quickly such as population numbers, and make numbers easier to read and remember.

Numbers can be rounded to any identified digit place value, including the first digit of a number (rounding all the way). To round whole numbers, use the following three steps:

Rounding Whole Numbers

Step 1 Identify the place value of the digit you want to round.

Step 2 If the digit to the right of the identified digit in Step 1 is 5 or more, increase the identified digit by 1 (round up). If the digit to the right is less than 5, do not change the identified digit.

Step 3 Change all digits to the right of the rounded identified digit to zeros.

'Avengers' Posts Record $1.2 Billion Opening

By Erich Schwartzel

LOS ANGELES–Walt Disney Co.'s superhero epic "Avengers: Endgame" became the first movie to gross more than $1 billion in its debut at the world-wide box office.

The Marvel Studios block-buster, powered by record-setting hauls in the U.S. and China, collected an estimated $1.2 billion in its first five days of release. An estimated $350 million of that total came from the U.S. and Canada, an amount that blew past the previous opening-weekend record.

Schwartzel, Erich. "'Avengers' Posts Record $1.2 Billion Opening." *The Wall Street Journal* (April 29, 2020).

Cartoon Collections

Example 1: Round 9,362 to the nearest hundred.

Step 1 9,362   The digit 3 is in the hundreds place value.

Step 2   The digit to the right of 3 is 5 or more (6). Thus, 3, the identified digit in Step 1, is now rounded to 4. You change the identified digit only if the digit to the right is 5 or more.

9,462

Step 3   9,400   Change digits 6 and 2 to zeros, since these digits are to the right of 4, the rounded number.

By rounding 9,362 to the nearest hundred, you can see that 9,362 is closer to 9,400 than to 9,300.

Next, we show you how to round to the nearest thousand.

Example 2: Round 67,951 to the nearest thousand.

Step 1 67,951   The digit 7 is in the thousands place value.

Step 2   The digit to the right of 7 is 5 or more (9). Thus, 7, the identified digit in Step 1, is now rounded to 8.

68,951

Step 3   68,000   Change digits 9, 5, and 1 to zeros, since these digits are to the right of 8, the rounded number.

By rounding 67,951 to the nearest thousand, you can see that 67,951 is closer to 68,000 than to 67,000.

Now let's look at **rounding all the way**. To round a number all the way, you round to the first digit of the number (the leftmost digit) and have only one nonzero digit remaining in the number.

Example 3: Round 7,843 all the way.

Step 1 7,843   Identified leftmost digit is 7.

Step 2   Digit to the right of 7 is greater than 5, so 7 becomes 8.

8,843

Step 3   8,000   Change all other digits to zeros.

Rounding 7,843 all the way gives 8,000.

Remember that rounding a digit to a specific place value depends on the degree of accuracy you want in your estimate. For example, in the *Wall Street Journal* article "'Avengers' Posts Record $1.2 Billion Opening," 1.2 billion rounded all the way would be 1 billion. Note the digit to the right of the identified digit is less than 5 so the identified digit (1) is kept at 1.

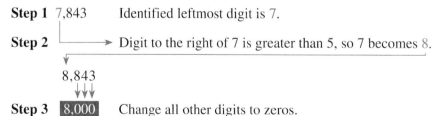

Learn: How to Dissect and Solve a Word Problem

As a student, your author found solving word problems difficult. Not knowing where to begin after reading the word problem caused the difficulty. Today, students still struggle with word problems as they try to decide where to begin.

Solving word problems involves *organization* and *persistence*. Recall how persistent you were when you learned to ride a two-wheel bike. Do you remember the feeling of success you experienced when you rode the bike without help? Apply this persistence to word problems. Do not be discouraged. Each person learns at a different speed. Your goal must be to FINISH THE RACE and experience the success of solving word problems with ease.

To be organized in solving word problems, you need a plan of action that tells you where to begin—a blueprint aid. Like a builder, you will refer to this blueprint aid constantly until you know the procedure. The blueprint aid for dissecting and solving a word problem appears below. Note that the blueprint aid serves an important function—**it decreases your math anxiety.**

Aha!

Remember to RTDQ2: Read the darn question and then read it again before trying to solve it.

Blueprint Aid for Dissecting and Solving a Word Problem

	The facts	Solving for?	Steps to take	Key points
Blueprint				

Now let's study this blueprint aid. The first two columns require that you *read* the word problem slowly. Think of the third column as the basic information you must know or calculate before solving the word problem. Often this column contains formulas that provide the foundation for the step-by-step problem solution. The last column reinforces the key points you should remember.

It's time now to try your skill at using the blueprint aid for dissecting and solving a word problem.

The Word Problem On the 100th anniversary of Tootsie Roll Industries, the company reported sharply increased sales and profits. Sales reached one hundred ninety-four million dollars and a record profit of twenty-two million, five hundred fifty-six thousand dollars. The company president requested that you round the sales and profit figures all the way.

Study the following blueprint aid and note how we filled in the columns with the information in the word problem. You will find the organization of the blueprint aid most helpful. Be persistent! You *can* dissect and solve word problems! When you are finished with the word problem, make sure the answer seems reasonable.

	The facts	Solving for?	Steps to take	Key points
Blueprint	*Sales:* One hundred ninety-four million dollars. *Profit:* Twenty-two million, five hundred fifty-six thousand dollars.	Sales and profit rounded all the way.	Express each verbal form in numeric form. Identify leftmost digit in each number.	Rounding all the way means only the leftmost digit will remain. All other digits become zeros.

Steps to solving problem

1. Convert verbal to numeric.

One hundred ninety-four million dollars ⟶ $194,000,000
Twenty-two million, five hundred fifty-six thousand dollars ⟶ $ 22,556,000

2. Identify leftmost digit of each number.

$194,000,000 $22,556,000

3. Round.

$200,000,000 $200,000,000

Note that in the final answer, $200,000,000 and $20,000,000 have only one nonzero digit.

Remember that you cannot round numbers expressed in verbal form. You must convert these numbers to numeric form.

Now you should see the importance of the information in the third column of the blueprint aid. When you complete your blueprint aids for word problems, do not be concerned if the order of the information in your boxes does not follow the order given in the text boxes. Often you can dissect a word problem in more than one way.

Your first Practice Quiz follows. Be sure to study the paragraph that introduces the Practice Quiz.

Practice Quiz

Complete this Practice Quiz to see how you are doing.

At the end of each learning unit, you can check your progress with a Practice Quiz. If you had difficulty understanding the unit, the Practice Quiz will help identify your area of weakness. Work the problems on scrap paper. Check your answers with the worked-out solutions that follow the quiz. Ask your instructor about specific assignments and the videos available in Connect for each unit Practice Quiz.

1. Write in verbal form:

   a. 7,948   **b.** 48,775   **c.** 814,410,335,414

2. Round the following numbers as indicated:

Nearest ten	Nearest hundred	Nearest thousand	Rounded all the way
a. 92	**b.** 745	**c.** 8,341	**d.** 4,752

3. Kellogg's reported its sales as five million, one hundred eighty-one thousand dollars. The company earned a profit of five hundred two thousand dollars. What would the sales and profit be if each number were rounded all the way? (*Hint:* You might want to draw the blueprint aid since we show it in the solution.)

✓ Solutions

1. **a.** Seven thousand, nine hundred forty-eight

   b. Forty-eight thousand, seven hundred seventy-five

   c. Eight hundred fourteen billion, four hundred ten million, three hundred thirty-five thousand, four hundred fourteen

2. **a.** 90   **b.** 700   **c.** 8,000   **d.** 5,000

3. Kellogg's sales and profit:

	The facts	Solving for?	Steps to take	Key points
Blueprint	*Sales:* Five million, one hundred eighty-one thousand dollars. *Profit:* Five hundred two thousand dollars.	Sales and profit rounded all the way.	Express each verbal form in numeric form. Identify leftmost digit in each number.	Express each verbal form in numeric form. Identify leftmost digit in each number.

Steps to solving problem

1. Convert verbal to numeric.

   Five million, one hundred eighty-one thousand ⟶ $5,181,000

   Five hundred two thousand ⟶ $ 502,000

2. Identify leftmost digit of each number.

   $5,181,000 $502,000

3. Round. $5,000,000 $500,000

Learning Unit 1–2:
Adding and Subtracting Whole Numbers

In the *Wall Street Journal* clip "'Avengers' Posts Record $1.2 Billion Opening" on the following page reprinted from Learning Unit 1–1, note 'Avengers' was the first movie to gross over $1 billion in its debut. The amount of gross sales outside the U.S. and Canada was:

5-day sales	$1,200,000,000
U.S. and Canada	−350,000,000
	$850,000,000

This unit teaches you how to manually add and subtract whole numbers. When you least expect it, you will catch yourself automatically using this skill.

Learn: Addition of Whole Numbers

To add whole numbers, you unite two or more numbers called **addends** to make one number called a **sum,** *total,* or *amount.* The numbers are arranged in a column according to their place values—units above units, tens above tens, and so on. Then, you add the columns of numbers from top to bottom. To check the result, you re-add the columns from bottom to top. This procedure is illustrated in the steps that follow.

'Avengers' Posts Record $1.2 Billion Opening

BY ERICH SCHWARTZEL

LOS ANGELES–**Walt Disney** Co.'s superhero epic "Avengers: Endgame" became the first movie to gross more than $1 billion in its debut at the world-wide box office.
The Marvel Studios block-buster, powered by record-setting hauls in the U.S. and China, collected an estimated $1.2 billion in its first five days of release. An estimated $350 million of that total came from the U.S. and Canada, an amount that blew past the previous opening-weekend record.

Schwartzel, Erich. "'Avengers' Posts Record $1.2 Billion Opening." *The Wall Street Journal* (April 29, 2020).

Adding Whole Numbers

Step 1 Align the numbers to be added in columns according to their place values, beginning with the units place at the right and moving to the left.

Step 2 Add the units column. Write the sum below the column. If the sum is more than 9, write the units digit and carry the tens digit.

Step 3 Moving to the left, repeat Step 2 until all place values are added.

Example:

Adding top to bottom	$\overset{2\ 11}{1,362}$ 5,913 8,924 + 6,594 **22,793**	Checking bottom to top	**Alternate check** Add each column as a separate total and then combine. The end result is the same.

$$\begin{array}{r} 1,362 \\ 5,913 \\ 8,924 \\ +\ 6,594 \\ \hline 13 \\ 18 \\ 26 \\ 20 \\ \hline \textbf{22,793} \end{array}$$

How to Quickly Estimate Addition by Rounding All the Way   In Learning Unit 1–1, you learned that rounding whole numbers all the way gives quick arithmetic estimates. Using the *Wall Street Journal* clip "Coronavirus Daily Update" shown on the left, note how you can round each number all the way and the total will

Rounded all the way

1,000,000 ←	Rounding all the way
4,000,000	means each number has
80,000	only one nonzero digit.
300,000	
200,000	*Note:* The final answer
+2,000,000	could have more than
7,580,000	one nonzero digit since
	the total is not rounded
	all the way.

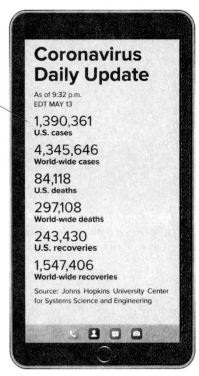

Coronavirus Daily Update

As of 9:32 p.m.
EDT MAY 13

1,390,361
U.S. cases

4,345,646
World-wide cases

84,118
U.S. deaths

297,108
World-wide deaths

243,430
U.S. recoveries

1,547,406
World-wide recoveries

Source: Johns Hopkins University Center for Systems Science and Engineering

"Coronavirus Daily Update."
The Wall Street Journal
(May 14, 2020).

At time of writing, deaths from covid have reached nearly 700,000.54% of population has been vacinated by the delta variant has cause a spike in new cases.

not be rounded all the way. Remember that rounding all the way does not replace actual computations, but it is helpful in making quick commonsense decisions.

Learn: Subtraction of Whole Numbers

Subtraction is the opposite of addition. Addition unites numbers; subtraction takes one number away from another number. In subtraction, the top (largest) number is the **minuend.** The number you subtract from the minuend is the **subtrahend,** which gives you the **difference** between the minuend and the subtrahend. The steps for subtracting whole numbers follow.

Subtracting Whole Numbers

Step 1 Align the minuend and subtrahend according to their place values.

Step 2 Begin the subtraction with the units digits. Write the difference below the column. If the units digit in the minuend is smaller than the units digit in the subtrahend, borrow 1 from the tens digit in the minuend. One tens digit is 10 units.

Step 3 Moving to the left, repeat Step 2 until all place values in the subtrahend are subtracted.

Example: The previous *Wall Street Journal* "Coronavirus Daily Update" clip illustrates the subtraction of whole numbers:

What is the difference between worldwide cases and U.S. cases? As shown below you can use subtraction to arrive at the 2,955,285 difference.

$$
\begin{array}{rl}
4,345,646 & \leftarrow \text{Minuend (larger number)} \\
-1,390,361 & \leftarrow \text{Subtrahend} \\
\hline
2,955,285 & \leftarrow \text{Difference}
\end{array}
$$

Check
$$
\begin{array}{r}
2,955,285 \\
+1,390,361 \\
\hline
4,345,646
\end{array}
$$

Checking subtraction requires adding the difference (2,955,285) to the subtrahend (1,390,361) to arrive at the minuend (4,345,646).

Learn: How to Dissect and Solve a Word Problem

Accurate subtraction is important in many business operations. In Chapter 4 we discuss the importance of keeping accurate subtraction in your checkbook balance. Now let's check your progress by dissecting and solving a word problem.

The Word Problem Hershey's produced 25 million Kisses in one day. The same day, the company shipped 4 million to Japan, 3 million to France, and 6 million throughout the United States. At the end of that day, what is the company's total inventory of Kisses? What is the inventory balance if you round the number all the way?

	The facts	Solving for?	Steps to take	Key points
Blueprint	*Produced:* 25 million. *Shipped:* Japan, 4 million; France, 3 million; United States, 6 million.	Total Kisses left in inventory. Inventory balance rounded all the way.	Total Kisses produced − Total Kisses shipped = Total Kisses left in inventory.	Minuend − Subtrahend = Difference. Rounding all the way means rounding to last digit on the left.

Steps to solving problem

1. Calculate the total Kisses shipped.

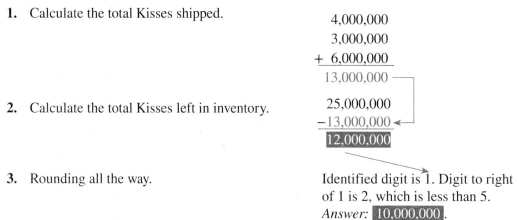

$$4,000,000$$
$$3,000,000$$
$$+\ 6,000,000$$
$$13,000,000$$

2. Calculate the total Kisses left in inventory.

$$25,000,000$$
$$-13,000,000$$
$$12,000,000$$

3. Rounding all the way.

Identified digit is 1. Digit to right of 1 is 2, which is less than 5. *Answer:* 10,000,000.

The Practice Quiz that follows will tell you how you are progressing in your study of Chapter 1.

Practice Quiz

Complete this Practice Quiz to see how you are doing.

1. Add by totaling each separate column:

   8,974
   6,439
   +6,941

2. Estimate by rounding all the way (do not round the total of estimate) and then do the actual computation:

   4,241
   8,794
   + 3,872

3. Subtract and check your answer:

   9,876
   − 4,967

4. Jackson Manufacturing Company projected its year 2022 furniture sales at $900,000. During 2022, Jackson earned $510,000 in sales from major clients and $369,100 in sales from the remainder of its clients. What is the amount by which Jackson over- or underestimated its sales? Use the blueprint aid, since the answer will show the completed blueprint aid.

✓ Solutions

1.　14
　　14
　　2 2
　　20
　　22,354

2.　**Estimate**　**Actual**
　　4,000　　4,241
　　9,000　　8,794
　　+ 4,000　+ 3,872
　　17,000　**16,907**

3.　$\overset{8\ \ 18\,6\,16}{\cancel{9},\cancel{8}\cancel{7}\cancel{6}}$ ←
　　−4,9 6 7
　　4,909

　　Check
　　4,909
　　+ 4,967
　　9,876

4. Jackson Manufacturing Company over- or underestimated sales:

The facts	Solving for?	Steps to take	Key points
Projected 2022 sales: $900,000. *Major clients:* $510,000. *Other clients:* $369,100.	How much were sales over- or underestimated?	Total projected sales − Total actual sales = Over- or underestimated sales.	Projected sales (minuend) − Actual sales (subtrahend) = Difference.

Blueprint

Steps to solving problem

1. Calculate total actual sales.

   $ 510,000
   + 369,100
   $ 879,100

2. Calculate overestimated or underestimated sales.

   $ 900,000
   − 879,100
   $ 20,900 (overestimated)

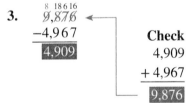

Learning Unit 1–3:
Multiplying and Dividing Whole Numbers

The *Wall Street Journal* clip on the following page reveals that Disney could lose $175 million if their parks remain closed for two more months. The $175 million figure divided by 2 averages an $87,500,000 loss per month.

This unit will sharpen your skills in two important arithmetic operations—multiplication and division. These two operations frequently result in knowledgeable business decisions.

Learn: Multiplication of Whole Numbers—Shortcut to Addition

From calculating the sales for 2 months you know that multiplication is a *shortcut to addition:*

$87,500,000 × 2 = $175,000,000

or

$87,500,000 + $87,500,000 = $175,000,000

Before learning the steps used to multiply whole numbers with two or more digits, you must learn some multiplication terminology.

Note in the following example that the top number (number we want to multiply) is the **multiplicand.** The bottom number (number doing the multiplying) is the **multiplier.** The final number (answer) is the **product.** The numbers between the multiplier and the product are **partial products.** Also note how we positioned the partial product 2090. This number is the result of multiplying 418 by 50 (the 5 is in the tens position). On each line in the partial products, we placed the first digit directly below the digit we used in the multiplication process.

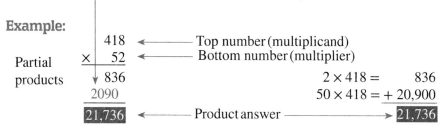

Example:

$$
\begin{array}{r}
418 \quad \longleftarrow \text{Top number (multiplicand)} \\
\times \quad 52 \quad \longleftarrow \text{Bottom number (multiplier)} \\
\hline
836 \\
2090 \\
\hline
21{,}736 \quad \longleftarrow \text{Product answer}
\end{array}
$$

Partial products

$$
\begin{aligned}
2 \times 418 &= 836 \\
50 \times 418 &= + 20{,}900 \\
\hline
&\, 21{,}736
\end{aligned}
$$

We can now give the following steps for multiplying whole numbers with two or more digits:

Multiplying Whole Numbers With Two Or More Digits

Step 1 Align the multiplicand (top number) and multiplier (bottom number) at the right. Usually, you should make the smaller number the multiplier.

Step 2 Begin by multiplying the right digit of the multiplier with the right digit of the multiplicand. Keep multiplying as you move left through the multiplicand. Your first partial product aligns at the right with the multiplicand and multiplier.

Step 3 Move left through the multiplier and continue multiplying the multiplicand. Your partial product right digit or first digit is placed directly below the digit in the multiplier that you used to multiply.

Step 4 Continue Steps 2 and 3 until you have completed your multiplication process. Then add the partial products to get the final product.

Checking and Estimating Multiplication We can check the multiplication process by reversing the multiplicand and multiplier and then multiplying. Let's first estimate 52 × 418 by rounding all the way.

Example:

$$
\begin{array}{rr}
50 \leftarrow & 52 \\
\times\ 400 \leftarrow \times\ 418 & \\
\hline
20{,}000 & 416 \\
& 52 \\
& \underline{20\ 8} \\
& \boxed{21{,}736}
\end{array}
$$

By estimating before actually working the problem, we know our answer should be about 20,000. When we multiply 52 by 418, we get the same answer as when we multiply 418 × 52—and the answer is about 20,000. Remember, if we had not rounded all the way, our estimate would have been closer. If we had used a calculator, the rounded estimate would have helped us check the calculator's answer. Our commonsense estimate tells us our answer is near 20,000—not 200,000.

Before you study the division of whole numbers, you should know (1) the multiplication shortcut with numbers ending in zeros and (2) how to multiply a whole number by a power of 10.

Multiplication Shortcut With Numbers Ending In Zeros

Step 1 When zeros are at the end of the multiplicand or the multiplier, or both, disregard the zeros and multiply.

Step 2 Count the number of zeros in the multiplicand and multiplier.

Step 3 Attach the number of zeros counted in Step 2 to your answer.

Example:

$$
\begin{array}{ccc}
65{,}000 & 65 & 3\ \text{zeros} \\
\times\ 420 & \times\ 42 & +\ 1\ \text{zero} \\
\cline{2-3}
& 1\ 30 & 4\ \text{zeros} \\
& \underline{26\ 0} & \\
& 27{,}300{,}000 &
\end{array}
$$

No need to multiply rows of zeros

$$
\begin{array}{r}
65{,}000 \\
\times\ \quad 420 \\
\hline
00\ 000 \\
1\ 300\ 00 \\
\underline{26\ 000\ 0} \\
\boxed{27{,}300{,}000}
\end{array}
$$

Multiplying A Whole Number By A Power Of 10

Step 1 Count the number of zeros in the power of 10 (a whole number that begins with 1 and ends in one or more zeros such as 10, 100, 1,000, and so on).

Step 2 Attach that number of zeros to the right side of the other whole number to obtain the answer. Insert comma(s) as needed every three digits, moving from right to left.

Example:

99×10	$= 990$	$= 900$	← Add 1 zero
99×100	$= 9,900$	$= 9,900$	← Add 2 zeros
$99 \times 1,000$	$= 99,000$	$= 99,000$	← Add 3 zeros

When a zero is in the center of the multiplier, you can do the following:

Example:

$$\begin{array}{r} 658 \\ \times \quad 403 \\ \hline 1\ 974 \\ 263\ 2 \quad \\ \hline 265,174 \end{array}$$

$$\begin{array}{r} 3 \times 658 = \quad 1,974 \\ 400 \times 658 = +\ 263,200 \\ \hline 265,174 \end{array}$$

Learn: Division of Whole Numbers

Division is the reverse of multiplication and a time-saving shortcut related to subtraction. For example, in the introduction of this learning unit you determined in the Disney example that lost sales for 2 months resulted in $175,000,000. You multiplied $87,500,000 × 2 to get $175,000,000. Since division is the reverse of multiplication you can also say that $175,000,000 ÷ 2 = $87,500,000.

Division can be indicated by the common symbols ÷ and ⌐, or by the bar — in a fraction and the forward slant / between two numbers, which means the first number is divided by the second number. Division asks how many times one number **(divisor)** is contained in another number **(dividend).** The answer, or result, is the **quotient.** When the divisor (number used to divide) doesn't divide evenly into the dividend (number we are dividing), the result is a **partial quotient,** with the leftover amount the **remainder** (expressed as fractions in later chapters). The following example reflecting how much is spent on coffee for 15 weeks illustrates *even division* (this is also an example of *long division* because the divisor has more than one digit).

Example:

$$\begin{array}{r} 18 \leftarrow \text{Quotient} \\ \text{Divisor} \longrightarrow 15\overline{)270} \leftarrow \text{Dividend} \\ \underline{15} \\ 120 \\ \underline{120} \end{array}$$

This example divides 15 into 27 once with 12 remaining. The 0 in the dividend is brought down to 12. Dividing 120 by 15 equals 8 with no remainder; that is, even division. The following example illustrates *uneven division with a remainder* (this is also an example of *short division* because the divisor has only one digit).

Example:

$$\begin{array}{r} 24\ \text{R1} \leftarrow \text{Remainder} \\ 7\overline{)169} \\ \underline{14} \\ 29 \\ \underline{28} \\ 1 \end{array}$$

Check

$$(7 \times 24) + 1 = 169$$

Divisor × Quotient + Remainder = Dividend

Note how doing the check gives you assurance that your calculation is correct. When the divisor has one digit (short division) as in this example, you can often calculate the division mentally as illustrated in the following examples:

Example:

$$\begin{array}{r} 108 \\ 8\overline{)864} \end{array} \qquad \begin{array}{r} 16\ R6 \\ 7\overline{)118} \end{array}$$

Next, let's look at the value of estimating division.

Estimating Division   Before actually working a division problem, estimate the quotient by rounding. This estimate helps you check the answer. The example that follows is rounded all the way. After you make an estimate, work the problem and check your answer by multiplication.

Example:

$$\begin{array}{r} 36\ R111 \\ 138\overline{)5,079} \\ 4\ 14 \\ \hline 939 \\ 828 \\ \hline 1\ 1\ 1 \end{array}$$

Estimate

$$\begin{array}{r} 50 \\ 100\overline{)5,000} \end{array}$$

Check

$$\begin{array}{r} 138 \\ \times\ 36 \\ \hline 828 \\ 4\ 14 \\ \hline 4,968 \\ +\ 111 \quad \longleftarrow \text{Add remainder} \\ \hline 5,079 \end{array}$$

Now let's turn our attention to division shortcuts with zeros.

Division Shortcuts with Zeros   The steps that follow show a shortcut that you can use when you divide numbers with zeros.

Division Shortcut With Numbers Ending In Zeros

Step 1   When the dividend and divisor have ending zeros, count the number of ending zeros in the divisor.

Step 2   Drop the same number of zeros in the dividend as in the divisor, counting from right to left.

Note the following examples of division shortcuts with numbers ending in zeros. Since two of the symbols used for division are ÷ and $\overline{)}\,$, our first examples show the zero shortcut method with the ÷ symbol.

Examples:

One ending zero

Dividend   Divisor

Drop 1 zero in dividend

$95,000 \div 10 \longrightarrow 95,00\underline{0} = \boxed{9,500}$

$95,000 \div 100 \longrightarrow 95,0\underline{00} = \boxed{950}$ Drop 2 zeros

$95,000 \div 1,000 \longrightarrow 95,\underline{000} = \boxed{95}$ Drop 3 zeros

In a long division problem with the $\overline{)}\,$ symbol, you again count the number of ending zeros in the divisor. Then drop the same number of ending zeros in the dividend and divide as usual.

Example:

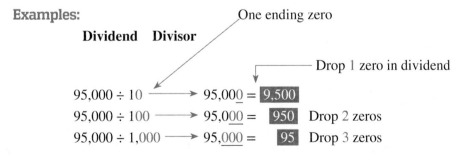

$6,5\underline{00}\overline{)88,0\underline{00}} \longleftarrow$ Drop 2 zeros $\boxed{13\ R35}$

$$\begin{array}{r} 13\ R35 \\ 65\overline{)880} \\ 65 \\ \hline 230 \\ 195 \\ \hline 35 \end{array}$$

$65\overline{)\,880} \longleftarrow$

Learn: How to Dissect and Solve a Word Problem

The blueprint aid presented in LU 1–1(3) will be your guide to dissecting and solving the following word problem.

The Word Problem Dunkin' Donuts sells to four different companies a total of $3,500 worth of doughnuts per week. What is the total annual sales to these companies? What is the yearly sales per company? (Assume each company buys the same amount.) Check your answer to show how multiplication and division are related.

	The facts	Solving for?	Steps to take	Key points
Blueprint	*Sales per week:* $3,500. *Companies:* 4.	Total annual sales to all four companies. Yearly sales per company.	Sales per week × Weeks in year (52) = Total annual sales. Total annual sales ÷ Total companies = Yearly sales per company.	Division is the reverse of multiplication.

Steps to solving problem

1. Calculate total annual sales. $3,500 × 52 weeks = $182,000

2. Calculate yearly sales per company, $182,000 ÷ 4 = $45,500

Check

$45,500 × 4 = $182,000

It's time again to check your progress with a Practice Quiz.

Practice Quiz
Complete this Practice Quiz to see how you are doing.

1. Estimate the actual problem by rounding all the way, work the actual problem, and check:

Actual	**Estimate**	**Check**
3,894		
× 18		

2. Multiply by shortcut method:

   77,000
   × 1,800

3. Multiply by shortcut method:

   $95 \times 10,000$

4. Divide by rounding all the way, complete the actual calculation, and check, showing remainder as a whole number.

   $26\overline{)5,325}$

5. Divide by shortcut method:

   $4,000\overline{)96,000}$

6. Assume General Motors produces 960 Chevrolets each workday (Monday through Friday). If the cost to produce each car is $6,500, what is General Motors' total cost for the year? Check your answer.

✓ Solutions

1. | **Estimate** | **Actual** | **Check** |
   |---|---|---|
   | 4,000 | 3,894 | $8 \times 3,894 =$ 31,152 |
   | ×   20 | ×   18 | $10 \times 3,894 = +$ 38,940 |
   | **80,000** | 31 152 | 70,092 |
   | | 38 94 | |
   | | **70,092** | |

2. $77 \times 18 = 1,386 + 5$ zeros $=$ **138,600,000**   3. $95 + 4$ zeros $=$ **950,000**

4. | **Rounding** | **Actual** | **Check** |
   |---|---|---|
   | **166 R20** | **204 R21** | $26 \times 204 = 5,304$ |
   | $30\overline{)5,000}$ | $26\overline{)5,325}$ | +   21 |
   | 3 0 | 5 2 | 5,325 |
   | 200 | 125 | |
   | 1 80 | 104 | |
   | 200 | 21 | |
   | 180 | | |
   | 20 | | |

5. Drop 3 zeros = **24**
   $4\overline{)96}$

6. General Motors' total cost per year:

	The facts	Solving for?	Steps to take	Key points
Blueprint	*Cars produced each workday:* 960. *Workweek:* 5 days. *Cost per car:* $6,500.	Total cost per year.	Cars produced per week × 52 = Total cars produced per year. Total cars produced per year × Total cost per car = Total cost per year.	Whenever possible, use multiplication and division shortcuts with zeros. Multiplication can be checked by division.

Steps to solving problem

1. Calculate total cars produced per week.

$5 \times 960 = 4,800$ cars produced per week

2. Calculate total cars produced per year.

$4,800$ cars $\times 52$ weeks $= 249,600$ total cars produced per year

3. Calculate total cost per year.

$249,600$ cars $\times \$6,500 = \boxed{\$1,622,400,000}$ (multiply $2,496 \times 65$ and add zeros)

Check

$\$1,622,400,000 \div 249,600 = \$6,500$ (drop 2 zeros before dividing)

Chapter 1 Review

Topic/Procedure/Formula	Example	You try it*
Reading and writing numeric and verbal whole numbers Placement of digits in a number gives the value of the digits (Figure 1.1). Commas separate every three digits, moving from right to left. Begin at left to read and write number in verbal form. Do not read zeros or use *and*. Hyphenate numbers twenty-one to ninety-nine. Reverse procedure to change verbal number to numeric.	462 $\longrightarrow$ Four hundred sixty-two 6,741 $\longrightarrow$ Six thousand, seven hundred forty-one	**Write in verbal form** 571 $\longrightarrow$ 7,943 $\longrightarrow$
Rounding whole numbers 1. Identify place value of the digit to be rounded. 2. If digit to the right is 5 or more, round up; if less than 5, do not change. 3. Change all digits to the right of rounded identified digit to zeros.	643 to nearest ten | 4 in tens place value	3 is not 5 or more | Thus, 643 rounds to 640.	**Round to nearest ten** 691
Rounding all the way Round to first digit of number. One nonzero digit remains. In estimating, you round each number of the problem to one nonzero digit. The final answer is not rounded.	468,451 $\longrightarrow$ 500,000 The 5 is the only nonzero digit remaining.	**Round all the way** 429,685 $\longrightarrow$
Adding whole numbers 1. Align numbers at the right. 2. Add units column. If sum is more than 9, carry tens digit. 3. Moving left, repeat Step 2 until all place values are added. Add from top to bottom. Check by adding bottom to top or adding each column separately and combining.	$\begin{array}{r} 1 \\ 65 \\ +47 \\ \hline 112 \end{array}$ $\begin{array}{r} 12 \\ +10 \\ \hline 112 \end{array}$ Checking sum of each digit	**Add** $\begin{array}{r} 76 \\ +38 \\ \hline \end{array}$

Chapter 1 Review (Continued)

Topic/Procedure/Formula	Example	You try it*
Subtracting whole numbers 1. Align minuend and subtrahend at the right. 2. Subtract units digits. If necessary, borrow 1 from tens digit in minuend. 3. Moving left, repeat Step 2 until all place values are subtracted. Minuend less subtrahend equals difference.	$\overset{5\ 18}{\cancel{6}\cancel{8}5}$ 193 $-\ 492$ $+\ 492$ $\boxed{193}$ 685	**Subtract** 629 -134
Multiplying whole numbers 1. Align multiplicand and multiplier at the right. 2. Begin at the right and keep multiplying as you move to the left. First partial product aligns at the right with multiplicand and multiplier. 3. Move left through multiplier and continue multiplying multiplicand. Partial product right digit or first digit is placed directly below digit in multiplier. 4. Continue Steps 2 and 3 until multiplication is complete. Add partial products to get final product.	223 $\times\ 32$ 446 $6\ 69$ $\boxed{7,136}$	**Multiply** 491 $\times\ 28$

Shortcuts: (a) When multiplicand or multiplier, or both, end in zeros, disregard zeros and multiply; attach same number of zeros to answer. If zero is in center of multiplier, no need to show row of zeros. (b) If multiplying by power of 10, attach same number of zeros to whole number multiplied.	a. 48,000 48 3 zeros 524
	   $\times$ 40 4 +1 zero $\times$ 206
	              $\boxed{1,920,000}$ $\leftarrow$ 4 zeros 3 144
	                                  104 8
	                                 $\boxed{107,944}$
	
	b. 14 $\times$ 10 = $\boxed{140}$ (attach 1 zero)
	    14 $\times$ 1,000 = $\boxed{14,000}$ (attach 3 zeros)
		Multiply by shortcut
		13 $\times$ 10 =
		13 $\times$ 1,000 =

Chapter 1 Review (Continued)

Topic/Procedure/Formula	Example	You try it*
Dividing whole numbers 1. When divisor is divided into the dividend, the remainder is less than divisor. 2. Drop zeros from dividend right to left by number of zeros found in the divisor. Even division has no remainder; uneven division has a remainder; divisor with one digit is short division; and divisor with more than one digit is long division.	1. $\begin{array}{r} \boxed{5\,R6} \\ 14\overline{)76} \\ \underline{70} \\ 6 \end{array}$ 2. $5{,}000 \div 100 = 50 \div 1 = \boxed{50}$ $\;\;\;5{,}000 \div 1{,}000 = 5 \div 1 = \boxed{5}$	**Divide** 1. $16\overline{)95}$ **Divide by shortcut** 2. $4{,}000 \div 100$ $\;\;\;4{,}000 \div 1{,}000$

Key Terms

Addends	Minuend	Quotient
Decimal point	Multiplicand	Remainder
Decimal system	Multiplier	Rounding all the way
Difference	Partial products	Subtrahend
Dividend	Partial quotient	Sum
Divisor	Product	Whole number

* Worked-out solutions are in Appendix A.

Critical Thinking Discussion Questions with Chapter Concept Check

1. List the four steps of the decision-making process. Do you think all companies should be required to follow these steps? Give an example.

2. Explain the three steps used to round whole numbers. Pick a whole number and explain why it should not be rounded.

3. How do you check subtraction? If you were to attend a movie, explain how you might use the subtraction check method.

4. Explain how you can check multiplication. If you visit a local supermarket, how could you show multiplication as a shortcut to addition?

5. Explain how division is the reverse of multiplication. Using the supermarket example in question 4, explain how division is a timesaving shortcut related to subtraction.

6. **Chapter Concept Check.** Using all the math you learned in Chapter 1, compare the number of COVID-19 cases in your state to the entire country.

End-of-Chapter Problems

Name _____   Date_____

Check figures for odd-numbered problems in Appendix A.

Drill Problems

Add the following: *LU 1–2(1)*

1–1.	90 +15	1–2.	900 + 250	1–3.	77 + 77	1–4.	88 + 75

1–5.	6,251 + 7,329	1–6.	59,481 51,411 + 70,821	1–7.	78,159 15,850 + 19,681

Subtract the following: *LU 1–2(2)*

1–8.	68 −19	1–9.	80 −42	1–10.	287 −199

1–11.	9,000 −5,400	1–12.	9,800 −8,900	1–13.	1,622 − 548

Multiply the following: *LU 1–3(1)*

1–14.	50 × 6	1–15.	510 × 61	1–16.	800 × 200

1–17.	677 × 503	1–18.	309 × 850	1–19.	450 × 280

Divide the following by short division: *LU 1–3(2)*

1–20.	4)1,600	1–21.	9)810	1–22.	4)164

Divide the following by long division. Show work and remainder.   *LU 1–3(2)*

1–23.	6)520	1–24.	62)8,915

Add the following without rearranging: *LU 1–2(1)*

1–25. 95 + 310

1–26. 1,055 + 88

1–27. 666 + 950

1–28. 1,011 + 17

1–29. Add the following and check by totaling each column individually without carrying numbers: *LU 1–2(1)*

Check

8,539
6,842
+ 9,495

Estimate the following by rounding all the way and then do actual addition: *LU 1–1(2), LU 1–2(1)*

	Actual	**Estimate**		**Actual**	**Estimate**
1–30.	7,700		**1–31.**	6,980	
	9,286			3,190	
	+ 3,900			+ 7,819	

Subtract the following without rearranging: *LU 1–2(2)*

1–32. 190 − 66

1–33. 950 − 870

1–34. Subtract the following and check answer: *LU 1–2(2)*

591,001
−375,956

Multiply the following horizontally: *LU 1–3(1)*

1–35. 19 × 7

1–36. 84 × 8

1–37. 27 × 8

1–38. 19 × 5 =

Divide the following and check by multiplication: *LU 1–2(2)*

Check

1–39. 45)876   **Check**

1–40. 46)1,950

Complete the following: *LU 1–2(2)*

1–41.	9,200	**1–42.**	3,000,000
	− 1,510		− 769,459
	− 700		− 68,541

End-of-Chapter Problems (Continued)

1–43. Estimate the following problem by rounding all the way and then do the actual multiplication: *LU 1–1(2), LU 1–3(1)*

Actual   **Estimate**
  870
× 81

Divide the following by the shortcut method: *LU 1–3(2)*

1–44. $1{,}000\overline{)950{,}000}$

1–45. $100\overline{)70{,}000}$

1–46. Estimate actual problem by rounding all the way and do actual division:
LU 1–1(2), LU 1–3(2)

Actual   **Estimate**

$695\overline{)8{,}950}$

Word Problems

1–47. *The Wall Street Journal* reported that the cost for lightbulbs over a 10-year period at a local Walmart parking lot in Kansas would be $248,134 if standard lightbulbs were used. If LED lightbulbs were used over the same period, the total cost would be $220,396. What would Walmart save by using LED bulbs?   *LU 1–2(2)*

1–48. An education can be the key to higher earnings. In a U.S. Census Bureau study, high school graduates earned $30,400 per year. Associate's degree graduates averaged $38,200 per year. Bachelor's degree graduates averaged $52,200 per year. Assuming a 50-year work-life, calculate the lifetime earnings for a high school graduate, associate's degree graduate, and bachelor's degree graduate. What's the lifetime income difference between a high school and associate's degree? What about the lifetime difference between a high school and bachelor's degree?   *LU 1–3(1), LU 1–2(2)*

1–49. Assume season-ticket prices in the lower bowl for the Buffalo Bills will rise from $480 for a 10-game package to $600. Fans sitting in the best seats in the upper deck will pay an increase from $440 to $540. Don Manning plans to purchase two season tickets for either lower bowl or upper deck. **(a)** How much more will two tickets cost for lower bowl? **(b)** How much more will two tickets cost for upper deck? **(c)** What will be his total cost for a 10-game package for lower bowl? **(d)** What will be his total cost for a 10-game package for upper deck? *LU 1–2(2), LU 1–3(1)*

1–50. Some ticket prices for *Lion King* on Broadway were $70, $95, $200, and $250. For a family of four, estimate the cost of the $95 tickets by rounding all the way and then do the actual multiplication: *LU 1–1(2), LU 1–3(1)*

1–51. Walt Disney World Resort and United Vacations got together to create a special deal. The air-inclusive package features accommodations for three nights at Disney's All-Star Resort, hotel taxes, and a four-day unlimited Magic Pass. Prices are $609 per person traveling from Washington, DC, and $764 per person traveling from Los Angeles. **(a)** What would be the cost for a family of four leaving from Washington, DC? **(b)** What would be the cost for a family of four leaving from Los Angeles? **(c)** How much more will it cost the family from Los Angeles? *LU 1–3(1)*

1–52. NTB Tires bought 910 tires from its manufacturer for $36 per tire. What is the total cost of NTB's purchase? If the store can sell all the tires at $65 each, what will be the store's gross profit, or the difference between its sales and costs (Sales − Costs = Gross profit)? *LU 1–3(1), LU 1–2(2)*

End-of-Chapter Problems (Continued)

1–53. What was the total average number of visits for these websites?   *LU 1–2(1), LU 1–3(2)*

Website	Average daily unique visitors
1. Orbitz.com	1,527,000
2. Mypoints.com	1,356,000
3. Americangreetings.com	745,000
4. Bizrate.com	503,000
5. Half.com	397,000

1–54. As of mid-September 2021, 229,552,716 worldwide cases of coronavirus were reported by www.worldometers.info. It was also reported 206,248,522 have recovered and 4,709,175 have died. How many cases are unaccounted for to date?

1–55. A report from the Center for Science in the Public Interest—a consumer group based in Washington, DC—released a study listing calories of various ice cream treats sold by six of the largest ice cream companies. The worst treat tested by the group was 1,270 total calories. People need roughly 2,200 to 2,500 calories per day. Using a daily average, how many additional calories should a person consume after eating ice cream?   *LU 1–2(1), LU 1–3(2)*

1–56. At Rose State College, Alison Wells received the following grades in her online accounting class: 90, 65, 85, 80, 75, and 90. Alison's instructor, Professor Clark, said he would drop the lowest grade. What is Alison's average?   *LU 1–2(1)*

1–57. The Bureau of Transportation's list of the 10 most expensive U.S. airports and their average fares is given below. Please use this list to answer the questions that follow.   *LU 1–2(1, 2)*

1. Houston, TX	$477
2. Huntsville, AL	473
3. Newark, NJ	470
4. Cincinnati, OH	466

5. Washington, DC	465	
6. Charleston, SC	460	
7. Memphis, TN	449	
8. Knoxville, TN	449	
9. Dallas–Fort Worth, TX	431	
10. Madison, WI	429	

a. What is the total of all the fares?

b. What would the total be if all the fares were rounded all the way?

c. How much does the actual number differ from the rounded estimate?

1–58. Ron Alf, owner of Alf's Moving Company, bought a new truck. On Ron's first trip, he drove 1,200 miles and used 80 gallons of gas. How many miles per gallon did Ron get from his new truck? On Ron's second trip, he drove 840 miles and used 60 gallons. What is the difference in miles per gallon between Ron's first trip and his second trip? *LU 1–3(2)*

1–59. For the first time in eight years, monthly credit card debt in the United States has dropped an average of 14% despite COVID-19, as reported by Experian. In 2019 the average individual's monthly credit card balance was $6,194. In 2020, this fell to $5,315. How much did the average monthly credit card balance decrease?

1–60. Assume BarnesandNoble.com has 289 business math texts in inventory. During one month, the online bookstore ordered and received 1,855 texts; it also sold 1,222 on the web. What is the bookstore's inventory at the end of the month? If each text costs $59, what is the end-of-month inventory cost? *LU 1–2(1), LU 1–2(2)*

1–61. Assume Cabot Company produced 2,115,000 cans of paint in August. Cabot sold 2,011,000 of these cans. If each can cost $18, what were Cabot's ending inventory of paint cans and its total ending inventory cost? *LU 1–2(2), LU 1–3(1)*

1–62. A local community college has 20 faculty members in the business department, 40 in psychology, 26 in English, and 140 in all other departments. What is the total number of faculty at this college? If each faculty member advises 25 students, how many students attend the local college? *LU 1–2(1), LU 1–3(1)*

End-of-Chapter Problems (Continued)

1–63. Hometown Buffet had 90 customers on Sunday, 70 on Monday, 65 on Tuesday, and a total of 310 on Wednesday to Saturday. How many customers did Hometown Buffet serve during the week? If each customer spends $9, what were the total sales for the week?   *LU 1–2(1), LU 1–3(1)*

If Hometown Buffet had the same sales each week, what were the sales for the year?

1–64. A good credit utilization ratio, measuring your credit card debt divided by your credit card limits, is 30% or less, according to Forbes.com. Surprisingly, the credit utilization ratio fell from 29% in 2019 to 25% in 2020 despite the coronavirus pandemic, as stated by one of the credit agencies, Experian. How many percentage points did the credit utilization ratio fall?

1–65. Ryan Seary works at US Airways and earned $71,000 last year before tax deductions. From Ryan's total earnings, his company subtracted $1,388 for federal income taxes, $4,402 for Social Security, and $1,030 for Medicare taxes. What was Ryan's actual, or net, pay for the year?   *LU 1–2(1, 2)*

1–66. CompareCards.com lists credit card offers by such categories as low interest, no annual fee, cash back, and so on. A top card offers no interest payments for 18 months. If 11 credit card companies make this offer and 25,652 people are approved, on average how many new customers does each credit card company gain?   *LU 1–3(2)*

1–67. Roger Company produces beach balls and operates three shifts. Roger produces 5,000 balls per shift on shifts 1 and 2. On shift 3, the company can produce 6 times as many balls as on shift 1. Assume a 5-day workweek. How many beach balls does Roger produce per week and per year?   *LU 1–2(1), LU 1–3(1)*

1–68. Assume 6,000 children go to Disneyland today. How much additional revenue will Disneyland receive if it raises the cost of admission from $31 to $41?   *LU 1–2(1), LU 1–3(1)*

1–69. Moe Brink has a $900 balance in his checkbook. During the week, Moe wrote the following checks: rent, $350; telephone, $44; food, $160; and entertaining, $60. Moe also made a $1,200 deposit. What is Moe's new checkbook balance? *LU 1–2(1, 2)*

1–70. A local Dick's Sporting Store, an athletic sports shop, bought and sold the following merchandise: *LU 1–2(1, 2)*

	Cost	Selling price
Tennis rackets	$2,900	$3,999
Tennis balls	70	210
Bowling balls	1,050	2,950
Sneakers	+ 8,105	+ 14,888

What was the total cost of the merchandise bought by Dick's Sporting Store? If the shop sold all its merchandise, what were the sales and the resulting gross profit (Sales − Costs = Gross profit)?

excel 1–71. Rich Engel, the bookkeeper for Engel's Real Estate, and his manager are concerned about the company's telephone bills. Last year the company's average monthly phone bill was $32. Rich's manager asked him for an average of this year's phone bills. Rich's records show the following: *LU 1–2(1), LU 1–3(2)*

January	$34	July	$28
February	60	August	23
March	20	September	29
April	25	October	25
May	30	November	22
June	59	December	41

What is the average of this year's phone bills? Did Rich and his manager have a justifiable concern?

excel 1–72. On Monday, a local True Value Hardware sold 15 paint brushes at $3 each, six wrenches at $5 each, seven bags of grass seed at $3 each, four lawn mowers at $119 each, and 28 cans of paint at $8 each. What were True Value's total dollar sales on Monday? *LU 1–2(1), LU 1–3(1)*

End-of-Chapter Problems (Continued)

1-73. While redecorating, Lee Owens went to Carpet World and bought 150 square yards of commercial carpet. The total cost of the carpet was $6,000. How much did Lee pay per square yard? *LU 1–3(2)*

excel 1-74. Washington Construction built 12 ranch houses for $115,000 each. From the sale of these houses, Washington received $1,980,000. How much gross profit (Sales − Costs = Gross profit) did Washington make on the houses? *LU 1–2(2), LU 1–3(1, 2)*

The four partners of Washington Construction split all profits equally. How much will each partner receive?

Challenge Problems

1-75. A mall in Lexington has 18 stores. The following is a breakdown of what each store pays for rent per month. The rent is based on square footage.

5 department/computer stores $1,250		2 bakeries	$ 500
5 restaurants	860	2 drugstores	820
3 bookstores	750	1 supermarket	1,450

Calculate the total rent that these stores pay annually. What would the answer be if it were rounded all the way? How much more each year do the drugstores pay in rent compared to the bakeries? *LU 1–2(2), LU 1–3(1)*

1-76. Paula Sanchez is trying to determine her 2022 finances. Paula's actual 2021 finances were as follows: *LU 1–1, LU 1–2, LU 1–3*

2021			
Income:		Assets:	
Gross income	$69,000	Checking account	$ 1,950
Interest income	450	Savings account	8,950
Total	$69,450	Automobile	1,800
Expenses:		Personal property	14,000
Living	$24,500	Total	$26,700
Insurance premium	350	Liabilities:	
Taxes	14,800	Note to bank	4,500
Medical	585	Net worth	$22,200
Investment	4,000		($26,700 − $4,500)
Total	$44,235		

Net worth = Assets − Liabilities
(own)   (owe)

Paula believes her gross income will double in 2022 but her interest income will decrease $150. She plans to reduce her 2022 living expenses by one-half. Paula's insurance company wrote a letter announcing that her insurance premiums would triple in 2022. Her accountant estimates her taxes will decrease $250 and her medical costs will increase $410. Paula also hopes to cut her investments expenses by one-fourth. Paula's accountant projects that her savings and checking accounts will each double in value. On January 2, 2022, Paula sold her automobile and began to use public transportation. Paula forecasts that her personal property will decrease by one-seventh. She has sent her bank a $375 check to reduce her bank note. Could you give Paula an updated list of her 2022 finances? If you round all the way each 2021 and 2022 asset and liability, what will be the difference in Paula's net worth?

Summary Practice Test

Do you need help? Connect videos have step-by-step worked-out solutions.

1. Translate the following verbal forms to numbers and add.   *LU 1–1(1), LU 1–2(1)*
   a. Four thousand, eight hundred thirty-nine
   b. Seven million, twelve
   c. Twelve thousand, three hundred ninety-two

2. Express the following number in verbal form.   *LU 1–1(1)*
   9,622,364

3. Round the following numbers.   *LU 1–1(2)*

Nearest ten	Nearest hundred	Nearest thousand	Round all the way
a. 68	b. 888	c. 8,325	d. 14,821

4. Estimate the following actual problem by rounding all the way, work the actual problem, and check by adding each column of digits separately.   *LU 1–1(2), LU 1–2(1)*

   Actual   **Estimate**   **Check**

   $$\begin{array}{r} 1,886 \\ 9,411 \\ +\ 6,395 \\ \hline \end{array}$$

5. Estimate the following actual problem by rounding all the way and then do the actual multiplication.   *LU 1–1(2), LU 1–3(1)*

   Actual   **Estimate**

   $$\begin{array}{r} 8,843 \\ \times\quad 906 \\ \hline \end{array}$$

6. Multiply the following by the shortcut method.   *LU 1–3(1)*
   $829,412 \times 1,000$

7. Divide the following and check the answer by multiplication.   *LU 1–3(1, 2)*

   Check

   $39\overline{)14,800}$

8. Divide the following by the shortcut method.   *LU 1–3(2)*
   $6,000 \div 60$

9. Ling Wong bought a $299 pair of Bluetooth earbuds that was reduced to $205. Ling gave the clerk three $100 bills. What change will Ling receive?   *LU 1–2(2)*

10. Sam Song plans to buy a $16,000 Ford Focus with an interest charge of $4,000. Sam figures he can afford a monthly payment of $400. If Sam must pay 40 equal monthly payments, can he afford the Ford Focus?   *LU 1–2(1), LU 1–3(2)*

11. Lester Hal has the oil tank at his business filled 20 times per year. The tank has a capacity of 200 gallons. Assume **(a)** the price of oil fuel is $3 per gallon and **(b)** the tank is completely empty each time Lester has it filled. What is Lester's average monthly oil bill? Complete the following blueprint aid for dissecting and solving the word problem.   *LU 1–3(1, 2)*

	The facts	Solving for?	Steps to take	Key points
Blueprint				

Steps to solving problem

MY MONEY

🔍 My Money = My Plan

 What I need to know

"If you fail to plan, then plan to fail" is a quote you may have heard before. This saying is applicable in many situations including your finances. A solid financial plan will guide your current and future financial decisions. Creating and following good financial habits now will pay huge dividends in the future and prepare you for the financial ups and downs you will most likely experience. There are many sources of information on the topic of financial planning. It is important you seek out assistance matching up with the financial goals you have set for yourself. Any assistance you can find is only helpful if you are able to apply the knowledge to your unique financial situation. Therefore, personal financial planning is not a one-size-fits-all strategy but needs to be customized to your unique situation and finances.

 What I need to do

Create a monthly budget, build an emergency savings of six times your monthly expenses, and save and invest 20% of all monies received. Manage your money. Pay off all credit cards monthly. Make certain you are earning rewards for your credit card spending. Maximize your 401k. Diversify your investments. Pay at least one additional month's mortgage payment each year. Plan for your retirement, bearing in mind inflation and long-term care needs. Plan your estate. Have a will, a general power of attorney, a medical power of attorney, and a living trust.

Obtaining financial excellence is possible and realistic. Budgeting is important no matter your income level. Establishing a personal budget based on current earnings will develop skills you need to continue budgeting in the future. Identify and document your current income and expenses. Be realistic with your budget amounts and review regularly to make any adjustments. Be sure your budget includes dollars directed to an emergency savings account for those unexpected events. It is acceptable to start small with your savings contributions and increase over time as your income allows. Establishing a habit of saving is more important in the beginning than the dollar amount you can save at first.

Make smart decisions to manage your money. If your credit card balances are more than you can pay off every month, reduce your credit card spending. Use credit cards offering rewards (e.g., airline miles, cash back, etc.) on your purchases and pay off the entire balance monthly to avoid costly interest. Begin investing early in life to maximize the benefit of time. Starting your investments early gives your money more time to work in your favor. Time will also help to even out the volatility you will experience with your investment accounts.

 Steps I need to take

1. Create a realistic budget for yourself and commit to following it.
2. Provide for yourself by saving and investing for your future.
3. Protect your loved ones through planning your estate.

 Resources I can use

- Mint: Personal Finance & Money (mobile app)—create your personal budget and track your progress.
- https://www.thebalance.com/five-steps-to-an-effective-financial-plan-2386045—steps for creating a personal financial plan.

MY MONEY ACTIVITY

- Record your current expenses and income by tracking all activity for one month.
- Create a personal budget for the next month then year using the monthly figures.
- Track actual versus estimated amounts and make any adjustments.

PERSONAL FINANCE

A KIPLINGER APPROACH

"WHAT YOU'LL PAY TO ADOPT A DOG." Kiplinger's. November 2020.

WHAT YOU'LL PAY TO ADOPT A DOG

The up-front cost of adopting a dog in 2020 ranges from $610 to $2,350, according to Rover.com. And that's just the initial tab. The annual costs of owning a dog can range from $650 to $2,115. The most budget-minded dog owners spend less than $1,000 per year, but nearly half of owners spend about $3,400 on their dogs annually, according to a survey by Rover.

Adoption fees	$50 – $500
Spay/neuter	$35 – $400
Other medical care	$250 – $275
Flea and tick protection	$40 – $200
Toys	$10 – $200
Bed	$5 – $200
Crate	$30 – $150
Puppy vaccinations	$75 – $100
Potty pads	$10 – $50
Food & water bowls	$10 – $50
Collar/harness/leash	$14 – $90
Microchip	$45
Shampoo & brush	$5 – $20
Miscellaneous costs	$30 – $60

Business Math Issue

Pet insurance is a must for dog owners.

1. List the key points of the article and information to support your position.

2. Write a group defense of your position using math calculations to support your view. If you are in an online course, post to a discussion board.

Chapter 2
Fractions

Amazon Prepares to Retrain A Third of Its U.S. Workforce

By Chip Cutter

U.S. companies are increasingly paying up to retrain workers as new technologies transform the workplace and companies struggle to recruit talent in one of the hottest job markets in decades.

Amazon.com Inc. is the latest example of a large employer committing to help its workers gain new skills. The online retailer said Thursday it plans to spend $700 million over about six years to retrain a third of its U.S. workforce as automation, machine learning and other technology upends the way many of its employees do their jobs.

Companies as varied as AT&T Inc., Walmart Inc., JP-Morgan Chase & Co. and Accenture PLC have embarked on efforts to prepare workers for new roles. At a time of historically low unemployment, coupled with rapid digital transformation that requires hightech job skills, more U.S. companies said they want to help their employees transition to new positions–and they have their bottom line squarely in focus.

Many have concluded that they must coach existing staff to take on different types of work, or face a dire talent shortage, said Ryan Carson, founder and chief executive of Treehouse, a firm that pairs tech apprentices, often from underrepresented groups, with employers and helps train them.

"It's the beginning of the flood," Mr. Carson said. "We're basically just going back to a time where companies would invest in their own workforces."

Cutter, Chip. "Amazon Prepares to Retrain a Third of Its U.S. Workforce." *The Wall Street Journal,* July 12, 2019.

Learning Unit Objectives

LU 2–1: Types of Fractions and Conversion Procedures

1. Recognize the three types of fractions.
2. Convert improper fractions to whole or mixed numbers and mixed numbers to improper fractions.
3. Convert fractions to lowest and highest terms.

LU 2–2: Adding and Subtracting Fractions

1. Add like and unlike fractions.
2. Find the least common denominator by inspection and prime numbers.
3. Subtract like and unlike fractions.
4. Add and subtract mixed numbers with the same or different denominators.

LU 2–3: Multiplying and Dividing Fractions

1. Multiply and divide proper fractions and mixed numbers.
2. Use the cancellation method in the multiplication and division of fractions.

⊖ Essential Question

How can I use fractions to understand the business world and make my life easier?

🌐 Math Around the World

The *Wall Street Journal* chapter opener clip "Amazon Prepares to Retrain a Third of Its U.S. Workforce" illustrates the use of a fraction. From the clipping you learn that Amazon plans to spend $700 million over the next six years to retrain $\frac{1}{3}$ of its workforce.

Now let's look at Milk Chocolate M&M'S® candies as another example of using fractions.

As you know, M&M'S® candies come in different colors. Do you know how many of each color are in a bag of M&M'S®? If you go to the M&M'S® website, you learn that a typical bag of M&M'S® contains approximately 17 brown, 11 yellow, 11 red, and 5 each of orange, blue, and green M&M'S®.[1]

The 1.69-ounce bag of M&M'S® shown on the next page contains 55 M&M'S®. In this bag, you will find the following colors:

18 yellow	9 blue	6 brown
10 red	7 orange	5 green

The number of yellow candies in a bag might suggest that yellow is the favorite color of many people. Since this is a business math text, however, let's look at the 55 M&M'S® in terms of fractional arithmetic.

Of the 55 M&M'S® in the 1.69-ounce bag, 5 of these M&M'S® are green, so we can say that 5 parts of 55 represent green candies. We could also say that 1 out of 11 M&M'S® is green. Are you confused?

For many people, fractions are difficult. If you are one of these people, this chapter is for you. First you will review the types of fractions and the fraction conversion procedures. Then you will gain a clear understanding of the addition, subtraction, multiplication, and division of fractions.

[1] Off 1 due to rounding.

55 pieces in the bag

Learning Unit 2–1:
Types of Fractions and Conversion Procedures

Note in the *Wall Street Journal* clip "Top Billing" that nearly $\frac{2}{3}$ of all product clicks are for Amazon's private-label products.

This chapter explains the parts of whole numbers called **fractions.** With fractions you can divide any object or unit—a whole—into a definite number of equal parts. For example, the bag of 55 M&M'S® described above contains 6 brown candies. If you eat only the brown M&M'S®, you have eaten 6 parts of 55, or 6 parts of the whole bag of M&M'S®. We can express this in the following fraction:

Top Billing

When people search for products on Amazon; nearly two-thirds of all product clicks come from the first page of results...

First page	Row 1	2	3	4 5 6 7 8
Other pages				

0   20   40   60%

...so the proliferation of Amazon's private-label products on the first page makes it more likely people choose those items.

"Top Billing." *The Wall Street Journal,* September 17, 2019.

$$\frac{6}{55}$$

6 is the **numerator,** or top of the fraction. The numerator describes the number of equal parts of the whole bag that you ate.

55 is the **denominator,** or bottom of the fraction. The denominator gives the total number of equal parts in the bag of M&M'S®.

Before reviewing the arithmetic operations of fractions, you must recognize the three types of fractions described in this unit. You must also know how to convert fractions to a workable form.

Learn: Types of Fractions

There are three types of fractions: proper fractions, improper fractions, and mixed numbers.

Proper Fractions

A **proper fraction** has a value less than 1; its numerator is smaller than its denominator.

Examples:

$$\frac{2}{3}, \frac{1}{4}, \frac{1}{2}, \frac{1}{10}, \frac{1}{12}, \frac{1}{3}, \frac{4}{7}, \frac{9}{10}, \frac{12}{13}, \frac{18}{55}, \frac{499}{1,000}, \frac{501}{1,000}$$

Improper Fractions

An **improper fraction** has a value equal to or greater than 1; its numerator is equal to or greater than its denominator.

Examples:

$$\frac{15}{15}, \frac{9}{8}, \frac{15}{14}, \frac{22}{19}$$

Mixed Numbers

A **mixed number** is the sum of a whole number greater than zero and a proper fraction.

Examples:

$$7\frac{1}{8}, 5\frac{9}{10}, 8\frac{7}{8}, 33\frac{5}{6}, 139\frac{9}{11}$$

Mother Goose & Grimm © 2019 Grimmy Inc., Dist. by King Features Syndicate

Learn: Conversion Procedures

In Chapter 1 we worked with two of the division symbols ($\div$ and $\overline{)}$). The horizontal line (or the diagonal) that separates the numerator and the denominator of a fraction also indicates division. The numerator, like the dividend, is the number we are dividing into. The denominator, like the divisor, is the number we use to divide. Then, referring to the 6 brown M&M'S® in the bag of 55 M&M'S® $\left(\frac{6}{55}\right)$ shown at the beginning of this unit, we can say that we are dividing 55 into 6, or 6 is divided by 55. Also, in the fraction $\frac{3}{4}$, we can say that we are dividing 4 into 3, or 3 is divided by 4. *Remember "The top dog gets the hat" when converting proper fractions to decimals. For example, in the fraction $\frac{3}{4}$, the 3 is the top dog. The division sign is the hat. Put the hat over the 3 and divide:* $4\overline{)3}$ = .75.

Aha!

Working with the smaller numbers of simple fractions such as $\frac{3}{4}$ is easier, so we often convert fractions to their simplest terms. In this unit we show how to convert improper fractions to whole or mixed numbers, mixed numbers to improper fractions, and fractions to lowest and highest terms.

Converting Improper Fractions to Whole or Mixed Numbers

Business situations often make it necessary to change an improper fraction to a whole number or mixed number. You can use the following steps to make this conversion:

Converting Improper Fractions to Whole or Mixed Numbers

Step 1 Divide the numerator of the improper fraction by the denominator.

Step 2 **a.** If you have no remainder, the quotient is a whole number.

b. If you have a remainder, the whole number part of the mixed number is the quotient. The remainder is placed over the original denominator as the proper fraction of the mixed number.

Examples:

$$\frac{15}{15} = 1 \qquad \frac{16}{5} = 3\frac{1}{5}$$

$$\begin{array}{r} 3\ R1 \\ 5\overline{)16} \\ \underline{15} \\ 1 \end{array}$$

Converting Mixed Numbers to Improper Fractions   By reversing the procedure of converting improper fractions to mixed numbers, we can change mixed numbers to improper fractions.

Converting Mixed Numbers to Improper Fractions

Step 1 Multiply the denominator of the fraction by the whole number.

Step 2 Add the product from Step 1 to the numerator of the original fraction.

Step 3 Place the total from Step 2 over the denominator of the original fraction to get the improper fraction.

Example:

$$6\frac{1}{8} = \frac{(8 \times 6) + 1}{8} = \frac{49}{8}$$

Note that the denominator stays the same.

Converting (Reducing) Fractions to Lowest Terms

When solving fraction problems, you always reduce the fractions to their lowest terms. This reduction does not change the value of the fraction. For example, in the bag of M&M'S®, 5 out of 55 were green. The fraction for this is $\frac{5}{55}$. If you divide the top and bottom of the fraction by 5, you have reduced the fraction to $\frac{1}{11}$ without changing its value. Remember, we said in the chapter introduction that 1 out of 11 M&M'S® in the bag of 55 M&M'S® represents green candies. Now you know why this is true.

To reduce a fraction to its lowest terms, begin by inspecting the fraction, looking for the largest whole number that will divide into both the numerator and the denominator without leaving a remainder. This whole number is the **greatest common divisor,** which cannot be zero. When you find this largest whole number, you have reached the point where the fraction is reduced to its **lowest terms.** At this point, no number (except 1) can divide evenly into both parts of the fraction.

Reducing Fractions to Lowest Terms by Inspection

Step 1 By inspection, find the largest whole number (greatest common divisor) that will divide evenly into the numerator and denominator (does not change the fraction value).

Step 2 Divide the numerator and denominator by the greatest common divisor. Now you have reduced the fraction to its lowest terms, since no number (except 1) can divide evenly into the numerator and denominator.

Example:

$$\frac{24}{30} = \frac{24 \div 6}{30 \div 6} = \boxed{\frac{4}{5}}$$

Using inspection, you can see that the number 6 in the above example is the greatest common divisor. When you have large numbers, the greatest common divisor is not so obvious. For large numbers, you can use the following step approach to find the greatest common divisor:

Step Approach for Finding Greatest Common Divisor

Step 1 Divide the smaller number (numerator) of the fraction into the larger number (denominator).

Step 2 Divide the remainder of Step 1 into the divisor of Step 1.

Step 3 Divide the remainder of Step 2 into the divisor of Step 2. Continue this division process until the remainder is a 0, which means the last divisor is the greatest common divisor.

Example:

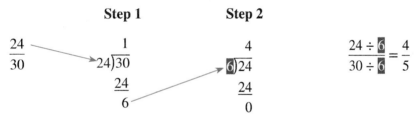

Reducing a fraction by inspection is to some extent a trial-and-error method. Sometimes you are not sure what number you should divide into the top (numerator) and bottom (denominator) of the fraction. The following reference table on divisibility tests will be helpful. Note that to reduce a fraction to lowest terms might result in more than one division.

	2	3	4	5	5	10
Will divide evenly into a number if the	Last digit is 0, 2, 4, 6, 8.	Sum of the digits is divisible by 3.	Last two digits can be divided by 4.	Last digit is 0 or 5.	The number is even and 3 will divide into the sum of digits.	The last digit is 0.
Examples	$\dfrac{12}{14} = \dfrac{6}{7}$	$\dfrac{36}{69} = \dfrac{12}{23}$ $3 + 6 = 9 \div 3 = 3$ $6 + 9 = 15 \div 3 = 5$	$\dfrac{140}{160} = \dfrac{1(40)}{1(60)}$ $= \dfrac{35}{40} = \dfrac{7}{8}$	$\dfrac{15}{20} = \dfrac{3}{4}$	$\dfrac{12}{18} = \dfrac{2}{3}$	$\dfrac{90}{100} = \dfrac{9}{10}$

Converting (Raising) Fractions to Higher Terms   Later, when you add and subtract fractions, you will see that sometimes fractions must be raised to **higher terms.** Recall that when you reduced fractions to their lowest terms, you looked for the largest whole number (greatest common divisor) that would divide evenly into both the numerator and the denominator. When you raise fractions to higher terms, you do the opposite and multiply the numerator and the denominator by the same whole number. For example, if you want to raise the fraction $\frac{1}{4}$, you can multiply the numerator and denominator by 2.

Example:

$$\frac{1}{4} \times \frac{2}{2} = \frac{2}{8}$$

The fractions $\frac{1}{4}$ and $\frac{2}{8}$ are **equivalent** in value. By converting $\frac{1}{4}$ to $\frac{2}{8}$, you only divided it into more parts.

Let's suppose that you have eaten $\frac{4}{7}$ of a pizza. You decide that instead of expressing the amount you have eaten in 7ths, you want to express it in 28ths.
How would you do this?

To find the new numerator when you know the new denominator (28), use the steps that follow.

Raising Fractions to Higher Terms When Denominator is Known

Step 1   Divide the *new* denominator by the *old* denominator to get the common number that raises the fraction to higher terms.

Step 2   Multiply the common number from Step 1 by the old numerator and place it as the new numerator over the new denominator.

Example:

$$\frac{4}{7} = \frac{?}{28}$$

Step 1 Divide 28 by 7 = 4.

Step 2 Multiply 4 by the numerator 4 = 16.
Result:

$$\frac{4}{7} = \frac{16}{28} \quad \left(Note\text{: This is the same as multiplying } \frac{4}{7} \times \frac{4}{4}. \right)$$

Note that $\frac{4}{7}$ and $\frac{16}{28}$ are equivalent in value, yet they are different fractions.

Now try the following Practice Quiz to check your understanding of this unit.

Practice Quiz

Complete this Practice Quiz to see how you are doing.

1. Identify the type of fraction—proper, improper, or mixed:

   a. $\dfrac{4}{5}$ **b.** $\dfrac{6}{5}$ **c.** $19\dfrac{1}{5}$ **d.** $\dfrac{20}{20}$

2. Convert to a mixed number:

   $\dfrac{160}{9}$

3. Convert the mixed number to an improper fraction:

   $9\dfrac{5}{8}$

4. Find the greatest common divisor by the step approach and reduce to lowest terms:

   a. $\dfrac{24}{40}$ **b.** $\dfrac{91}{156}$

5. Convert to higher terms:

   a. $\dfrac{14}{20} = \dfrac{}{200}$ **b.** $\dfrac{8}{10} = \dfrac{}{60}$

✓ Solutions

1. **a.** Proper
    b. Improper
    c. Mixed
    d. Improper

2. $17\frac{7}{9}$

    $$9\overline{)160}$$
    $$\underline{9}$$
    $$70$$
    $$\underline{63}$$
    $$7$$

3. $\dfrac{(9 \times 8) + 5}{8} = \dfrac{77}{8}$

4. **a.**

    $$\overset{1}{24\overline{)40}} \quad \overset{1}{16\overline{)24}} \quad \overset{1}{8\overline{)\,16}}$$
    $$\underline{24} \qquad \underline{16} \qquad \underline{16}$$
    $$16 \qquad\quad 8 \qquad\quad 0$$

    8 is greatest common divisor.

    $\dfrac{24 \div 8}{40 \div 8} = \dfrac{3}{5}$

    b.

    $$\overset{1}{91\overline{)156}} \quad \overset{1}{65\overline{)91}} \quad \overset{2}{26\overline{)65}} \quad \overset{2}{13\overline{)26}}$$
    $$\underline{91} \qquad \underline{65} \qquad \underline{52} \qquad \underline{26}$$
    $$65 \qquad\quad 26 \qquad\quad 13 \qquad\quad 0$$

    13 is greatest common divisor.

    $\dfrac{91 \div 13}{156 \div 13} = \dfrac{7}{12}$

5. **a.** $\overset{10}{20\overline{)200}}$ $\quad 10 \times 14 = 140 \quad$ $\dfrac{14}{20} = \dfrac{140}{200}$

    b. $\overset{6}{10\overline{)60}}$ $\quad 6 \times 8 = 48 \quad$ $\dfrac{8}{10} = \dfrac{48}{60}$

Learning Unit 2–2:

Adding and Subtracting Fractions

As a result of the pandemic, more teachers are using online video-sharing sites that are modeled after Google Inc.'s YouTube. As you can see in the blackboard illustration, these fractions can be added because the fractions have the same denominator. These are called *like fractions*.

In this unit you learn how to add and subtract fractions with the same denominators **(like fractions)** and fractions with different denominators **(unlike fractions).** We have also included how to add and subtract mixed numbers.

Learn: Addition of Fractions

When you add two or more quantities, they must have the same name or be of the same denomination. You cannot add 6 quarts and 3 pints unless you change the denomination of one or both quantities. You must either make the quarts into pints or the pints into quarts. The same principle also applies to fractions. That is, to add two or more fractions, they must have a **common denominator.**

Adding Like Fractions

Earlier we stated that because the fractions had the same denominator, or a common denominator, they were *like fractions*. Adding like fractions is similar to adding whole numbers.

Adding Like Fractions

Step 1 Add the numerators and place the total over the original denominator.

Step 2 If the total of your numerators is the same as your original denominator, convert your answer to a whole number; if the total is larger than your original denominator, convert your answer to a mixed number.

Example:

$$\frac{1}{7} + \frac{4}{7} = \boxed{\frac{5}{7}}$$

The denominator, 7, shows the number of pieces into which some whole was divided. The two numerators, 1 and 4, tell how many of the pieces you have. So if you add 1 and 4, you get 5, or $\frac{5}{7}$.

Adding Unlike Fractions   Since you cannot add *unlike fractions* because their denominators are not the same, you must change the unlike fractions to *like fractions*—fractions with the same denominators. To do this, find a denominator that is common to all the fractions you want to add. Then look for the **least common denominator (LCD).**[2] The LCD is the smallest nonzero whole number into which all denominators will divide evenly. You can find the LCD by inspection or with prime numbers.

[2] Often referred to as the *lowest common denominator.*

Finding the Least Common Denominator (LCD) by Inspection The example that follows shows you how to use inspection to find an LCD (this will make all the denominators the same).

Example:

$$\frac{3}{7} + \frac{5}{21}$$

Inspection of these two fractions shows that the smallest number into which denominators 7 and 21 divide evenly is 21. Thus, **21** is the LCD.

You may know that 21 is the LCD of $\frac{3}{7} + \frac{5}{21}$, but you cannot add these two fractions until you change the denominator of $\frac{3}{7}$ to 21. You do this by building (raising) the equivalent of $\frac{3}{7}$, as explained in Learning Unit 2–1. You can use the following steps to find the LCD by inspection:

Step 1 Divide the new denominator (21) by the old denominator (7): $21 \div 7 = 3$.

Step 2 Multiply the 3 in Step 1 by the old numerator (3): $3 \times 3 = 9$. The new numerator is 9.

Result:

$$\frac{3}{7} = \frac{9}{21}$$

Now that the denominators are the same, you add the numerators.

$$\frac{9}{21} + \frac{5}{21} = \frac{14}{21} = \frac{2}{3}$$

Note that $\frac{14}{21}$ is reduced to its lowest terms $\frac{2}{3}$. Always reduce your answer to its lowest terms.

You are now ready for the following general steps for adding proper fractions with different denominators. These steps also apply to the following discussion on finding LCD by prime numbers.

Adding Unlike Fractions

Step 1 Find the LCD.

Step 2 Change each fraction to a like fraction with the LCD.

Step 3 Add the numerators and place the total over the LCD.

Step 4 If necessary, reduce the answer to lowest terms.

Finding the Least Common Denominator (LCD) by Prime Numbers When you cannot determine the LCD by inspection, you can use the prime number method. First you must understand prime numbers.

Prime Numbers

A **prime number** is a whole number greater than 1 that is only divisible by itself and 1. The number 1 is not a prime number.

Source: FRAZZ ©2010 Jef Mallett/Andrews McMeel Syndication

Example:

$$2, 3, 5, 7, 11, 13, 17, 19, 23, 29, 31, 37, 41, 43$$

Note that the number 4 is not a prime number. Not only can you divide 4 by 1 and by 4, but you can also divide 4 by 2. A whole number that is greater than 1 and is only divisible by itself and 1 has become a source of interest to some people.

Example:

$$\frac{1}{3} + \frac{1}{8} + \frac{1}{9} + \frac{1}{12}$$

Step 1 Copy the denominators and arrange them in a separate row.

    3   8   9   12

Step 2 Divide the denominators in Step 1 by prime numbers. Start with the smallest number that will divide into at least two of the denominators. Bring down any number that is not divisible. Keep in mind that the lowest prime number is 2.

$$2\ \overline{)\ 3\quad 8\quad 9\quad 12}$$
$$\ 3\quad 4\quad 9\quad 6$$

Note: The 3 and 9 were brought down, since they were not divisible by 2.

Step 3 Continue Step 2 until no prime number will divide evenly into at least two numbers.

Note: The 3 is used, since 2 can no longer divide evenly into at least two numbers.

$$2\ \overline{)\ 3\quad 8\quad 9\quad 12}$$
$$2\ \overline{)\ 3\quad 4\quad 9\quad 6}$$
$$3\ \overline{)\ 3\quad 2\quad 9\quad 3}$$
$$\ 1\quad 2\quad 3\quad 1$$

Step 4 To find the LCD, multiply all the numbers in the divisors (2, 2, 3) and in the last row (1, 2, 3, 1).

$$\boxed{2 \times 2 \times 3} \times \boxed{1 \times 2 \times 3 \times 1} = \boxed{72}\,(\text{LCD})$$

Divisors   ×   Last row

Step 5 Raise each fraction so that each denominator will be 72 and then add fractions.

$$\frac{24}{72} + \frac{9}{72} + \frac{8}{72} + \frac{6}{72} = \frac{47}{72}$$

$$\frac{1}{3} = \frac{?}{72} \qquad 72 \div 3 = 24$$
$$24 \times 1 = 24$$

$$\frac{1}{8} = \frac{?}{72} \qquad 72 \div 8 = 9$$
$$9 \times 1 = 9$$

The above five steps used for finding LCD with prime numbers are summarized as follows:

Finding Lcd for Two or More Fractions

Step 1 Copy the denominators and arrange them in a separate row.

Step 2 Divide the denominators by the smallest prime number that will divide evenly into at least two numbers.

Step 3 Continue until no prime number divides evenly into at least two numbers.

Step 4 Multiply all the numbers in divisors and last row to find the LCD.

Step 5 Raise all fractions so each has a common denominator and then complete the computation.

Adding Mixed Numbers   The following steps will show you how to add mixed numbers:

Adding Mixed Numbers

Step 1 Add the fractions (remember that fractions need common denominators, as in the previous section).

Step 2 Add the whole numbers.

Step 3 Combine the totals of Steps 1 and 2. Be sure you do not have an improper fraction in your final answer. Convert the improper fraction to a whole or mixed number. Add the whole numbers resulting from the improper fraction conversion to the total whole numbers of Step 2. If necessary, reduce the answer to lowest terms.

Example:

$$4\frac{7}{20} \qquad 4\frac{7}{20}$$
$$6\frac{3}{5} \qquad 6\frac{12}{20}$$
$$+7\frac{1}{4} \qquad +7\frac{5}{20}$$

$$\frac{3}{5} = \frac{?}{20}$$
$$20 \div 5 = \quad 4$$
$$\times 3$$
$$12$$

Using prime numbers to find LCD of example

$$\begin{array}{r} 2\,\underline{/20 \quad 5 \quad 4} \\ 2\,\underline{/10 \quad 5 \quad 2} \\ 5\,\underline{/5 \quad 5 \quad 1} \\ 1 \quad 1 \quad 1 \end{array}$$

$$2 \times 2 \times 5 = 20 \text{ LCD}$$

Step 1 $\longrightarrow$ $\dfrac{24}{20} = 1\dfrac{4}{20}$

Step 2 $+\,\underline{17} \qquad (4 + 6 + 7)$

Step 3 $\longrightarrow$ $= 18\dfrac{4}{20} = \boxed{18\dfrac{1}{5}}$

Learn: Subtraction of Fractions

The subtraction of fractions is similar to the addition of fractions. This section explains how to subtract like and unlike fractions and how to subtract mixed numbers.

Subtracting Like Fractions   To subtract like fractions, use the steps that follow.

Subtracting Like Fractions

Step 1 Subtract the numerators and place the answer over the common denominator.

Step 2 If necessary, reduce the answer to lowest terms.

Example:

$$\frac{9}{10} - \frac{1}{10} = \frac{8 \div 2}{10 \div 2} = \frac{4}{5}$$

$$\uparrow \qquad \uparrow$$
$$\textbf{Step} \quad \textbf{1} \quad \textbf{Step} \quad \textbf{2}$$

Subtracting Unlike Fractions   Now let's learn the steps for subtracting unlike fractions.

Subtracting Unlike Fractions

Step 1 Find the LCD.

Step 2 Raise the fraction to its equivalent value.

Step 3 Subtract the numerators and place the answer over the LCD.

Step 4 If necessary, reduce the answer to lowest terms.

Example:

$$
\begin{array}{cc}
\dfrac{5}{8} & \dfrac{40}{64} \\[2ex]
-\dfrac{2}{64} & -\dfrac{2}{64} \\[2ex]
\hline
& \dfrac{38}{64} = \boxed{\dfrac{19}{32}}
\end{array}
$$

By inspection, we see that LCD is 64. Thus $64 \div 8 = 8 \times 5 = 40$.

Subtracting Mixed Numbers   When you subtract whole numbers, sometimes borrowing is not necessary. At other times, you must borrow. The same is true of subtracting mixed numbers.

Subtracting Mixed Numbers

When Borrowing Is Not Necessary	*When Borrowing Is Necessary*
Step 1 Subtract fractions, making sure to find the LCD.	**Step 1** Make sure the fractions have the LCD.
Step 2 Subtract whole numbers.	**Step 2** Borrow from the whole number of the minuend (top number).
Step 3 Reduce the fraction(s) to lowest terms.	**Step 3** Subtract the whole numbers and fractions.
	Step 4 Reduce the fraction(s) to lowest terms.

Example: Where borrowing is not necessary: Find LCD of 2 and 8. LCD is 8.

$$6\frac{1}{2}$$

$$-\frac{3}{8}$$

$$6\frac{4}{8}$$
$$-\frac{3}{8}$$
$$6\frac{1}{8}$$

Example: Where borrowing is necessary:

$$3\frac{1}{2} = \quad 3\frac{2}{4} = \quad 2\frac{6}{4}\left(\frac{4}{4}+\frac{2}{4}\right)$$

$$-1\frac{3}{4} = \quad -1\frac{3}{4} = \quad -1\frac{3}{4}$$

LCD is 4. $$1\frac{3}{4}$$

Since $\frac{3}{4}$ is larger than $\frac{2}{4}$, we must borrow 1 from the 3. This is the same as borrowing $\frac{4}{4}$. A fraction with the same numerator and denominator represents a whole. When we add $\frac{4}{4}$ + $\frac{2}{4}$, we get $\frac{6}{4}$. Note how we subtracted the whole number and fractions, being sure to reduce the final answer if necessary.

Learn: How to Dissect and Solve a Word Problem

Let's now look at how to dissect and solve a word problem involving fractions.

The Word Problem Albertsons grocery store has $550\frac{1}{4}$ total square feet of floor space. Albertsons' meat department occupies $115\frac{1}{2}$ square feet, and its deli department occupies $145\frac{7}{8}$ square feet. If the remainder of the floor space is for groceries, what square footage remains for groceries?

The facts	Solving for?	Steps to take	Key points
Total square footage: $550\frac{1}{4}$ sq. ft. *Meat department:* $115\frac{1}{2}$ sq. ft. *Deli department:* $145\frac{7}{8}$ sq. ft.	Total square footage for groceries.	Total floor space – Total meat and deli floor space = Total grocery floor space.	Denominators must be the same before adding or subtracting fractions. $\frac{8}{8} = 1$ Never leave improper fraction as final answer.

Blueprint

Steps to solving problem

1. Calculate total square footage of the meat and deli departments.

    Meat: $\quad 115\dfrac{1}{2} = \quad 115\dfrac{4}{8}$

    Deli: $\quad +145\dfrac{7}{8} = +145\dfrac{7}{8}$

    $\qquad\qquad\qquad 260\dfrac{11}{8} = 261\dfrac{3}{8}\text{ sq. ft.}$

2. Calculate total grocery square footage.

    $550\dfrac{1}{4} = \quad 550\dfrac{2}{8} = \quad 549\dfrac{10}{8}$ $\qquad$ **Check** $\qquad\qquad 261\dfrac{3}{8}$

    $-261\dfrac{3}{8} = -261\dfrac{3}{8} = -\;261\dfrac{3}{8} \quad \left(\dfrac{2}{8}+\dfrac{8}{8}\right)$ $\qquad\qquad +288\dfrac{7}{8}$

    $\qquad\qquad\qquad\qquad\quad 288\dfrac{7}{8}\text{ sq. ft.}$ $\qquad\qquad 549\dfrac{10}{8} = 550\dfrac{2}{8} = 550\dfrac{1}{4}\text{ sq. ft.}$

Note how the above blueprint aid helped to gather the facts and identify what we were looking for. To find the total square footage for groceries, we first had to sum the areas for meat and deli. Then we could subtract these areas from the total square footage. Also note that in Step 1 above, we didn't leave the answer as an improper fraction. In Step 2, we borrowed from the 550 so that we could complete the subtraction.

It's your turn to check your progress with a Practice Quiz.

Practice Quiz

Complete this Practice Quiz to see how you are doing.

1. Find LCD by the division of prime numbers:

   12, 9, 6, 4

2. Add and reduce to lowest terms if needed:

   a. $\dfrac{3}{40} + \dfrac{2}{5}$ **b.** $2\dfrac{3}{4} + 6\dfrac{1}{20}$

3. Subtract and reduce to lowest terms if needed:

   a. $\dfrac{6}{7} - \dfrac{1}{4}$ **b.** $8\dfrac{1}{4} - 3\dfrac{9}{28}$ **c.** $4 - 1\dfrac{3}{4}$

4. Computerland has $660\frac{1}{4}$ total square feet of floor space. Three departments occupy this floor space: hardware, $201\frac{1}{8}$ square feet; software, $242\frac{1}{4}$ square feet; and customer service, _____ square feet. What is the total square footage of the customer service area? You might want to try a blueprint aid, since the solution will show a completed blueprint aid.

✓ Solutions

1.
```
2 /12   9   6   4
2 / 6   9   3   2
3 / 3   9   3   1
    1   3   1   1
```
$\text{LCD} = 2 \times 2 \times 3 \times 1 \times 3 \times 1 \times 1 = \boxed{36}$

2. **a.** $\dfrac{3}{40} + \dfrac{2}{5} = \dfrac{3}{40} + \dfrac{16}{40} = \boxed{\dfrac{19}{40}}$ $\left(\dfrac{2}{5} = \dfrac{?}{40} \atop 40 \div 5 = 8 \times 2 = 16 \right)$

   b.
$$\begin{array}{r} 2\dfrac{3}{4} \quad 2\dfrac{15}{20} \\[2mm] +6\dfrac{1}{20} \quad +6\dfrac{1}{20} \\[2mm] \hline 8\dfrac{16}{20} = \boxed{8\dfrac{4}{5}} \end{array}$$

   $\dfrac{3}{4} = \dfrac{?}{20}$ $20 \div 4 = 5 \times 3 = 15$

3. **a.**
$$\begin{array}{r} \dfrac{6}{7} = \dfrac{24}{28} \\[2mm] -\dfrac{1}{4} = -\dfrac{7}{28} \\[2mm] \hline \boxed{\dfrac{17}{28}} \end{array}$$

   b.
$$\begin{array}{r} 8\dfrac{1}{4} = 8\dfrac{7}{28} = 7\dfrac{35}{28} \\[2mm] -3\dfrac{9}{28} = -3\dfrac{9}{28} = -3\dfrac{9}{28} \\[2mm] \hline 4\dfrac{26}{28} = \boxed{4\dfrac{13}{14}} \end{array}$$
   $\longleftarrow \left(\dfrac{28}{28} + \dfrac{7}{28} \right)$

   c. Note how we showed the 4 as $3\dfrac{4}{4}$.
$$\begin{array}{r} 3\dfrac{4}{4} \\[2mm] -1\dfrac{3}{4} \\[2mm] \hline \boxed{2\dfrac{1}{4}} \end{array}$$

Practice Quiz *Continued*

4. Computerland's total square footage for customer service:

The facts	Solving for?	Steps to take	Key points
Blueprint *Total square footage:* $660\frac{1}{4}$ sq. ft. *Hardware:* $201\frac{1}{8}$ sq. ft. *Software:* $242\frac{1}{4}$ sq. ft.	Total square footage for customer service.	Total floor space − Total hardware and software floor space = Total customer service floor space.	Denominators must be the same before adding or subtracting fractions.

Steps to solving problem

1. Calculate the total square footage of hardware and software.

$$201\frac{1}{8} = \quad 201\frac{1}{8} \text{ (hardware)}$$
$$+242\frac{1}{4} = +242\frac{2}{8} \text{ (software)}$$
$$\overline{\hphantom{+242\frac{1}{4}} 443\frac{3}{8}}$$

2. Calculate the total square footage for customer service.

$$660\frac{1}{4} = \quad 660\frac{2}{8} = \quad 659\frac{10}{8} \text{ (total square footage)}$$
$$-443\frac{3}{8} = -443\frac{3}{8} = -443\frac{3}{8} \text{ (hardware plus software)}$$
$$216\frac{7}{8} \text{ sq. ft. (customer service)}$$

Multiplying and Dividing Fractions

The following recipe for Coconutty "M&M'S"® Brand Brownies makes 16 brownies. What would you need if you wanted to triple the recipe and make 48 brownies?

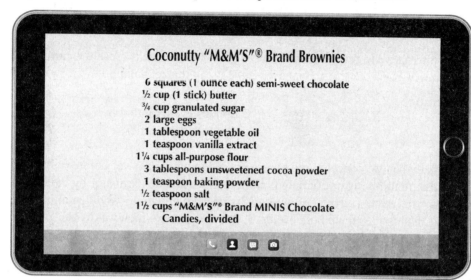

Coconutty "M&M'S"® Brand Brownies

6 squares (1 ounce each) semi-sweet chocolate
½ cup (1 stick) butter
¾ cup granulated sugar
2 large eggs
1 tablespoon vegetable oil
1 teaspoon vanilla extract
1¼ cups all-purpose flour
3 tablespoons unsweetened cocoa powder
1 teaspoon baking powder
½ teaspoon salt
1½ cups "M&M'S"® Brand MINIS Chocolate
Candies, divided

Source: Adapted from Mars, Inc.

In this unit you learn how to multiply and divide fractions.

Learn: Multiplication of Fractions

Multiplying fractions is easier than adding and subtracting fractions because you do not have to find a common denominator. This section explains the multiplication of proper fractions and the multiplication of mixed numbers.

Multiplying Proper Fractions[3]

Step 1 Multiply the numerators and the denominators.

Step 2 Reduce the answer to lowest terms or use the cancellation method.

First let's look at an example that results in an answer that we do not have to reduce.

Example:

$$\frac{1}{7} \times \frac{5}{8} = \boxed{\frac{5}{56}}$$

In the next example, note how we reduce the answer to lowest terms.

Example:

$$\frac{5}{1} \times \frac{1}{6} \times \frac{4}{7} = \frac{20}{42} = \boxed{\frac{10}{21}} \qquad \text{Keep in mind } \frac{5}{1} \text{ is equal to 5.}$$

[3] You would follow the same procedure to multiply improper fractions.

We can reduce $\frac{20}{42}$ by the step approach as follows:

$$\begin{array}{r} 2 \\ 20\overline{)42} \\ \underline{40} \\ 2 \end{array} \qquad \begin{array}{r} 10 \\ 2\overline{)20} \\ \underline{20} \\ 0 \end{array}$$

We could also have found the greatest common divisor by inspection.

$$\frac{20 \div 2}{42 \div 2} = \boxed{\frac{10}{21}}$$

As an alternative to reducing fractions to lowest terms, we can use the **cancellation** technique. Let's work the previous example using this technique.

Example:

$$\frac{5}{1} \times \frac{1}{\cancel{6}_{3}} \times \frac{\cancel{4}^{2}}{7} = \boxed{\frac{10}{21}}$$

2 divides evenly into 4 twice and into 6 three times.

Note that when we cancel numbers, we are reducing the answer before multiplying. We know that multiplying or dividing both numerator and denominator by the same number gives an equivalent fraction. So we can divide both numerator and denominator by any number that divides them both evenly. It doesn't matter which we divide first. Note that this division reduces $\frac{10}{21}$ to its lowest terms.

Multiplying Mixed Numbers The following steps explain how to multiply mixed numbers:

Multiplying Mixed Numbers

Step 1 Convert the mixed numbers to improper fractions.

Step 2 Multiply the numerators and denominators.

Step 3 Reduce the answer to lowest terms or use the cancellation method.

Example:

$$2\frac{1}{3} \times 1\frac{1}{2} = \frac{7}{\cancel{3}} \times \frac{\cancel{3}^{1}}{2} = \frac{7}{2} = \boxed{3\frac{1}{2}}$$

Step 1   **Step 2**   **Step 3**

Before we look at dividing fractions, reference the article below from the *Wall Street Journal*, "Seeing is Believing," showing research of the brain and its relationship to your fingers and math skills.

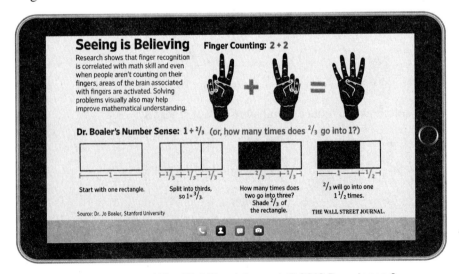

Reprinted by permission of *The Wall Street Journal*, © 2016 Dow Jones & Company, Inc. All rights reserved worldwide.

Learn: Division of Fractions

When you studied whole numbers in Chapter 1, you saw how multiplication can be checked by division. The multiplication of fractions can also be checked by division, as you will see in this section on dividing proper fractions and mixed numbers.

Dividing Proper Fractions   The division of proper fractions introduces a new term—the **reciprocal.** To use reciprocals, we must first recognize which fraction in the problem is the divisor—the fraction that we divide by. Let's assume the problem we are to solve is $\frac{1}{8} \div \frac{2}{3}$. We read this problem as "$\frac{1}{8}$ divided by $\frac{2}{3}$." The divisor is the fraction after the division sign (or the second fraction). The steps that follow show how the divisor becomes a reciprocal.

Jef Mallett dated 2/27/2013/Andrews McMeel Syndication

Dividing Proper Fractions

Step 1   Invert (turn upside down) the divisor (the second fraction). The inverted number is the *reciprocal.*

Step 2   Multiply the fractions.

Step 3   Reduce the answer to lowest terms or use the cancellation method.

Do you know why the inverted fraction number is a reciprocal? Reciprocals are two numbers that when multiplied give a product of 1. For example, 2 (which is the same as $\frac{2}{1}$) and $\frac{1}{2}$ are reciprocals because multiplying them gives 1.

Example:
$$\frac{1}{8} \div \frac{2}{3} \qquad \frac{1}{8} \times \frac{3}{2} = \frac{3}{16}$$

Dividing Mixed Numbers   Now you are ready to divide mixed numbers by using improper fractions.

Dividing Mixed Numbers

Step 1 Convert all mixed numbers to improper fractions.

Step 2 Invert the divisor (take its reciprocal) and multiply. If your final answer is an improper fraction, reduce it to lowest terms. You can do this by finding the greatest common divisor or by using the cancellation technique.

Example:

$$8\frac{3}{4} \div 2\frac{5}{6}$$

Step 1 $\dfrac{35}{4} \div \dfrac{17}{6}$

Step 2 $\dfrac{35}{\underset{2}{\cancel{4}}} \times \dfrac{\overset{3}{\cancel{6}}}{17} = \dfrac{105}{34} = 3\frac{3}{34}$ Here we used the cancellation technique.

Learn: How to Dissect and Solve a Word Problem

The Word Problem Jamie ordered $5\frac{1}{2}$ cords of oak. The cost of each cord is $150. He also ordered $2\frac{1}{4}$ cords of maple at $120 per cord. Jamie's neighbor, Al, said that he would share the wood and pay him $\frac{1}{5}$ of the total cost. How much did Jamie receive from Al?

Note how we filled in the blueprint aid columns. We first had to find the total cost of all the wood before we could find Al's share—$\frac{1}{5}$ of the total cost.

	The facts	Solving for?	Steps to take	Key points
Blueprint	*Cords ordered:* $5\frac{1}{2}$ at $150 per cord; $2\frac{1}{4}$ at $120 per cord. *Al's cost share:* $\frac{1}{5}$ the total cost.	What will Al pay Jamie?	Total cost of wood $\times \frac{1}{5}$ = Al's cost.	Convert mixed numbers to improper fractions when multiplying. Cancellation is an alternative to reducing fractions.

Money Tip Make good buying decisions. Do not spend more money than you make. In fact, remember to pay yourself first by putting away money each paycheck for your retirement—even $10 each paycheck adds up.

Steps to solving problem

1. Calculate the cost of oak.

$$5\frac{1}{2} \times \$150 = \frac{11}{\underset{1}{\cancel{2}}} \times \overset{\$75}{\cancel{\$150}} = \$825$$

2. Calculate the cost of maple.

$$2\frac{1}{4} \times \$120 = \frac{9}{\underset{1}{\cancel{4}}} \times \overset{\$30}{\cancel{\$120}} = \$270$$

$$\overline{\$1,095} \text{ (total cost of wood)}$$

3. What Al pays.

$$\frac{1}{\underset{1}{\cancel{5}}} \times \overset{\$219}{\cancel{\$1,095}} = \boxed{\$219}$$

You should now be ready to test your knowledge of the final unit in the chapter.

Practice Quiz

Complete this Practice Quiz to see how you are doing.

1. Multiply (use cancellation technique):

   a. $\dfrac{4}{8} \times \dfrac{4}{6}$ b. $35 \times \dfrac{4}{7}$

2. Multiply (do not use canceling; reduce by finding the greatest common divisor):

   $\dfrac{14}{15} \times \dfrac{7}{10}$

3. Complete the following. Reduce to lowest terms as needed.

   a. $\dfrac{1}{9} \div \dfrac{5}{6}$ b. $\dfrac{51}{5} \div \dfrac{5}{9}$

4. Jill Estes bought a mobile home that was $8\frac{1}{8}$ times as expensive as the home her brother bought. Jill's brother paid $16,000 for his mobile home. What is the cost of Jill's new home?

✓ Solutions

1. a. $\dfrac{\cancelto{1}{\cancel{4}}}{\underset{2}{\cancel{8}}} \times \dfrac{\cancelto{1}{\cancel{4}}}{\underset{3}{\cancel{6}}} = \boxed{\dfrac{1}{3}}$ b. $\overset{5}{\cancel{35}} \times \dfrac{4}{\underset{1}{\cancel{7}}} = \boxed{20}$

2. $\dfrac{14}{15} \times \dfrac{7}{10} = \dfrac{98 \div 2}{150 \div 2} = \boxed{\dfrac{49}{75}}$

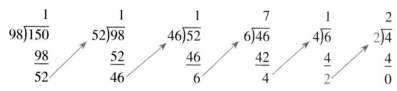

$$
\begin{array}{cccccc}
\overset{1}{98\overline{)150}} & \overset{1}{52\overline{)98}} & \overset{1}{46\overline{)52}} & \overset{7}{6\overline{)46}} & \overset{1}{4\overline{)6}} & \overset{2}{2\overline{)4}} \\
\underline{98} & \underline{52} & \underline{46} & \underline{42} & \underline{4} & \underline{4} \\
52 & 46 & 6 & 4 & 2 & 0
\end{array}
$$

3. a. $\dfrac{1}{9} \times \dfrac{6}{5} = \dfrac{6 \div 3}{45 \div 3} = \boxed{\dfrac{2}{15}}$ b. $\dfrac{51}{5} \times \dfrac{9}{5} = \dfrac{459}{25} = \boxed{18\dfrac{9}{25}}$

4. Total cost of Jill's new home:

	The facts	Solving for?	Steps to take	Key points
Blueprint	Jill's mobile home: $8\frac{1}{8}$ as expensive as her brother's. *Brother paid:* $16,000.	Total cost of Jill's new home.	$8\frac{1}{8} \times$ Total cost of Jill's brother's mobile home = Total cost of Jill's new home.	Canceling is an alternative to reducing.

Steps to solving problem

1. Convert $8\frac{1}{8}$ to a mixed number. $\dfrac{65}{8}$

2. Calculate the total cost of Jill's home. $\dfrac{65}{\underset{1}{\cancel{8}}} \times \overset{\$2,000}{\cancel{\$16,000}} = \boxed{\$130,000}$

Chapter 2 Review

Topic/Procedure/Formula	Example	You try it*
Types of fractions *Proper:* Value less than 1; numerator smaller than denominator. *Improper:* Value equal to or greater than 1; numerator equal to or greater than denominator. *Mixed:* Sum of whole number greater than zero and a proper fraction.	$\dfrac{3}{5}, \dfrac{7}{9}, \dfrac{8}{15}$ $\dfrac{14}{14}, \dfrac{19}{18}$ $6\dfrac{3}{8}, 9\dfrac{8}{9}$	**Identify type of fraction** $\dfrac{3}{10}, \dfrac{9}{8}, 1\dfrac{4}{5}$
Fraction conversions *Improper to whole or mixed:* Divide numerator by denominator; place remainder over old denominator. *Mixed to improper:* $\dfrac{\text{Whole number} \times \text{Denominator} + \text{Numerator}}{\text{Old denominator}}$	$\dfrac{17}{4} = 4\dfrac{1}{4}$ $4\dfrac{1}{8} = \dfrac{32+1}{8} = \dfrac{33}{8}$	**Convert to mixed number** $\dfrac{18}{7}$ **Convert to improper fraction** $5\dfrac{1}{7}$
Reducing fractions to lowest terms 1. Divide numerator and denominator by largest possible divisor (does not change fraction value). 2. When reduced to lowest terms, no number (except 1) will divide evenly into both numerator and denominator.	$\dfrac{18 \div 2}{46 \div 2} = \dfrac{9}{23}$	**Reduce to lowest terms** $\dfrac{16}{24}$
Step approach for finding greatest common denominator 1. Divide smaller number of fraction into larger number. 2. Divide remainder into divisor of Step 1. Continue this process until no remainder results. 3. The last divisor used is the greatest common divisor.	$\dfrac{15}{65} \longrightarrow \begin{array}{r} 4 \\ 15\overline{)65} \\ \underline{60} \\ 5 \end{array} \quad \begin{array}{r} 3 \\ 5\overline{)15} \\ \underline{15} \\ 0 \end{array}$ **5** is greatest common divisor.	**Find greatest common denominator** $\dfrac{20}{50}$
Raising fractions to higher terms Multiply numerator and denominator by same number. Does not change fraction value.	$\dfrac{15}{41} = \dfrac{?}{410}$ $410 \div 40 = 10 \times 15 = \boxed{150}$	**Raise to higher terms** $\dfrac{16}{31} = \dfrac{?}{310}$

Chapter 2 Review (Continued)

Topic/Procedure/Formula	Example	You try it*
Adding and subtracting like and unlike fractions When denominators are the same (like fractions), add (or subtract) numerators, place total over original denominator, and reduce to lowest terms. When denominators are different (unlike fractions), change them to like fractions by finding LCD using inspection or prime numbers. Then add (or subtract) the numerators, place total over LCD, and reduce to lowest terms.	$\dfrac{4}{9} + \dfrac{1}{9} = \boxed{\dfrac{5}{9}}$ $\dfrac{4}{9} - \dfrac{1}{9} = \dfrac{3}{9} = \boxed{\dfrac{1}{3}}$ $\dfrac{4}{5} + \dfrac{2}{7} = \dfrac{28}{35} + \dfrac{10}{35} = \dfrac{38}{35} = \boxed{1\dfrac{3}{35}}$	**Add** $\dfrac{3}{7} + \dfrac{2}{7}$ **Subtract** $\dfrac{5}{7} - \dfrac{2}{7}$ **Add** $\dfrac{5}{8} + \dfrac{3}{40}$
Prime numbers Whole numbers larger than 1 that are only divisible by itself and 1.	2, 3, 5, 7, 11	**List the next two prime numbers after 11**
LCD by prime numbers 1. Copy denominators and arrange them in a separate row. 2. Divide denominators by smallest prime number that will divide evenly into at least two numbers. 3. Continue until no prime number divides evenly into at least two numbers. 4. Multiply all the numbers in the divisors and last row to find LCD. 5. Raise fractions so each has a common denominator and complete computation.	$\dfrac{1}{3} + \dfrac{1}{6} + \dfrac{1}{8} + \dfrac{1}{12} + \dfrac{1}{9}$ $2\,\underline{/\,3\quad 6\quad 8\quad 12\quad 9}$ $2\,\underline{/\,3\quad 3\quad 4\quad 6\quad 9}$ $3\,\underline{/\,3\quad 3\quad 2\quad 3\quad 9}$ $\,1\quad 1\quad 2\quad 1\quad 3$ $2 \times 2 \times 3 \times 1 \times 1 \times 2 \times 1 \times 3 = \boxed{72}$	**Find LCD** $\dfrac{1}{2} + \dfrac{1}{4} + \dfrac{1}{5}$
Adding mixed numbers 1. Add fractions. 2. Add whole numbers. 3. Combine totals of Steps 1 and 2. If denominators are different, a common denominator must be found. Answer cannot be left as improper fraction.	$1\dfrac{4}{7} + 1\dfrac{3}{7}$ Step 1: $\dfrac{4}{7} + \dfrac{3}{7} = \dfrac{7}{7}$ Step 2: $1 + 1 = 2$ Step 3: $2\dfrac{7}{7} = \boxed{3}$	**Add mixed numbers** $2\dfrac{1}{4} + 3\dfrac{3}{4}$
Subtracting mixed numbers 1. Subtract fractions. 2. If necessary, borrow from whole numbers. 3. Subtract whole numbers and fractions if borrowing was necessary. 4. Reduce fractions to lowest terms. If denominators are different, a common denominator must be found.	$12\dfrac{2}{5} - 7\dfrac{3}{5}$ $11\dfrac{7}{5} - 7\dfrac{3}{5}$ $= 4\dfrac{4}{5}$ Due to borrowing $\dfrac{5}{5}$ from number 12 $\dfrac{5}{5} + \dfrac{2}{5} = \dfrac{7}{5}$ The whole number is now 11.	**Subtract mixed numbers** $11\dfrac{1}{3}$ $-2\dfrac{2}{3}$

Chapter 2 Review (Continued)

Topic/Procedure/Formula	Example	You try it*
Multiplying proper fractions 1. Multiply numerators and denominators. 2. Reduce answer to lowest terms or use cancellation method.	$\dfrac{4}{\cancel{7}_{1}} \times \dfrac{\cancel{7}^{1}}{9} = \dfrac{4}{9}$	**Multiply and reduce** $\dfrac{4}{5} \times \dfrac{25}{26}$
Multiplying mixed numbers 1. Convert mixed numbers to improper fractions. 2. Multiply numerators and denominators. 3. Reduce answer to lowest terms or use cancellation method.	$1\dfrac{1}{8} \times 2\dfrac{5}{8}$ $\dfrac{9}{8} \times \dfrac{21}{8} = \dfrac{189}{64} = 2\dfrac{61}{64}$	**Multiply and reduce** $2\dfrac{1}{4} \times 3\dfrac{1}{4}$
Dividing proper fractions 1. Invert divisor. 2. Multiply. 3. Reduce answer to lowest terms or use cancellation method.	$\dfrac{1}{4} \div \dfrac{1}{8} = \dfrac{1}{\cancel{4}_{1}} \times \dfrac{\cancel{8}^{2}}{1} = 2$	**Divide** $\dfrac{1}{8} \div \dfrac{1}{4}$
Dividing mixed numbers 1. Convert mixed numbers to improper fractions. 2. Invert divisor and multiply. If final answer is an improper fraction, reduce to lowest terms by finding greatest common divisor or using the cancellation method.	$1\dfrac{1}{2} \div 1\dfrac{5}{8} = \dfrac{3}{2} \div \dfrac{13}{8}$ $= \dfrac{3}{\cancel{2}_{1}} \times \dfrac{\cancel{8}^{4}}{13}$ $= \dfrac{12}{13}$	**Dividing mixed numbers** $3\dfrac{1}{4} \div 1\dfrac{4}{5}$

Key Terms

Cancellation	Higher terms	Mixed numbers
Common denominator	Improper fraction	Numerator
Denominator	Least common denominator (LCD)	Prime numbers
Equivalent		Proper fraction
Fraction	Like fractions	Reciprocal
Greatest common divisor	Lowest terms	Unlike fractions

*Worked-out solutions are in Appendix A.

Chapter 2 Review (Continued)

Critical Thinking Discussion Questions with Chapter Concept Check

1. What are the steps to convert improper fractions to whole or mixed numbers? Give an example of how you could use this conversion procedure when you eat at Pizza Hut.

2. What are the steps to convert mixed numbers to improper fractions? Show how you could use this conversion procedure when you order doughnuts at Dunkin' Donuts.

3. What is the greatest common divisor? How could you use the greatest common divisor to write an advertisement showing that 35 out of 60 people prefer MCI to AT&T?

4. Explain the step approach for finding the greatest common divisor. How could you use the MCI–AT&T example in question 3 to illustrate the step approach?

5. Explain the steps of adding or subtracting unlike fractions. Using a ruler, measure the heights of two different-size cans of food and show how to calculate the difference in height.

6. What is a prime number? Using the two cans in question 5, show how you could use prime numbers to calculate the LCD.

7. Explain the steps for multiplying proper fractions and mixed numbers. Assume you went to Staples (a stationery superstore). Give an example showing the multiplying of proper fractions and mixed numbers.

8. **Chapter Concept Check.** Using all the information you have learned about fractions, search the web to find out how many cars are produced in the United States in a year and what fractional part represents cars produced by foreign-owned firms. Finally, present calculations using fractions.

9. Explain how you can use fractions to summarize the pandemic in the United States.

End-of-Chapter Problems

Name _____   Date _____

Check figures for odd-numbered problems in Appendix A.

Drill Problems

Identify the following types of fractions: *LU 2-1(1)*

2–1. $\dfrac{9}{10}$

2–2. $\dfrac{12}{11}$

2–3. $\dfrac{25}{13}$

Convert the following to mixed numbers: *LU 2-1(2)*

2–4. $\dfrac{91}{10}$

2–5. $\dfrac{921}{15}$

Convert the following to improper fractions: *LU 2-1(2)*

2–6. $8\dfrac{7}{8}$

2–7. $19\dfrac{2}{3}$

Reduce the following to the lowest terms. Show how to calculate the greatest common divisor by the step approach.   *LU 2-1(3)*

2–8. $\dfrac{16}{38}$

2–9. $\dfrac{44}{52}$

Convert the following to higher terms: *LU 2-1(3)*

2–10. $\dfrac{9}{10} = \dfrac{}{70}$

Determine the LCD of the following (a) by inspection and (b) by division of prime numbers: *LU 2-2(2)*

2–11. $\dfrac{3}{4}, \dfrac{7}{12}, \dfrac{5}{6}, \dfrac{1}{5}$ **Check**

  Inspection

2–12. $\dfrac{5}{6}, \dfrac{7}{18}, \dfrac{5}{9}, \dfrac{2}{72}$ **Check**

  Inspection

2–13. $\dfrac{1}{4}, \dfrac{3}{32}, \dfrac{5}{48}, \dfrac{1}{8}$ **Check**

  Inspection

Add the following and reduce to lowest terms: *LU 2-2(1), LU 2-1(3)*

2–14. $\dfrac{3}{9} + \dfrac{3}{9}$

2–15. $\dfrac{3}{7} + \dfrac{4}{21}$

2–16. $6\dfrac{1}{8} + 4\dfrac{3}{8}$

2–17. $6\dfrac{3}{8} + 9\dfrac{1}{24}$

2–18. $9\dfrac{9}{10} + 6\dfrac{7}{10}$

Subtract the following and reduce to lowest terms: *LU 2-2(3), LU 2-1(3)*

2–19. $\dfrac{11}{12} - \dfrac{1}{12}$

2–20. $14\dfrac{3}{8} - 10\dfrac{5}{8}$

2–21. $12\dfrac{1}{9} - 4\dfrac{2}{3}$

Multiply the following and reduce to lowest terms. Do not use the cancellation technique for these problems.   *LU 2-3(1), LU 2-1(3)*

2–22. $17 \times \dfrac{4}{2}$

2–23. $\dfrac{5}{6} \times \dfrac{3}{8}$

2–24. $8\dfrac{7}{8} \times 64$

Multiply the following. Use the cancellation technique.   *LU 2-3(1), LU 2-1(2)*

2–25. $\dfrac{4}{10} \times \dfrac{30}{60} \times \dfrac{6}{10}$

2–26. $3\dfrac{3}{4} \times \dfrac{8}{9} \times 4\dfrac{9}{12}$

Divide the following and reduce to lowest terms. Use the cancellation technique as needed. *LU 2-3(2), LU 2-1(2)*

2–27. $\dfrac{12}{9} \div 4$

2–28. $18 \div \dfrac{1}{5}$

2–29. $4\dfrac{2}{3} \div 12$

2–30. $3\dfrac{5}{6} \div 3\dfrac{1}{2}$

Word Problems

2–31. Michael Wittry has been investing in his Roth IRA retirement account for 20 years. Two years ago, his account was worth \$215,658. After losing $\frac{1}{3}$ of its original value, it then gained $\frac{1}{2}$ of its new value back. What is the current value of his Roth IRA?   *LU 2-3(1)*

End-of-Chapter Problems (Continued)

2–32. Delta pays Pete Rose $180 per day to work in the maintenance department at the airport. Pete became ill on Monday and went home after $\frac{1}{6}$ of a day. What did he earn on Monday? Assume no work, no pay. *LU 2-3(1)*

2–33. The Spanish flu infected $\frac{1}{3}$ of the worldwide population in 1918–1919. If the worldwide population was 1,500,000, how many people contracted the disease?

2–34. Joy Wigens, who works at Putnam Investments, received a check for $1,600. She deposited $\frac{1}{4}$ of the check in her Citibank account. How much money does Joy have left after the deposit? *LU 2-3(1)*

2–35. Lee Jenkins worked the following hours as a manager for a local Pizza Hut: $14\frac{1}{4}, 5\frac{1}{4}, 8\frac{1}{2}$ and $7\frac{1}{4}$. How many total hours did Lee work? *LU 2-2(1)*

2–36. Lester bought a piece of property in Vail, Colorado. The sides of the land measure $115\frac{1}{2}$ feet, $66\frac{1}{4}$ feet, $106\frac{1}{8}$ feet, and $110\frac{1}{4}$ feet. Lester wants to know the perimeter (sum of all sides) of his property. Can you calculate the perimeter for Lester? *LU 2-2(1)*

2–37. Tiffani Lind got her new weekly course schedule from Roxbury Community College in Boston. Following are her classes and their length: Business Math, $2\frac{1}{2}$ hours; Introduction to Business, $1\frac{1}{2}$ hours; Microeconomics, $1\frac{1}{2}$ hours; Spanish, $2\frac{1}{4}$ hours; Marketing, $1\frac{1}{4}$ hours; and Business Statistics, $1\frac{3}{4}$ hours. How long will she be in class each week? *LU 2-2(1)*

2–38. Seventy-seven million people were born between 1946 and 1964. The U.S. Census classifies this group of individuals as baby boomers. It is said that today and every day for the next 18 years, 10,000 baby boomers will reach 65. If $\frac{1}{4}$ of the 65 and older age group uses e-mail, $\frac{1}{5}$ obtains the news from the Internet, and $\frac{1}{6}$ searches the Internet, find the LCD and determine total technology usage for this age group as a fraction. *LU 2-2(1, 2)*

2–39. At a local Walmart store, a Coke dispenser held $19\frac{1}{4}$ gallons of soda. During working hours, $12\frac{3}{4}$ gallons were dispensed. How many gallons of Coke remain?   *LU 2-2(2, 3)*

2–40. If two coronavirus vaccines have been administered to a total of 398,675,414 people in the United States, and the Pfizer-BioNTech vaccine was administered to $\frac{9}{17}$ of the U.S. population, how many people received the Moderna vaccine? Round to the nearest whole person.

2–41. A local garden center charges $250 per cord of wood. If Logan Grace orders $3\frac{1}{2}$ cords, what will the total cost be?   *LU 2-3(1)*

2–42. A local Target store bought 90 pizzas at Pizza Hut for its holiday party. Each guest ate $\frac{1}{6}$ of a pizza and there was no pizza left over. How many guests did Target have for the party?   *LU 2-3(1)*

2–43. Marc, Steven, and Daniel entered into a Subway sandwich shop partnership. Marc owns $\frac{1}{9}$ of the shop and Steven owns $\frac{1}{4}$. What part does Daniel own?   *LU 2-2(1, 2)*

2–44. Lionel Sullivan works for Burger King. He is paid time and one-half for Sundays. If Lionel works on Sunday for 6 hours at a regular pay of $8 per hour, what does he earn on Sunday?   *LU 2-3(1)*

2–45. DaveRamsey.com's "Baby Step 3" out of "7 Baby Steps" for financial health recommends a $1,000 emergency fund if you have debt; and, once you are free of debt, he recommends a fully funded emergency fund of at least six months of monthly expenses depending on your job situation. If your starting goal is to have a 4-month emergency fund and your monthly expenses total $2,750, how much more do you have to save if you currently have $\frac{2}{3}$ of your fund saved? Round to the nearest dollar.

excel 2–46. A trip to the White Mountains of New Hampshire from Boston will take you $2\frac{3}{4}$ hours. Assume you have traveled $\frac{1}{11}$ of the way. How much longer will the trip take?
LU 2-3(1, 2)

End-of-Chapter Problems (Continued)

excel **2–47.** Andy, who loves to cook, makes apple cobbler for his family. The recipe (serves 6) calls for $1\frac{1}{2}$ pounds of apples, $3\frac{1}{4}$ cups of flour, $\frac{1}{4}$ cup of margarine, $2\frac{3}{8}$ cups of sugar, and 2 teaspoons of cinnamon. Since guests are coming, Andy wants to make a cobbler that will serve 15 (or increase the recipe $2\frac{1}{2}$ times). How much of each ingredient should Andy use? *LU 2-3(1, 2)*

2–48. Mobil allocates $1,692\frac{3}{4}$ gallons of gas per month to Jerry's Service Station. The first week, Jerry sold $275\frac{1}{2}$ gallons; second week, $280\frac{1}{4}$ gallons; and third week, $189\frac{1}{8}$ gallons. If Jerry sells $582\frac{1}{2}$ gallons in the fourth week, how close is Jerry to selling his allocation? *LU 2-2(4)*

2–49. A marketing class at North Shore Community College conducted a viewer preference survey. The survey showed that $\frac{5}{6}$ of the people surveyed preferred Apple's iPhone over the Blackberry. Assume 2,400 responded to the survey. How many favored using a Blackberry? *LU 2-3(1, 2)*

2–50. The price of a used Toyota LandCruiser has increased to $1\frac{1}{4}$ times its earlier price. If the original price of the LandCruiser was $30,000, what is the new price? *LU 2-3(1, 2)*

2–51. Tempco Corporation has a machine that produces $12\frac{1}{2}$ baseball gloves each hour. In the last 2 days, the machine has run for a total of 22 hours. How many baseball gloves has Tempco produced? *LU 2-3(2)*

2–52. Alicia, an employee of Dunkin' Donuts, receives $23\frac{1}{4}$ days per year of vacation time. So far this year she has taken $3\frac{1}{8}$ days in January, $5\frac{1}{2}$ days in May, $6\frac{1}{4}$ days in July, and $4\frac{1}{4}$ days in September. How many more days of vacation does Alicia have left? *LU 2-2(1, 2, 3)*

excel 2–53. A Hamilton multitouch watch was originally priced at $600. At a closing of the Alpha Omega Jewelry Shop, the watch is being reduced by $\frac{1}{4}$. What is the new selling price? *LU 2-3(1)*

2–54. Shelly Van Doren hired a contractor to refinish her kitchen. The contractor said the job would take $49\frac{1}{2}$ hours. To date, the contractor has worked the following hours:

Monday	$4\frac{1}{4}$
Tuesday	$9\frac{1}{8}$
Wednesday	$4\frac{1}{4}$
Thursday	$3\frac{1}{2}$
Friday	$10\frac{5}{8}$

How much longer should the job take to be completed? *LU 2-2(4)*

2–55. An issue of *Taunton's Fine Woodworking* included plans for a hall stand. The total height of the stand is $81\frac{1}{2}$ inches. If the base is $36\frac{5}{16}$ inches, how tall is the upper portion of the stand? *LU 2-2(4)*

2–56. Albertsons grocery planned a big sale on apples and received 750 crates from the wholesale market. Albertsons will bag these apples in plastic. Each plastic bag holds $\frac{1}{9}$ of a crate. If Albertsons has no loss to perishables, how many bags of apples can be prepared? *LU 2-3(1)*

End-of-Chapter Problems (Continued)

2–57. Frank Puleo bought 6,625 acres of land in ski country. He plans to subdivide the land into parcels of $13\frac{1}{4}$ acres each. Each parcel will sell for $125,000. How many parcels of land will Frank develop? If Frank sells all the parcels, what will be his total sales? *LU 2-3(1)*

If Frank sells $\frac{3}{5}$ of the parcels in the first year, what will be his total sales for the year?

2–58. A local Papa Gino's conducted a food survey. The survey showed that $\frac{1}{9}$ of the people surveyed preferred eating pasta to hamburger. If 5,400 responded to the survey, how many actually favored hamburger? *LU 2-3(1)*

2–59. Tamara, Jose, and Milton entered into a partnership that sells men's clothing on the web. Tamara owns $\frac{3}{8}$ of the company and Jose owns $\frac{1}{4}$. What part does Milton own? *LU 2-2(1, 3)*

2–60. *Quilters Newsletter Magazine* gave instructions on making a quilt. The quilt required $4\frac{1}{2}$ yards of white-on-white print, 2 yards blue check, $\frac{1}{2}$ yard blue-and-white stripe, $2\frac{3}{4}$ yards blue scraps, $\frac{3}{4}$ yard yellow scraps, and $4\frac{7}{8}$ yards lining. How many total yards are needed? *LU 2-2(1, 2)*

2–61. A trailer carrying supplies for a Krispy Kreme from Virginia to New York will take $3\frac{1}{4}$ hours. If the truck traveled $\frac{1}{5}$ of the way, how much longer will the trip take? *LU 2-3(1, 2)*

2–62. Land Rover has increased the price of a FreeLander by $\frac{1}{5}$ from the original price. The original price of the FreeLander was $30,000. What is the new price? *LU 2-3(1, 2)*

Challenge Problems

2–63. *Woodsmith* magazine gave instructions on how to build a pine cupboard. Lumber will be needed for two shelves $10\frac{1}{4}$ inches long, two base sides $12\frac{1}{2}$ inches long, and two door stiles $29\frac{1}{8}$ inches long. Your lumber comes in 6 foot lengths. **(a)** How many feet of lumber will you need? **(b)** If you want $\frac{1}{2}$ a board left over, is this possible with two boards?
LU 2-2(1, 2, 3, 4)

End-of-Chapter Problems (Continued)

2–64. Jack MacLean has entered into a real estate development partnership with Bill Lyons and June Reese. Bill owns $\frac{1}{4}$ of the partnership, while June has a $\frac{1}{5}$ interest. The partners will divide all profits on the basis of their fractional ownership. The partnership bought 900 acres of land and plans to subdivide each lot into $2\frac{1}{4}$ acres. Homes in the area have been selling for $240,000. By time of completion, Jack estimates the price of each home will increase by $\frac{1}{3}$ of the current value. The partners sent a survey to 12,000 potential customers to see whether they should heat the homes with oil or gas. One-fourth of the customers responded by indicating a 5-to-1 preference for oil. From the results of the survey, Jack now plans to install a 270-gallon oil tank at each home. He estimates that each home will need five fills per year. The current price of home heating fuel is $1 per gallon. The partnership estimates its profit per home will be $\frac{1}{8}$ the selling price of each home. From the above, please calculate the following:
LU 2-1(1, 2, 3), LU 2-2(1, 2, 3, 4), LU 2-3(1, 2)

a. Number of homes to be built.   **b.** Selling price of each home.

c. Number of people responding to survey.   **d.** Number of people desiring oil.

e. Average monthly cost per house to heat using oil.

f. Amount of profit Jack will receive from the sale of homes.

Summary Practice Test

Do you need help? Connect videos have step-by-step worked-out solutions.

Identify the following types of fractions. *LU 2-1(1)*

1. $5\dfrac{1}{8}$

2. $\dfrac{2}{7}$

3. $\dfrac{20}{19}$

4. Convert the following to a mixed number. *LU 2-1(2)*

$$\dfrac{163}{9}$$

5. Convert the following to an improper fraction. *LU 2-1(2)*

$$8\dfrac{1}{8}$$

6. Calculate the greatest common divisor of the following by the step approach and reduce to lowest terms. *LU 2-2(1, 2)*

$$\dfrac{63}{90}$$

7. Convert the following to higher terms. *LU 2-1(3)*

$$\dfrac{16}{94} = \dfrac{?}{376}$$

8. Find the LCD of the following by using prime numbers. Show your work. *LU 2-2(2)*

$$\dfrac{1}{8} + \dfrac{1}{3} + \dfrac{1}{2} + \dfrac{1}{12}$$

9. Subtract the following. *LU 2-2(4)*

$$15\dfrac{4}{5}$$
$$-8\dfrac{19}{20}$$

Complete the following using the cancellation technique. *LU 2-3(1, 2)*

10. $\dfrac{3}{4} \times \dfrac{2}{4} \times \dfrac{6}{9}$

11. $7\dfrac{1}{9} \times \dfrac{6}{7}$

12. $\dfrac{3}{7} \div 6$

13. A trip to Washington from Boston will take you $5\frac{3}{4}$ hours. If you have traveled $\frac{1}{3}$ of the way, how much longer will the trip take? *LU 2-3(1)*

Summary Practice Test (Continued)

14. Quiznos produces 640 rolls per hour. If the oven runs $12\frac{1}{4}$ hours, how many rolls will the machine produce? *LU 2-3(1, 2)*

15. A taste-testing survey of Zing Farms showed that $\frac{2}{3}$ of the people surveyed preferred the taste of veggie burgers to regular burgers. If 90,000 people were in the survey, how many favored veggie burgers? How many chose regular burgers? *LU 2-3(1)*

16. Jim Janes, an employee of Enterprise Co., worked $9\frac{1}{4}$ hours on Monday, $4\frac{1}{2}$ hours on Tuesday, $9\frac{1}{4}$ hours on Wednesday, $7\frac{1}{2}$ hours on Thursday, and 9 hours on Friday. How many total hours did Jim work during the week? *LU 2-2(1, 2)*

17. JCPenney offered a $\frac{1}{3}$ rebate on its $39 hair dryer. Joan bought a JCPenney hair dryer. What did Joan pay after the rebate? *LU 2-3(1)*

Q Landing Your Dream Job!

 What I need to know

As you work through your college courses, you aspire to attaining a degree or credential to get you the job, promotion, or career move you are seeking. As you prepare for that eventual career it is important to understand and anticipate the level of salary you will earn upon your graduation. Additionally, assess the types of employment matching closest to your career aspirations. The potential job market has changed due to the coronavirus pandemic. As you consider your career options you may discover your ideal position is not what you had originally anticipated. This newfound flexibility can be used to your advantage to pursue a variety of career options.

What I need to do

Research expected salaries before committing to a desired course of study. This information is available through your institution and is provided based upon the program of study you pursue. Although only a range may be given from your college, it will give you a rough estimate from which to determine your educational path. Compare the cost you will incur to attain your degree to the expected salary upon graduation to determine the cost effectiveness of each degree option you are considering. Ultimately you will want to determine whether a career field will fit into your financial plans. How does this salary range compare to your financial goals and will you be able to meet these goals with such earnings? Place the salary expectation against your budget to see how it will meet your expenses. Furthermore, determine what salary range will allow for spending opportunities outside of your expenses such as investments, savings, entertainment, etc.

Seek out advice from professionals in the field you are considering for some insights on the career. Ask these professionals about their personal experiences within this career field. What do they like best about their chosen profession? What do they see as the future opportunities within this career? Are there other factors to consider outside of just salary such as benefits, personal growth, contribution to a greater cause, etc.? If these professionals had it to do all over again, what might they do differently as it relates to career preparedness? The knowledge you gain will assist you in selecting an educational field of study to achieve your desired career.

 Steps I need to take

1. Be flexible with your job search and open yourself to a variety of career options.
2. Know the financial impact of your chosen career to determine best fit.
3. Learn from others by gaining valuable insights from within your desired career field.

 Resources I can use

- Indeed Job Search (mobile app)—find your next career and express your interest directly with employers.
- https://www.payscale.com/—salary expectations by education level, job title and much more.
- https://www.themuse.com/advice/job-search-coronavirus—tips for how your job search has changed because of the coronavirus pandemic.

MY MONEY ACTIVITY

- Search for job openings related to your degree
- Compare the expected salaries to obtain a range for this position.
- How does the expected salary fit into your financial goals?

PERSONAL FINANCE

A KIPLINGER APPROACH

"Farewell to the Office." Kiplinger's. February 2021.

FROM <u>THE</u> EDITOR
MARK SOLHEIM

FAREWELL TO THE OFFICE

We can all list numerous ways, good and bad, that the pandemic has altered our lives. It gets trickier to identify changes that will last long after the risk of COVID has faded. But for many people, working remotely is here to stay.

My wife, Allyson, and I have been working remotely since mid March—initially from Michigan and lately from our Washington, D.C., home. In our northern Michigan house, we each have our own home office. In D.C., I have ceded the spare bedroom/home office to Allyson because, unlike me, she typically has back-to-back video calls. My workday finds me perched on a stool at the kitchen island or slouched in one of the chairs next to the fireplace, having conversations with Alexa.

In theory, we get to recapture time we used to spend commuting. But I find myself rolling out of bed and turning on the computer, and it takes discipline to ignore work e-mails. We do appreciate the more-relaxed wardrobe demands. My go-to garb includes T-shirts, a couple of sweaters and a pair of blue jeans (the bit of stretchiness accommodates the extra pounds I attribute to the 5 P.M. drinks and hors d'oeuvres).

I miss the ergonomics of the office and the structure that working there gave to my day. I miss seeing colleagues face to face rather than in two dimensions in a Hollywood Squares grid. And I particularly worry about our younger staffers missing out on the collaboration and camaraderie that can be a crucial part of their professional development.

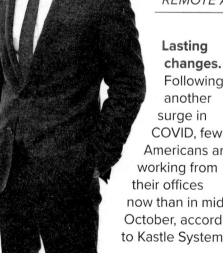

Poon Watchara-Amphaiwan

Lasting changes. Following another surge in COVID, fewer Americans are working from their offices now than in mid October, according to Kastle Systems, the security firm that provides badges to swipe into offices. In late November, fewer than 18% of workers had returned to the office, compared with 27% a month earlier.

That more or less jibes with what we found in a nationwide survey Kiplinger commissioned, in partnership with Personal Capital, in early November (see page 58 for highlights). The survey asked how retirement savers were doing in the wake of coronavirus, and it accompanies our cover story on getting your nest egg back on track (see page 48). We sprinkled in a few questions about

---

A SURVEY OF HIRING MANAGERS FOUND THAT ONE-FIFTH OF THE WORKFORCE COULD BE ENTIRELY REMOTE AFTER THE PANDEMIC.

---

remote work. More than 70% of respondents confirmed they were working from another location besides their normal workplace: 56% from their primary residence, 10% from a second home, and 5% from "another location." Two-thirds said they "love it," and one-third said it was okay but they wanted to return to the office to see colleagues. Only 6% said they don't like it at all.

According to a recent survey of hiring managers by the global freelancing platform Upwork, one-fifth of the workforce could be entirely remote after the pandemic. And that trend has far-reaching, long-term ramifications, not only for retailers selling business clothes but also for real estate and transportation. White collar and tech workers in job-rich urban areas have been fleeing cramped, high-rent apartments and pricey condos for the suburbs and exurbs. Many offices sit almost empty. Mass transit systems are struggling. Here in D.C., Metro announced that it may have to curtail bus routes and completely end weekend subway service. That leaves lower-paid, front-line workers in the lurch, or at least the ones who are left with jobs as the service, travel and hospitality industries go through upheaval. Many downtown coffee shops, bars and restaurants are endangered species.

For workers who are telecommuting from another state—or even from abroad—we offer advice on taxes, health care and other considerations starting on page 38. The tax rules are hellishly complex. We'll be on the hook for Michigan income taxes for the time we worked there, but D.C. offers a credit. This may be the year I break a lifetime string of DIY tax returns and recruit an accountant. ■

Mark Solheim

MARK SOLHEIM, EDITOR
MARK_SOLHEIM@KIPLINGER.COM
TWITTER: @MARKSOLHEIM

Business Math Issue

Post Pandemic means $\frac{3}{5}$ of workers will still work.

1. List the key points of the article and information to support your position.

2. Write a group defense of your position using math calculations to support your view. If you are in an online course, post to a discussion board.

Chapter 3
Decimals

Learning Unit Objectives

LU 3–1: Rounding Decimals; Fraction and Decimal Conversions

1. Explain the place values of whole numbers and decimals; round decimals.
2. Convert decimal fractions to decimals, proper fractions to decimals, mixed numbers to decimals, and pure and mixed decimals to decimal fractions.

LU 3–2: Adding, Subtracting, Multiplying, and Dividing Decimals

1. Add, subtract, multiply, and divide decimals.
2. Complete decimal applications in foreign currency.
3. Multiply and divide decimals by shortcut methods.

ⓔ Essential Question

How can I use decimals to understand the business world and make my life easier?

🌐 Math Around the World

In the chapter opener, "Lyft Records Severe Revenue, Ridership Drop in Pandemic", Lyft has lost many riders due to the pandemic. Let's look at how many riders were lost.

$$
\begin{array}{r}
21.8 \text{ million} \\
-8.7 \text{ million} \\
\hline
13.1 \text{ million riders lost}
\end{array}
$$

This chapter is divided into two learning units. The first unit discusses rounding decimals, converting fractions to decimals, and converting decimals to fractions. The second unit shows you how to add, subtract, multiply, and divide decimals, along with some shortcuts for multiplying and dividing decimals. Added to this unit is a global application of decimals dealing with foreign exchange rates. One of the most common uses of decimals occurs when we spend dollars and cents, which is a *decimal number*.

Decimals are decimal numbers with digits to the right of a *decimal point*, indicating that decimals, like fractions, are parts of a whole that are less than one. Thus, we can interchange the terms *decimals* and *decimal numbers*. Remembering this will avoid confusion between the terms *decimal, decimal number,* and *decimal point*.

Chapter 2 introduced the 1.69-ounce bag of M&M'S®. In Table 3.1, the six colors in that 1.69-ounce bag of M&M'S® are given in fractions and their values expressed in decimal equivalents that are rounded to the nearest hundredths.

Lyft Records Severe Revenue, Ridership Drop in Pandemic

BY PREETIKA RANA

Lyft Inc. reported a dramatic drop in riders and revenue in the second quarter, as rising coronavirus infections in the U.S. and prolonged shutdowns weighed on its results.

The San Francisco-based ride-hailing company said it had 8.7 million active riders in the three months ended June 30, compared with 21.8 million in the year-earlier quarter and 21.2 million in the previous quarter, when the pandemic hit the U.S., keeping people at home and disrupting travel.

Rana, Preetika. "Lyft Records Severe Revenue, Ridership Drop in Pandemic." *The Wall Street Journal*, August 13, 2020.

Table 3-1 Analyzing a bag of M&M'S®

Color*	Fraction	Decimal
Yellow	$\frac{18}{55}$	.33
Red	$\frac{10}{55}$	.18
Blue	$\frac{9}{55}$	.16
Orange	$\frac{7}{55}$	.13
Brown	$\frac{6}{55}$	.11
Green	$\frac{5}{55}$	.09
Total	$\frac{55}{55} = 1$	1.00

*The color ratios currently given are a sample used for educational purposes. They do not represent the manufacturer's color ratios.

Learning Unit 3–1:

Rounding Decimals; Fraction and Decimal Conversions

In Chapter 1 we stated that the **decimal point** is the center of the decimal numbering system. So far we have studied the whole numbers to the left of the decimal point and the parts of whole numbers called fractions. We also learned that the position of the digits in a whole number gives the place values of the digits (Figure 1.1). Now we will study the position (place values) of the digits to the right of the decimal point (Figure 3.1). Note that the words to the right of the decimal point end in *ths*.

You should understand why the decimal point is the center of the decimal system. If you move a digit to the left of the decimal point by place (ones, tens, and so on), *you increase its value 10 times for each place (power of 10).* If you move a digit to the right of the decimal point by place (tenths, hundredths, and so on), *you decrease its value 10 times for each place.*

Examples: $.06 \longleftarrow$ The 6 is in the hundred*ths* place value.

        1.527 $\longrightarrow$ The 5 is in the ten*ths* place value.

        2.8394 $\longrightarrow$ The 4 is in the ten thousand*ths* place value.

        .33 $\longrightarrow$ The thirty-three hundred*ths* represents the yellow M&M'S® in our M&M'S® bag of 55 M&M'S®.

        1.69 oz. $\longrightarrow$ The one ounce and sixty-nine hundred*ths* of another ounce is the weight of our bag of M&M'S®.

Do you recall from Chapter 1 how you used a place-value chart to read or write whole numbers in verbal form? To read or write decimal numbers, you read or write the decimal number as if it were a whole number. Then you use the name of the decimal place of the last digit as given in Figure 3.1. For example, you would read or write the decimal .0796 as seven hundred ninety-six ten thousandths (the last digit, 6, is in the ten thousandths place).

FIGURE 3.1 Decimal place-value chart

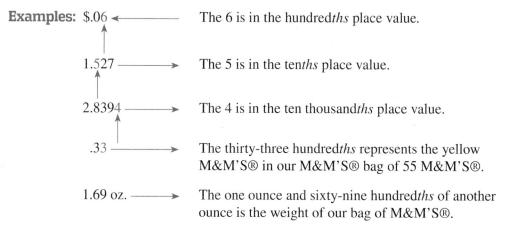

	Whole Number Groups				Decimal Place Values				
Thousands	Hundreds	Tens	Ones (units)	Decimal point (and)	Tenths	Hundredths	Thousandths	Ten thousandths	Hundred thousandths
1,000	100	10	1	and	$\frac{1}{10}$	$\frac{1}{100}$	$\frac{1}{1,000}$	$\frac{1}{10,000}$	$\frac{1}{100,000}$

To read a decimal with four or fewer whole numbers, you can also refer to Figure 3.1. For larger whole numbers, refer to the whole number place-value chart in Chapter 1 (Figure 1.1). For example, from Figure 3.1 you would read the number 126.2864 as one hundred twenty-six and two thousand eight hundred sixty-four ten thousandths. *Remember to read the decimal point as* "and."

Now let's round decimals. Rounding decimals is similar to the rounding of whole numbers that you learned in Chapter 1.

Learn: Rounding Decimals

From Table 3.1, you know that the 1.69-ounce bag of M&M'S® introduced in Chapter 2 contained $\frac{18}{55}$, or .33, yellow M&M'S®. The .33 was rounded to the nearest hundredth. **Rounding decimals** involves the following steps:

Rounding Decimals to a Specified Place Value

Step 1 Identify the place value of the digit you want to round.

Step 2 If the digit to the right of the identified digit in Step 1 is 5 or more, increase the identified digit by 1. If the digit to the right is less than 5, do not change the identified digit.

Step 3 Drop all digits to the right of the identified digit.

Let's practice rounding by using the $\frac{18}{55}$ yellow M&M'S® that we rounded to .33 in Table 3.1. Before we rounded $\frac{18}{55}$ to .33, the number we rounded was .32727. This is an example of a **repeating decimal** since the 27 repeats itself.

Cartoon Collections

Example: Round .3272727 to the nearest hundredth.

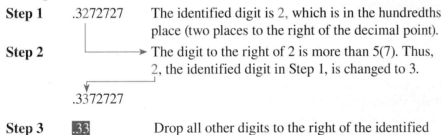

Step 1 .3272727 The identified digit is 2, which is in the hundredths place (two places to the right of the decimal point).

Step 2 The digit to the right of 2 is more than 5(7). Thus, 2, the identified digit in Step 1, is changed to 3.

.3372727

Step 3 .33 Drop all other digits to the right of the identified digit 3.

We could also round the .3272727 M&M'S® to the nearest tenth or thousandth as follows:

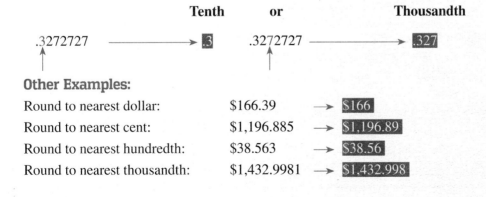

	Tenth	**or**		**Thousandth**
.3272727	→ .3	.3272727	→	.327

Other Examples:

Round to nearest dollar:	$166.39	→ $166
Round to nearest cent:	$1,196.885	→ $1,196.89
Round to nearest hundredth:	$38.563	→ $38.56
Round to nearest thousandth:	$1,432.9981	→ $1,432.998

The rules for rounding can differ with the situation in which rounding is used. For example, have you ever bought one item from a supermarket produce department that was marked "3 for $1" and noticed what the cashier charged you? One item marked "3 for $1" would not cost you $33\frac{1}{3}$ cents rounded to 33 cents. You will pay 34 cents. Many retail stores round to the next cent even if the digit following the identified digit is less than $\frac{1}{2}$ of a penny. In this text we round on the concept of 5 or more.

Learn: Fraction and Decimal Conversions

In business operations we must frequently convert fractions to decimal numbers and decimal numbers to fractions. This section begins by discussing three types of fraction-to-decimal conversions. Then we discuss converting pure and mixed decimals to decimal fractions.

2014 Jef Mallett/Andrews McMeel Syndication

Converting Decimal Fractions to Decimals   From Figure 3.1 you can see that a **decimal fraction** (expressed in the digits to the right of the decimal point) is a fraction with a denominator that has a power of 10, such as $\frac{1}{10}$, $\frac{17}{100}$, and $\frac{23}{1,000}$. To convert a decimal fraction to a decimal, follow these steps:

Converting Decimal Fractions to Decimals

Step 1 Count the number of zeros in the denominator.

Step 2 Place the numerator of the decimal fraction to the right of the decimal point the same number of places as you have zeros in the denominator. (The number of zeros in the denominator gives the number of digits your decimal has to the right of the decimal point.) Do not go over the total number of denominator zeros.

Now let's change $\frac{3}{10}$ and its higher multiples of 10 to decimals.

Examples:

Verbal form	Decimal fraction	Decimal[1]	Number of decimal places to right of decimal point
a. Three tenths	$\dfrac{3}{10}$	.3	1
b. Three hundredths	$\dfrac{3}{100}$	.03	2
c. Three thousandths	$\dfrac{3}{1,000}$	.003	3
d. Three ten thousandths	$\dfrac{3}{10,000}$	.0003	4

Note how we show the different values of the decimal fractions above in decimals. The zeros after the decimal point and before the number 3 indicate these values. If you add zeros after the number 3, you do not change the value. Thus, the numbers .3, .30, and .300 have the same value. So 3 tenths of a pizza, 30 hundredths of a pizza, and 300 thousandths of a pizza are the same total amount of pizza. The first pizza is sliced into 10 pieces. The second pizza is sliced into 100 pieces. The third pizza is sliced into 1,000 pieces. Also, we don't need to place a zero to the left of the decimal point.

Converting Proper Fractions to Decimals   Recall from Chapter 2 that proper fractions are fractions with a value less than 1. That is, the numerator of the fraction is smaller than its denominator. How can we convert these proper fractions to decimals? Since proper fractions are a form of division, it is possible to convert proper fractions to decimals by carrying out the division.

Converting Proper Fractions to Decimals

Step 1   Divide the numerator of the fraction by its denominator. (If necessary, add a decimal point and zeros to the number in the numerator.)

Step 2   Round as necessary.

Examples:

$$\frac{3}{4} = \begin{array}{r} .75 \\ 4\overline{)3.00} \\ \underline{2\,8} \\ 20 \\ \underline{20} \end{array} \qquad \frac{3}{8} = \begin{array}{r} .375 \\ 8\overline{)3.000} \\ \underline{2\,4} \\ 60 \\ \underline{56} \\ 40 \\ \underline{40} \end{array} \qquad \frac{1}{3} = \begin{array}{r} .33\overline{3} \\ \overline{)1.000} \\ \underline{9} \\ 10 \\ \underline{9} \\ 10 \\ \underline{9} \\ 1 \end{array}$$

Note that in the last example $\frac{1}{3}$, the 3 in the quotient keeps repeating itself (never ends). The short bar over the last 3 means that the number endlessly repeats.

Learn: Converting Mixed Numbers to Decimals

A mixed number, you will recall from Chapter 2, is the sum of a whole number greater than zero and a proper fraction. To convert mixed numbers to decimals, use the following steps:

[1] From .3 to .0003, the values get smaller and smaller, but if you go from .3 to .3000, the values remain the same.

Converting Mixed Numbers to Decimals

Step 1 Convert the fractional part of the mixed number to a decimal (as illustrated in the previous section).

Step 2 Add the converted fractional part to the whole number.

Example:

$$8\frac{2}{5} = \textbf{(Step 1)} \quad 5\overline{)2.0} \quad \begin{array}{r} .4 \\ \underline{2.0} \end{array} \qquad \textbf{(Step 2)} = \begin{array}{r} 8.00 \\ +\ .40 \\ \hline \boxed{8.40} \end{array}$$

Now that we have converted fractions to decimals, let's convert decimals to fractions.

Converting Pure and Mixed Decimals to Decimal Fractions A **pure decimal** has no whole number(s) to the left of the decimal point (.43, .458, and so on). A **mixed decimal** is a combination of a whole number and a decimal. An example of a mixed decimal follows:

Example: 737.592 = Seven hundred thirty-seven and five hundred ninety-two thousandths

Note the following conversion steps for converting pure and mixed decimals to decimal fractions:

Converting Pure and Mixed Decimals to Decimal Fractions

Step 1 Place the digits to the right of the decimal point in the numerator of the fraction. Omit the decimal point. (For a decimal fraction with a fractional part, see examples **c** and **d** below.)

Step 2 Put a 1 in the denominator of the fraction.

Step 3 Count the number of digits to the right of the decimal point. Add the same number of zeros to the denominator of the fraction. For mixed decimals, add the fraction to the whole number.

Examples:		Step 1	Step 2	Places	Step 3
a.	.3	$\dfrac{3}{}$	$\dfrac{3}{1}$	1	$\boxed{\dfrac{3}{10}}$
b.	.24	$\dfrac{24}{}$	$\dfrac{24}{1}$	2	$\boxed{\dfrac{24}{100}}$
c.	$.24\frac{1}{2}$	$\dfrac{245}{}$	$\dfrac{245}{1}$	3	$\boxed{\dfrac{245}{1,000}}$

If desired, you can reduce the fractions in Step 3.

Money Tip Formula for Financial Success: Reduce Spending + Decrease Debt + Increase Savings (Investing) = Healthy Net Worth

Before completing Step 1 in example **c,** we must remove the fractional part, convert it to a decimal ($\frac{1}{2} = .5$), and multiply it by .01 ($.5 \times .01 = .005$). We use .01 because the 4 of .24 is in the hundredths place. Then we add $.005 + .24 = .245$ (three places to right of the decimal) and complete Steps 1, 2, and 3.

		Step 1	Step 2	Places	Step 3
d.	$.07\frac{1}{4}$	$\dfrac{725}{}$	$\dfrac{725}{1}$	4	$\boxed{\dfrac{725}{10,000}}$

In example **d,** be sure to convert $\frac{1}{4}$ to .25 and multiply by .01. This gives .0025. Then add .0025 to .07, which is .0725 (four places), and complete Steps 1, 2, and 3.

		Step 1	Step 2	Places	Step 3	
e.	17.45	$\dfrac{45}{}$	$\dfrac{45}{1}$	2	$\dfrac{45}{100}$	$= \boxed{17\dfrac{45}{100}}$

Example **e** is a mixed decimal. Since we substitute *and* for the decimal point, we read this mixed decimal as seventeen and forty-five hundredths. Note that after we converted the .45 of the mixed decimal to a fraction, we added it to the whole number 17.

The Practice Quiz that follows will help you check your understanding of this unit.

Practice Quiz

Complete this Practice Quiz to see how you are doing.

Write the following as a decimal number.

1. Four hundred eight thousandths

Name the place position of the identified digit:

2. 6.8241   3. 9.3942

Round each decimal to place indicated:

	Tenth	Thousandth
4. .62768	a.	b.
5. .68341	a.	b.

Convert the following to decimals:

6. $\dfrac{9}{10,000}$ 7. $\dfrac{14}{100,000}$

Convert the following to decimal fractions (do not reduce):

8. .819   9. 16.93   10. $.05\dfrac{1}{4}$

Convert the following fractions to decimals and round answer to nearest hundredth:

11. $\dfrac{1}{6}$ 12. $\dfrac{3}{8}$ 13. $12\dfrac{1}{8}$

✓ Solutions

1. **.408** (3 places to right of decimal)

2. Hundredths   3. Thousandths

4. a. **.6** (identified digit 6—digit to right less than 5)   b. **.628** (identified digit 7—digit to right greater than 5)

5. a. **.7** (identified digit 6—digit to right greater than 5)   b. **.683** (identified digit 3—digit to right less than 5)

6. **.0009** (4 places)   7. **.00014** (5 places)

8. $\dfrac{819}{1,000}\left(\dfrac{819}{1 + 3 \text{ zeros}}\right)$ 9. $16\dfrac{93}{100}$

10. $\dfrac{525}{10,000}\left(\dfrac{525}{1 + 4 \text{ zeros}} \quad \dfrac{1}{4}\times.01 = .0025 + .05 = .0525\right)$

11. $.16666 = $ **.17**   12. $.375 = $ **.38**   13. $12.125 = $ **12.13**

Adding, Subtracting, Multiplying, and Dividing Decimals

Monthly cost of streaming plans

Netflix (standard) — $12.99

Hulu (ad free) — 11.99

YouTube Premium — 11.99

CBS All Access — 9.99

Netflix (basic) — 8.99

*Disney+ — 6.99

*Disney+ (under 3-year plan) — 3.92

*Disney service launches in November
Source: the companies

"Monthly Cost of Streaming Services." *The Wall Street Journal*.

The *Wall Street Journal* clip "Monthly cost of streaming plans", comparing monthly streaming plan costs uses decimals showing the price of Netflix (standard) at $12.99 and Disney+ (under 3-year plan) at $3.92. Netflix monthly plan costs $9.07 more per month.

$$\begin{array}{r} \$12.99 \\ -\ 3.92 \\ \hline \$\ 9.07 \end{array}$$

This learning unit shows you how to add, subtract, multiply, and divide decimals. You also make calculations involving decimals, including decimals used in foreign currency.

Learn: Addition and Subtraction of Decimals

Since you know how to add and subtract whole numbers, to add and subtract decimal numbers you have only to learn about the placement of the decimals. The following steps will help you:

Adding And Subtracting Decimals

Step 1 Vertically write the numbers so that the decimal points align. You can place additional zeros to the right of the decimal point if needed without changing the value of the number.

Step 2 Add or subtract the digits starting with the right column and moving to the left.

Step 3 Align the decimal point in the answer with the above decimal points.

Examples: Add $4 + 7.3 + 36.139 + .0007 + 8.22$.

Whole number to the right of the last digit is assumed to have a decimal. $\longrightarrow$

$$\begin{array}{r} 4.0000 \\ 7.3000 \\ 36.1390 \\ .0007 \\ 8.2200 \\ \hline 55.6597 \end{array}$$

Extra zeros have been added to make calculation easier.

Subtract $45.3 - 15.273$.

$$\begin{array}{r} {}^{2\,9\,10}\\ 45.\cancel{300} \\ -15.273 \\ \hline 30.027 \end{array}$$

Subtract $7 - 6.9$.

$$\begin{array}{r} {}^{6\ 10}\\ 7.\cancel{0} \\ -6.9 \\ \hline .1 \end{array}$$

Learn: Multiplication of Decimals

The multiplication of decimal numbers is similar to the multiplication of whole numbers except for the additional step of placing the decimal in the answer (product). The steps that follow simplify this procedure.

Multiplying Decimals

Step 1 Multiply the numbers as whole numbers, ignoring the decimal points.

Step 2 Count and total the number of decimal places in the multiplier and multiplicand.

Step 3 Starting at the right in the product, count to the left the number of decimal places totaled in Step 2. Place the decimal point so that the product has the same number of decimal places as totaled in Step 2. If the total number of places is greater than the places in the product, insert zeros in front of the product.

Examples:

$$\begin{array}{r} 8.52 \\ \times\ 6.7 \\ \hline 5\,964 \\ 5112 \\ \hline 57.084 \end{array}$$
(2 decimal places)
Step 1
(1 decimal place) ← Step 2
Step 3
= 3 decimal places

$$\begin{array}{r} 2.36 \\ \times\ .016 \\ \hline 1416 \\ 236 \\ \hline .03776 \end{array}$$
(2 places)
(3 places)
Need to add zero
= 5 decimal places

Learn: Division of Decimals

If the divisor in your decimal division problem is a whole number, first place the decimal point in the quotient directly above the decimal point in the dividend. Then divide as usual. If the divisor has a decimal point, complete the steps that follow.

Dividing Decimals

Step 1 Make the divisor a whole number by moving the decimal point to the right.

Step 2 Move the decimal point in the dividend to the right the same number of places that you moved the decimal point in the divisor (Step 1). If there are not enough places, add zeros to the right of the dividend.

Step 3 Place the decimal point in the quotient above the new decimal point in the dividend. Divide as usual.

Example

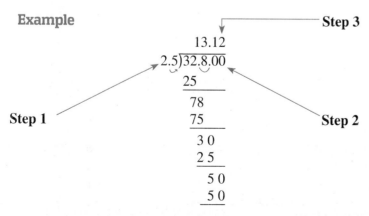

Step 3

$$\begin{array}{r} 13.12 \\ 2.5\overline{)32.8.00} \\ 25 \\ \hline 78 \\ 75 \\ \hline 3\,0 \\ 2\,5 \\ \hline 5\,0 \\ 5\,0 \\ \hline \end{array}$$

Step 1   Step 2

Stop a moment and study the above example. Note that the quotient does not change when we multiply the divisor and the dividend by the same number. This is why we can move the decimal point in division problems and always divide by a whole number.

Learn: Decimal Applications in Foreign Currency

Example:

Hanna Lind, who lives in Canada, wanted to buy a new Apple iPad. She went on eBay and found that the cost would be $800 in U.S. dollars. Wanting to know how much this would cost in Canadian dollars, Hanna consulted the following *Wall Street Journal* currency table and found that a Canadian dollar was worth $.7490 in U.S. dollars. Therefore, for each Canadian dollar it would cost $1.3352 to buy a U.S. good.

Using this information, Hanna completed the following calculation to determine what an Apple iPad would cost her:

$800   × $1.3352   = $1,068.16
(cost of the iPad in   (cost of the iPad
   U.S. dollars)   in Canadian dollars)

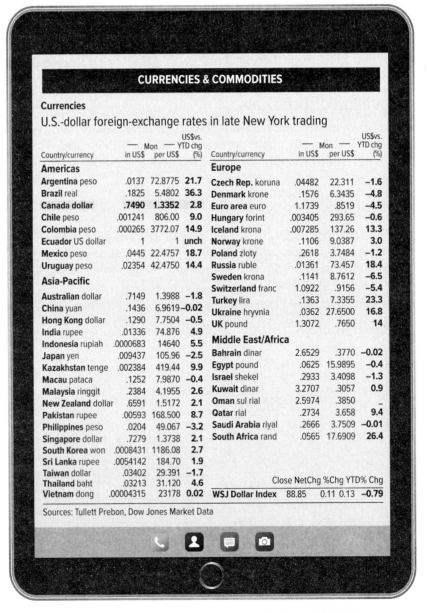

CURRENCIES & COMMODITIES

Currencies
U.S.-dollar foreign-exchange rates in late New York trading

Country/currency	in US$	Mon per US$	US$vs. YTD chg (%)
Americas			
Argentina peso	.0137	72.8775	21.7
Brazil real	.1825	5.4802	36.3
Canada dollar	.7490	1.3352	2.8
Chile peso	.001241	806.00	9.0
Colombia peso	.000265	3772.07	14.9
Ecuador US dollar	1	1	unch
Mexico peso	.0445	22.4757	18.7
Uruguay peso	.02354	42.4750	14.4
Asia-Pacific			
Australian dollar	.7149	1.3988	−1.8
China yuan	.1436	6.9619	−0.02
Hong Kong dollar	.1290	7.7504	−0.5
India rupee	.01336	74.876	4.9
Indonesia rupiah	.0000683	14640	5.5
Japan yen	.009437	105.96	−2.5
Kazakhstan tenge	.002384	419.44	9.9
Macau pataca	.1252	7.9870	−0.4
Malaysia ringgit	.2384	4.1955	2.6
New Zealand dollar	.6591	1.5172	2.1
Pakistan rupee	.00593	168.500	8.7
Philippines peso	.0204	49.067	−3.2
Singapore dollar	.7279	1.3738	2.1
South Korea won	.0008431	1186.08	2.7
Sri Lanka rupee	.0054142	184.70	1.9
Taiwan dollar	.03402	29.391	−1.7
Thailand baht	.03213	31.120	4.6
Vietnam dong	.00004315	23178	0.02

Country/currency	in US$	Mon per US$	US$vs. YTD chg (%)
Europe			
Czech Rep. koruna	.04482	22.311	−1.6
Denmark krone	.1576	6.3435	−4.8
Euro area euro	1.1739	.8519	−4.5
Hungary forint	.003405	293.65	−0.6
Iceland krona	.007285	137.26	13.3
Norway krone	.1106	9.0387	3.0
Poland zloty	.2618	3.7484	−1.2
Russia ruble	.01361	73.457	18.4
Sweden krona	.1141	8.7612	−6.5
Switzerland franc	1.0922	.9156	−5.4
Turkey lira	.1363	7.3355	23.3
Ukraine hryvnia	.0362	27.6500	16.8
UK pound	1.3072	.7650	14
Middle East/Africa			
Bahrain dinar	2.6529	.3770	−0.02
Egypt pound	.0625	15.9895	−0.4
Israel shekel	.2933	3.4098	−1.3
Kuwait dinar	3.2707	.3057	0.9
Oman sul rial	2.5974	.3850	–
Qatar rial	.2734	3.658	9.4
Saudi Arabia riyal	.2666	3.7509	−0.01
South Africa rand	.0565	17.6909	26.4

	Close	NetChg	%Chg	YTD% Chg
WSJ Dollar Index	88.85	0.11	0.13	−0.79

Sources: Tullett Prebon, Dow Jones Market Data

""Currencies" Infographic." *The Wall Street Journal*, August 11, 2020.

To check her findings, Hanna did the following calculation:

$1,068.16	× $.7490	= $800.05 (off due to rounding)
(cost of the Apple iPad in Canadian dollars)	(what the Canadian dollar is worth against the U.S. dollar)	(U.S. selling price)

Learn: Multiplication and Division Shortcuts for Decimals

The shortcut steps that follow show how to solve multiplication and division problems quickly involving multiples of 10 (10, 100, 1,000, 10,000, etc.).

Shortcuts for Multiples of 10

Multiplication

Step 1 Count the zeros in the multiplier.

Step 2 Move the decimal point in the multiplicand the same number of places to the right as you have zeros in the multiplier.

Division

Step 1 Count the zeros in the divisor.

Step 2 Move the decimal point in the dividend the same number of places to the left as you have zeros in the divisor.

In multiplication, the answers are *larger* than the original number.

Example: If Toyota spends $60,000 for magazine advertising, what is the total value if it spends this same amount for 10 years? What would be the total cost?

$$\$60,000 \times 10 = \boxed{\$600,000} \qquad \text{(1 place to the right)}$$

Other Examples:

$$6.89 \times 10 = \boxed{68.9} \qquad \text{(1 place to the right)}$$

$$6.89 \times 100 = \boxed{689.} \qquad \text{(2 places to the right)}$$

$$6.89 \times 1,000 = \boxed{6.890.} \qquad \text{(3 places to the right)}$$

In division, the answers are *smaller* than the original number.

Examples:

$$6.89 \div 10 = \boxed{.689} \qquad \text{(1 place to the left)}$$

$$6.89 \div 100 = \boxed{.0689} \qquad \text{(2 places to the left)}$$

$$6.89 \div 1,000 = \boxed{.00689} \qquad \text{(3 places to the left)}$$

$$6.89 \div 10,000 = \boxed{.000689} \qquad \text{(4 places to the left)}$$

Next, let's dissect and solve a word problem.

Learn: How to Dissect and Solve a Word Problem

The Word Problem May O'Mally went to Sears to buy wall-to-wall carpet. She needs 101.3 square yards for downstairs, 16.3 square yards for the upstairs bedrooms, and 6.2 square yards for the halls. The carpet cost $14.55 per square yard. The padding cost $3.25 per square yard. Sears quoted an installation charge of $6.25 per square yard. What was May O'Mally's total cost?

By completing the following blueprint aid, we will slowly dissect this word problem. Note that before solving the problem, we gather the facts, identify what we are solving for, and list the steps that must be completed before finding the final answer, along with any key points we should remember. Let's go to it!

	The facts	Solving for?	Steps to take	Key points
Blueprint	*Carpet needed:* 101.3 sq. yd.; 16.3 sq. yd.; 6.2 sq. yd. *Costs:* Carpet, $14.55 per sq. yd.; padding, $3.25 per sq. yd.; installation, $6.25 per sq. yd.	Total cost of carpet, padding, and installation.	Total square yards × Cost per square yard = Total cost.	Align decimals. Round answer to nearest cent.

Steps to solving problem

1. Calculate the total number of square yards.

   101.3
   16.3
   6.2
   ___
   123.8 square yards

2. Calculate the total cost per square yard.

   $14.55
   3.25
   6.25
   ___
   $24.05

3. Calculate the total cost of carpet, padding, and installation.

   $123.8 × $24.05 = $2,977.39$

It's time to check your progress.

Practice Quiz

Complete this Practice Quiz to see how you are doing.

1. Rearrange vertically and add:

   14, .642, 9.34, 15.87321

2. Rearrange and subtract:

   28.1549 − .885

3. Multiply and round the answer to the nearest tenth:

   28.53 × 17.4

4. Divide and round to the nearest hundredth:

   2,182 ÷ 2.83

Complete by the shortcut method:

5. 14.28 × 100   6. 9,680 ÷ 1,000   7. 9,812 ÷ 10,000

8. Could you help Mel decide which product is the "better buy"?

   Dog food A: $9.01 for 64 ounces   **Dog food B:** $7.95 for 50 ounces

Round to the nearest cent as needed:

9. At Avis Rent-A-Car, the cost per day to rent a medium-size car is $39.99 plus 29 cents per mile. What will it cost to rent this car for 2 days if you drive 602.3 miles? Since the solution shows a completed blueprint, you might use a blueprint also.

10. A trip to Mexico cost 6,000 pesos. What would this be in U.S. dollars? Check your answer.

✓ Solutions

1.
```
   14.00000
     .64200
    9.34000
   15.87321
   39.85521
```

2.
```
    7 101414
   28.1549
   − .8850
   27.2699
```

3.
```
      28.53
    ×  17.4
      11 412
     199 71
     285 3
     496.422  = 496.4
```

4.
```
           771.024 = 771.02
     2.83)218200.000
          1981
          2010
          1981
           290
           283
            7 00
            5 66
            1 340
            1 132
```

5. 14.28 = 1,428   6. 9,680 = 9.680   7. 9,812 = .9812

8. **A:** $9.01 ÷ 64 = $.14   **B:** $7.95 ÷ 50 = $.16   Buy A.

Practice Quiz *Continued*

9. Avis Rent-A-Car total rental charge:

	The facts	Solving for?	Steps to take	Key points
Blueprint	Cost per day, $39.99. 29 cents per mile. Drove 602.3 miles. 2-day rental.	Total rental charge.	Total cost for 2 days' rental + Total cost of driving = Total rental charge.	In multiplication, count the number of decimal places. Starting from right to left in the product, insert decimal in appropriate place. Round to nearest cent.

Steps to solving problem

1. Calculate total cost for 2 days' rental. $39.99 \times 2 = $79.98

2. Calculate the total cost of driving. $.29 \times 602.3 = $174.667 = $174.67

3. Calculate the total rental charge.

$$\begin{array}{r} \$\ 79.98 \\ +174.67 \\ \hline \$254.65 \end{array}$$

10. $6,000 \times $.0445 = $267

   Check $267 \times 22.4757 = 6,001.01$ pesos due to rounding

Chapter 3 Review

Topic/Procedure/Formula	Example	You try it*
Identifying place value $$10, 1, \frac{1}{10}, \frac{1}{100}, \frac{1}{1,000}, \text{etc.}$$	.439 in thousandths place value	**Identify place value** .8256
Rounding decimals 1. Identify place value of digit you want to round. 2. If digit to right of identified digit in Step 1 is 5 or more, increase identified digit by 1; if less than 5, do not change identified digit. 3. Drop all digits to right of identified digit.	.875 rounded to nearest tenth = .9 Identified digit	**Round to nearest tenth** .841
Converting decimal fractions to decimals 1. Decimal fraction has a denominator with multiples of 10. Count number of zeros in denominator. 2. Zeros show how many places are in the decimal.	$$\frac{8}{1,000} = .008$$ $$\frac{6}{10,000} = .0006$$	**Convert to decimal** $$\frac{9}{1,000}$$ $$\frac{3}{10,000}$$
Converting proper fractions to decimals 1. Divide numerator of fraction by its denominator. 2. Round as necessary.	$$\frac{1}{3} \text{ (to nearest tenth)} = .3$$	**Convert to decimal (to nearest tenth)** $$\frac{1}{7}$$
Converting mixed numbers to decimals 1. Convert fractional part of the mixed number to a decimal. 2. Add converted fractional part to whole number.	$$6\frac{1}{4} \quad \frac{1}{4} = .25 + 6 = 6.25$$	**Convert to decimal** $$5\frac{4}{5}$$
Converting pure and mixed decimals to decimal fractions 1. Place digits to right of decimal point in numerator of fraction. 2. Put 1 in denominator. 3. Add zeros to denominator, depending on decimal places of original number. For mixed decimals, add fraction to whole number.	.984 (3 places) 1. $\underline{984}$ 2. $\dfrac{984}{1}$ 3. $\dfrac{984}{1,000}$	**Convert to fraction** .865

Chapter 3 Review (Continued)

Topic/Procedure/Formula	Example	You try it*
Adding and subtracting decimals 1. Vertically write and align numbers on decimal points. 2. Add or subtract digits, starting with right column and moving to the left. 3. Align decimal point in answer with above decimal points.	Add $1.3 + 2 + .4$ $\begin{array}{r} 1.3 \\ 2.0 \\ .4 \\ \hline \boxed{3.7} \end{array}$ Subtract $5 - 3.9$ $\begin{array}{r} \overset{4\ \ 10}{\cancel{5}.\cancel{0}} \\ -3.9 \\ \hline \boxed{1.1} \end{array}$	**Add** $1.7 + 3 + .8$ **Subtract** $6 - 4.1$
Multiplying decimals 1. Multiply numbers, ignoring decimal points. 2. Count and total number of decimal places in multiplier and multiplicand. 3. Starting at right in the product, count to the left the number of decimal places totaled in Step 2. Insert decimal point. If number of places greater than space in answer, add zeros.	$\begin{array}{r} 2.48\ \ \text{(2 places)} \\ \times\ .018\ \ \text{(3 places)} \\ \hline 1984 \\ 248 \\ \hline \boxed{.04464} = 5\ \text{places} \end{array}$	**Multiply** $\begin{array}{r} 3.49 \\ \times\ .015 \end{array}$
Dividing a decimal by a whole number 1. Place decimal point in quotient directly above the decimal point in dividend. 2. Divide as usual.	$\begin{array}{r} \boxed{1.1} \\ 42\overline{)46.2} \\ \underline{42} \\ 42 \\ \underline{42} \end{array}$	**Divide (to nearest tenth)** $33\overline{)49.5}$
Dividing if the divisor is a decimal 1. Make divisor a whole number by moving decimal point to the right. 2. Move decimal point in dividend to the right the same number of places as in Step 1. 3. Place decimal point in quotient above decimal point in dividend. Divide as usual.	$\begin{array}{r} \boxed{14.3} \\ 2.9\overline{)41.39} \\ \underline{29} \\ 123 \\ \underline{116} \\ 79 \\ \underline{58} \\ 21 \end{array}$	**Divide (to nearest tenth)** $3.2\overline{)1.48}$

Chapter 3 Review (Continued)

Topic/Procedure/Formula	Example	You try it*
Shortcuts on multiplication and division of decimals When multiplying by 10, 100, 1,000, and so on, move decimal point in multiplicand the same number of places to the right as you have zeros in multiplier. For division, move decimal point to the left.	$4.85 \times 100 = 485$ $4.85 \div 100 = .0485$	**Multiply by shortcut** 6.92×100 **Divide by shortcut** $6.92 \div 100$

Key Terms

Decimals	Mixed decimal	Rounding decimals
Decimal fraction	Pure decimal	
Decimal point	Repeating decimal	

* Answers are now found in Appendix A.

Critical Thinking Discussion Questions with Chapter Concept Check

1. What are the steps for rounding decimals? Federal income tax forms allow the taxpayer to round each amount to the nearest dollar. Do you agree with this?

2. Explain how to convert fractions to decimals. If 1 out of 20 people buys a Land Rover, how could you write an advertisement in decimals?

3. Explain why .07, .70, and .700 are not equal. Assume you take a family trip to Disney World that covers 500 miles. Show that $\frac{8}{10}$ of the trip, or .8 of the trip, represents 400 miles.

4. Explain the steps in the addition or subtraction of decimals. Visit a car dealership and find the difference between two sticker prices. Be sure to check each sticker price for accuracy. Should you always pay the sticker price?

5. **Chapter Concept Check.** Go to the Lyft website and compare its ridership today versus the data from the chapter opener.

End-of-Chapter Problems

Name _____   Date_____

Check figures for odd-numbered problems in Appendix A.

Drill Problems

Identify the place value for the following: *LU 3-1(1)*

3–1. 7.5328   **3–2.** 229.448

       ↑   ↑

Round the following as indicated: *LU 3-1(1)*

	Tenth	**Hundredth**	**Thousandth**
3–3. .7391			
3–4. 6.8629			
3–5. 5.8312			
3–6. 6.8415			
3–7. 6.5555			
3–8. 75.9913			

Round the following to the nearest cent: *LU 3-1(1)*

3–9. $4,822.775   **3–10.** $4,892.046

Convert the following types of decimal fractions to decimals (round to nearest hundredth as needed):   *LU 3-1(2)*

3–11. $\dfrac{8}{100}$ **3–12.** $\dfrac{3}{10}$ **3–13.** $\dfrac{61}{1,000}$ **3–14.** $\dfrac{610}{1,000}$

3–15. $\dfrac{91}{100}$ **3–16.** $\dfrac{979}{1,000}$ **3–17.** $16\dfrac{61}{100}$

Convert the following decimals to fractions. Do not reduce to lowest terms.   *LU 3-1(2)*

3–18. .9   **3–19.** .71   **3–20.** .009   **3–21.** .0125

3–22. .609   **3–23.** .825   **3–24.** .9999   **3–25.** .7065

Convert the following to mixed numbers. Do not reduce to the lowest terms.   *LU 3-1(2)*

3–26. 7.1   **3–27.** 28.48   **3–28.** 6.025

Write the decimal equivalent of the following: *LU 3-1(2)*

3–29. Five thousandths   **3–30.** Three hundred three and two hundredths

3–31. Eighty-five ten thousandths   **3–32.** Seven hundred seventy-five thousandths

Rearrange the following and add: *LU 3-2(1)*

3–33. .115, 10.8318, 4.7, 802.4811

3–34. .005, 2,002.181, 795.41, 14.0, .184

Rearrange the following and subtract: *LU 3-2(1)*

3–35. 9.2 − 5.8 =

3–36. 7 − 2.0815 =

3–37. 3.4 − 1.08 =

Estimate by rounding all the way and multiply the following (do not round final answer): *LU 3-2(1)*

3–38. 6.24 × 3.9 =

Estimate

3–39. .413 × 3.07 =

Estimate

3–40. 675 × 1.92 =

Estimate

3–41. 4.9 × .825 =

Estimate

Divide the following and round to the nearest hundredth: *LU 3-2(1)*

3–42. .8931 ÷ 3 =

3–43. 29.432 ÷ .0012 =

3–44. .0065 ÷ .07 =

3–45. 7,742.1 ÷ 48 =

3–46. 8.95 ÷ 1.81 =

3–47. 2,600 ÷ 381 =

Convert the following to decimals and round to the nearest hundredth: *LU 3-1(2)*

3–48. $\dfrac{1}{8}$

3–49. $\dfrac{1}{25}$

3–50. $\dfrac{5}{6}$

3–51. $\dfrac{5}{8}$

Complete these multiplications and divisions by the shortcut method (do not do any written calculations):   *LU 3-2(3)*

3–52. 96.7 ÷ 10 =

3–53. 258 ÷ 100 =

3–54. 8.51 × 1,000 =

3–55. .86 ÷ 100 =

3–56. 9.015 × 100 =

3–57. 48.6 × 10 =

3–58. 750 × 10 =

3–59. 3,950 ÷ 1,000 =

3–60. 8.45 ÷ 10 =

3–61. 7.9132 × 1,000 =

Word Problems

As needed, round answers to the nearest cent.

3–62. A Chevy Volt costs $30,000 in the United States. Using the exchange rate given in the WSJ currency table on page 3-10, what would it cost in Canada? Check your answer.   *LU 3-2(2)*

End-of-Chapter Problems (Continued)

3–63. Dustin Pedroia got 7 hits out of 12 at bats. What was his batting average to the nearest thousandths place? *LU 3-1(2)*

3–64. Pete Ross read in a *Wall Street Journal* article that the cost of parts and labor to make an Apple iPhone 4S were as follows: *LU 3-2(1)*

Display	$37.00	Wireless	$23.54
Memory	$28.30	Camera	$17.60
Labor	$ 8.00	Additional items	$81.56

Assuming Pete pays $649 for an iPhone 4S, how much profit does the iPhone generate?

3–65. At the Party Store, JoAnn Greenwood purchased 21.50 yards of ribbon. Each yard cost 91 cents. What was the total cost of the ribbon? Round to the nearest cent. *LU 3-2(1)*

3–66. Douglas Noel went to Home Depot and bought four doors at $42.99 each and six bags of fertilizer at $8.99 per bag. What was the total cost to Douglas? If Douglas had $300 in his pocket, what does he have left to spend? *LU 3-2(1)*

3–67. The stock of Intel has a high of $48.50 today. It closed at $47.75. How much did the stock drop from its high? *LU 3-2(1)*

3–68. If you net $14.25 per hour and work 40 hours a week, 4 weeks per month, and have monthly expenses of: rent $825.50, car payment $458.79, utilities $110, food $150, gas $105, phone $125.25, savings $225, and insurance $118.36, what do you have left to invest for your retirement? *LU 3-2(1)*

3–69. Mark Ogara rented a truck from Avis Rent-A-Car for the weekend (2 days). The base rental price was $29.95 per day plus $14\frac{1}{2}$ cents per mile. Mark drove 410.85 miles. How much does Mark owe? *LU 3-2(1)*

3–70. Nursing home costs are on the rise as consumeraffairs.com reports in its quarterly newsletter. The average cost is around $192 a day with an average length of stay of 2.5 years. Calculate the cost of the average nursing home stay.   *LU 3-2(1)*

3–71. Bob Ross bought a smartphone on the web for $89.99. He saw the same smartphone in the mall for $118.99. How much did Bob save by buying on the web?   *LU 3-2(1)*

3–72. Russell is preparing the daily bank deposit for his coffee shop. Before the deposit, the coffee shop had a checking account balance of $3,185.66. The deposit contains the following checks:

No. 1 $ 99.50   No. 3 $8.75

No. 2   110.35   No. 4   6.83

Russell included $820.55 in currency with the deposit. What is the coffee shop's new balance, assuming Russell writes no new checks?   *LU 3-2(1)*

excel 3–73. The United Nations claims India will overtake China as the world's most populous country within seven years. If China has 1.436 billion people and India has 1.345 billion people, what is the difference in population?   *LU 3-2(1)*

3–74. Randi went to Lowe's to buy wall-to-wall carpeting. She needs 110.8 square yards for downstairs, 31.8 square yards for the halls, and 161.9 square yards for the bedrooms upstairs. Randi chose a shag carpet that costs $14.99 per square yard. She ordered foam padding at $3.10 per square yard. The carpet installers quoted Randi a labor charge of $3.75 per square yard. What will the total job cost Randi?   *LU 3-2(1)*

3–75. Paul Rey bought four new Dunlop tires at Goodyear for $95.99 per tire. Goodyear charged $3.05 per tire for mounting, $2.95 per tire for valve stems, and $3.80 per tire for balancing. If Paul paid no sales tax, what was his total cost for the four tires?   *LU 3-2(1)*

End-of-Chapter Problems (Continued)

excel 3–76. Shelly is shopping for laundry detergent, mustard, and canned tuna. She is trying to decide which of two products is the better buy. Using the following information, can you help Shelly?   *LU 3-2(1)*

Laundry detergent A	**Mustard A**	**Canned tuna A**
$2.00 for 37 ounces	$.88 for 6 ounces	$1.09 for 6 ounces

Laundry detergent B	**Mustard B**	**Canned tuna B**
$2.37 for 38 ounces	$1.61 for $12\frac{1}{2}$ ounces	$1.29 for $8\frac{3}{4}$ ounces

excel 3–77. Roger bought season tickets for weekend professional basketball games. The cost was $945.60. The season package included 36 home games. What is the average price of the tickets per game? Round to the nearest cent. Marcelo, Roger's friend, offered to buy four of the tickets from Roger. What is the total amount Roger should receive?   *LU 3-2(1)*

3–78. A nurse was to give each of her patients a 1.32-unit dosage of a prescribed drug. The total remaining units of the drug at the hospital pharmacy were 53.12. The nurse has 38 patients. Will there be enough dosages for all her patients?   *LU 3-2(1)*

3–79. Jill Horn went to Japan and bought an animation cel of Spongebob. The price was 25,000 yen. Using the WSJ currency table, what is the price in U.S. dollars? Round your answer to the nearest cent. Check your answer.   *LU 3-2(2)*

3–80. Cryptocurrency, such as bitcoin, functions like a check, arcade token, or casino chip, as it allows you to buy goods and services. According to CoinMarketCap, the cryptocurrency market as of early 2021 was more than $897.3 billion with bitcoin being the largest trading cryptocurrency at $563.8 billion of that market. If the next largest trading cryptocurrency, Ethereum, is $142.9 billion, how much larger is bitcoin?   *LU 3-2(1)*

3–81. According to the Labor Department, a record 20.5 million jobs were lost in the USA in April 2020, sending the unemployment rate to 14.7%, the highest since the Great Depression. Pre-pandemic weekly unemployment claims were around 200,000. During the week of March 28, 2020, 6.9 million filed a claim. How many more claims were filed during the week of March 28, 2020 than the typical weekly claim?   *LU 3-2(1)*

3–82. Morris Katz bought four new tires at Goodyear for $95.49 per tire. Goodyear also charged Morris $2.50 per tire for mounting, $2.40 per tire for valve stems, and $3.95 per tire for balancing. Assume no tax. What was Morris's total cost for the four tires? *LU 3-2(1)*

3–83. yahoo!finance reported findings about the effectiveness of the top coronavirus vaccines. The Pfizer (PFE) vaccine has the best results with a 95% effectiveness rate while the Moderna (MRNA) vaccine is at 94.1% and Johnson & Johnson (JNJ) at 66%. What is the difference between the effectiveness rates of the top two vaccines, Pfizer and Moderna? *LU 3-2(1)*

3–84. If your car gets 28 miles per gallon and you travel 30 miles round-trip to work five days a week, how much do you pay each 4-week month if gas is $3.05 a gallon? Round each calculation to the hundredth. *LU 3-2(1)*

3–85. Gracie went to Home Depot to buy wall-to-wall carpeting for her house. She needs 104.8 square yards for downstairs, 17.4 square yards for halls, and 165.8 square yards for the upstairs bedrooms. Gracie chose a shag carpet that costs $13.95 per square yard. She ordered foam padding at $2.75 per square yard. The installers quoted Gracie a labor cost of $5.75 per square yard in installation. What will the total job cost Gracie? *LU 3-2(1)*

Challenge Problems

excel **3–86.** Fred and Winnie O'Callahan have put themselves on a very strict budget. Their goal at the end of the year is to buy a car for $14,000 in cash. Their budget includes the following per dollar:

$.40 food and lodging

.20 entertainment

.10 educational

Fred earns $2,000 per month and Winnie earns $2,500 per month. After 1 year will Fred and Winnie have enough cash to buy the car? *LU 3-2(1)*

End-of-Chapter Problems (Continued)

3–87. Jill and Frank decided to take a long weekend in New York. City Hotel has a special getaway weekend for $79.95. The price is per person per night, based on double occupancy. The hotel has a minimum two-night stay. For this price, Jill and Frank will receive $50 credit toward their dinners at City's Skylight Restaurant. Also included in the package is a $3.99 credit per person toward breakfast for two each morning.

Since Jill and Frank do not own a car, they plan to rent a car. The car rental agency charges $19.95 a day with an additional charge of $.22 a mile and $1.19 per gallon of gas used. The gas tank holds 24 gallons.

From the following facts, calculate the total expenses of Jill and Frank (round all answers to nearest hundredth or cent as appropriate). Assume no taxes.   *LU 3-2(1)*

Car rental (2 days):		Dinner cost at Skylight	$182.12
Beginning odometer reading	4,820	Breakfast for two:	
Ending odometer reading	4,940	Morning No. 1	24.17
Beginning gas tank: $\frac{3}{4}$ full		Morning No. 2	26.88
Gas tank on return: $\frac{1}{2}$ full		Hotel room	79.95
Tank holds 24 gallons			

Summary Practice Test

Do you need help? Connect videos have step-by-step worked-out solutions.

1. Add the following by translating the verbal form to the decimal equivalent.
   LU 3-1(1), LU 3-2(1)

   Three hundred thirty-eight and seven hundred five thousandths

   Nineteen and fifty-nine hundredths

   Five and four thousandths

   Seventy-five hundredths

   Four hundred three and eight tenths

Convert the following decimal fractions to decimals. *LU 3-1(2)*

2. $\dfrac{7}{10}$ **3.** $\dfrac{7}{100}$ **4.** $\dfrac{7}{1,000}$

Convert the following to proper fractions or mixed numbers. Do not reduce to the lowest terms. *LU 3-1(2)*

5. .9   **6.** 6.97   **7.** .685

Convert the following fractions to decimals (or mixed decimals) and round to the nearest hundredth as needed. *LU 3-1(2)*

8. $\dfrac{2}{7}$ **9.** $\dfrac{1}{8}$ **10.** $4\dfrac{4}{7}$ **11.** $\dfrac{1}{13}$

12. Rearrange the following decimals and add. *LU 3-2(1)*

    5.93,   11.862,   284.0382,   88.44

13. Subtract the following and round to the nearest tenth. *LU 3-2(1)*

    $13.111 - 3.872$

14. Multiply the following and round to the nearest hundredth. *LU 3-2(1)*

    7.4821×15.861

15. Divide the following and round to the nearest hundredth. *LU 3-2(1)*

    $203,942 \div 5.88$

Complete the following by the shortcut method. *LU 3-2(3)*

16. $62.94 \times 1,000$

17. $8,322,249.821 \times 100$

18. The average pay of employees is $795.88 per week. Lee earns $820.44 per week. How much is Lee's pay over the average? *LU 3-2(1)*

Summary Practice Test

19. Lowes reimburses Ron $.49 per mile. Ron submitted a travel log for a total of 1,910.81 miles. How much will Lowes reimburse Ron? Round to the nearest cent.   *LU 3-2(1)*

20. Lee Chin bought two new car tires from Michelin for $182.11 per tire. Michelin also charged Lee $3.99 per tire for mounting, $2.50 per tire for valve stems, and $4.10 per tire for balancing. What is Lee's final bill?   *LU 3-2(1)*

21. Could you help Judy decide which of the following products is cheaper per ounce?
LU 3-2(1)

Canned fruit A

$.37 for 3 ounces

Canned fruit B

$.58 for $3\frac{3}{4}$ ounces

22. Paula Smith bought a computer tablet for 400 shekels in Israel. Using the WSJ currency table, what is this price in U.S. dollars? Round your answer to the nearest cent.   *LU 3-2(2)*

23. Google stock traded at a high of $522.00 and closed at $518.55. How much did the stock fall from its high?   *LU 3-2(1)*

🔍 Nailing the Interview!

 What I need to know

Interviews can be intimidating and cause anxiety on the part of the interviewee. Reduce this anxiety by preparing yourself for the interview. The interview is an opportunity for both the interviewer and interviewee to determine whether there is a good fit. The interview itself takes a relatively short time so making sure you are well prepared will allow you to make a good first impression. You want to show the potential employer you are the best candidate for the job and match well to the type of candidate they are seeking.

 What I need to do

"Practice makes perfect" applies directly to the interview process. The more experience you have with interviewing the more comfortable and effective you will become. Participate in mock interviews offered by your institution. Many of these mock interviews utilize local employers who can provide vital feedback on your interviewing skills. Ask friends and family to conduct a mock interview for you or put you into contact with their employers who may be able to offer a practice interview. The feedback you receive from this practice will be beneficial for you as you sharpen your interviewing skills.

Another key interviewing issue for college students is the lack of experience students have with the new career they are about to enter. As a student studying for a professional career, you most likely have worked in jobs not directly related to your new career path. Therefore, you need to find a way to transfer the experiences you have gained to the job you desire. For example, if you worked in a retail environment during college you had many interactions with customers and these skills can transfer to a new job in which customer service is a key requirement. By identifying your transferable skills you show an employer you understand the job requirements and you can demonstrate how you meet these expectations along with how your experiences will benefit the organization.

In an interview many questions are asked to assess the ability of the candidate to perform the job's required tasks. Common questions appear in many interviews as noted in the link below and you should practice your responses to these common questions. Additionally, you should have questions to ask of the employer during the interview. These questions allow you the opportunity to determine whether this employer is a good fit for you. Asking questions also lets your interviewer know that you are serious about the position and have a genuine interest in the job and the organization.

 Steps I need to take

1. Practice your interviewing skills and get feedback on your performance.
2. Show how your existing skill set matches to the expectations of the employer.
3. Be prepared for common interview questions and develop questions you can ask.

 Resources I can use

- Glassdoor Job Search (mobile app)—search and preparation tools to find your desired career.
- https://zety.com/blog/job-interview-questions-and-answers—job interview questions to prepare for as well as questions you should be prepared to ask the interviewer.

MY MONEY ACTIVITY

- Select five questions from the top interview questions link above.
- Find a partner and take turns interviewing each other and providing feedback.

PERSONAL FINANCE

A KIPLINGER APPROACH

"How to Surf the Net More Safely." Kiplinger's. May 2020.

TECH

HOW TO SURF THE NET MORE SAFELY

To help you fight identity theft, consider adding a VPN.

BY MONITORING YOUR INTERNET browsing activity, identity thieves can steal personal information and use it to hack into your financial accounts. And ID theft has spiked during the pandemic. The number of cases more than doubled from 2019 to 2020, to nearly 1.4 million, according to a recent report by the Federal Trade Commission.

One way to protect yourself is by using a virtual private network. With a VPN, your connection is encrypted, which means anyone trying to monitor it will see nothing more than an unintelligible sequence of letters and numbers. A VPN also hides your IP address from the websites you visit. An IP address is a unique online identifier, and it can help hackers stitch together information about you.

Choose wisely. It's difficult for the average user to navigate their way through the marketing claims made by the many VPN services. But you can see what experts say at sites such as Top10VPN.com and CNET.

com. Although there are a lot of options, ExpressVPN, SurfShark and NordVPN are ranked highly by both Top10VPN and CNET, and all three offer money-back guarantees after 30 days. ExpressVPN costs $8.32 per month for a 12-month plan, but it has been offering a free three-month trial and 15-month plans for $6.67 per month. NordVPN costs $3.71 per month for a two-year plan, and SurfShark costs only $2.49 per month for a two-year plan. To install a VPN, you simply download it and follow a short, step-by-step installation wizard. You'll need to enter an activation code that's supplied when you subscribe or create log-in credentials.

Why spend the money? Because free VPNs are notoriously unreliable and can expose users to more risk than not using one at all, says Simon Migliano, head of research at Top10VPN.com. Free VPN services can make data harvesting even easier. They are rife with advertising, showing you targeted ads based on your browsing behavior, and

they do little to protect your data. And it's hard to be sure whether a free VPN will keep your browsing history private from your internet service provider.

Caveats. "It's important to remember that a VPN

isn't some all-singing, all-dancing solution to all your privacy and security problems," says Migliano. However, if used properly, it's significantly safer and more private to use a VPN than to go online without one, he says.

Because they mask your IP address, VPNs can also occasionally confuse sites you're signing into, such as bank or brokerage accounts, which will often flag login attempts from multiple IP addresses as potential fraud, says Carrie Kerskie, author of

Protect Your Identity, a Step-by-Step Guide. Given how commonplace it is to access accounts via mobile devices on various cellular and Wi-Fi networks, this shouldn't be a problem for most people. But consider asking a potential VPN

service provider to confirm that their service works with your financial service of choice, and make sure the VPN service offers a money-back guarantee before signing up.

Even with the added protection of a VPN, always take care to manage your passwords wisely. Avoid saving passwords in your web browser, says Kerskie, and change them regularly. It's also a good idea to invest in a password manager, such as LastPass or 1Password; both cost about $36 per year.

Also beware of phishing and malware. More and more scammers are using e-mail, texts and phone calls to try to pry your personal information from you. Never click on links in e-mails that seem suspicious, and be cautious when responding to e-mails from addresses you don't recognize.

EMMA PATCH
Emma_Patch@kiplinger.com

Business Math Issue

Signing up with VPN makes surfing always safe.

1. List the key points of the article and information to support your position.

2. Write a group defense of your position using math calculations to support your view. If you are in an online course, post to a discussion board.

Cumulative Review

A Word Problem Approach—Chapters 1, 2, 3

1. The Waldorf Astoria New York rate per night is $754 compared to the Ritz Carlton in Boston which is $730. If John spends 9 days at one of these hotels, how much can he save if he stays at the Ritz?   *LU 1-2(2), LU 1-3(1)*

2. Robert Half Placement Agency was rated best by 4 to 1 in an independent national survey. If 250,000 responded to the survey, how many rated Robert Half the best?   *LU 2-3(1)*

3. Of the 63.2 million people who watch professional football, only $\frac{1}{5}$ watch the commercials. How many viewers do not watch the commercials?   *LU 2-3(1)*

4. AT&T advertised a 500-minute prepaid domestic calling card for $25. Diamante sells a 500-minute prepaid domestic calling card for $12.25. Assuming Bill Splat needs two 500-minute cards, how much could he save by buying Diamante's?   *LU 3-2(1)*

5. A square foot of rental space in New York City, Boston, and Providence costs as follows: New York City, $6.25; Boston, $5.75; and Providence, $3.75. If Hewlett Packard wants to rent 112,500 square feet of space, what will Hewlett Packard save by renting in Providence rather than Boston?   *LU 3-2(1)*

6. American Airlines has a frequent-flier program. Coupon brokers who buy and sell these awards pay between 1 and $1\frac{1}{2}$ cents for each mile earned. Fred Dietrich earned a 50,000-mile award (worth two free tickets to any city). If Fred decided to sell his award to a coupon broker, approximately how much would he receive?   *LU 3-2(1)*

7. Lillie Wong bought four new Firestone tires at $82.99 each. Firestone also charged $2.80 per tire for mounting, $1.95 per tire for valves, and $3.15 per tire for balancing. Lillie turned her four old tires in to Firestone, which charged $1.50 per tire to dispose of them. What was Lillie's final bill?   *LU 3-2(1)*

8. Tootsie Roll Industries bought Charms Company for $65 million. Some analysts believe that in 4 years the purchase price could rise to three times as much. If the analysts are right, how much did Tootsie Roll save by purchasing Charms immediately?   *LU 1-3(1)*

9. Today the average business traveler will spend $47.73 a day on food. The breakdown is dinner, $22.26; lunch, $10.73; breakfast, $6.53; tips, $6.23; and tax, $1.98. If Clarence Donato, an executive for Kroger, spends only .33 of the average, what is Clarence's total cost for food for the day? If Clarence wanted to spend $\frac{1}{3}$ more than the average on the next day, what would be his total cost on the second day? Round to the nearest cent.   *LU 2-3(1), LU 3-2(1)*

Be sure you use the fractional equivalent in calculating $.3\overline{3}$.

Chapter 4
Banking

Learning Unit Objectives

LU 4–1: The Checking Account

1. Define and state the purpose of signature cards, checks, deposit slips, check stubs, check registers, and endorsements.
2. Correctly prepare deposit slips and write checks.

LU 4–2: Bank Statement and Reconciliation Process; Using Mobile and Online Banking

1. Explain trends in the banking industry.
2. Define and state the purpose of the bank statement.
3. Complete a check register and a bank reconciliation.
4. Using mobile and online banking.

Banking App Acceptance

How They Are Used

Consumers who use mobile-banking apps like them for different things, with some performing only a couple of tasks. In percent of U.S. adults with a bank account and a mobile phone:

- General user
- Task-only user*

Check balance or transactions
Receive alerts
Locate ATM or branch
Transfer money between accounts
Pay bills
Deposit a check electronically
Send money to relatives or friends in the U.S.

0% 5 10 15 20 25 30 35 40

Top Reasons for Not Mobile Banking, by Generation

Concerns about security were foremost in the minds of respondents to a 2016 international survey who said they weren't likely to use mobile-banking services.

- Generation Z (ages 13–20)
- Millennials (21–34)
- Generation X (35–49)
- Baby boomers (50–64)
- Silent generation (65+)

I'm concerned about security
I prefer to visit a physical location/bank
I don't need it
Screen too small

0% 10 20 30 40 50 60

*Respondents who reported doing one or more mobile banking tasks as opposed to being a 'general user' of mobile banking
Sources: Federal Reserve 2015 Mobile Survey (Use); Nielsen Co. (Reasons Not)

""Banking App Acceptance" infographic." *The Wall Street Journal,* September 23, 2019.

@ Essential Question

How can I use banking to understand the business world and make my life easier?

🌐 Math Around the World

The *Wall Street Journal* clip "Banking App Acceptance" on the chapter opener shows how banks are offering new ways to conduct mobile banking. *The Wall Street Journal* clip "Overheard" describes how the pandemic has affected ATMs.

In this chapter we will first look at how to do banking transactions manually, then look at mobile and online banking.

An important fixture in today's banking is the **automatic teller machine (ATM).** The ability to get instant cash is a convenience many bank customers enjoy.

The effect of using an ATM card is the same as using a **debit card**—both transactions result in money being immediately deducted from your checking account balance. As a result, debit cards have been called enhanced ATM cards or *check cards.* Often banks charge fees for these card transactions. The frequent complaints of bank customers have made many banks offer their ATMs as a free service, especially if customers use an ATM in the same network as their bank. Some banks charge fees for using another bank's ATM.

Remember that the use of debit cards involves planning. As *check cards,* you must be aware of your bank balance every time you use a debit card. Also, if you use a credit card instead of a debit card, you can only be held responsible for $50 of illegal charges; and during the time the credit card company investigates the illegal charges, they are removed from your account. However, with a debit card, this legal limit only applies if you report your card lost or stolen within two business days. Payment apps, such as Venmo, grew in popularity during the pandemic because they allow for virtual peer-to-peer transfer of money 24/7 by linking a credit card or bank account to one's Venmo account.

This chapter begins with a discussion of the checking account. You will follow Molly Kate as she opens a checking account for Gracie's Natural Superstore and performs her banking transactions. Pay special attention to the procedure used by Gracie's to reconcile its checking account and bank statement. This information will help you reconcile your checkbook records with the bank's record of your account. The chapter concludes by discussing the use of mobile and online banking.

The Checking Account

A **check** or **draft** is a written order instructing a bank, credit union, or savings and loan institution to pay a designated amount of your money on deposit to a person or an organization. Checking accounts are offered to individuals and businesses. Note that the business checking account usually receives more services than the personal checking account but may come with additional fees.

Most small businesses depend on a checking account for efficient record keeping. In this learning unit you will follow the checking account procedures of a newly organized small business. You can use many of these procedures in your personal check writing. You will also learn about e-checks.

Learn: Opening the Checking Account

Molly Kate, treasurer of Gracie's Natural Superstore, went to Ipswich Bank to open a business checking account. The bank manager gave Molly a **signature card.** The signature card contained space for the company's name and address, references, type of account, and the signature(s) of the person(s) authorized to sign checks. If necessary, the bank will use the signature card to verify that Molly signed the checks. Some companies authorize more than one person to sign checks or require more than one signature on a check.

Molly then lists on a **deposit slip** (or deposit ticket) the checks and/or cash she is depositing in her company's business account. The bank gave Molly a temporary checkbook to use until the company's printed checks arrived. Molly also will receive *preprinted* checking account deposit slips like the one shown in Figure 4.1. Since the deposit slips are in duplicate, Molly can keep a record of her deposit. Note that the increased use of ATM machines has made it more convenient for people to make their deposits.

FIGURE 4.1 Deposit slip

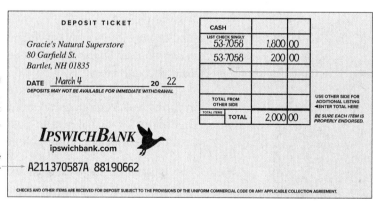

FIGURE 4.2 The structure of a check

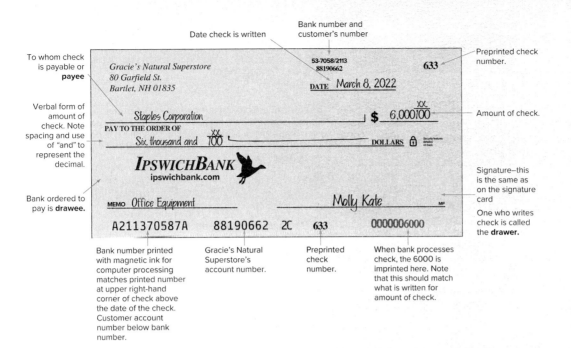

Check Stub

It should be completed before the check is written.

Writing business checks is similar to writing personal checks. Before writing any checks, however, you must understand the structure of a check and know how to write a check. Carefully study Figure 4.2. Note that the verbal amount written in the check should match the figure amount. If these two amounts are different, by law the bank uses the verbal amount. Also, note the bank imprint on the bottom right section of the check. When processing the check, the bank imprints the check's amount. This makes it easy to detect bank errors.

Learn: Using the Checking Account

Once the check is written, the writer must keep a record of the check. Knowing the amount of your written checks and the amount in the bank should help you avoid writing a bad check. Business checkbooks usually include attached **check stubs** to keep track of written checks. The sample check stub in the margin shows the information that the check writer will want to record. Some companies use a **check register** to keep their check records instead of check stubs. Figure 4.6 later in the chapter shows a check register with a ✓ column that is often used in balancing the checkbook with the bank statement (Learning Unit 4–2).

Gracie's Natural Superstore has had a busy week, and Molly must deposit its checks in the company's checking account. However, before she can do this, Molly must **endorse,** or sign, the back left side of the checks. Figure 4.3 explains the three types of check endorsements: **blank endorsement, full endorsement,** and **restrictive endorsement.** These endorsements transfer Gracie's ownership to the bank, which collects the money from the person or company issuing the check. Federal Reserve regulation limits all endorsements to the top $1\frac{1}{2}$ inches of the trailing edge on the back left side of the check.

After the bank receives Molly's deposit slip, shown in Figure 4.1, it increases (or credits) Gracie's account by $2,000. Often Molly leaves the deposit in a locked bag in a night depository. Then the bank credits (increases) Gracie's account when it processes the deposit on the next working day.

Money Tip Conduct an annual check of your bank's interest rates and fees. You may find better rates and lower fees at a credit union.

FIGURE 4.3

Types of common
endorsements

A. Blank Endorsement

**Gracie's Natural Superstore
88190662**

The company stamp or a signature alone on the back left side of a check legally makes the check payable to anyone holding the check. It can be *further* endorsed. This is not a safe type of endorsement.

B. Full Endorsement

Pay to the order of
Ipswich Bank
**Gracie's Natural Superstore
88190662**

Safer type of endorsement since Gracie's Natural Superstore indicates the name of the company or person to whom the check is to be payable. Only the person or company named in the endorsement can transfer the check to someone else.

C. Restrictive Endorsement

Pay to the order of
Ipswich Bank

For deposit only
**Gracie's Natural Superstore
88190662**

Safest endorsement for businesses. Gracie's stamps the back of the check so that this check must be deposited in the firm's bank account. This limits any further negotiation of the check.

Let's check your understanding of this unit.

Practice Quiz

Complete this Practice Quiz to see how you are doing.

Complete the following check and check stub for Long Company. Note the $9,500.60 balance brought forward on check stub No. 113. You must make a $690.60 deposit on May 3. Sign the check for Roland Small.

Date	Check no.	Amount	Payable to	For
June 5, 2022	113	$83.76	Angel Corporation	Rent

No. 113   **$** _____
_____ 20 _____
To _____
For _____

	DOLLARS	CENTS
BALANCE	9,500	60
AMT. DEPOSITED		
TOTAL		
AMT. THIS CHECK		
BALANCE FORWARD		

Long Company
22 Aster Rd.
Salem, MA 01970

No. 113

PAY
TO THE
ORDER
OF _____   _____ 20 _____   5-13/110

$ _____

_____ DOLLARS

IPSWICHBANK
ipswichbank.com

MEMO_____

A011000138A   14 0380 113

✓ Solution

No. 113   **$** 83.76
June 5 20 22
To Angel Corp.
For Rent

	DOLLARS	CENTS
BALANCE	9,500	60
AMT. DEPOSITED	690	60
TOTAL	10,191	20
AMT. THIS CHECK	83	76
BALANCE FORWARD	10,107	44

Long Company
22 Aster Rd.
Salem, MA 01970

No. 113

PAY
TO THE
ORDER
OF Angel Corporation   June 5 20 22   5-13/110

$ 83 $\frac{76}{100}$

Eighty-three and $\frac{76}{100}$ _____ DOLLARS

IPSWICHBANK
ipswichbank.com

Roland Small

MEMO   Rent

A011000138A   14 0380 113

Learning Unit 4–2:

Bank Statement and Reconciliation Process; Using Mobile and Online Banking

This learning unit is divided into two sections: (1) bank statement and reconciliation process, and (2) using mobile and online banking. The bank statement discussion will teach you why it was important for Gracie's Natural Superstore to reconcile its checkbook balance with the balance reported on its bank statement. Note that you can also use this reconciliation process in reconciling your personal checking account to avoid the expensive error of an overdrawn account.

Learn: Bank Statement and Reconciliation Process

Each month, Ipswich Bank sends Gracie's Natural Superstore a **bank statement** (Figure 4.4). We are interested in the following:

1. Beginning bank balance.

2. Total of all the account increases. Each time the bank increases the account amount, it *credits* the account.

FIGURE 4.4 Bank statement

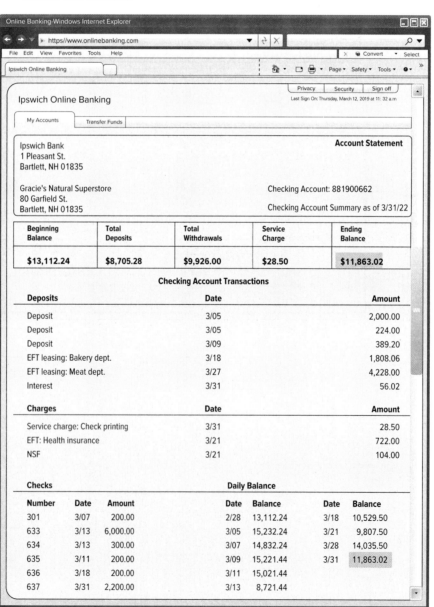

Beginning Balance	Total Deposits	Total Withdrawals	Service Charge	Ending Balance
$13,112.24	$8,705.28	$9,926.00	$28.50	$11,863.02

Checking Account Transactions

Deposits	Date	Amount
Deposit	3/05	2,000.00
Deposit	3/05	224.00
Deposit	3/09	389.20
EFT leasing: Bakery dept.	3/18	1,808.06
EFT leasing: Meat dept.	3/27	4,228.00
Interest	3/31	56.02

Charges	Date	Amount
Service charge: Check printing	3/31	28.50
EFT: Health insurance	3/21	722.00
NSF	3/21	104.00

Checks — **Daily Balance**

Number	Date	Amount		Date	Balance		Date	Balance
301	3/07	200.00		2/28	13,112.24		3/18	10,529.50
633	3/13	6,000.00		3/05	15,232.24		3/21	9,807.50
634	3/13	300.00		3/07	14,832.24		3/28	14,035.50
635	3/11	200.00		3/09	15,221.44		3/31	11,863.02
636	3/18	200.00		3/11	15,021.44			
637	3/31	2,200.00		3/13	8,721.44			

Checkbook balance		Bank balance
+ EFT (electronic funds transfer)	− NSF check	+ Deposits in transit
+ Interest earned	− Online fees	− Outstanding checks
+ Notes collected^	− Automatic payments*	± Bank errors
+ Credit memo		
+ Direct deposits	− Overdrafts†	
− ATM withdrawals	− Service charges	
− Automatic withdrawals	− Stop payments‡	
− Debit memo	± Book errors§	

^Notes (money) collected by a financial institution on behalf of a customer.

*Preauthorized payments for utility bills, mortgage payments, insurance, etc.

†**Overdrafts** occur when the customer has no overdraft protection and a check bounces back to the company or person who received the check because the customer has written a check without enough money in the bank to pay for it.

‡A stop payment is issued when the writer of the check does not want the receiver to cash the check.

§If a $60 check is recorded at $50, the checkbook balance must be decreased by $10.

FIGURE 4.5
Reconciling checkbook with bank statement

3. Total of all account decreases. Each time the bank decreases the account amount, it *debits* the account.

4. Final ending balance.

Due to differences in timing, the bank balance on the bank statement frequently does not match the customer's checkbook balance. Also, the bank statement can show transactions that have not been entered in the customer's checkbook. Figure 4.5 tells you what to look for when comparing a checkbook balance with a bank balance.

Aha!

Be vigilant in safeguarding business and personal information to bank safely electronically.

Gracie's Natural Superstore is planning to offer to its employees the option of depositing their checks directly into each employee's checking account. This is accomplished through the **electronic funds transfer (EFT)**—a computerized operation that electronically transfers funds among parties without the use of paper checks. Gracie's, which sublets space in the store, receives rental payments by EFT. Gracie's also has the bank pay the store's health insurance premiums by EFT.

To reconcile the difference between the amount on the bank statement and in the checkbook, the customer should complete a **bank reconciliation.** Today, many companies and home computer owners are using software such as Quicken and QuickBooks to complete their bank reconciliation. Also, we have mentioned the increased use of **banking apps** available to customers. However, you should understand the following steps for manually reconciling a bank statement.

Money Tip Use effective passwords and guard them carefully. Be aware passwords consisting of 8 numbers is instantly hackable but using a combination of 11 characters including numbers, upper and lowercase letters and symbols will take 400 years to hack according to Hive Systems and www.howsecureismy password.net.

Reconciling a Bank Statement

Step 1 Identify the outstanding checks (checks written but not yet processed by the bank). You can use the ✓ column in the check register (Figure 4.6) to check the canceled checks listed in the bank statement against the checks you wrote in the check register. The unchecked checks are the outstanding checks.

Step 2 Identify the deposits in transit (deposits made but not yet processed by the bank), using the same method in Step 1.

Step 3 Analyze the bank statement for transactions not recorded in the check stubs or check registers (like EFT).

Step 4 Check for recording errors in checks written, in deposits made, or in subtraction and addition.

Step 5 Compare the adjusted balances of the checkbook and the bank statement. If the balances are not the same, repeat Steps 1–4.

Molly uses a check register (Figure 4.6) to keep a record of Gracie's checks and deposits. By looking at Gracie's check register, you can see how to complete Steps 1 and 2 above. The explanation that follows for the first four bank statement reconciliation steps will help you understand the procedure.

FIGURE 4.6 Gracie's Natural Superstore check register

		RECORD ALL CHARGES OR CREDITS THAT AFFECT YOUR ACCOUNT						BALANCE	
NUMBER	DATE 2019	DESCRIPTION OF TRANSACTION	PAYMENT/DEBIT (−)	√	FEE (IF ANY) (−)	DEPOSIT/CREDIT (+)	$	12,912	24
	3/04	Deposit	$		$	$ 2,000 00	+ 2,000	00
							14,912	24
	3/04	Deposit				224 00	+ 224	00
							15,136	24
633	3/08	Staples Company	6,000 00	✓			− 6,000	00
							9,136	24
634	3/09	Health Foods Inc.	1,020 00	✓			− 1,020	00
							8,116	24
	3/09	Deposit				389 20	+ 389	20
							8,505	44
635	3/10	Liberty Insurance	200 00	✓			− 200	00
							8,305	44
636	3/18	Ryan Press	200 00	✓			− 200	00
							8,105	44
637	3/29	Logan Advertising	2,200 00	✓			− 2,200	00
							5,905	44
	3/30	Deposit				3,383 26	+ 3,383	26
							9,288	70
638	3/31	Sears Roebuck	572 00				− 572	00
							8,716	70
639	3/31	Flynn Company	638 94				− 638	94
							8,077	76
640	3/31	Lynn's Farm	166 00				− 166	00
							7,911	76
641	3/31	Ron's Wholesale	406 28				− 406	28
							7,505	48
642	3/31	Grocery Natural, Inc.	917 06				− 917	06
							$6,588	42

REMEMBER TO RECORD AUTOMATIC PAYMENTS/DEPOSITS ON DATE AUTHORIZED.

Money Tip Always review your monthly bank statement to ensure there are no errors. The earlier you catch an error, the easier it is to remedy.

Step 1. Identify Outstanding Checks **Outstanding checks** are checks that Gracie's Natural Superstore has written but Ipswich Bank has not yet recorded for payment when it sends out the bank statement. Gracie's treasurer identifies the following checks written on 3/31 as outstanding:

No. 638	$572.00
No. 639	638.94
No. 640	166.00
No. 641	406.28
No. 642	917.06

Step 2. Identify Deposits in Transit **Deposits in transit** are deposits that did not reach Ipswich Bank by the time the bank prepared the bank statement. The March 30 deposit of $3,383.26 did not reach Ipswich Bank by the bank statement date. You can see this by comparing the company's bank statement with its check register.

Step 3. Analyze Bank Statement for Transactions Not Recorded in Check Stubs or Check Register   The bank statement of Gracie's Natural Superstore (Figure 4.4) begins with the deposits, or increases, made to Gracie's bank account. Increases to accounts are known as credits. These are the result of a **credit memo (CM).** Gracie's received the following increases or credits in March:

1. *EFT leasing:* $1,808.06 and $4,228.00.

    Each month the bakery and meat departments pay for space they lease in the store.

2. *Interest credited:* $56.02.

    Gracie's has a checking account that pays interest; the account has earned $56.02.

When Gracie's has charges against its bank account, the bank decreases, or debits, Gracie's account for these charges. Banks usually inform customers of a debit transaction by a **debit memo (DM).** The following items will result in debits to Gracie's account:

1. *Service charge:* $28.50.

    The bank charged $28.50 for printing Gracie's checks.

2. *EFT payment:* $722.

    The bank made a health insurance payment for Gracie's.

3. *NSF check:* $104.

    One of Gracie's customers wrote Gracie's a check for $104. Gracie's deposited the check, but the check bounced for **nonsufficient funds (NSF).** Thus, Gracie's has $104 less than it figured.

Step 4. Check for Recording Errors   The treasurer of Gracie's Natural Superstore, Molly Kate, recorded check No. 634 for the wrong amount—$1,020 (see the check register in Figure 4.6). The bank statement showed that check No. 634 cleared for $300. To reconcile Gracie's checkbook balance with the bank balance, Gracie's must add $720 to its checkbook balance. Neglecting to record a deposit also results in an error in the company's checkbook balance. As you can see, reconciling the bank's balance with a checkbook balance is a necessary part of business and personal finance.

Step 5. Completing the Bank Reconciliation   Now we can complete the bank reconciliation on the back side of the bank statement as shown in Figure 4.7. This form is usually on the back of a bank statement. If necessary, however, the person reconciling the bank statement can construct a bank reconciliation form similar to Figure 4.8.

FIGURE 4.7
Reconciliation
process

CHECKS OUTSTANDING (NOT YET CHARGED TO ACCOUNT)		
NUMBER OR DATE	DOLLARS	CENTS
638	$572	00
639	638	94
640	166	00
641	406	28
642	977	06
TOTAL	$2,700	28

CHECKING ACCOUNT RECONCILEMENT

Enter the new balance shown on the other side of this statement. $ 11,863.02

ADD
Deposits not shown } 3,383.26

ADD
Advances and Transfers to Checking not shown }

SUBTOTAL 15,246.28

DEDUCT
Checks outstanding 2,700.28

SUBTOTAL 12,546.00

DEDUCT
Transfers from Checking not shown _____

This amount should agree with the balance in your checkbook register. $ 12,546.00

Checkbook balance $6,588.42

+ EFT leasing $6,036.06
+ Interest   56.02
+ Checkbook error   720.00   6,812.08
− Service charge $ 28.50
− EFT: health insurance 722.00
− NSF   104.00   854.50

Ending checkbook balance $12,546.00

FIGURE 4.8 Bank
reconciliation

GRACIE'S NATURAL SUPERSTORE Bank Reconciliation as of March 31, 2022					
Checkbook balance			**Bank balance**		
Gracie's checkbook balance		$6,588.42	Bank balance		$11,863.02
Add:			Add:		
EFT leasing: Bakery dept.	$1,808.06		Deposit in transit, 3/30		3,383.26
EFT leasing: Meat dept.	4,228.00				$15,246.28
Interest	56.02				
Error: Overstated check No. 634	720.00	$ 6,812.08			
		$13,400.50			
Deduct:			Deduct:		
Service charge	$ 28.50		Outstanding checks:		
NSF check	104.00		No. 638	$572.00	
EFT health insurance payment	722.00	854.50	No. 639	638.94	
			No. 640	166.00	
			No. 641	406.28	
			No. 642	917.06	2,700.28
Reconciled balance		$12,546.00	Reconciled balance		$12,546.00

Learn: Using Mobile and Online Banking

Mobile and online banking technologies have transformed banking for many users throughout the last decade. Gone are the days of having to wait until the bank or credit union opens to conduct financial business. The Federal Reserve reports that 71 out of 100 of customers use online banking and 43 out of 100 use mobile banking. Online banking is typically conducted on the internet using the financial institution's website, requiring an internet connection and a computer. Mobile banking conveniently uses a smartphone or tablet. Many functions historically conducted by a teller can now be accomplished online or with mobile devices. Viewing transactions, inspecting current or past statements, transferring money, managing and paying bills, and myriad other capabilities have blossomed from when online banking commenced back in the late 1980s.

Mobile and online banking provide several benefits. Mobile apps and online banking websites are easy to use; convenient, with 24-7 accessibility; save time; and have higher security such as a two-step verification process and one-time password sent to a registered device before permitting transactions or account management. Apps used by financial institutions for mobile banking use industry-standard encryption making them, in some instances, even safer than conducting business directly with a teller.

Pepper ... And Salt

THE WALL STREET JOURNAL

STATE BANK

"Does every penny have to come out exactly even?"

Used by permission of Cartoon Features Syndicate

The Check Clearing for the 21st Century Act permits banks to accept for deposit the legal equivalent of a physical check, known as a substitute check. Online-only banks and mobile apps have taken advantage of this Act. Online-only banks, such as Axos Bank and Varo Bank, are able to pass on lower overhead savings to members via reduced fees and higher interest rates paid. Mobile apps make it possible to deposit "substitute checks" into a U.S. bank account, in addition to other banking transactions, using the camera on mobile devices and the financial institution's app. Once a mobile check is deposited, the information is encrypted and transmitted to the financial institution following standard steps for processing and clearing.

Mobile check depositing requires following a simple process of endorsing the check, opening the financial institution's app and following the steps for making a deposit, taking a picture of the front and back of the check, verifying the information, submitting, and finally, awaiting the confirmation. It's that easy and that quick. Remote deposit capture technology or the "bank-check selfie" does all the rest.

Mobile Check Deposit

Step 1 Endorse the check

Step 2 Open the bank or credit union app and register or login

Step 3 Locate mobile check deposit in the menu and follow the steps for making a deposit

Step 4 Identify which account you want the check deposited to

Step 5 Enter the check amount

Step 6 Take a picture of the front and back of the check ensuring all four corners are included in the photo and the photo is crisp and clear

Step 7 Verify the information and submit

Step 8 Wait for a confirmation

Money Tip Set up a low-balance alert with your financial institution for text or email notifications to reduce your risk of making an overdraft saving you time and money.

"It's cash. It's kind of like a debit card."

Below are some tips for using mobile check deposit:

- Do not destroy the check deposited via mobile deposit until you have verified the check has been deposited into your account.

- Date and mark on the check that it has been deposited and retain it for 30–60 days for documentation. Then shred the check to avoid accidentally redepositing it and incurring a return-deposit fee.

- Check with your financial institution to determine if there are any remote deposit fees, any limits to quantity of checks or deposit limits for mobile deposits, what the cut-off time is for mobile deposits, and when your mobile deposit funds will be available.

- Use direct deposit instead of mobile deposit to avoid any delays in funding or fees being assessed when having access to the funds immediately is important.

- When receiving a check from someone you don't know, verify it with the bank it is drawn upon by finding the bank's number online (don't call the number printed on the check) and calling it requesting verification of the account and if the check will clear at that moment in time. Cashing the check at the check writer's bank is a best practice.

Aha! *ALWAYS avoid the use of public WiFi for banking or for any transaction or inquiry using sensitive information.*

The ease and safety of using banking technologies has created dedicated users. These technologies also provided hands-free banking during the COVID-19 pandemic. With bank lobbies closing and minimal access to tellers, mobile, online and online-only banking provided what was needed. Banking technologies have continued to improve the mobile and online experience resulting in more and more people choosing these methods for their banking needs.

The *Wall Street Journal* clip "Four Apps to Help You Fight Inaction", below shows four apps to help you with your saving and banking needs.

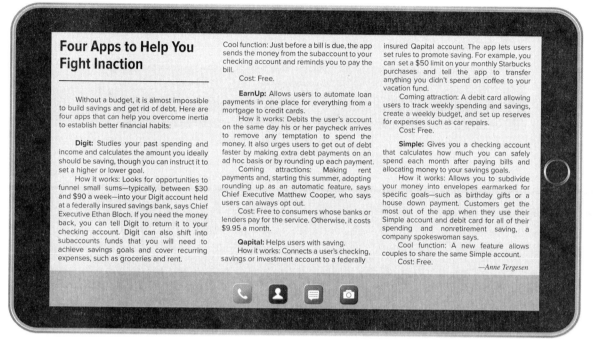

Four Apps to Help You Fight Inaction

Without a budget, it is almost impossible to build savings and get rid of debt. Here are four apps that can help you overcome inertia to establish better financial habits:

Digit: Studies your past spending and income and calculates the amount you ideally should be saving, though you can instruct it to set a higher or lower goal.

How it works: Looks for opportunities to funnel small sums—typically, between $30 and $90 a week—into your Digit account held at a federally insured savings bank, says Chief Executive Ethan Bloch. If you need the money back, you can tell Digit to return it to your checking account. Digit can also shift into subaccounts funds that you will need to achieve savings goals and cover recurring expenses, such as groceries and rent.

Cool function: Just before a bill is due, the app sends the money from the subaccount to your checking account and reminds you to pay the bill.

Cost: Free.

EarnUp: Allows users to automate loan payments in one place for everything from a mortgage to credit cards.

How it works: Debits the user's account on the same day his or her paycheck arrives to remove any temptation to spend the money. It also urges users to get out of debt faster by making extra debt payments on an ad hoc basis or by rounding up each payment.

Coming attractions: Making rent payments and, starting this summer, adopting rounding up as an automatic feature, says Chief Executive Matthew Cooper, who says users can always opt out.

Cost: Free to consumers whose banks or lenders pay for the service. Otherwise, it costs $9.95 a month.

Qapital: Helps users with saving.

How it works: Connects a user's checking, savings or investment account to a federally insured Qapital account. The app lets users set rules to promote saving. For example, you can set a $50 limit on your monthly Starbucks purchases and tell the app to transfer anything you didn't spend on coffee to your vacation fund.

Coming attraction: A debit card allowing users to track weekly spending and savings, create a weekly budget, and set up reserves for expenses such as car repairs.

Cost: Free.

Simple: Gives you a checking account that calculates how much you can safely spend each month after paying bills and allocating money to your savings goals.

How it works: Allows you to subdivide your money into envelopes earmarked for specific goals—such as birthday gifts or a house down payment. Customers get the most out of the app when they use their Simple account and debit card for all of their spending and nonretirement saving, a company spokeswoman says.

Cool function: A new feature allows couples to share the same Simple account.

Cost: Free.

—*Anne Tergesen*

Check your knowledge of this learning unit by completing the following Practice Quiz.

Practice Quiz

Complete this Practice Quiz to see how you are doing.

Rosa Garcia received her February 3, 2022, bank statement showing a balance of $212.80. Rosa's checkbook has a balance of $929.15. The bank statement showed that Rosa had an ATM fee of $12.00 and a deposited check returned fee of $20.00. Rosa earned interest of $1.05. She had three outstanding checks: No. 300, $18.20; No. 302, $38.40; and No. 303, $68.12. A deposit for $810.12 was not on her bank statement. Prepare Rosa Garcia's bank reconciliation.

✓ **Solution**

ROSA GARCIA
Bank Reconciliation as of February 3, 2022

Checkbook balance			Bank balance		
Rosa's checkbook balance		$ 929.15	Bank balance		$ 212.80
Add:			Add:		
Interest		1.05	Deposit in transit		810.12
		$930.20			$1,022.92
Deduct:			Deduct:		
Deposited check			Outstanding checks:		
returned fee	$20.00		No. 300	$18.20	
ATM	12.00	32.00	No. 302	38.40	
			No. 303	68.12	124.72
Reconciled balance		$898.20	Reconciled balance		$ 898.20

Chapter 4 Review

Topic/Procedure/Formula	Example	You try it*
Types of endorsements *Blank:* Not safe; can be further endorsed. *Full:* Only person or company named in endorsement can transfer check to someone else. *Restrictive:* Check must be deposited. Limits any further negotiation of the check.	Jones Co. 21-333-9 Pay to the order of Regan Bank Jones Co. 21-333-9 Pay to the order of Regan Bank. For deposit only. Jones Co. 21-333-9	**Write a sample of a blank, full, and restrictive endorsement.** Use Pete Co. Acct. # 24-111-9

Bank reconciliation
 	
Checkbook balance
+ EFT (electronic funds transfer)
+ Interest earned
+ Notes collected
+ Direct deposits
− ATM withdrawals
− NSF check
− Online fees
− Automatic withdrawals
− Overdrafts
− Service charges
− Stop payments
± Book errors (see note, below)
CM—adds to balance
DM—deducts from balance
 	
Bank balance
+ Deposits in transit
− Outstanding checks
± Bank errors
	Checkbook balance
	
	Balance $800
	  − NSF _40_
	    $760
	  − Service charge _4_
	    $756
	
	Bank balance
	
	Balance $ 632
	  + Deposits in transit _416_
	    $1,048
	  − Outstanding checks _292_
	    $ 756
		Calculate ending checkbook balance
		
		1. Beg. checkbook bal.: $300
		2. NSF: $50
		3. Deposit in transit: $100
		4. Outstanding check: $60
		5. ATM service charge: $20

Chapter 4 Review *(Continued)*

Key Terms

Automatic teller machine (ATM)	Debit card	Full endorsement
	Debit memo (DM)	Mobile banking
Bank reconciliation	Deposit slip	Nonsufficient funds (NSF)
Bank statement	Deposits in transit	
Banking apps	Draft	Outstanding checks
Blank endorsement	Drawee	Overdrafts
Check	Drawer	Payee
Check register	Electronic funds transfer (EFT)	Restrictive endorsement
Check stub		
Credit memo (CM)	Endorse	Signature card

Note: If a $60 check is recorded as $50, we must decrease checkbook balance by $10.
*Worked-out solutions are in Appendix A.

Critical Thinking Discussion Questions with Chapter Concept Check

1. Explain the structure of a check. The trend in bank statements is not to return the canceled checks. Do you think this is fair?

2. List the three types of endorsements. Endorsements are limited to the top $1\frac{1}{2}$ inches of the trailing edge on the back left side of your check. Why do you think the Federal Reserve made this regulation?

3. List the steps in reconciling a bank statement. Today, many banks charge a monthly fee for certain types of checking accounts. Do you think all checking accounts should be free? Please explain.

4. What is mobile and online banking? Will we become a cashless society in which all transactions are made with some type of credit card or cryptocurrency?

5. Should banks be allowed to fail? Why or why not?

6. **Chapter Concept Check.** Create your own company and provide needed data to prepare a bank reconciliation. Then go to a bank website and explain how you would use the bank's app versus the manual system of banking.

7. **Chapter Concept Check.** Use one mobile and/or online banking function from the chapter and report on its effectiveness.

End-of-Chapter Problems

Name _____   Date_____

Check figures for odd-numbered problems in Appendix A.

Drill Problems

4–1. Fill out the check register that follows with this information for August 2022:   *LU 4-1(1)*

Aug 7	Check No. 959	AT&T	$143.50
15	Check No. 960	Staples	66.10
19	Deposit		800.00
20	Check No. 961	West Electric	451.88
24	Check No. 962	Bank of America	319.24
29	Deposit		400.30

		RECORD ALL CHARGES OR CREDITS THAT AFFECT YOUR ACCOUNT							BALANCE	
NUMBER	DATE 2019	DESCRIPTION OF TRANSACTION	PAYMENT/DEBIT (−)	√	FEE (IF ANY) (−)	DEPOSIT/CREDIT (+)		$	4,500	75
			$		$	$				

4–2. November 1, 2022, Payroll.com, an Internet company, has a $10,481.88 checkbook balance. Record the following transactions for Payroll.com by completing the two checks and check stubs provided. Sign the checks Garth Scholten, controller.   *LU 4-1(2)*

a. November 8, 2022, deposited $688.10

b. November 8, check No. 190 payable to Staples for office supplies—$766.88

c. November 15, check No. 191 payable to Best Buy for computer equipment—$3,815.99.

No. _____ $ _____	PAYROLL.COM	No. 190
_____ 20 _____	1 LEDGER RD.	
To _____	ST. PAUL, MN 55113	
For _____		

	DOLLARS	CENTS
BALANCE		
AMT. DEPOSITED		
TOTAL		
AMT. THIS CHECK		
BALANCE FORWARD		

PAY TO THE ORDER OF _____ $ _____

_____ DOLLARS

IPSWICHBANK
ipswichbank.com

MEMO_____

A011000138A   25 11103 190

No. _____ $ _____	PAYROLL.COM	No. 191
_____ 20 _____	1 LEDGER RD.	
To _____	ST. PAUL, MN 55113	
For _____		

```
No. _____ $ _____          PAYROLL.COM                              No. 191
                                   1 LEDGER RD.
              20                   ST. PAUL, MN 55113
To _____
For _____                      PAY                          _____ 20 _____   5-13/110
         DOLLARS   CENTS           TO THE
BALANCE                            ORDER                                 $ _____
                                   OF _____
AMT. DEPOSITED
                                   _____  DOLLARS

                                   IPSWICHBANK
TOTAL                              ipswichbank.com        _____
AMT. THIS CHECK                    MEMO_____
BALANCE FORWARD                      A011000138A      25  11103  191
```

4–3. Using the check register in Problem 4–1 and the following bank statement, prepare a bank reconciliation for Lee.com. *LU 4-2(3)*

BANK STATEMENT			
Date	**Checks**	**Deposits**	**Balance**
8/1 balance			$4,500.75
8/18	$143.50		4,357.25
8/19		$800.00	5,157.25
8/26	319.24		4,838.01
8/30	15.00 SC		4,823.01

Word Problems

4–4. The World Bank forecasted growth of world trade to be 2.9%, up from 2.7%. This change caused Galapagos Islands Resort to analyze its current financial situation, beginning with reconciling its accounts. Galapagos Islands Resort received its bank statement showing a balance of $8,788. Its checkbook balance is $15,252. Deposits in transit are $3,450 and $6,521. There is a service charge of $45 and interest earned of $3. Notes collected total $1,575. Outstanding checks are No. 1021 for $1,260 and No. 1022 for $714. All numbers are in U.S. dollars. Help Galapagos Islands Resort reconcile its balances. *LU 4-2(3)*

4–5. The U.S. Chamber of Commerce provides a free monthly bank reconciliation template at business.uschamber.com/tools/bankre_m.asp. Riley Whitelaw just received her bank statement notice online. She wants to reconcile her checking account with her bank statement and has chosen to reconcile her accounts manually. Her checkbook shows a balance of $698. Her bank statement reflects a balance of $1,348. Checks outstanding are No. 2146, $25; No. 2148, $58; No. 2152, $198; and No. 2153, $464. Deposits in transit are $100 and $50. There is a $15 service charge and $5 ATM charge in addition to notes collected of $50 and $25. Reconcile Riley's balances. *LU 4-2(3)*

End-of-Chapter Problems (Continued)

4–6. A local bank began charging $2.50 each month for returning canceled checks. The bank also has an $8.00 "maintenance" fee if a checking account slips below $750. Donna Sands likes to have copies of her canceled checks for preparing her income tax returns. She has received her bank statement with a balance of $535.85. Donna received $2.68 in interest and has been charged for the canceled checks and the maintenance fee. The following checks were outstanding: No. 94, $121.16; No. 96, $106.30; No. 98, $210.12; and No. 99, $64.84. A deposit of $765.69 was not recorded on Donna's bank statement. Her checkbook shows a balance of $806.94. Prepare Donna's bank reconciliation.
LU 4-2(3)

e)cel 4–7. Ben Luna received his bank statement with a $27.04 fee for a bounced check (NSF). He has an $815.75 monthly mortgage payment paid through his bank. There was also a $3.00 teller fee and a check printing fee of $3.50. His ATM card fee was $6.40. There was also a $530.50 deposit in transit. The bank shows a balance of $119.17. The bank paid Ben $1.23 in interest. Ben's checkbook shows a balance of $1,395.28. Check No. 234 for $80.30 and check No. 235 for $28.55 were outstanding. Prepare Ben's bank reconciliation.
LU 4-2(3)

4–8. Kameron Gibson's bank statement showed a balance of $717.72. Kameron's checkbook had a balance of $209.50. Check No. 104 for $110.07 and check No. 105 for $15.55 were outstanding. A $620.50 deposit was not on the statement. He has his payroll check electronically deposited to his checking account—the payroll check was for $1,025.10. There was also a $4 teller fee and an $18 service charge. Prepare Kameron Gibson's bank reconciliation.   *LU 4-2(3)*

excel **4–9.** Banks are finding more ways to charge fees, such as a $25 overdraft fee. Sue McVickers has an account in Fayetteville; she has received her bank statement with this $25 charge. Also, she was charged a $6.50 service fee; however, the good news is she earned $5.15 interest. Her bank statement's balance was $315.65, but it did not show the $1,215.15 deposit she had made. Sue's checkbook balance shows $604.30. The following checks have not cleared: No. 250, $603.15; No. 253, $218.90; and No. 254, $130.80. Prepare Sue's bank reconciliation.
LU 4-2(3)

End-of-Chapter Problems (Continued)

4–10. Carol Stokke receives her April 6 bank statement showing a balance of $859.75; her checkbook balance is $954.25. The bank statement shows an ATM charge of $25.00, NSF fee of $27.00, earned interest of $2.75, and Carol's $630.15 refund check, which was processed by the IRS and deposited to her account. Carol has two checks that have not cleared—No. 115 for $521.15 and No. 116 for $205.50. There is also a deposit in transit for $1,402.05. Prepare Carol's bank reconciliation.   *LU 4-2(3)*

4–11. Lowell Bank reported the following checking account fees: $2 to see a real-live teller, $20 to process a bounced check, and $1 to $3 if you need an original check to prove you paid a bill or made a charitable contribution. This past month you had to transact business through a teller six times—a total $12 cost to you. Your bank statement shows a $305.33 balance; your checkbook shows a $1,009.76 balance. You received $1.10 in interest. An $801.15 deposit was not recorded on your statement. The following checks were outstanding: No. 413, $28.30; No. 414, $18.60; and No. 418, $60.72. Prepare your bank reconciliation.   *LU 4-2(3)*

excel **4–12.** According to the *Portland Business Journal,* Jim Houser, a Portland auto specialist, landed a key Small Business Administration appointment. Help Jim reconcile Remington's Auto Clinic's checkbook and bank balance according to the following: bank statement balance, $18,769; checkbook balance, $22,385,015; interest earned, $3,948; deposits in transit, $100,656 and $22,375,000; ATM card fees, $150; outstanding checks—No. 10189, $55,678; No. 10192, $15,287; No. 10193, $22,350; and No. 10194, $12,297.   *LU 4-2(3)*

4–13. Identity Theft Resource Center (ITRC) provides consumer and victim support, public education, and advice. Marlena's grandmother is concerned her identity has been stolen. Help Marlena reconcile her grandmother's checkbook and bank statement. The checkbook reflects a balance of $1,245. The bank statement shows a balance of $207. Notes collected were $100 and $210. The bank charged a $25 service fee. Outstanding checks were No. 255, $985; No. 261, $233; and No. 262, $105. There is a deposit in transit of $2,646.   *LU 4-2(2)*

End-of-Chapter Problems (*Continued*)

4–14. Family Loved Labs', out of Sandstone, Minnesota, bookkeeper wanted to reconcile the business bank statement with their checkbook. One 8-puppy litter was recently sold at $2,200 per puppy and the total deposit in transit was $17,600. The bank statement balance was $22,456 and the checkbook balance was $33,299.18. Outstanding checks were No. 975, $58.10; No, 978, $1,276.34; No. 982, $899.11; and No. 986, $4,525.80. The bank statement showed a service fee of $25, check printing charge of $15.75, and interest earned of $38.22.

4–15. Chris Montgomery was concerned about his checking account. He had been laid off work during the pandemic and was relying on the stimulus checks to be able to pay bills. Chris's checkbook balance was a negative $1,352.37. His bank statement reflected a balance of $1,567.45. The Cares Act stimulus checks of $1,200 and $600 were not reflected in Chris's checkbook. The bank statement showed interest earned of $5.92 and a service charge of $15.00. An EFT for health insurance for $575 and notes collected of $272.90 were included. Outstanding checks were No. 367, $575; and No. 368, $856. Help Chris reconcile his bank statement and checkbook balance.

Challenge Problems

4–16. Carolyn Crosswell, who banks in New Jersey, wants to balance her checkbook, which shows a balance of $985.20. The bank shows a balance of $1,430.33. The following transactions occurred: $135.20 automatic withdrawal to the gas company, $6.50 ATM fee, $8.00 service fee, and $1,030.05 direct deposit from the IRS. Carolyn used her debit card five times and was charged 45 cents for each transaction; she was also charged $3.50 for check printing. A $931.08 deposit was not shown on her bank statement. The following checks were outstanding: No. 235, $158.20; No. 237, $184.13; No. 238, $118.12; and No. 239, $38.83. Carolyn received $2.33 interest. Prepare Carolyn's bank reconciliation. *LU 4-2(3)*

excel 4–17. Melissa Jackson, bookkeeper for Kinko Company, cannot prepare a bank reconciliation. From the following facts, can you help her complete the June 30, 2020, reconciliation? The bank statement showed a $2,955.82 balance. Melissa's checkbook showed a $3,301.82 balance. Melissa placed a $510.19 deposit in the bank's night depository on June 30. The deposit did not appear on the bank statement. The bank included two DMs and one CM with the returned checks: $690.65 DM for NSF check, $8.50 DM for service charges, and $400.00 CM (less $10 collection fee) for collecting a $400.00 non-interest-bearing note. Check No. 811 for $110.94 and check No. 912 for $82.50, both written and recorded on June 28, were not with the returned checks. The bookkeeper had correctly written check No. 884, $1,000, for a new cash register, but she recorded the check as $1,069. The May bank reconciliation showed check No. 748 for $210.90 and check No. 710 for $195.80 outstanding on April 30. The June bank statement included check No. 710 but not check No. 748.   *LU 4-2(3)*

Summary Practice Test

Do you need help? Connect videos have step-by-step worked-out solutions.

1. Walgreens has a $12,925.55 beginning checkbook balance. Record the following transactions in the check stubs provided.   *LU 4-1(2)*

   a. November 4, 2022, check No. 180 payable to Ace Medical Corporation, $1,700.88 for drugs.

   b. $5,250 deposit—November 24.

   c. November 24, 2022, check No. 181 payable to John's Wholesale, $825.55 merchandise.

No. _____ $ _____		
_____ 20 ____		
To _____		
For _____		
	DOLLARS	CENTS
BALANCE		
AMT. DEPOSITED		
TOTAL		
AMT. THIS CHECK		
BALANCE FORWARD		

No. _____ $ _____		
_____ 20 ____		
To _____		
For _____		
	DOLLARS	CENTS
BALANCE		
AMT. DEPOSITED		
TOTAL		
AMT. THIS CHECK		
BALANCE FORWARD		

2. On April 1, 2022, Lester Company received a bank statement that showed a balance of $8,950. Lester showed an $8,000 checking account balance. The bank did not return check No. 115 for $750 or check No. 118 for $370. A $900 deposit made on March 31 was in transit. The bank charged Lester $20 for check printing and $250 for NSF checks. The bank also collected a $1,400 note for Lester. Lester forgot to record a $400 withdrawal at the ATM. Prepare a bank reconciliation.   *LU 4-2(3)*

3. Felix Babic banks at Role Federal Bank. Today he received his March 31, 2022, bank statement showing a $762.80 balance. Felix's checkbook shows a balance of $799.80. The following checks have not cleared the bank: No. 140, $130.55; No. 149, $66.80; and No. 161, $102.90. Felix made an $820.15 deposit that is not shown on the bank statement. He has his $617.30 monthly mortgage payment paid through the bank. His $1,100.20 IRS refund check was mailed to his bank. Prepare Felix Babic's bank reconciliation.   *LU 4-2(3)*

4. On June 30, 2022, Wally Company's bank statement showed a $7,500.10 bank balance. Wally has a beginning checkbook balance of $9,800.00. The bank statement also showed that it collected a $1,200.50 note for the company. A $4,500.10 June 30 deposit was in transit. Check No. 119 for $650.20 and check No. 130 for $381.50 are outstanding. Wally's bank charges $.40 cents per check. This month, 80 checks were processed. Prepare a reconciled statement. *LU 4-2(3)*

MY MONEY

Q Breaking (down) the Bank

 What I need to know

There are many banking options available and you should find the one that best fits your needs. When you enter your chosen career field your earnings will likely increase. The bank you have been using may no longer be the best option for you going forward. It is in your best interest to consider how a bank can satisfy your needs and assist you in meeting your financial goals in the future. Not all banks are created equal. Some will offer more incentives and financial services than others and some will even pay you interest on your money! Understanding the banking options available will help you in finding the best bank for you.

 What I need to do

Consider the types of services you may be interested in when choosing where to bank. It is helpful to reference your financial goals and find a bank aligned with these goals. For instance, if you have plans to someday purchase a house it may be important to select a bank that will be able to provide financing for home purchases. Additionally, you may want to consider whether mobile banking is of importance to you and to what level you want to access and complete transactions with your bank via a mobile device. Electronic bill pay is another feature that provides considerable convenience and ease of use in managing your expenses.

Maybe a small local bank is right for you if you desire a personal approach to your finances and help in meeting your financial goals. A credit union is established to benefit its members through low fees and low interest rates while emphasizing a personal touch. Smaller local banks will also strive to take a personal interest in you and your financial plans. On the other hand, a large national bank has the benefit of accessibility through multiple physical locations and Automated Teller Machines (ATMs) having wide spread availability.

Research the details of each account type offered by your bank. You will find there are variations in the amount of interest you can earn on your money from different banks. Additionally, you need to understand the fees associated with various types of transactions on your account. You do not want to lose money by not paying attention to transactions with associated fees. For instance, there may be a fee for using a debit card, exceeding a certain number of withdrawals, and using an ATM not affiliated with your bank.

 Steps I need to take

1. Identify the banking options and services important to you.
2. Determine the level of personal service you desire with your banking.
3. Look for opportunities to grow your money through interest-bearing accounts and avoiding fees.

 Resources I can use

- https://www.gobankingrates.com/banking/banks/10-steps-finding-better-bank/—important steps to consider in finding the right bank for your needs.

MY MONEY ACTIVITY

- Visit https://www.findabetterbank.com/ and search for banks in your local area. Consider online only banks, too. www.nerdwallet.com/article/banking/best-online-banks
- Choose the features that are important to you in your bank choice.
- Compare results to the offerings of your current bank.

PERSONAL FINANCE

A KIPLINGER APPROACH

"IF YOU'RE USING CASH LESS OFTEN, YOU'RE PART OF A TREND." Kiplinger's. August 2020.

INTERVIEW

IF YOU'RE USING CASH LESS OFTEN, YOU'RE PART OF A TREND

The pandemic is speeding up the use of digital payments.

Dayna Ford is senior director-analyst for Gartner, a market research firm. She focuses on digital wallets and other forms of electronic payment.

Electronic payments have soared since the pandemic began. Do you expect that trend to continue after the crisis is over? I do, though the trend toward digital payments due to the pandemic has taken different forms: shopping online, paying digitally while doing a physical pickup or using contactless methods of payment, such as digital wallets. All were existing trends that had been steadily climbing over the past couple of years but have accelerated since the pandemic began. After the crisis is over, the rate of digital payments will drop, but not to what it was before.

Apple, Google and other providers offer apps designed to eliminate the physical wallet in favor of a digital one, but they're not widely used. Do you think they'll become more popular now? In places such as Asia, the use of digital wallets is very pervasive. A key reason we don't have more adoption here is consumer inertia. Consumers in the U.S. and Europe are accustomed to using plastic credit and debit cards. But concerns about health and hygiene could get consumers to try digital wallets. Once they try them, some will continue to use them.

What about contactless, peer-to-peer payment systems? We will continue to see the growth of peer-to-peer systems such as Venmo. You also have programs such as Zelle, which caters more to consumers who use traditional banks. The pandemic adds momentum because some consumers don't want to touch cash. Concerns about hygiene will get some people over the inertia that prevented them from trying these systems.

How can consumers protect themselves from identity theft when using these products? In some ways, digital wallets are more secure than credit and debit cards. For example, when you make online purchases using a digital wallet, only the wallet provider sees your credit or debit card information. Consumers should follow best practices for protecting data on their phones—use passwords and biometric authentication, auto-lock the screen—to keep their phones protected. And make sure you choose a reputable digital wallet provider, because you are storing your payment credentials with them.

Other countries, such as Sweden, are essentially cashless. How far is the U.S. from becoming a cashless society? I don't think cash is going away anytime soon. We've seen a number of legislative initiatives to protect cash, such as banning cashless stores in San Francisco, New Jersey and Philadelphia because they're seen as discriminating against consumers who don't have a bank account. Plastic cards may go away before cash. I do think we will see some reduction in the amount of cash in circula tion due to the pandemic.

What are the prospects for the Federal Reserve introducing a digital form of the U.S. dollar? The U.S. Federal Reserve, along with central banks in Europe and other regions, has considered developing its own digital currency. Other countries, such as China, are farther along this path. It seems likely that the U.S. will watch to see how these initiatives fare in other regions before it makes a decision. **SANDRA BLOCK**

Business Math Issue

Using Cash less will result in banks adding more branches.

1. List the key points of the article and information to support your position.

2. Write a group defense of your position using math calculations to support your view. If you are in an online course, post to a discussion board.

Chapter 5

Solving for the Unknown: A How-to Approach for Solving Equations

Learning Unit Objectives

LU 5–1: Solving Equations for the Unknown

1. Explain the basic procedures used to solve equations for the unknown.
2. List the five rules and the mechanical steps used to solve for the unknown in seven situations; know how to check the answers.

LU 5–2: Solving Word Problems for the Unknown

1. List the steps for solving word problems.
2. Complete blueprint aids to solve word problems; check the solutions.

⊖ Essential Question

How can I use solving equations in the business world and make my life easier?

🌐 Math Around the World

The Wall Street Journal chapter opener, "Follow a formula for retirement saving", shows the need to plan for retirement. Formulas play an important role in the planning process. Note the *Wall Street Journal* clip in the margin showing how Amazon is using math to increase profit.

Learning Unit 5–1 explains how you can solve for unknowns in equations. In Learning Unit 5–2 you learn how to solve for unknowns in word problems. When you complete these learning units, you will not have to memorize as many formulas to solve business and personal math applications.

Pepper... And Salt

THE WALL STREET JOURNAL

"OK, everybody set for that Zoom call?"

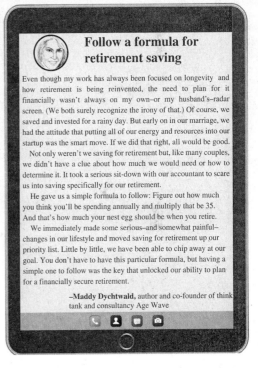

Follow a formula for retirement saving

Even though my work has always been focused on longevity and how retirement is being reinvented, the need to plan for it financially wasn't always on my own–or my husband's–radar screen. (We both surely recognize the irony of that.) Of course, we saved and invested for a rainy day. But early on in our marriage, we had the attitude that putting all of our energy and resources into our startup was the smart move. If we did that right, all would be good.

Not only weren't we saving for retirement but, like many couples, we didn't have a clue about how much we would need or how to determine it. It took a serious sit-down with our accountant to scare us into saving specifically for our retirement.

He gave us a simple formula to follow: Figure out how much you think you'll be spending annually and multiply that be 35. And that's how much your nest egg should be when you retire.

We immediately made some serious–and somewhat painful–changes in our lifestyle and moved saving for retirement up our priority list. Little by little, we have been able to chip away at our goal. You don't have to have this particular formula, but having a simple one to follow was the key that unlocked our ability to plan for a financially secure retirement.

–**Maddy Dychtwald,** author and co-founder of think tank and consultancy Age Wave

Amazon.com Inc. has adjusted its product-search system to more prominently feature listings that are more profitable for the company, said people who worked on the project—a move, contested internally, that could favor Amazon's own brands.

Late last year, these people said, Amazon optimized the secret algorithm that ranks listings so that instead of showing customers mainly the most-relevant and best-selling listings when they search.

Mattioli, Dana. "Amazon Search Change Boosts Its Own Products." *The Wall Street Journal*, September 9, 2019.

Learning Unit 5–1:
Solving Equations for the Unknown

The following Rose Smith letter is based on a true story. Note how Rose states that the blueprint aids, the lesson on repetition, and the chapter organizers were important factors in the successful completion of her business math course.

Many of you are familiar with the terms *variables* and *constants*. If you are planning to prepare for your retirement by saving only what you can afford each year, your saving is a *variable;* if you plan to save the same amount each year, your saving is a *constant.* Now you can also say that you cannot buy clothes by size because of the many variables involved. This unit explains the importance of mathematical variables and constants when solving equations.

Rose Smith
15 Locust Street
Lynn, MA 01915

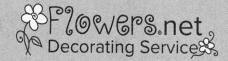

Dear Professor Slater,

    Thank you for helping me get through your Business Math class. When I first started, my math anxiety level was real high. I felt I had no head for numbers. When you told us we would be covering the chapter on solving equations, I'll never forget how I started to shake. I started to panic. I felt I could never solve a word problem. I thought I was having an algebra attack.

    Now that it's over (90 on the chapter on unknowns), I'd like to tell you what worked for me so you might pass this on to other students. It was your blueprint aids. Drawing boxes helped me to think things out. They were a tool that helped me more clearly understand how to dissect each word problem. They didn't solve the problem for me, but gave me the direction I needed. Repetition was the key to my success. At first I got them all wrong but after the third time, things started to click. I felt more confident. Your chapter organizers at the end of the chapter were great. Thanks for your patience – your repetition breeds success – now students are asking me to help them solve a word problem. Can you believe it!

Best,

Rose

Rose Smith

Learn: Basic Equation-Solving Procedures

Do you know the difference between a mathematical expression, equation, and formula? A mathematical **expression** is a meaningful combination of numbers and letters called *terms.* Operational signs (such as $+$ or $-$) within the expression connect the terms to show a relationship between them. For example, $6 + 2$ and $6A - 4A$ are mathematical expressions. An **equation** is a mathematical statement with an equals sign showing that a mathematical expression on the left equals the mathematical expression on the right. An equation has an equals sign; an expression does not have an equals sign. A **formula** is an equation that expresses in symbols a general fact, rule, or principle. Formulas are shortcuts for expressing a word concept. For example, in Chapter 10 you will learn that the formula for simple interest is Interest (I) = Principal (P) × Rate (R) × Time (T). This means that when you see $I = P \times R \times T$, you recognize the simple interest formula. Now let's study basic equations.

As a mathematical statement of equality, equations show that two numbers or groups of numbers are equal. For example, $6 + 4 = 10$ shows the equality of an equation. Equations also use letters as symbols that represent one or more numbers. These symbols, usually a letter of the alphabet, are **variables** that stand for a number. We can use a variable even though we may not know what it represents. For example, $A + 2 = 6$. The variable A represents the number or **unknown** (4 in this example) for which we are solving. We distinguish variables from numbers, which have a fixed value. Numbers such as 3 or -7 are **constants** or **knowns,** whereas A and $3A$ (this means 3 times the variable A) are variables. So we can now say that variables and constants are *terms of mathematical expressions.*

Usually in solving for the unknown, we place variable(s) on the left side of the equation and constants on the right. The following rules for variables and constants are important.

Variables and Constants Rules

Step 1 If no number is in front of a letter, it is a 1: $B = 1B$; $C = 1C$.

Step 2 If no sign is in front of a letter or number, it is a +: $C = +C$; $4 = +4$.

You should be aware that in solving equations, the meaning of the symbols $+$, $-$, $\times$, and $\div$ has not changed. However, some variations occur. For example, you can also write $A \times B$ (A times B) as $A \cdot B$, $A(B)$, or AB. Also, A divided by B is the same as A/B. Remember that to solve an equation, you must find a number that can replace the unknown in the equation and make it a true statement. Now let's take a moment to look at how we can change verbal statements into variables.

Assume Dick Hersh, an employee of Nike, is 50 years old. Let's assign Dick Hersh's changing age to the symbol A. The symbol A is a variable.

Verbal statement	Variable A (age)
Dick's age 8 years ago	$A - 8$
Dick's age 8 years from today	$A + 8$
Four times Dick's age	$4A$
One-fifth Dick's age	$A/5$

To visualize how equations work, think of the old-fashioned balancing scale shown in Figure 5.1. The pole of the scale is the equals sign. The two sides of the equation are the two pans of the scale. In the left pan or left side of the equation, we have $A + 8$; in the right pan or right side of the equation, we have 58. To solve for the unknown (Dick's present age), we isolate or place the unknown (variable) on the left side and the numbers on the right. We will do this soon. For now, remember that to keep an equation (or scale) in balance, we must perform mathematical operations (addition, subtraction, multiplication, and division) to *both* sides of the equation.

FIGURE 5.1 Equality in equations

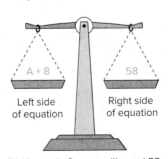

Dick's age in 8 years will equal 58.

Learn: How to Solve for Unknowns in Equations

This section presents seven drill situations and the rules that will guide you in solving for unknowns in these situations. We begin with two basic rules—the opposite process rule and the equation equality rule.

Opposite Process Rule

If an equation indicates a process such as addition, subtraction, multiplication, or division, solve for the unknown or variable by using the opposite process. For example, if the equation process is addition, solve for the unknown by using subtraction.

Equation Equality Rule

You can add the same quantity or number to both sides of the equation and subtract the same quantity or number from both sides of the equation without affecting the equality of the equation. You can also divide or multiply both sides of the equation by the same quantity or number *(except zero)* without affecting the equality of the equation.

To check your answer(s), substitute your answer(s) for the letter(s) in the equation. The sum of the left side should equal the sum of the right side.

Drill Situation 1: Subtracting Same Number from Both Sides of Equation

Example

$A + 8 = 58$

Dick's age A plus 8 equals 58.

Mechanical steps

$$
\begin{array}{rcr}
A + 8 &=& 58 \\
-8 & & -8 \\
\hline
A &=& \boxed{50}
\end{array}
$$

Explanation

8 is subtracted from *both* sides of the equation to isolate variable A on the left.

Check

$50 + 8 = 58$

$58 = 58$

Note: Since the equation process used *addition,* we use the opposite process rule and solve for variable *A* with *subtraction.* We also use the equation equality rule when we subtract the same quantity from both sides of the equation.

Drill Situation 2: Adding Same Number to Both Sides of Equation

Example

$B - 50 = 80$

Some number *B* less 50 equals 80.

Mechanical steps

$$B - 50 = \quad 80$$
$$\underline{+50 \quad\quad +50}$$
$$B \quad\quad = \quad 130$$

Explanation

50 is added to *both* sides to isolate variable *B* on the left.

Check

$$130 - 50 = 80$$
$$80 = 80$$

Note: Since the equation process used *subtraction,* we use the opposite process rule and solve for variable *B* with *addition.* We also use the equation equality rule when we add the same quantity to both sides of the equation.

Drill Situation 3: Dividing Both Sides of Equation by Same Number

Example

$7G = 35$

Some number *G* times 7 equals 35.

Mechanical steps

$$7G = 35$$
$$\frac{7G}{7} = \frac{35}{7}$$
$$G = 5$$

Explanation

By dividing both sides by 7, *G* equals 5.

Check

$$7(5) = 35$$
$$35 = 35$$

Note: Since the equation process used *multiplication,* we use the opposite process rule and solve for variable *G* with *division.* We also use the equation equality rule when we divide both sides of the equation by the same quantity.

Drill Situation 4: Multiplying Both Sides of Equation by Same Number

Example

$\dfrac{V}{5} = 70$

Some number *V* divided by 5 equals 70.

Mechanical steps

$$\frac{V}{5} = 70$$
$$5\left(\frac{V}{5}\right) = 70(5)$$
$$V = 350$$

Explanation

By multiplying both sides by 5, *V* is equal to 350.

Check

$$\frac{350}{5} = 70$$
$$70 = 70$$

Note: Since the equation process used *division,* we use the opposite process rule and solve for variable *V* with *multiplication.* We also use the equation equality rule when we multiply both sides of the equation by the same quantity.

Drill Situation 5: Equation That Uses Subtraction and Multiplication to Solve for Unknown

Multiple Processes Rule

When solving for an unknown that involves more than one process, do the addition and subtraction before the multiplication and division.

Example

$$\frac{H}{4} + 2 = 5$$

When we divide unknown H by 4 and add the result to 2, the answer is 5.

Mechanical steps

$$\frac{H}{4} + 2 = 5$$

$$\frac{H}{4} + 2 = 5$$
$$\underline{\quad -2 \qquad -2\quad}$$
$$\frac{H}{4} = 3$$

$$\cancel{4}\left(\frac{H}{\cancel{4}}\right) = 4(3)$$

$$H = \boxed{12}$$

Explanation

1. Move constant to right side by subtracting 2 from both sides.

2. To isolate H, which is divided by 4, we do the opposite process and multiply 4 times *both* sides of the equation.

Check

$$\frac{12}{4} + 2 = 5$$
$$3 + 2 = 5$$
$$5 = 5$$

Drill Situation 6: Using Parentheses in Solving for Unknown

Parentheses Rule

When equations contain parentheses (which indicate grouping together), you solve for the unknown by first multiplying each item inside the parentheses by the number or letter just outside the parentheses. Then you continue to solve for the unknown with the opposite process used in the equation. Do the additions and subtractions first; then the multiplications and divisions.

Example

$5(P - 4) = 20$

The unknown P less 4, multiplied by 5 equals 20.

Mechanical steps

$$5(P - 4) = 20$$
$$5P - 20 = 20$$
$$\underline{\quad +20 \qquad +20\quad}$$
$$\frac{\cancel{5}P}{\cancel{5}} = \frac{40}{5}$$
$$P = \boxed{8}$$

Explanation

1. Parentheses tell us that everything inside parentheses is multiplied by 5. Multiply 5 by P and 5 by -4.

2. Add 20 to both sides to isolate $5P$ on left.

3. To remove 5 in front of P, divide both sides by 5 to result in P equals 8.

Check

$$5(8 - 4) = 20$$
$$5(4) = 20$$
$$20 = 20$$

Drill Situation 7: Combining Like Unknowns

Like Unknowns Rule

To solve equations with like unknowns, you first combine the unknowns and then solve with the opposite process used in the equation.

Example

$4A + A = 20$

Mechanical steps

$$4A + A = 20$$

$$\frac{\cancel{5}A}{\cancel{5}} = \frac{20}{5}$$

$$A = 4$$

Explanation

To solve this equation:
$4A + 1A = 5A$. Thus,
$5A = 20$. To solve for A,
divide both sides by 5,
leaving A equals 4.

Check

$$4(4) + 4 = 20$$

$$20 = 20$$

Aha! *Always, always, always do a logic check on your answer. Reread the word problem to see if your answer makes sense. If it does, move on. If it does not, review your strategy for solving the problem and make any needed adjustments.*

Before you go to Learning Unit 5–2, let's check your understanding of this unit.

Learning Unit 5–1

Practice Quiz

Complete this Practice Quiz to see how you are doing.

1. Write equations for the following (use the letter Q as the variable). Do not solve for the unknown.

   a. Nine less than one-half a number is fourteen.

   b. Eight times the sum of a number and thirty-one is fifty.

   c. Ten decreased by twice a number is two.

   d. Eight times a number less two equals twenty-one.

   e. The sum of four times a number and two is fifteen.

   f. If twice a number is decreased by eight, the difference is four.

2. Solve the following:

   a. $B + 24 = 60$ b. $D + 3D = 240$ c. $12B = 144$

   d. $\dfrac{B}{6} = 50$ e. $\dfrac{B}{4} + 4 = 16$ f. $3(B - 8) = 18$

✓ Solutions

1. a. $\dfrac{1}{2}Q - 9 = 14$ b. $8(Q + 31) = 50$ c. $10 - 2Q = 2$

   d. $8Q - 2 = 21$ e. $4Q + 2 = 15$ f. $2Q - 8 = 4$

2. a.
$$
\begin{aligned}
B + 24 &= 60 \\
-24 \quad &-24 \\
\hline
B \quad &= \boxed{36}
\end{aligned}
$$

   b.
$$
\frac{\cancel{4}D}{\cancel{4}} = \frac{240}{4} \\
D = \boxed{60}
$$

   c.
$$
\frac{\cancel{12}B}{\cancel{12}} = \frac{144}{12} \\
B = \boxed{12}
$$

   d.
$$
\cancel{6}\left(\frac{B}{\cancel{6}}\right) = 50(6) \\
B = \boxed{300}
$$

   e.
$$
\begin{aligned}
\frac{B}{4} + 4 &= 16 \\
-4 \quad &-4 \\
\hline
\frac{B}{4} &= 12 \\
\cancel{4}\left(\frac{B}{\cancel{4}}\right) &= 12(4) \\
B &= \boxed{48}
\end{aligned}
$$

   f.
$$
\begin{aligned}
3(B - 8) &= 18 \\
3B - 24 &= 18 \\
+24 \quad &+24 \\
\hline
\frac{\cancel{3}B}{\cancel{3}} &= \frac{42}{3} \\
B &= \boxed{14}
\end{aligned}
$$

Solving Word Problems for the Unknown

When you buy a candy bar such as a Snickers, you should turn the candy bar over and carefully read the ingredients and calories contained on the back of the candy bar wrapper. For example, on the back of the Snickers wrapper you will read that there are "170 calories per piece." You could misread this to mean that the entire Snickers bar has 170 calories. However, look closer and you will see that the Snickers bar is divided into three pieces, so if you eat the entire bar, instead of consuming 170 calories, you will consume 510 calories. Making errors like this could result in a weight gain that you cannot explain.

$$\frac{1}{3}S = 170 \text{ calories}$$

$$\cancel{3}\left(\frac{1}{\cancel{3}}S\right) = 170 \times 3$$

$$S = \boxed{510} \text{ calories per bar}$$

In this unit, we use blueprint aids in six different situations to help you solve for unknowns. Be patient and *persistent*. Remember that the more problems you work, the easier the process becomes. Do not panic! Repetition is the key. Study the five steps that follow. They will help you solve for unknowns in word problems.

Solving Word Problems For Unknowns

Step 1 Carefully read the entire problem. You may have to read it several times.

Step 2 Ask yourself: What is the problem looking for?

Step 3 When you are sure what the problem is asking, let a variable represent the unknown. If the problem has more than one unknown, represent the second unknown in terms of the same variable. For example, if the problem has two unknowns, Y is one unknown. The second unknown is $4Y$—4 times the first unknown.

Step 4 Visualize the relationship between unknowns and variables. Then set up an equation to solve for the unknown(s).

Step 5 Check your result to see if it is accurate.

The clip from *The Wall Street Journal,* "How to Ace That Test," may also help you in the process of solving word problems.

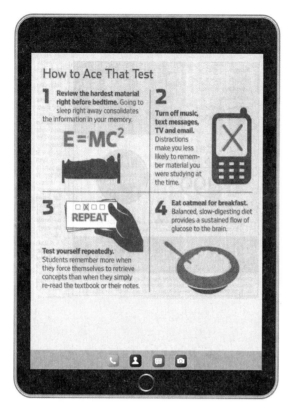

Word Problem Situation 1: Number Problems Today on sale at a local Stop & Shop supermarket, the price of a bag of Dunkin' Donuts coffee is $9.99. This is a $2 savings. What was the original price of the bag of coffee?

Mechanical steps

$$P - 2 = \$\ 9.99$$
$$\underline{+2 \quad +2}$$
$$P \quad = \boxed{\$11.99}$$

	Unknown(s)	Variable(s)	Relationship*
Blueprint	Original price of Dunkin' Donuts	P	$P - \$2 =$ New price

*This column will help you visualize the equation before setting up the actual equation.

Explanation

The original price less $2 = \$9.99$. Note that we added $2 to both sides to isolate P on the left. Remember, $1P = P$.

Check

$\$11.99 - 2 = \$9.99$

$\$9.99\quad = \$9.99$

Word Problem Situation 2: Finding the Whole When Part Is Known A local Burger King budgets $\frac{1}{8}$ of its monthly profits on salaries. Salaries for the month were $12,000. What were Burger King's monthly profits?

	Unknown(s)	Variable(s)	Relationship
Blueprint	Monthly profits	p	$\frac{1}{8}P$ Salaries = \$12,000

Mechanical steps

$$\frac{1}{8}P = \$12,000$$

$$8\left(\frac{P}{8}\right) = \$12,000(8)$$

$$P = \boxed{\$96,000}$$

Explanation

$\frac{1}{8}P$ represents Burger King's monthly salaries. Since the equation used division, we solve for P by multiplying both sides by $8.$

Check

$\frac{1}{8}(\$96,000) = \$12,000$

$\$12,000 = \$12,000$

Word Problem Situation 3: Difference Problems ICM Company sold 4 times as many computers as Ring Company. The difference in their sales is 27. How many computers of each company were sold?

	Unknown(s)	Variable(s)	Relationship
Blueprint	ICM Ring	$4C$ C	$4C$ $-C$ $\overline{27}$

Note: If problem has two unknowns, assign the variable to smaller item or one who sells less. Then assign the other unknown using the same variable. *Use the same letter.*

Mechanical steps

$$4C - C = 27$$

$$\frac{3C}{3} = \frac{27}{3}$$

$$C = \boxed{9}$$

Ring $= \boxed{9}$ computers

ICM $= 4(9)$

$\quad = \boxed{36}$ computers

Explanation

The variables replace the names ICM and Ring. We assigned Ring the variable C, since it sold fewer computers. We assigned ICM $4C$, since it sold 4 times as many computers.

Check

36 computers

$\underline{-9}$

27 computers

Word Problem Situation 4: Calculating Unit Sales Together Barry Sullivan and Mitch Ryan sold a total of 300 homes for Regis Realty. Barry sold 9 times as many homes as Mitch. How many did each sell?

	Unknown(s)	Variable(s)	Relationship
Blueprint	*Homes sold:* B. Sullivan M. Ryan	 $9H$ $H*$	 $9H$ $+H$ $\overline{300\ \text{homes}}$

*Assign H to Ryan since he sold less.

Mechanical steps

$$9H + H = 300$$

$$\frac{10H}{10} = \frac{300}{10}$$

$$H = \boxed{30}$$

Ryan: $\boxed{30}$ homes

Sullivan: $9(30) = \boxed{270}$ homes

Explanation

We assigned Mitch H, since he sold fewer homes. We assigned Barry $9H$, since he sold 9 times as many homes. Together Barry and Mitch sold 300 homes.

Check

$30 + 270 = 300$

Word Problem Situation 5: Calculating Unit and Dollar Sales (Cost per Unit) When Total Units Are Not Given Andy sold watches ($9) and alarm clocks ($5) at a flea market. Total sales were $287. People bought 4 times as many watches as alarm clocks. How many of each did Andy sell? What were the total dollar sales of each?

<table>
<thead>
<tr><th></th><th>Unknown(s)</th><th>Variable(s)</th><th>Price</th><th>Relationship</th></tr>
</thead>
<tbody>
<tr><td rowspan="4">Blueprint</td><td>*Unit sales:*</td><td></td><td></td><td></td></tr>
<tr><td>Watches</td><td>$4C$</td><td>$9</td><td>$36C$</td></tr>
<tr><td>Clocks</td><td>C</td><td>5</td><td>$+ 5C$</td></tr>
<tr><td></td><td></td><td></td><td>$287 total sales</td></tr>
</tbody>
</table>

Mechanical steps

$$36C + 5C = 287$$
$$\frac{41C}{41} = \frac{287}{41}$$
$$C = 7$$

7 clocks
$4(7) = 28$ watches

Explanation

Number of watches times $9 sales price plus number of alarm clocks times $5 sales price equals $287 total sales.

Check

$7($5) + 28($9) = 287
$$\$35 + \$252 = \$287$$
$$\$287 = \$287$$

Word Problem Situation 6: Calculating Unit and Dollar Sales (Cost per Unit) When Total Units Are Given Andy sold watches ($9) and alarm clocks ($5) at a flea market. Total sales for 35 watches and alarm clocks were $287. How many of each did Andy sell? What were the total dollar sales of each?

<table>
<thead>
<tr><th></th><th>Unknown(s)</th><th>Variable(s)</th><th>Price</th><th>Relationship</th></tr>
</thead>
<tbody>
<tr><td rowspan="4">Blueprint</td><td>*Unit sales:*</td><td></td><td></td><td></td></tr>
<tr><td>Watches</td><td>W*</td><td>$9</td><td>$9W$</td></tr>
<tr><td>Clocks</td><td>$35 - W$</td><td>5</td><td>$+ 5(35 - W)$</td></tr>
<tr><td></td><td></td><td></td><td>$287 total sales</td></tr>
</tbody>
</table>

*The more expensive item is assigned to the variable first only for this situation to make the mechanical steps easier to complete.

Mechanical steps

$$9W + 5(35 - W) = \quad 287$$
$$9W + 175 - 5W = \quad 287$$
$$4W + 175 \quad\quad = \quad 287$$
$$\underline{-175 \quad\quad\quad\quad -175}$$
$$\frac{4W}{4} = \frac{112}{4}$$
$$W = \quad 28$$

Watches $= 28$
Clocks $= 35 - 28 = 7$

Explanation

Number of watches (W) times price per watch plus number of alarm clocks times price per alarm clock equals $287. Total units given was 35.

Why did we use $35 - W$? Assume we had 35 pizzas (some cheese, others meatball). If I said that I ate all the meatball pizzas (5), how many cheese pizzas are left? Thirty? Right, you subtract 5 from 35. Think of $35 - W$ as meaning one number.

Note in Word Problem Situations 5 and 6 that the situation is the same. In Word Problem Situation 5, we were not given total units sold (but we were told which sold better). In Word Problem Situation 6, we were given total units sold, but we did not know which sold better.

Now try these six types of word problems in the Practice Quiz. Be sure to complete blueprint aids and the mechanical steps for solving the unknown(s).

Check

$28($9) + 7($5) = 287
$$\$252 + \$35 = \$287$$
$$\$287 = \$287$$

Money Tip Go green. Recycle plastics, paper, cardboard, glass, steel, aluminum cans and foil, plastic bags, motor oil, tires, batteries, computer printers, compost, household cleaners, and so on. Reduce the volume of packaging you require by using a canvas bag. Do your part to help solve the unknown environmental challenges of the future.

Practice Quiz

Complete this Practice Quiz to see how you are doing.

Situations

1. An L. L. Bean sweater was reduced $30. The sale price was $90. What was the original price?

2. Kelly Doyle budgets $\frac{1}{8}$ of her yearly salary for entertainment. Kelly's total entertainment bill for the year is $6,500. What is Kelly's yearly salary?

3. Micro Knowledge sells 5 times as many computers as Morse Electronics. The difference in sales between the two stores is 20 computers. How many computers did each store sell?

4. Susie and Cara sell stoves at Elliott's Appliances. Together they sold 180 stoves in January. Susie sold 5 times as many stoves as Cara. How many stoves did each sell?

5. Pasquale's Pizza sells meatball pizzas ($6) and cheese pizzas ($5). In March, Pasquale's total sales were $1,600. People bought 2 times as many cheese pizzas as meatball pizzas. How many of each did Pasquale's sell? What were the total dollar sales of each?

6. Pasquale's Pizza sells meatball pizzas ($6) and cheese pizzas ($5). In March, Pasquale's sold 300 pizzas for $1,600. How many of each did Pasquale's sell? What was the dollar sales price of each?

✓ **Solutions**

1.

Blueprint	Unknown(s)	Variable(s)	Relationship
	Original price	P*	P − $30 = Sale price Sale price = $90

*p = Original price.

Mechanical steps

$$P - \$30 = \$90$$
$$\underline{+\,30 \qquad +\,30}$$
$$P \qquad = \boxed{\$120}$$

2.

Blueprint	Unknown(s)	Variable(s)	Relationship
	Yearly salary	S*	$\frac{1}{8}S$ Entertainment = $6,500

*S = Salary.

Mechanical steps

$$\frac{1}{8}S = \$6,500$$
$$8\left(\frac{S}{8}\right) = \$6,500(8)$$
$$S = \boxed{\$52,000}$$

3.

Blueprint	Unknown(s)	Variable(s)	Relationship
	Micro	5C*	5C
	Morse	C	− C 20 computers

*C = Computers.

Mechanical steps

$$5C - C = 20$$
$$\frac{\cancel{4}C}{\cancel{4}} = \frac{20}{4}$$
$$C = \boxed{5} \text{ (Morse)}$$
$$5C = \boxed{25} \text{ (Micro)}$$

Practice Quiz *Continued*

4.

Unknown(s)	Variable(s)	Relationship
Stoves sold:		
Susie	5S*	5S
Cara	S	+ S
		180 stoves

Mechanical steps

$$5S + S = 180$$
$$\frac{\cancel{6}S}{\cancel{6}} = \frac{180}{6}$$
$$S = \boxed{30} \text{ (Cara)}$$
$$5S = \boxed{150} \text{ (Susie)}$$

*S = Stoves.

5.

Unknown(s)	Variable(s)	Price	Relationship
Meatball	M	$6	6M
Cheese	2M	5	+ 10M
			$1,600 total sales

Check

$$(100 \times \$6) + (200 \times \$5) = \$1,600$$
$$\$600 + \$1,000 = \$1,600$$
$$\$1,600 = \$1,600$$

Mechanical steps

$$6M + 10M = 1,600$$
$$\frac{\cancel{16}M}{\cancel{16}} = \frac{1,600}{16}$$
$$M = \boxed{100} \text{ (meatball)}$$
$$2M = \boxed{200} \text{ (cheese)}$$

6.

Unknown(s)	Variable(s)	Price	Relationship*
Unit sales:	M*	$6	6M
Meatball	300 − M	5	+ 5(300 − M)
Cheese			$1,600 total sales

*We assign the variable to the most expensive item to make the mechanical steps easier to complete.

Check

$$100(\$6) + 200(\$5) = \$600 + \$1,000$$
$$= \$1,600$$

Mechanical steps

$$6M + 5(300 - M) = 1,600$$
$$6M + 1,500 - 5M = 1,600$$
$$M + 1,500 = 1,600$$
$$\underline{-1,500 \qquad\qquad -1,500}$$
$$M = \boxed{100}$$

Meatball = $\boxed{100}$
Cheese = 300 − 100 = $\boxed{200}$

Chapter 5 Review

Solving for unknowns from basic equations	Mechanical steps to solve unknowns	Key point(s)	You try it*
Situation 1: Subtracting same number from both sides of equation	$D + 10 = 12$ $\underline{-10 \quad -10}$ $D \quad = \boxed{2}$	Subtract 10 from both sides of equation to isolate variable D on the left. Since equation used addition, we solve by using opposite process—subtraction.	**Solve** $E + 15 = 14$
Situation 2: Adding same number to both sides of equation	$L - 24 = 40$ $\underline{+24 \quad +24}$ $L \quad = \boxed{64}$	Add 24 to both sides to isolate unknown L on left. We solve by using opposite process of subtraction—addition.	**Solve** $B - 40 = 80$
Situation 3: Dividing both sides of equation by same number	$6B = 24$ $\dfrac{\cancel{6}B}{\cancel{6}} = \dfrac{24}{\cancel{6}}$ $B = \boxed{4}$	To isolate B on the left, divide both sides of the equation by 6. Thus, the 6 on the left cancels—leaving B equal to 4. Since equation used multiplication, we solve unknown by using opposite process—division.	**Solve** $5C = 75$
Situation 4: Multiplying both sides of equation by same number	$\dfrac{R}{3} = 15$ $\cancel{3}\left(\dfrac{R}{\cancel{3}}\right) = 15(3)$ $R = \boxed{45}$	To remove denominator, multiply both sides of the equation by 3—the 3 on the left side cancels, leaving R equal to 45. Since equation used division, we solve unknown by using opposite process—multiplication.	**Solve** $\dfrac{A}{6} = 60$
Situation 5: Equation that uses subtraction and multiplication to solve for unknown	$\dfrac{B}{3} + 6 = 13$ $\underline{-6 \quad -6}$ $\dfrac{B}{3} = 7$ $\cancel{3}\left(\dfrac{B}{\cancel{3}}\right) = 7(3)$ $B = \boxed{21}$	1. Move constant 6 to right side by subtracting 6 from both sides. 2. Isolate B on left by multiplying both sides by 3.	**Solve** $\dfrac{C}{4} + 10 = 17$

Solving for unknowns from basic equations	Mechanical steps to solve unknowns	Key point(s)	You try it*
Situation 6: Using parentheses in solving for unknown	$6(A - 5) = 12$ $6A - 30 = 12$ $\underline{+30 \quad +30}$ $\dfrac{6A}{6} = \dfrac{42}{6}$ $A = \boxed{7}$	Parentheses indicate multiplication. Multiply 6 times A and 6 times -5. Result is $6A - 30$ on left side of the equation. Now add 30 to both sides to isolate $6A$ on left. To remove 6 in front of A, divide both sides by 6, to result in A equal to 7. Note that when deleting parentheses, we did not have to multiply the right side.	**Solve** $7(B - 10) = 35$
Situation 7: Combining like unknowns	$6A + 2A = 64$ $\dfrac{8A}{8} = \dfrac{64}{8}$ $A = \boxed{8}$	$6A + 2A$ combine to $8A$. To solve for A, we divide both sides by 8.	**Solve** $5B + 3B = 16$

Solving for unknowns from word problems	Blueprint aid		Mechanical steps to solve unknown with check	You try it*
Situation 1: Number problems U.S. Air reduced its airfare to California by $60. The sale price was $95. What was the original price?	**Blueprint** 	Unknown(s): Original price Variable(s): P Relationship: $P - \$60 = $ Sale price Sale price $= \$95$	$P - \$60 = \$\ 95$ $\underline{+60 \qquad +60}$ $P \qquad = \boxed{\$155}$ **Check** $\$155 - \$60 = \$95$ $\$95 = \$95$	**Solve** U.S. Air reduced its airfare to California by $53. The sale price was $110. What was the original price?
Situation 2: Finding the whole when part is known K. McCarthy spends ⅛ of her budget for school. What is the total budget if school costs $5,000?	**Blueprint** 	Unknown(s): Total budget Variable(s): B Relationship: ⅛B Sale price $= \$5,000$	$\dfrac{1}{8}B = \$5,000$ $8\left(\dfrac{B}{8}\right) = \$5,000(8)$ $B = \boxed{\$40,000}$ **Check** $\dfrac{1}{8}(\$40,000) = \$5,000$ $\$5,000 = \$5,000$	**Solve** K. McCarthy spends ½ of her budget for school. What is the total budget if school costs $6,000?

Solving for unknowns from word problems	Blueprint aid			Mechanical steps to solve unknown with check	You try it*

Situation 3: Difference problems

Moe sold 8 times as many suitcases as Bill. The difference in their sales is 280 suitcases. How many suitcases did each sell?

Blueprint

Unknown(s)	Variable(s)	Relationship
Suitcases sold:	$8S$	$8S$
Moe	S	$-S$
Bill		280 suitcases

$8S - S = 280$

$$\frac{7S}{7} = \frac{280}{7}$$

$S = \boxed{40}$ (Bill)

$8(40) = \boxed{320}$ (Moe)

Check

$320 - 40 = 280$

$280 = 280$

Solve

Moe sold 9 times as many suitcases as Bill. The difference in their sales is 640 suitcases. How many suitcases did each sell?

Situation 4: Calculating unit sales

Moe sold 8 times as many suitcases as Bill. Together they sold a total of 360. How many did each sell?

Blueprint

Unknown(s)	Variable(s)	Relationship
Suitcases sold:	$8S$	$8S$
Moe	S	$+S$
Bill		360 suitcases

$8S + S = 360$

$$\frac{9S}{9} = \frac{360}{9}$$

$S = \boxed{40}$ (Bill)

$8(40) = \boxed{320}$ (Moe)

Check

$320 + 40 = 360$

$360 = 360$

Solve

Moe sold 9 times as many suitcases as Bill. Together they sold a total of 640. How many did each sell?

Situation 5: Calculating unit and dollar sales (cost per unit) when *total units not given*

Blue Furniture Company ordered sleepers ($300) and nonsleepers ($200) that cost $8,000. Blue expects sleepers to outsell nonsleepers 2 to 1. How many units of each were ordered? What were the dollar costs of each?

Blueprint

Unknown(s)	Variable(s)	Price	Relationship
Sleepers	$2N$	$300	$600N$
Nonsleepers	N	200	$+200N$
			$8,000 total cost

$600N + 200N = 8,000$

$$\frac{800N}{800} = \frac{8,000}{800}$$

$N = \boxed{10}$ (nonsleepers)

$2N = \boxed{20}$ (sleepers)

Check

$10 \times \$200 = \$2,000$

$20 \times \$300 = \underline{6,000}$

$= \$8,000$

Solve

Blue Furniture Company ordered sleepers ($400) and nonsleepers ($300) that cost $15,000. Blue expects sleepers to outsell nonsleepers 3 to 1. How many units of each were ordered? What were the dollar costs of each?

Solving for unknowns from word problems	Blueprint aid	Mechanical steps to solve unknown with check	You try it*
Situation 6: Calculating unit and dollar sales (cost per unit) when *total units given* **Blue Furniture Company ordered 30 sofas (sleepers and nonsleepers) that cost $8,000. The wholesale unit cost was $300 for the sleepers and $200 for the nonsleepers. How many units of each were ordered? What were the dollar costs of each?**	(blueprint table below)	(mechanical steps below)	**Solve** Blue Furniture Company ordered 40 sofas (sleepers and nonsleepers) that cost $15,000. The wholesale unit cost was $400 for the sleepers and $300 for the nonsleepers. How many units of each were ordered? What were the dollar costs of each?

Blueprint aid table:

	Unknown(s)	Variable(s)	Price	Relationship
Blueprint	Unit costs			
	Sleepers	S	$300	$300S$
	Nonsleepers	$30 - S$	200	$+\ 200(30 - S)$
				$\overline{\$8{,}000 \text{ total cost}}$

Note: When the total units are given, the higher-priced item (sleepers) is assigned to the variable first. This makes the mechanical steps easier to complete.

Mechanical steps to solve unknown with check:

$$300S + 200(30 - S) = 8{,}000$$
$$300S + 6{,}000 - 200S = 8{,}000$$
$$100S + 6{,}000 = 8{,}000$$
$$\underline{-\ 6{,}000 \qquad\qquad -6{,}000}$$
$$\frac{\cancel{100}S}{\cancel{100}} = \frac{2{,}000}{100}$$
$$S = 20$$
$$\text{Nonsleepers} = 30 - 20$$
$$= 10$$

Check

$$20(\$300) + 10(\$200) = \$8{,}000$$
$$\$6{,}000 + \$2{,}000 = \$8{,}000$$
$$\$8{,}000 = \$8{,}000$$

Key Terms

Constants   Formula   Variables

Equation   Knowns

Expression   Unknown

* Worked-out solutions are in Appendix A.

Critical Thinking Discussion Questions with Chapter Concept Check

1. Explain the difference between a variable and a constant. What would you consider your monthly car payment—a variable or a constant?

2. How does the opposite process rule help solve for the variable in an equation? If a Mercedes costs 3 times as much as a Saab, how could the opposite process rule be used? The selling price of the Mercedes is $60,000.

3. What is the difference between Word Problem Situations 5 and 6 in Learning Unit 5–2? Show why the more expensive item in Word Problem Situation 6 is assigned to the variable first.

4. **Chapter Concept Check.** Go to a weight-loss website and create several equations on how to lose weight. Be sure to create a word problem and specify the steps you need to take to solve this weight-loss problem.

End-of-Chapter Problems

Name _____ Date_____

Check figures for odd-numbered problems in Appendix A.

Drill Problems (First of Three Sets)

Solve the unknown from the following equations: *LU 5-1(2)*

5–1. $X - 40 = 400$ **5–2.** $A + 64 = 98$ **5–3.** $Q + 100 = 400$ **5–4.** $Q - 60 = 850$

5–5. $5Y = 75$ **5–6.** $\dfrac{P}{6} = 92$ **5–7.** $8Y = 96$ **5–8.** $\dfrac{N}{16} = 5$

5–9. $4(P - 9) = 64$ **5–10.** $3(P - 3) = 27$

Word Problems (First of Three Sets)

5–11. Lee and Fred are elementary school teachers. Fred works for a charter school in Pacific Palisades, California, where class size reduction was a goal for the school year. Lee works for a noncharter school where funds do not allow for class size reduction policies. Lee's fifth-grade class has 1.4 times as many students as Fred's. If there are a total of 60 students, how many students does Fred's class have? How many students does Lee's class have? *LU 5-2(2)*

excel 5–12. A car that originally cost $3,668 in 1955 is valued today at $62,125 if in excellent condition, which is $1\frac{3}{4}$ times as much as a car in very nice condition—if you can find an owner willing to part with one for any price. What would be the value of the car in very nice condition? *LU 5-2(2)*

End-of-Chapter Problems (Continued)

5–13. Jessica and Josh are selling Entertainment Books to raise money for the art room at their school. One book sells for $15. Jessica received the prize for selling the most books in the school. Jessica sold 15 times more books than Josh. Together they sold 256 books. How many did each one of them sell? *LU 5-2(1)*

excel **5–14.** Nanda Yueh and Lane Zuriff sell homes for Equity Real Estate. Over the past 6 months they sold 120 homes. Nanda sold 3 times as many homes as Lane. How many homes did each sell? *LU 5-2(2)*

5–15. Dots sells T-shirts ($2) and shorts ($4). In April, total sales were $600. People bought 4 times as many T-shirts as shorts. How many T-shirts and shorts did Dots sell? Check your answer. *LU 5-2(2)*

5–16. Dots sells a total of 250 T-shirts ($2) and shorts ($4). In April, total sales were $600. How many T-shirts and shorts did Dots sell? Check your answer. *Hint:* Let S = Shorts.   *LU 5-2(2)*

Drill Problems (Second of Three Sets)

Solve the unknown from the following equations: *LU 5-1(2)*

5–17. $7B = 490$

5–18. $7(A - 5) = 63$

5–19. $\dfrac{N}{9} = 7$

5–20. $18(C - 3) = 162$ **5–21.** $9Y - 10 = \ 53$ **5–22.** $7B + 5 = \ 26$

Word Problems (Second of Three Sets)

5–23. On a flight from Boston to San Diego, American reduced its Internet price by $190.00. The new sale price was $420.99. What was the original price? *LU 5-2(2)*

5–24. Jill, an employee at Old Navy, budgets $\frac{1}{5}$ of her yearly salary for clothing. Jill's total clothing bill for the year is $8,000. What is her yearly salary? *LU 5-2(2)*

excel 5–25. Bill's Roast Beef sells 5 times as many sandwiches as Pete's Deli. The difference between their sales is 360 sandwiches. How many sandwiches did each sell? *LU 5-2(2)*

5–26. Kathy and Mark Smith believe investing in their retirement is critical. Kathy began investing 20% of each paycheck in a retirement account when she was 20 years old. She has saved four times more than Mark, who began saving when he was 35. If their total retirement savings equals $1,450,000, how much is Kathy's investment worth? *LU 5-2(2)*

5–27. During the pandemic, it was not uncommon for toilet paper and wet wipes to be unavailable because people were hording these items as stores were regularly sold out. King Soopers grocers had a shipment arrive allowing them to briefly restock their shelves. They sold toilet paper for $3 a roll and wet wipes for $5 a dispenser. If customers bought 5 times as many toilet paper rolls as wet wipes and total sales were $960, how many of each did King Soopers sell? Check your answer. *LU 5-2(2)*

End-of-Chapter Problems (Continued)

5–28. Maleri Designs sells cartons of cloth face masks ($10) and cartons of hand-sanitizer ($4) on eBay. One of their customers, Mod World, purchased 24 cartons for $210. How many of cartons of each did Mod World purchase? Check your answer. *Hint:* Let F = Face masks.   *LU 5-2(2)*

Drill Problems (Third of Three Sets)

Solve the unknown from the following equations: *LU 5-1(2)*

5–29. $A + 90 - 15 = \ \ 210$

5–30. $5Y + 15(Y + 1) = \ \ 35$

5–31. $3M + 20 = \ \ 2M + 80$

5–32. $20(C - 50) = \ \ 19{,}000$

5–33. If Colorado Springs, Colorado, has 1.2 times as many days of sunshine as Boston, Massachusetts, how many days of sunshine does each city have if there are a total of 464 days of sunshine between the two in a year? (Round to the nearest day.)   *LU 5-2(2)*

5–34. Ben and Jerry's sells 4 times more ice cream cones ($3) than shakes ($8). If last month's sales totaled $4,800, how many of each were sold? Check your answer.   *LU 5-2(1)*

Drill Problems (Third of Three Sets)

5–35. Ivy Corporation gave 84 people a bonus. If Ivy had given 2 more people bonuses, Ivy would have rewarded $\frac{2}{3}$ of the workforce. How large is Ivy's workforce?   *LU 5-2(2)*

5–36. Jim Murray and Phyllis Lowe received a total of $50,000 from a deceased relative's estate. They decided to put $10,000 in a trust for their nephew and divide the remainder. Phyllis received $\frac{3}{4}$ of the remainder; Jim received $\frac{1}{4}$. How much did Jim and Phyllis receive?
LU 5-2(2)

5–37. Active dry yeast and bread machine yeast were not available in stores during the coronavirus pandemic. Fleischmann's attempted to staff two shifts to produce enough active dry yeast to put some yeast back on the market. If the first shift produced $1\frac{1}{2}$ times as many yeast packets as the second shift, and 5,600 packets of yeast were produced during the month, how many packets of yeast were produced on each shift?   *LU 5-2(2)*

5–38. Banana Republic at the Orlando, Florida, outlet store sells casual jeans for $40 and dress jeans for $60. If customers bought 5 times more casual than dress jeans and last month's sales totaled $6,500, how many of each type of jeans were sold? Check your answer.
LU 5-2(2)

End-of-Chapter Problems (Continued)

5–39. Lowe's sells boxes of wrenches ($100) and hammers ($300). Howard ordered 40 boxes of wrenches and hammers for $8,400. How many boxes of each are in the order? Check your answer. *LU 5-2(2)*

5–40. The Susan Hansen Group in St. George, Utah, sells $16,000,000 of single-family homes and townhomes a year. If single-family homes, with an average selling price of $250,000, sell 3.5 times more often than townhomes, with an average selling price of $190,000, how many of each are sold? (Round to nearest whole.) *LU 5-2(2)*

5–41. Want to donate to a better cause? Consider micro-lending. Micro-lending is a process where you lend directly to entrepreneurs in developing countries. You can lend starting at $25. Kiva.org boasts a 99% repayment rate. The average loan to an entrepreneur is $388.44 and the average loan amount is $261.14. With a total amount loaned of $283,697,150, how many people are lending money if the average number of loans per lender is 8? (Round final answer to nearest whole lender.) *LU 5-2(2)*

Challenge Problems

5–42. Myron Corporation is sponsoring a walking race at its company outing. Leona Jackson and Sam Peterson love to walk. Leona walks at the rate of 5 miles per hour. Sam walks at the rate of 6 miles per hour. Assume they start walking from the same place and walk in a straight line. Sam starts $\frac{1}{2}$ hour after Leona. Answer the questions that follow. *Hint:* Distance = Rate × Time.   *LU 5-2(2)*

    a. How long will it take Sam to meet Leona?

    b. How many miles would each have walked?

    c. Assume Leona and Sam meet in Lonetown Station where two buses leave along parallel routes in opposite directions. The bus traveling east has a 60 mph speed. The bus traveling west has a 40 mph speed. In how many hours will the buses be 600 miles apart?

5–43. Bessy has 6 times as much money as Bob, but when each earns $6, Bessy will have 3 times as much money as Bob. How much does each have before and after earning the $6?   *LU 5-2(2)*

Summary Practice Test

Do you need help? Connect videos have step-by-step worked-out solutions.

1. Delta reduced its round-trip ticket price from Portland to Boston by $140. The sale price was $401.90. What was the original price? *LU 5-2(2)*

2. David Role is an employee of Google. He budgets $\frac{1}{7}$ of his salary for clothing. If David's total clothing for the year is $12,000, what is his yearly salary? *LU 5-2(2)*

3. A local Best Buy sells 8 times as many iPads as Target. The difference between their sales is 490 iPads. How many iPads did each sell? *LU 5-2(2)*

4. Working at Staples, Jill Reese and Abby Lee sold a total of 1,200 calculators. Jill sold 5 times as many calculators as Abby. How many did each sell? *LU 5-2(2)*

5. Target sells sets of pots ($30) and dishes ($20) at the local store. On the July 4 weekend, Target's total sales were $2,600. People bought 6 times as many pots as dishes. How many of each did Target sell? Check your answer. *LU 5-2(2)*

6. A local Dominos sold a total of 1,600 small pizzas ($9) and pasta dinners ($13) during the Super Bowl. How many of each did Dominos sell if total sales were $15,600? Check your answer.   *LU 5-2(2)*

MY MONEY

🔍 Goal Setting for Success

 What I need to know

Goal setting is powerful. Through a simple process you can produce amazing results by channeling your focus. The time you invest in setting specific, measurable, attainable, rewarding goals will be time well spent. Reviewing your goals and accomplishments regularly throughout the year provides you with timely, motivating feedback propelling you forward as you work your way to achieving your desired results. Goal setting is a continual process, and it is important to complete the process on a yearly basis.

Financial goals direct where you would like to be in the future when it comes to your finances. There may be certain purchases you would like to make in the future such as buying a home or a new car. Or you may want to travel during your retirement. In addition, it is important to set personal goals for yourself. You might want to increase your physical activity to improve your overall health or pursue a degree to advance your career and expand your knowledge base. Goals set a course for your life toward a sense of achievement and success.

 What I need to do

Guidelines for goal setting include making the goals realistic, measurable, and a challenge for you to achieve. A realistic goal gives you the belief this can be achieved and helps maintain your motivation toward reaching the goal. You also need to be able to measure your goal so you know whether progress is being made. Finally, set challenging goals to push yourself toward accomplishment. An example of a goal using these guidelines could be saving $2,000 toward a car purchase in 18 months. This goal is measurable in both the amount of money to be saved and the time frame in which to complete it. To reach this goal you would need to save just over $100 per month ($2,000/18) which provides you a challenge as you adjust your budget.

It may be helpful to incorporate visuals as part of your goal-setting process. Creating a physical space to display reminders, motivators, progress, notes of encouragement, and other visuals is often referred to as a *vision board*. This may provide the inspiration you need to stay focused on achieving your goals. For instance, if you are saving money for a trip to Hawaii, post a picture from a Hawaiian beach to help you visualize why you are making the sacrifices to save money. Upon returning from your Hawaiian vacation, place a beach photo of yourself on your vision board as a reminder of your success.

 Steps I need to take

1. Set both personal and financial goals for yourself every January. Write them down.
2. Create goals which are realistic, measurable, and challenging.
3. Review goals informally throughout the year and formally, in writing, at the end of December and celebrate your accomplishments. Save your annual goals and accomplishments and review them every five years to motivate you (and amaze you!) to continue goal setting.

 Resources I can use

- Strides: Goal & Habit Tracker (mobile app)—set and track your goals along with reminders and progress charting to keep you motivated.
- https://www.lifehack.org/864439/vision-board-ideas—ideas and helpful tips to create your own vision board for goal setting.

MY MONEY ACTIVITY

- Set one financial and one personal goal for yourself for the next month.
- Track your progress and make notes about how you performed.
- Apply what you learned and set some yearly goals for yourself.

PERSONAL FINANCE

A KIPLINGER APPROACH

"Buy Now, Pay Later Isn't a Slam Dunk." Kiplinger's. May 2021.

MILLENNIAL MONEY   Emma Patch

Buy Now, Pay Later Isn't A Slam Dunk

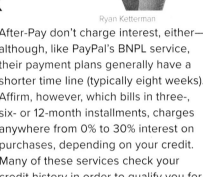

Ryan Ketterman

Once upon a time (I am told), it was common practice to walk into a store and put an item on layaway. You'd put down a deposit, make payments over time and collect your item once you paid it off. But now, a new service has turned layaway on its head. With Buy Now, Pay Later (BNPL), you don't have to wait to bring home something you pay for in installments. Instead, a third party offers you a loan at checkout to cover your purchase, in some cases with no interest or additional fees. More than one-third of U.S. consumers have used such a service at least once, according to research by The Ascent, a subsidiary of The Motley Fool. And BNPL is the fastest-growing e-commerce payment method globally, says Worldpay, a unit of payments processor FIS.

When shopping online, I've noticed BNPL offerings for clothes and shoes, but the most popular spending category for BNPL services is consumer electronics, according to a survey by Coupon Follow, a consumer savings engine that markets popular coupons. But these point-of-sale loans are on the increase for everything from furniture to travel. For example, VRBO, the online marketplace for vacation rentals, and travel provider

Expedia have partnered with Affirm, one of the largest BNPL services.

The industry is young, and the pandemic has certainly been a catalyst for its growth. Affirm went public in January, and its stock price has more than doubled. BNPL's market share in North America is expected to triple in the next three years, says Greg Fisher, chief marketing officer at Affirm. Klarna, whose Super Bowl commercial featured actor Maya Rudolph on a mission to buy a fabulous pair of boots, services more than 15 million customers in the U.S., says David Sykes, head of Klarna US. PayPal recently rolled out several new "pay later" products in the U.S., the U.K. and France.

> *BREAKING UP PAYMENTS CAN MAKE A BIG PURCHASE SEEM CHEAPER, WHICH CAN TEMPT YOU TO OVERSPEND.*

Weigh all the costs. BNPL services offer a way to cover the cost of something you need right away, says Linda Sherry, director of national priorities for Consumer Action. But it's important to thoroughly vet the service you choose. Keep an eye out for late fees, interest and whether you'll pay more for the product or service than you would by paying up front.

The services' interest policies vary. PayPal's U.S. BNPL service charges no interest on purchases as long as you spend at least $99. Klarna and

After-Pay don't charge interest, either—although, like PayPal's BNPL service, their payment plans generally have a shorter time line (typically eight weeks). Affirm, however, which bills in three-, six- or 12-month installments, charges anywhere from 0% to 30% interest on purchases, depending on your credit. Many of these services check your credit history in order to qualify you for the loan.

Even if the price is the same, breaking up payments can make a big purchase seem cheaper, which can tempt you to overspend. Also bear in mind that if you need to return an item, you might end up waiting longer for a refund than if you had paid in full up front. With some services, any interest you pay is non-refundable. And though making BNPL payments on time won't help you build credit, missing payments could hurt your credit score.

I love finding good deals, but BNPL feels like a perfect invitation to live above my means. So even though I would love to book a VRBO in Aspen for my 24th birthday and pay it off later, this Gen Zer/cusp millennial is going to save for 25 instead. ■

TO SHARE THIS COLUMN, GO TO KIPLINGER. COM/LINKS/MILLENNIALS. FOR QUESTIONS OR COMMENTS, PLEASE SEND AN E-MAIL TO EMMA_PATCH@KIPLINGER.COM.

Business Math Issue

'Buy now, pay later' will always result in constant problems for the consumer.

1. List the key points of the article and information to support your position.

2. Write a group defense of your position using math calculations to support your view. If you are in an online course, post to a discussion board.

Chapter 6

Percents and Their Applications

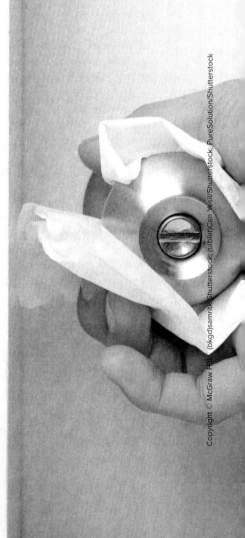

Clorox and Lysol Can't Catch Up on Wipes

By Sharon Terlep

Toilet paper is back on shelves and hand sanitizer is easier to find, but one item remains as elusive as ever: disinfectant wipes.

Clorox Co. and Lysol maker **Reckitt Benckiser Group PLC** have seen sales of sanitizing wipes more than double in the past two months amid the coronavirus pandemic, according to Nielsen. But while makers of other hard-to-find staples are catching up with demand, the producers of Clorox, Lysol and private-label wipes are as far behind as ever. Clorox said it doesn't expect to catch up until summer, while Reckitt Benckiser said it is unsure when supplies will replenish.

"We're shipping canisters of wipes every day to our customers, and within 30-45 minutes they're gone from shelves," Clorox finance chief Kevin Jacobsen said in an interview. "Demand has outstripped what anybody could have imagined."

Clorox has increased production of disinfectant products by 40%, but sales have stretched to five times the normal level at times during the spread of Covid-19, Mr. Jacobsen said. U.S. sales of disinfectant wipes were up 146% for the eight-week period ended March 25 compared with a year ago, according to Nielsen.

While wipes are in short supply, disinfectant sprays, surface cleaners and other coronavirus-fighting cleaning products are selling out as well. Many household and personal-care products, from paper towels to cold medicine to baby wipes, are also in high demand,

Terlep, Sharon. "Clorox and Lysol Can't Catch Up On Wipes." *The Wall Street Journal,* May 8, 2020.

Learning Unit Objectives

LU 6–1: Conversions

1. Convert decimals to percents (including rounding percents), percents to decimals, and fractions to percents.
2. Convert percents to fractions.

LU 6–2: Application of Percents—Portion Formula

1. List and define the key elements of the portion formula.
2. Solve for one unknown of the portion formula when the other two key elements are given.
3. Calculate the rate of percent increases and decreases.

ⓔ Essential Question

How can I use percentages to understand the business world and make my life easier?

🌐 Math Around the World

The chapter opening *Wall Street Journal* clip, "Clorox and Lysol Can't Catch Up on Wipes", shows how the COVID-19 pandemic affected sales of disinfectant wipes. Sales of disinfectant wipes were up 146% for an eight-week period. The *Wall Street Journal* clip "Prepare Your Kids" shows that many people have trouble managing their finances.

To understand percents, you should first understand the conversion relationship between decimals, percents, and fractions as explained in Learning Unit 6–1. Then, in Learning Unit 6–2, you will be ready to apply percents to personal and business events.

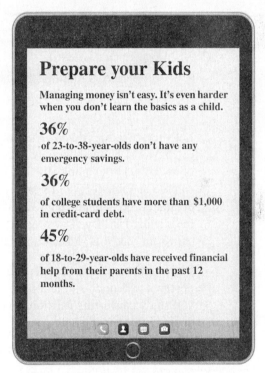

Prepare your Kids

Managing money isn't easy. It's even harder when you don't learn the basics as a child.

36%
of 23-to-38-year-olds don't have any emergency savings.

36%
of college students have more than $1,000 in credit-card debt.

45%
of 18-to-29-year-olds have received financial help from their parents in the past 12 months.

""Prepare Your Kids" Infographic." *The Wall Street Journal,* January 10, 2019.

Learning Unit 6–1:
Conversions

When we described parts of a whole in previous chapters, we used fractions and decimals. Percents also describe parts of a whole. The word *percent* means per 100. The percent symbol (%) indicates hundredths (division by 100). **Percents** are the result of expressing numbers as part of 100.

Rhymes With Orange © 2019 RWO Studios, Dist. By King Features Syndicate

Let's return to the M&M'S® example from earlier chapters. In Table 6.1, we use our bag of 55 M&M'S® to show how fractions, decimals, and percents can refer to the same parts of a whole. For example, the bag of 55 M&M'S® contains 18 yellow M&M'S®. As you can see in Table 6.1, the 18 candies in the bag of 55 can be expressed as a fraction $\left(\frac{18}{55}\right)$, decimal (.33), and percent (32.73%). If you visit the M&M'S® website, you will see that the standard is 11 yellow M&M'S®. The clip (below) "What Colors Come in Your Bag?" shows an M&M'S® Milk Chocolate Candies Color Chart.

In this unit we discuss converting decimals to percents (including rounding percents), percents to decimals, fractions to percents, and percents to fractions. You will see when you study converting fractions to percents why you should first learn how to convert decimals to percents.

Learn: Converting Decimals to Percents

Note in the *Wall Street Journal* clip to the left that 60% of consumers check their phone 52 times a day. If the clipping had stated the 60% as a decimal (.60), could you give its equivalent in percent? The decimal .60 in decimal fraction is $\frac{60}{100}$. As you know, percents are the result of expressing numbers as part of 100, so $60\% = \frac{60}{100}$. You can now conclude that $.60 = \frac{60}{100} = 60\%$.

More than 60% of consumers say they look at their phone within 15 minutes of waking and check their phones about 52 times a day, according to Deloitte.

Byron, Ellen. "What's the Rush? The Power of Slow Mornings." *The Wall Street Journal,* January 8, 2019.

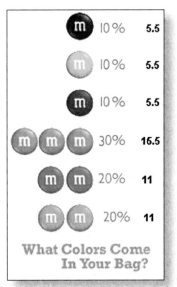

What Colors Come In Your Bag?

Information adapted from http://us.mms.com/us/about/products/milkchocolate/

TABLE 6.1 Analyzing a bag of M&M'S®

Color	Fraction	Decimal (hundredth)	Percent (hundredth)
Yellow	$\frac{18}{55}$	.33	32.73%
Red	$\frac{10}{55}$	.18	18.18
Blue	$\frac{9}{55}$	.16	16.36
Orange	$\frac{7}{55}$	.13	12.73
Brown	$\frac{6}{55}$	.11	10.91
Green	$\frac{5}{55}$	.09	9.09
Total	$\frac{55}{55} = 1$	1.00	100.00%

The steps for converting decimals to percents are as follows:

Converting Decimals to Percents

Step 1 Move the decimal point two places to the right. You are multiplying by 100. If necessary, add zeros. This rule is also used for whole numbers and mixed decimals.

Step 2 Add a percent symbol at the end of the number.

Examples:

$.60 = .60. = \boxed{60\%}$ $.8 = .80. = \boxed{80\%}$ $8 = 8.00. = \boxed{800\%}$

Add 1 zero to make two places.   Add 2 zeros to make two places.

$.425 = .42.5 = \boxed{42.5\%}$ $.007 = .00.7 = \boxed{.7\%}$ $2.51 = 2.51. = \boxed{251\%}$

Caution: One percent means 1 out of every 100. Since .7% is less than 1%, it means $\frac{7}{10}$ of 1%—a very small amount. Less than 1% is less than .01. To show a number less than 1%, you must use more than two decimal places and add 2 zeros. Example: .7% = .007.

Aha! *Use "D2P" to help you remember how to change a decimal to a percent. "D" stands for "decimal," "2" tells you to move the decimal two places, and "P" stands for "percent." Since P is to the right of D in D2P, we move the decimal two places to the right and add a percent sign:*

$.159 = 15.9\%$

Learn: Rounding Percents

When necessary, percents should be rounded. Rounding percents is similar to rounding whole numbers. Use the following steps to round percents:

""Inside Track" Infographic."
The Wall Street Journal,
May 2, 2020.

For example, Table 6.1 shows that the 18 yellow M&M'S® rounded to the nearest hundredth percent is 32.73% of the bag of 55 M&M'S®. Let's look at how we arrived at this figure.

Step 1 $\frac{18}{55} = .3272727 = 32.72727\%$ Note that the number is in percent! Identify the hundredth percent digit.

Step 2 32.73727% Digit to the right of the identified digit is greater than 5, so the identified digit is increased by 1.

Step 3 32.73%   Delete digits to the right of the identified digit.

Learn: Converting Percents to Decimals

During the pandemic consumers really changed their buying habits. Note in the *Wall Street Journal* clip "Inside Track" that Delta's U.S. passenger volume dropped 95%.

To convert percents to decimals, you reverse the process used to convert decimals to percents:

Aha! *Remember our D2P trick for converting decimals to percents? Well good news! It works in reverse for converting percents to decimals. Because we are changing a percent to a decimal, read D2P from right to left, that is, P2D. Start by removing the percent sign and then move the decimal two places to the left because P is to the left of D in P2D.*

$$95\% = 95 = .95$$

Examples:

Note that when a percent is less than 1%, the decimal conversion has at least two leading zeros before the number .004.

$.4\% = .00.4 = \boxed{.004}$

Add 2 zeros to make two places.

$2\% = .02. = \boxed{.02}$

Add 1 zero to make two places.

$.83\% = .00.83 = \boxed{.0083}$

Add 2 zeros to make two places.

$49\% = .49. = \boxed{.49}$

$54.5\% = .54.5 = \boxed{.545}$

$824.4\% = 8.24.4 = \boxed{8.244}$

Now we must explain how to change fractional percents such as $\frac{1}{5}\%$ to a decimal. Remember that fractional percents are values less than 1%. For example, $\frac{1}{5}\%$ is $\frac{1}{5}$ of 1%. Fractional percents can appear singly or in combination with whole numbers. To convert them to decimals, use the following steps:

Converting Fractional Percents to Decimals

Step 1 Convert a single fractional percent to its decimal equivalent by dividing the numerator by the denominator. If necessary, round the answer.

Step 2 If a fractional percent is combined with a whole number (mixed fractional percent), convert the fractional percent first. Then combine the whole number and the fractional percent.

Step 3 Drop the percent symbol; move the decimal point two places to the left (this divides the number by 100).

Examples:

$$\frac{1}{5}\% = .20\% = .00.20 = \boxed{.0020}$$

$$\frac{1}{4}\% = .25\% = .00.25 = \boxed{.0025}$$

$$7\frac{3}{4}\% = 7.75\% = .07.75 = \boxed{.0775}$$

$$6\frac{1}{2}\% = 6.5\% = .06.5 = \boxed{.065}$$

Think of $7\frac{3}{4}\%$ as

$$7\% = \quad .07$$
$$+\frac{3}{4}\% = \quad +.0075$$
$$\overline{}$$
$$7\frac{3}{4}\% = \quad .0775$$

Learn: Converting Fractions to Percents

When fractions have denominators of 100, the numerator becomes the percent. Other fractions must be first converted to decimals; then the decimals are converted to percents.

Converting Fractions to Percents

Step 1 Divide the numerator by the denominator to convert the fraction to a decimal.

Step 2 Move the decimal point two places to the right; add the percent symbol.

Examples:

$$\frac{3}{4} = .75 = .75. = \boxed{75\%} \qquad \frac{1}{5} = .20 = .20. = \boxed{20\%} \qquad \frac{1}{20} = .05 = .05. = \boxed{5\%}$$

Learn: Converting Percents to Fractions

Using the definition of percent, you can write any percent as a fraction whose denominator is 100. Thus, when we convert a percent to a fraction, we drop the percent symbol and write the number over 100, which is the same as multiplying the number by $\frac{1}{100}$. This method of multiplying by $\frac{1}{100}$ is also used for fractional percents.

Converting a Whole Percent (or a Fractional Percent) to a Fraction

Step 1 Drop the percent symbol.

Step 2 Multiply the number by $\frac{1}{100}$.

Step 3 Reduce to lowest terms.

Examples:

$$76\% = 76 \times \frac{1}{100} = \frac{76}{100} = \boxed{\frac{19}{25}} \qquad \frac{1}{8}\% = \frac{1}{8} \times \frac{1}{100} = \boxed{\frac{1}{800}}$$

$$156\% = 156 \times \frac{1}{100} = \frac{156}{100} = 1\frac{56}{100} = \boxed{1\frac{14}{25}}$$

Money Tip Nearly half, 47%, of adult Americans have no life insurance coverage. Consider the impact on survivors. At a minimum, carry burial insurance and a letter of last instruction stating your burial wishes.

Sometimes a percent contains a whole number and a fraction such as $12\frac{1}{2}\%$ or 22.5%. Extra steps are needed to write a mixed or decimal percent as a simplified fraction.

Converting a Mixed or Decimal Percent to a Fraction

Step 1 Drop the percent symbol.

Step 2 Change the mixed percent to an improper fraction.

Step 3 Multiply the number by $\frac{1}{100}$.

Step 4 Reduce to lowest terms.

Note: If you have a mixed or decimal percent, change the decimal portion to its fractional equivalent and continue with Steps 1 to 4.

Examples:

$$12\frac{1}{2}\% = \frac{25}{2} \times \frac{1}{100} = \frac{25}{200} = \boxed{\frac{1}{8}}$$

$$12.5\% = 12\frac{1}{2}\% = \frac{25}{2} \times \frac{1}{100} = \frac{25}{200} = \boxed{\frac{1}{8}}$$

$$22.5\% = 22\frac{1}{2}\% = \frac{45}{2} \times \frac{1}{100} = \frac{45}{200} = \boxed{\frac{9}{40}}$$

It's time to check your understanding of Learning Unit 6–1.

Practice Quiz

Complete this Practice Quiz to see how you are doing.

Convert to percents (round to the nearest tenth percent as needed):

1. .6666 _____ **2.** .832 _____

3. .004 _____ **4.** 8.94444 _____

Convert to decimals (remember, decimals representing less than 1% will have at least 2 leading zeros before the number):

5. $\frac{1}{4}\%$ _____ **6.** $6\frac{3}{4}\%$ _____

7. 87% _____ **8.** 810.9% _____

Convert to percents (round to the nearest hundredth percent):

9. $\frac{1}{7}$ _____ **10.** $\frac{2}{9}$ _____

Convert to fractions (remember, if it is a mixed number, first convert to an improper fraction):

11. 19% _____ **12.** $71\frac{1}{2}\%$ _____ **13.** 130% _____

14. $\frac{1}{2}\%$ _____ **15.** 19.9% _____

✓ Solutions

1. $.66.66 = \boxed{66.7\%}$ **2.** $.83.2 = \boxed{83.2\%}$

3. $.00.4 = \boxed{.4\%}$ **4.** $8.94.444 = \boxed{894.4\%}$

5. $\frac{1}{4}\% = .25\% = \boxed{.0025}$ **6.** $6\frac{3}{4}\% = 6.75\% = \boxed{.0675}$

7. $87\% = .87. = \boxed{.87}$ **8.** $810.9\% = 8.10.9 = \boxed{8.109}$

9. $\frac{1}{7} = .14.285 = \boxed{14.29\%}$ **10.** $\frac{2}{9} = .22.2\overline{2} = \boxed{22.22\%}$

11. $19\% = 19 \times \frac{1}{100} = \frac{19}{100}$ **12.** $71\frac{1}{2}\% = \frac{143}{2} \times \frac{1}{100} = \boxed{\frac{143}{200}}$

13. $130\% = 130 \times \frac{1}{100} = \frac{130}{100} = 1\frac{30}{100} = \boxed{1\frac{3}{10}}$ **14.** $\frac{1}{2}\% = \frac{1}{2} \times \frac{1}{100} = \boxed{\frac{1}{200}}$

15. $19\frac{9}{10}\% = \frac{199}{10} \times \frac{1}{100} = \boxed{\frac{199}{1,000}}$

Learning Unit 6–2:

Application of Percents—Portion Formula

The bag of M&M'S® we have been studying contains Milk Chocolate M&M'S®. M&M/Mars also makes Peanut M&M'S® and some other types of M&M'S®. To study the application of percents to problems involving M&M'S®, we make two key assumptions:

1. Total sales of Milk Chocolate M&M'S®, Peanut M&M'S®, and other M&M'S® chocolate candies are $400,000.

2. Eighty percent of M&M'S® sales are Milk Chocolate M&M'S®. This leaves the Peanut and other M&M'S® chocolate candies with 20% of sales (100% − 80%).

80% M&M'S® Milk Chocolate M&M'S®	+	20% M&M'S® Peanut and other chocolate candies	=	100% Total sales ($400,000)

Before we begin, you must understand the meaning of three terms—*base, rate,* and *portion.* These terms are the key elements in solving percent problems.

- Base (**B**). The **base** is the beginning whole quantity or value (100%) with which you will compare some other quantity or value. Often the problems give the base after the word *of.* For example, the whole (total) sales of M&M'S®—Milk Chocolate M&M'S, Peanut, and other M&M'S® chocolate candies—are $400,000.

- Rate (**R**). The **rate** is a percent, decimal, or fraction that indicates the part of the base that you must calculate. The percent symbol often helps you identify the rate. For example, Milk Chocolate M&M'S® currently account for 80% of sales. So the rate is 80%. Remember that 80% is also $\frac{4}{5}$, or .80.

- **Portion (P).** The **portion** is the amount or part that results from the base multiplied by the rate. For example, total sales of M&M'S® are $400,000 (base); $400,000 times .80 (rate) equals $320,000 (portion), or the sales of Milk Chocolate M&M'S®. *A key point to remember is that portion is a number and not a percent. In fact, the portion can be larger than the base if the rate is greater than 100%.*

Aha!

Learn: Solving Percents with the Portion Formula

In problems involving portion, base, and rate, we give two of these elements. You must find the third element. Remember the following key formula:

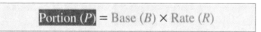

$$\text{Portion } (P) = \text{Base } (B) \times \text{Rate } (R)$$

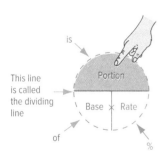

This line is called the dividing line

is — Portion — Base × Rate — of — %

To help you solve for the portion, base, and rate, this unit shows pie charts. The shaded area in each pie chart indicates the element that you must solve for. For example, since we shaded *portion* in the pie chart at the left, you must solve for portion. To use the pie charts, put your finger on the shaded area (in this case portion). The formula that remains tells you what to do. So in the pie chart at the left, you solve the problem by multiplying base by the rate. Note the circle around the pie chart is broken since we want to emphasize that portion can be larger than base if rate is greater than 100%. The horizontal line in the pie chart is called the dividing line, and we will use it when we solve for base or rate.

The following example summarizes the concept of base, rate, and portion. Assume that you received a small bonus check of $100. This is a gross amount—your company did not withhold any taxes. You will have to pay 20% in taxes.

Base: 100%—whole. Usually given after the word *of*—but not always.	Rate: Usually expressed as a percent but could also be a decimal or fraction.	Portion: A number—not a percent and not the whole.
$100 bonus check	20% taxes	$20 taxes

First decide what you are looking for. You want to know how much you must pay in taxes—the portion. How do you get the portion? From the portion formula Portion (P) = Base (B) × Rate (R), you know that you must multiply the base ($100) by the rate (20%). When you do this, you get $\$100 \times .20 = \$20.$ So you must pay $20 in taxes.

Let's try our first word problem by taking a closer look at the M&M'S® example to see how we arrived at the $320,000 sales of Milk Chocolate M&M'S® given earlier. We will be using blueprint aids to help dissect and solve each word problem.

Solving for Portion

The Word Problem Sales of Milk Chocolate M&M'S® are 80% of the total M&M'S® sales. Total M&M'S® sales are $400,000. What are the sales of Milk Chocolate M&M'S®?

	The facts	Solving for?	Steps to take	Key points
Blueprint	*Milk Chocolate M&M'S® sales: 80%.* *Total M&M'S® sales: $400,000.*	Sales of Milk Chocolate M&M'S®.	Identify key elements. *Base:* $400,000. *Rate:* .80. *Portion:* ? Portion = Base × Rate.	Amount or part of beginning Portion (?) Base × Rate ($400,000) (.80) Beginning whole quantity (often after "of") — Percent symbol or word (here we put into decimal) Portion and rate must relate to same piece of base.

Steps to solving problem

1. Set up the formula.

2. Calculate portion (sales of Milk Chocolate M&M'S®).

Portion = Base × Rate

$P = \$400,000 \times .80$

$P = \$320,000$

In the first column of the blueprint aid, we gather the facts. In the second column, we state that we are looking for sales of Milk Chocolate M&M'S®. In the third column, we identify each key element and the formula needed to solve the problem. Review the pie chart in the fourth column. *The portion and rate must relate to the same piece of the base.* In this word problem, we can see from the solution below the blueprint aid that sales of Milk Chocolate M&M'S® are $320,000. The $320,000 does indeed represent 80% of the base. Note here that the portion ($320,000) is less than the base of $400,000 since the rate is less than 100%.

Now let's work another word problem that solves for the portion.

The Word Problem Sales of Milk Chocolate M&M'S® are 80% of the total M&M'S® sales. Total M&M'S® sales are $400,000. What are the sales of Peanut and other M&M'S® chocolate candies?

	The facts	Solving for?	Steps to take	Key points
Blueprint	*Milk Chocolate M&M'S® sales: 80%.* Total M&M'S® sales: $400,000.	Sales of Peanut and other M&M'S® chocolate candies.	Identify key elements. *Base:* $400,000. *Rate:* .20 (100% − 80%). *Portion:* ? Portion = Base × Rate.	If 80% of sales are Milk Chocolate M&M'S, then 20% are Peanut and other M&M'S® chocolate candies. Portion (?) Base × Rate ($400,000) (.20) Portion and rate must relate to same piece of base.

Steps to solving problem

1. Set up the formula.

2. Calculate portion (sale of Peanut and other M&M'S® chocolate candies).

Portion = Base × Rate

$P = \$400,000 \times .20$

$P = \$80,000$

Aha! In the previous blueprint aid, *note that we must use a rate that agrees with the portion so the portion and rate refer to the same piece of the base.* Thus, if 80% of sales are Milk Chocolate M&M'S®, 20% must be Peanut and other M&M'S® chocolate candies (100% − 80% = 20%). So we use a rate of .20.

In Step 2, we multiplied $400,000 × .20 to get a portion of $80,000. This portion represents the part of the sales that were *not* Milk Chocolate M&M'S®. Note that the rate of .20 and the portion of $80,000 relate to the same piece of the base—$80,000 is 20% of $400,000. Also note that the portion ($80,000) is less than the base ($400,000) since the rate is less than 100%.

Take a moment to review the two blueprint aids in this section. Be sure you understand why the rate in the first blueprint aid was 80% and the rate in the second blueprint aid was 20%.

Solving for Rate

The Word Problem Sales of Milk Chocolate M&M'S® are $320,000. Total M&M'S® sales are $400,000. What is the percent of Milk Chocolate M&M'S® sales compared to total M&M'S® sales?

The facts	Solving for?	Steps to take	Key points
Blueprint *Milk Chocolate M&M'S® sales: $320,000.* *Total M&M'S® sales: $400,000.*	Percent of Milk Chocolate M&M'S® sales to total M&M'S® sales.	Identify key elements. Base: $400,000. Rate: ? Portion: $320,000 $$\text{Rate} = \frac{\text{Portion}}{\text{Base}}$$	Since portion is less than base, the rate must be less than 100% Portion ($320,000) Base ($400,000) × Rate (?) Portion and rate must relate to the same piece of base.

Steps to solving problem

1. Set up the formula.

$$\text{Rate} = \frac{\text{Portion}}{\text{Base}}$$

2. Calculate rate (percent of Milk Chocolate M&M'S® sales).

$$R = \frac{\$320,000}{\$400,000}$$

$$R = \boxed{80\%}$$

Note that in this word problem, the rate of 80% and the portion of $320,000 refer to the same piece of the base.

The Word Problem Sales of Milk Chocolate M&M'S® are $320,000. Total sales of Milk Chocolate M&M'S, Peanut, and other M&M'S® chocolate candies are $400,000. What percent of Peanut and other M&M'S® chocolate candies are sold compared to total M&M'S® sales?

The facts	Solving for?	Steps to take	Key points
Blueprint *Milk Chocolate M&M'S® sales: $320,000.* *Total M&M'S® sales: $400,000.*	Percent of Peanut and other M&M'S® chocolate candies sales compared to total M&M'S® sales.	Identify key elements. Base: $400,000. Rate: ? Portion: $80,000 ($400,000 − $320,000). $$\text{Rate} = \frac{\text{Portion}}{\text{Base}}$$	Represents sales of Peanut and other M&M'S® chocolate candies Portion ($80,000) Base ($400,000) × Rate (?) When portion becomes $80,000, the portion and rate now relate to same piece of base.

Steps to solving problem

1. Set up the formula.

$$\text{Rate} = \frac{\text{Portion}}{\text{Base}}$$

2. Calculate rate.

$$R = \frac{\$80,000}{\$400,000} \quad (\$400,000 - \$320,000)$$

$$R = \boxed{20\%}$$

The word problem asks for the rate of candy sales that are *not* Milk Chocolate M&M'S. Thus, $400,000 of total candy sales less sales of Milk Chocolate M&M'S® ($320,000) allows us to arrive at sales of Peanut and other M&M'S® chocolate candies ($80,000). The $80,000 portion represents 20% of total candy sales. The $80,000 portion and 20% rate refer to the same piece of the $400,000 base. Compare this blueprint aid with the blueprint aid for the previous word problem. Ask yourself why in the previous word problem the rate was 80% and in this word problem the rate is 20%. In both word problems, the portion was less than the base since the rate was less than 100%.

Now we go on to calculate the base. Remember to read the word problem carefully so that you match the rate and portion to the same piece of the base.

Solving for Base

The Word Problem Sales of Peanut and other M&M'S® chocolate candies are 20% of total M&M'S® sales. Sales of Milk Chocolate M&M'S® are $320,000. What are the total sales of all M&M'S®?

	The facts	Solving for?	Steps to take	Key points
Blueprint	*Peanut and other M&M'S® chocolate candies sales: 20%.* *Milk Chocolate M&M'S® sales: $320,000.*	Total M&M'S® sales.	Identify key elements. *Base: ?* *Rate: .80 (100% − 20%)* *Portion: $320,000* $\text{Base} = \dfrac{\text{Portion}}{\text{Rate}}$	Portion ($320,000) Base × Rate (?) (.80) (100% − 20%) Portion ($320,000) and rate (.80) do relate to the same piece of base.

Steps to solving problem

1. Set up the formula. $\text{Base} = \dfrac{\text{Portion}}{\text{Rate}}$

2. Calculate the base. $B = \dfrac{\$320,000}{.80}$ ⟵ $320,000 is 80% of base

$$B = \boxed{\$400,000}$$

Note that we could not use 20% for the rate. The $320,000 of Milk Chocolate M&M'S® represents 80% (100% − 20%) of the total sales of M&M'S®. We use 80% so that the portion and rate refer to same piece of the base. Remember that the portion ($320,000) is less than the base ($400,000) since the rate is less than 100%.

Learn: Calculating Percent Increases and Decreases

The following *Wall Street Journal* clip shows Procter & Gamble is keeping the price of Tide at $11.99 but reducing the number of loads the new product washes from 60 to only 48 loads. Let's calculate the cost per load (rounded to nearest cent) before and after the load change:

Before

$\dfrac{\$11.99}{60 \text{ loads}} = \$.20 \text{ per load}$

After

$\dfrac{\$11.99}{48 \text{ loads}} = \$.25 \text{ per load}$

Using this clip, let's look at how to calculate percent increases and decreases.

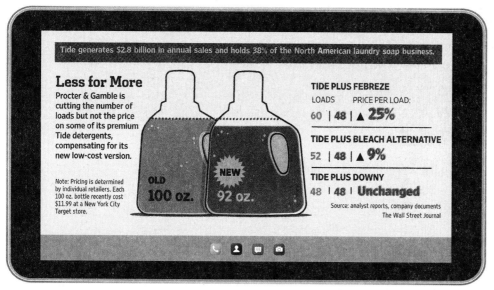

The Tide Example: Rate of Percent Increase in Price per Load   Assume: per load cost increase from $.20 to $.25:

$$\text{Rate} = \frac{\text{Portion}}{\text{Base}} \quad \begin{matrix}\leftarrow \text{ Difference between old and new price per load} \\ \leftarrow \text{ Old price per load}\end{matrix}$$

$$R = \frac{\$.05}{\$.20} \qquad (\$.25 - \$.20)$$

$$R = .25 = 25\% \text{ increase}$$

Let's prove the 25% with a pie chart.

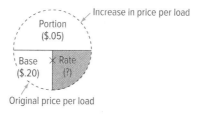

The formula for calculating **percent increase** is as follows:

Percent increase

$$\begin{array}{c}\text{Percent of} \\ \text{increase } (R) = \\ (25\%)\end{array} \frac{\text{Amount of price per load increase } (P)}{\text{Original price per load } (B)} \\ \quad (\$.05) \\ \quad (\$.20)$$

Now let's look at how to calculate the math for a decrease in price per load for Tide.

The Tide Example: Rate of Percent Decrease   Assume: Price of $11.99 but loads increase from 60 to 70. The first step is to calculate the price per load (rounded to nearest cent) before and after:

Before	After
$\dfrac{\$11.99}{60} = \$.20$ per load	$\dfrac{\$11.99}{70} = \$.17$ per load

$\text{Rate} = \dfrac{\text{Portion}}{\text{Base}}$ ←— Difference between old and new price per load
←— Old price per load

$R = \dfrac{\$.03}{\$.20}$ ($.20 − $.17 = $.03)

$R = \boxed{.15 = 15\%}$ decrease

Let's prove the 15% with a pie chart.

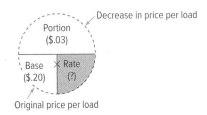

The formula for calculating **percent decrease** is as follows:

Percent decrease

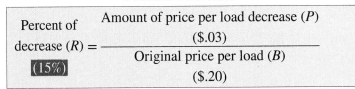

Percent of decrease (R) (15%) = $\dfrac{\text{Amount of price per load decrease } (P) \;(\$.03)}{\text{Original price per load } (B) \;(\$.20)}$

In conclusion, the following steps can be used to calculate percent increases and decreases:

Calculating Percent Increases and Decreases

Step 1 Find the difference between amounts (such as sales).

Step 2 Divide Step 1 by the original amount (the base): $R = P \div B$. Be sure to express your answer in percent.

Before concluding this chapter, we will show how to calculate a percent increase and decrease using M&M'S® (Figure 6.1).

Additional Examples Using M&M'S

The Word Problem Sheila Leary went to her local supermarket and bought the bag of M&M'S® shown in Figure 6.1. The bag gave its weight as 18.40 ounces, which was 15% more than a regular 1-pound bag of M&M'S®. Sheila, who is a careful shopper, wanted to check and see if she was actually getting a 15% increase. Let's help Sheila dissect and solve this problem.

	The facts	Solving for?	Steps to take	Key points
Blueprint	*New bag of M&M'S®:* 18.40 oz. 15% increase in weight. *Original bag of M&M'S®:* 16 oz. (1 lb.)	Checking percent increase of 15%.	Identify key elements. *Base:* 16 oz. *Rate:* ? *Portion:* 2.40 oz. $\left(\begin{array}{c} 18.40 \text{ oz.} \\ -16.00 \\ \hline 2.40 \text{ oz.} \end{array}\right)$ $\text{Rate} = \dfrac{\text{Portion}}{\text{Base}}$	Difference between base and new weight. Portion (2.40 oz.) Base × Rate (16 oz.) (?) Original amount sold

Steps to solving problem

1. Set up the formula. $\text{Rate} = \dfrac{\text{Portion}}{\text{Base}}$

2. Calculate the rate. $R = \dfrac{2.40 \text{ oz.}}{16.00 \text{ oz.}}$ ← Difference between base and new weight.
                                                                    ← Old weight equals 100%.

   $\boxed{R = 15\% \text{ increase}}$

The new weight of the bag of M&M'S® is really 115% of the old weight:

$$
\begin{array}{rcl}
16.00 \text{ oz.} & = & 100\% \\
+\ 2.40 & = & +\ 15 \\
\hline
18.40 \text{ oz.} & = & 115\% = 1.15
\end{array}
$$

FIGURE 6.1 Bag of 18.40-ounce M&M'S®

We can check this by looking at the following pie chart:

Portion = Base × Rate
$\boxed{18.40 \text{ oz.}}$ = 16 oz. × 1.15

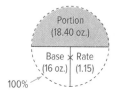

Portion (18.40 oz.) Base × Rate (16 oz.) (1.15) 100%

Why is the portion greater than the base? Remember that the portion can be larger than the base only if the rate is greater than 100%. Note how the portion and rate relate to the same piece of the base—18.40 oz. is 115% of the base (16 oz.).

Let's see what could happen if M&M/Mars has an increase in its price of sugar. This is an additional example to reinforce the concept of percent decrease.

Money Tip When planning for retirement, a rule of thumb is that you will need 70% of your preretirement pay to live comfortably. This number assumes your house is paid off and you are in good health. Automating your savings can be a huge factor in helping you reach your goals. So, begin planning early in life and start saving for a financially sound retirement.

The Word Problem The increase in the price of sugar caused the M&M/Mars company to decrease the weight of each 1-pound bag of M&M'S® to 12 ounces. What is the rate of percent decrease?

	The facts	Solving for?	Steps to take	Key points
Blueprint	*16-oz. bag of M&M'S®: reduced to 12 oz.*	Rate of percent decrease.	Identify key elements. *Base:* 16 oz. *Rate:* ? *Portion:* 4 oz. (16 oz. − 12 oz.) $\text{Rate} = \dfrac{\text{Portion}}{\text{Base}}$	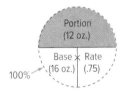

Steps to solving problem

1. Set up the formula. $\text{Rate} = \dfrac{\text{Portion}}{\text{Base}}$

2. Calculate the rate. $R = \dfrac{4 \text{ oz.}}{16.00 \text{ oz.}}$

$$R = .25 = 25\% \text{ decrease}$$

The new weight of the bag of M&M'S® is 75% of the old weight:

$$
\begin{array}{rcr}
16 \text{ oz.} = & & 100\% \\
- 4 & & - 25 \\
\hline
12 \text{ oz.} = & & 75\%
\end{array}
$$

We can check this by looking at the following pie chart:

$\text{Portion} = \text{Base} \times \text{Rate}$

$12 \text{ oz.} = 16 \text{ oz.} \times .75$

Note that the portion is smaller than the base because the rate is less than 100%. Also note how the portion and rate relate to the same piece of the base—12 ounces is 75% of the base (16 oz.).

After your study of Learning Unit 6–2, you should be ready for the Practice Quiz.

Practice Quiz

Complete this Practice Quiz to see how you are doing.

Solve for portion:

1. 38% of 900.   **2.** 60% of $9,000.

Solve for rate (round to the nearest tenth percent as needed):

3. 430 is _____% of 5,000.   **4.** 200 is _____% of 700.

Solve for base (round to the nearest tenth as needed):

5. 55 is 40% of _____.   **6.** 900 is $4\frac{1}{2}$% of _____.

Solve the following (blueprint aids are shown in the solution; you might want to try some on scrap paper):

7. Five out of 25 students in Professor Ford's class received an A grade. What percent of the class *did not* receive the A grade?

8. Abby Biernet has yet to receive 60% of her lobster order. Abby received 80 lobsters to date. What was her original order?

9. Assume in 2021 Dunkin' Donuts Company had $300,000 in doughnut sales. In 2022, sales were up 40%. What are Dunkin' Donuts sales for 2022?

10. The price of an Apple computer dropped from $1,600 to $1,200. What was the percent decrease?

11. In 1982, a ticket to the Boston Celtics cost $14. In 2022, a ticket cost $50. What is the percent increase to the nearest hundredth percent?

✓ Solutions

1.
$$\boxed{342} = 900 \times .38$$
$$(P) = (B) \times (R)$$

2.
$$\boxed{\$5,400} = \$9,000 \times .60$$
$$(P) \;\; = \;\; (B) \;\; \times (R)$$

3. $\dfrac{(P)430}{(B)5,000} = .086 = \boxed{8.6\% \; (R)}$

4. $\dfrac{(P)200}{(B)700} = .2857 = \boxed{28.6\% \; (R)}$

5. $\dfrac{(P)55}{(R).40} = \boxed{137.5 \; (B)}$

6. $\dfrac{(P)900}{(R).045} = \boxed{20,000 \; (B)}$

7. Percent of Professor Ford's class that did not receive an A grade:

The facts	Solving for?	Steps to take	Key points
5 As. 25 in class.	Percent that did not receive A.	Identify key elements. *Base:* 25 *Rate:* ? *Portion:* 20 (25 − 5) Rate = $\dfrac{\text{Portion}}{\text{Base}}$	Portion (20) Base × Rate (25) (?) The whole Portion and rate must relate to same piece of base.

Blueprint

Practice Quiz *Continued*

Steps to solving problem

1. Set up the formula. $\text{Rate} = \dfrac{\text{Portion}}{\text{Base}}$

2. Calculate the base rate. $R = \dfrac{20}{25}$

$$R = .80 = 80\%$$

8. Abby Biernet's original order:

The facts	Solving for?	Steps to take	Key points
60% of the order not in. 80 lobsters received.	Total order of lobsters.	Identify key elements. *Base:* ? *Rate:* .40 (100% − 60%) *Portion:* 80 $\text{Base} = \dfrac{\text{Portion}}{\text{Rate}}$	Portion (80) Base × Rate (?) (.40) 80 lobsters represent 40% of the order Portion and rate must relate to same piece of base.

(Blueprint)

Steps to solving problem

1. Set up the formula. $\text{Base} = \dfrac{\text{Portion}}{\text{Rate}}$ 80 lobsters is 40% of base.

2. Calculate the base rate. $B = \dfrac{80}{.40}$ ←

$$B = 200 \text{ lobsters}$$

9. Dunkin' Donuts Company sales for 2022:

The facts	Solving for?	Steps to take	Key points
2021: $300,000 sales. *2022:* Sales up 40% from 2021.	Sales for 2022.	Identify key elements. *Base:* $300,000. *Rate:* 1.40. Old year 100% New year + 40 140% *Portion:* ? Portion = Base × Rate.	2022 sales Portion (?) Base × Rate ($300,000) (1.40) 2021 sales When rate is greater than 100%, portion will be larger than base.

(Blueprint)

Steps to solving problem

1. Set up the formula. Portion = Base × Rate

2. Calculate the portion. $P = \$300,000 \times 1.40$

$$P = \$420,000$$

Practice Quiz *Continued*

10. Percent decrease in Apple computer price:

	The facts	Solving for?	Steps to take	Key points
Blueprint	Apple computer was $1,600; now, $1,200.	Percent decrease in price.	Identify key elements. *Base:* $1,600. *Rate:* ? *Portion:* $400 ($1,600 − $1,200). $$Rate = \frac{Portion}{Base}$$	Difference in price Portion ($400) Base ($1,600) × Rate (?) Original price

Steps to solving problem

1. Set up the formula. $\quad Rate = \dfrac{Portion}{Base}$

2. Calculate the rate. $\quad R = \dfrac{\$400}{\$1,600}$

$\qquad\qquad\qquad\qquad R = 25 = 25\%$

11. Percent increase in Boston Celtics ticket:

	The facts	Solving for?	Steps to take	Key points
Blueprint	$14 ticket (old). $50 ticket (new).	Percent increase in price.	Identify key elements. *Base:* $14 *Rate:* ? *Portion:* $36 ($50 − $14) $$Rate = \frac{Portion}{Base}$$	Difference in price Portion ($36) Base ($14) × Rate (?) Original price When portion is greater than base, rate will be greater than 100%.

Steps to solving problem

1. Set up the formula. $\quad Rate = \dfrac{Portion}{Base}$

2. Calculate the rate. $\quad R = \dfrac{\$36}{\$14}$

$\qquad\qquad\qquad R = 2.5714 = 257.14\%$

Chapter 6 Review

Topic/Procedure/Formula	Example	You try it*
Converting decimals to percents 1. Move decimal point two places to right. If necessary, add zeros. This rule is also used for whole numbers and mixed decimals. 2. Add a percent symbol at end of number.	$.81 = .81. = \boxed{81\%}$ $.008 = .00.8 = \boxed{.8\%}$ $4.15 = 4.15. = \boxed{415\%}$	**Convert to percent** .92 .009 5.46
Rounding percents 1. Answer must be in percent before rounding. 2. Identify specific digit. If digit to right is 5 or greater, round up. 3. Delete digits to right of identified digit.	Round to the nearest hundredth percent. $\dfrac{3}{7} = .4285714 = 42.85714 = \boxed{42.86\%}$	**Round to the nearest hundredth percent** $\dfrac{2}{9}$
Converting percents to decimals 1. Drop percent symbol. 2. Move decimal point two places to left. If necessary, add zeros. For fractional percents: 1. Convert to decimal by dividing numerator by denominator. If necessary, round answer. 2. If a mixed fractional percent, convert fractional percent first. Then combine whole number and fractional percent. 3. Drop percent symbol; move decimal point two places to left.	$.89\% = \boxed{.0089}$ $95\% = \boxed{.95}$ $195\% = \boxed{1.95}$ $8\dfrac{3}{4}\% = 8.75\% = \boxed{.0875}$ $\dfrac{1}{4}\% = .25\% = \boxed{.0025}$ $\dfrac{1}{5}\% = .20\% = \boxed{.0020}$	**Convert to decimal** .78% 96% 246% $7\dfrac{3}{4}\%$ $\dfrac{3}{4}\%$ $\dfrac{1}{2}\%$
Converting fractions to percents 1. Divide numerator by denominator. 2. Move decimal point two places to right; add percent symbol.	$\dfrac{4}{5} = .80 = \boxed{80\%}$	**Convert to percent** $\dfrac{3}{5}$

Chapter 6 Review (Continued)

Topic/Procedure/Formula	Example	You try it*
Converting percents to fractions Whole percent (or fractional percent) to a fraction: 1. Drop percent symbol. 2. Multiply number by $\frac{1}{100}$. 3. Reduce to lowest terms. Mixed or decimal percent to a fraction: 1. Drop percent symbol. 2. Change mixed percent to an improper fraction. 3. Multiply number by $\frac{1}{100}$. 4. Reduce to lowest terms. If you have a mixed or decimal percent, change decimal portion to fractional equivalent and continue with Steps 1 to 4.	$64\% \longrightarrow 64 \times \frac{1}{100} = \frac{64}{100} = \boxed{\frac{16}{25}}$ $\frac{1}{4}\% \longrightarrow \frac{1}{4} \times \frac{1}{100} = \boxed{\frac{1}{400}}$ $119\% \longrightarrow 119 \times \frac{1}{100} = \frac{119}{100} = \boxed{1\frac{19}{100}}$ $16\frac{1}{4}\% \longrightarrow \frac{65}{4} \times \frac{1}{100} = \frac{65}{400} = \boxed{\frac{13}{80}}$ $16.25\% \longrightarrow 16\frac{1}{4}\% = \frac{65}{4} \times \frac{1}{100}$ $\qquad\qquad = \frac{65}{400} = \boxed{\frac{13}{80}}$	**Convert to fractions** 74% $\frac{1}{5}\%$ 121% $17\frac{1}{5}\%$ 17.75%
Solving for portion 	10% of Mel's paycheck of $1,000 goes for food. What portion is deducted for food? $\boxed{\$100} = \$1,000 \times .10$ *Note:* If question was what amount does not go for food, the portion would have been: $\boxed{\$900} = \$1,000 \times .90$ $(100\% - 10\% = 90\%)$	**Find portion** Base $2,000 Rate 80%
Solving for rate 	Assume Mel spends $100 for food from his $1,000 paycheck. What percent of his paycheck is spent on food? $\dfrac{\$100}{\$1,000} = .10 = \boxed{10\%}$ *Note:* Portion is less than base since rate is less than 100%.	**Find rate** Base $2,000 Portion $500
Solving for base 	Assume Mel spends $100 for food, which is 10% of his paycheck. What is Mel's total paycheck? $\dfrac{\$100}{.10} = \boxed{\$1,000}$	**Find base** Rate 20% Portion $200

Chapter 6 Review (Continued)

Topic/Procedure/Formula	Example	You try it*
Calculating percent increases and decreases Amount of decrease or increase Portion Base × Rate (?) Original price	Stereo, $2,000 original price. Stereo, $2,500 new price. $\dfrac{\$500}{\$2,000} = .25 = \boxed{25\%}$ increase **Check** $2,000 × 1.25 = $2,500 *Note:* Portion is greater than base since rate is greater than 100%. Portion ($2,500) Base ($2,000) × Rate (1.25)	**Find percent increase** Old price $500 New price $600

Key Terms

Base   Percent increase   Portion

Percent decrease   Percents   Rate

Note: For how to dissect and solve a word problem, see Learning Unit 6–2.
*Worked-out solutions are in Appendix A.

Critical Thinking Discussion Questions with Chapter Concept Check

1. In converting from a percent to a decimal, when will you have at least 2 leading zeros before the whole number? Explain this concept, assuming you have 100 bills of $1.

2. Explain the steps in rounding percents. Count the number of students who are sitting in the back half of the room as a percent of the total class. Round your answer to the nearest hundredth percent. Could you have rounded to the nearest whole percent without changing the accuracy of the answer?

3. Define portion, rate, and base. Create an example using Walt Disney World to show when the portion could be larger than the base. Why must the rate be greater than 100% for this to happen?

4. How do we solve for portion, rate, and base? Create an example using Apple computer sales to show that the portion and rate do relate to the same piece of the base.

5. Explain how to calculate percent decreases or increases. Many years ago, comic books cost 10 cents a copy. Visit a bookshop or newsstand. Select a new comic book and explain the price increase in percent compared to the 10-cent comic. How important is the rounding process in your final answer?

6. **Chapter Concept Check.** Go to the Google or Facebook site and find out how many people the company employs. Assuming a 10% increase in employment this year, calculate the total number of new employees by the end of the year, and identify the base, rate, and portion. If, in the following year, the 10% increase in employment fell by 5%, what would the total number of current employees be?

7. **Chapter Concept Check.** Using percents, show how your buying habits changed during the pandemic.

End-of-Chapter Problems

Name _____ Date _____

Check figures for odd-numbered problems in Appendix A.

Drill Problems

Convert the following decimals to percents: *LU 6-1(1)*

6–1. .88   **6–2.** .384   **6–3.** .4

6–4. 8.00   **6–5.** 3.561   **6–6.** 6.006

Convert the following percents to decimals: *LU 6-1(1)*

6–7. 4%   **6–8.** 14%   **6–9.** $64\frac{3}{10}\%$

6–10. 75.9%   **6–11.** 119%   **6–12.** 89%

Convert the following fractions to percents (round to the nearest tenth percent as needed):   *LU 6-1(1)*

6–13. $\frac{1}{12}$ **6–14.** $\frac{1}{400}$

6–15. $\frac{7}{8}$ **6–16.** $\frac{11}{12}$

Convert the following percents to fractions and reduce to the lowest terms: *LU 6-1(2)*

6–17. 4%   **6–18.** $18\frac{1}{2}\%$

6–19. $31\frac{2}{3}\%$ **6–20.** $61\frac{1}{2}\%$

6–21. 6.75%   **6–22.** 182%

Solve for the portion (round to the nearest hundredth as needed):   *LU 6-2(2)*

excel 6–23. 7% of 150   **excel 6–24.** 125% of 4,320   **excel 6–25.** 25% of 410

excel 6–26. 119% of 128.9   **excel 6–27.** 17.4% of 900   **excel 6–28.** 11.2% of 85

6–29. $12\frac{1}{2}\%$ of 919   **6–30.** 45% of 300

6–31. 18% of 90   **6–32.** 30% of 2,000

Solve for the base (round to the nearest hundredth as needed):   *LU 6-2(2)*

6–33. 170 is 120% of _____   **6–34.** 36 is .75% of _____

6–35. 50 is .5% of _____   **6–36.** 10,800 is 90% of _____

6–37. 800 is $4\frac{1}{2}\%$ of _____

End-of-Chapter Problems (Continued)

Solve for rate (round to the nearest tenth percent as needed): *LU 6-2(2)*

6–38. _____ of 80 is 50

6–39. _____ of 85 is 92

6–40. _____ of 250 is 65

6–41. 110 is _____ of 100

6–42. .09 is _____ of 2.25

6–43. 16 is _____ of 4

Solve the following problems. Be sure to show your work. Round to the nearest hundredth or hundredth percent as needed: *LU 6-2(2)*

6–44. What is 180% of 310?

6–45. 66% of 90 is what?

6–46. 40% of what number is 20?

6–47. 770 is 70% of what number?

6–48. 4 is what percent of 90?

6–49. What percent of 150 is 60?

Complete the following table: *LU 6-2(3)*

| | | Selling price | | Amount of decrease | Percent change (to nearest |
	Product	2021	2022	or increase	hundredth percent as needed)
6–50.	Apple iPad	$650	$500		
6–51.	Smartphone	$100	$120		

Word Problems (First of Four Sets)

e\|cel **6–52.** At a local Dunkin' Donuts, a survey showed that out of 1,200 customers eating lunch, 240 ordered coffee with their meal. What percent of customers ordered coffee? *LU 6-2(2)*

e\|cel **6–53.** What percent of customers in Problem 6–52 did not order coffee? *LU 6-2(2)*

6–54. If the price of gas was on average $2.05 per gallon, and this was $1.20 cheaper than a year before, what is the price decrease? Round to the nearest hundredth percent.   *LU 6-2(3)*

6–55. Wally Chin, the owner of an ExxonMobil station, bought a used Ford pickup truck, paying $2,000 as a down payment. He still owes 80% of the selling price. What was the selling price of the truck?   *LU 6-2(2)*

6–56. Maria Fay bought four Dunlop tires at a local Goodyear store. The salesperson told her that her mileage would increase by 8%. Before this purchase, Maria was getting 24 mpg. What should her mileage be with the new tires? Round to the nearest hundredth.   *LU 6-2(2)*

excel **6–57.** The Social Security Administration announced the following rates to explain what percent of your Social Security benefits you will receive based on how old you are when you start receiving Social Security benefits.

Age	Percent of benefit
62	75
63	80
64	86.7
65	93.3
66	100

Assume Shelley Kate decides to take her Social Security at age 63. What amount of Social Security money will she receive each month, assuming she is entitled to $800 per month? *LU 6-2(2)*

excel **6–58.** Assume that in the year 2021, 800,000 people attended the Christmas Eve celebration at Walt Disney World. In 2022, attendance for the Christmas Eve celebration is expected to increase by 35%. What is the total number of people expected at Walt Disney World for this event?   *LU 6-2(2)*

Copyright © McGraw Hill

End-of-Chapter Problems (Continued)

6–59. Pete Smith found in his attic a Woody Woodpecker watch in its original box. It had a price tag on it for $4.50. The watch was made in 1949. Pete brought the watch to an antiques dealer and sold it for $35. What was the percent of increase in price? Round to the nearest hundredth percent. *LU 6-2(3)*

6–60. Christie's Auction sold a painting for $24,500. It charges all buyers a 15% premium of the final bid price. How much did the bidder pay Christie's? *LU 6-2(2)*

Word Problems (Second of Four Sets)

6–61. Out of 9,000 college students surveyed, 540 responded that they do not eat breakfast. What percent of the students do not eat breakfast? *LU 6-2(2)*

6–62. What percent of college students in Problem 6–61 eat breakfast? *LU 6-2(2)*

6–63. You are saving for an emergency fund totaling six months' worth of expenses as well as investing 20% of each paycheck for retirement. If your monthly expenses amount to $1,465 and you want to put aside 30% of each month's expenses for your emergency fund and you earn $2,650 each month, how much do you need to set aside each month for your emergency fund and retirement savings? Can you afford to do this? *LU 6-2(2)*

6–64. Rainfall for January in Fiji averages 12″ according to *World Travel Guide.* This year it rained 5% less. How many inches (to the nearest tenth) did it rain this year?   *LU 6-2(2)*

6–65. Jim and Alice Lange, employees at Walmart, have put themselves on a strict budget. Their goal at year's end is to buy a boat for $15,000 in cash. Their budget includes the following:

   40% food and lodging   20% entertainment   10% educational

   Jim earns $1,900 per month and Alice earns $2,400 per month. After 1 year, will Alice and Jim have enough cash to buy the boat?   *LU 6-2(2)*

6–66. The 1918 Spanish flu killed at least 50 million people with a fatality rate of around 2% worldwide. As of late 2021, COVID-19 had killed 4,866,952 people worldwide. With a worldwide population of 7,874,965,825, what percent of the worldwide population had COVID-19 killed? (Round final answer to nearest thousandth percent.)   *LU 6-2(2)*

6–67. The Museum of Science in Boston estimated that 64% of all visitors came from within the state. On Saturday, 2,500 people attended the museum. How many attended the museum from out of state?   *LU 6-2(2)*

6–68. Staples pays George Nagovsky an annual salary of $36,000. Today, George's boss informs him that he will receive a $4,600 raise. What percent of George's old salary is the $4,600 raise? Round to the nearest tenth percent.   *LU 6-2(2)*

End-of-Chapter Problems (Continued)

6–69. Assume in 2021, a local Dairy Queen had $550,000 in sales. In 2022, Dairy Queen's sales were up 35%. What were Dairy Queen's sales in 2022?   *LU 6-2(2)*

6–70. Blue Valley College has 600 female students. This is 60% of the total student body. How many students attend Blue Valley College?   *LU 6-2(2)*

6–71. Dr. Grossman was reviewing his total accounts receivable. This month, credit customers paid $44,000, which represented 20% of all receivables due (what customers owe). What was Dr. Grossman's total accounts receivable?   *LU 6-2(2)*

6–72. Your city has a sales tax rate of 8.25%. If you just spent $20 on sales tax, how much were your purchases?   *LU 6-2(2)*

6–73. The price of an antique doll increased from $600 to $800. What was the percent of increase? Round to the nearest tenth percent.   *LU 6-2(3)*

6–74. A local Barnes and Noble bookstore ordered 80 marketing books but received 60 books. What percent of the order was missing?   *LU 6-2(2)*

Word Problems (Third of Four Sets)

6–75. Early in the pandemic in April 2020, college-educated employees had an unemployment rate of 8.4% compared with noncollege-educated employees who experienced a 21.2% unemployment rate, as reported by the Congressional Research Service based on the Bureau of Labor Statistics data. The U.S. labor force had reached a high of 164.6 million persons right at the start of the pandemic in February 2020. Using that statistic, how many more noncollege-educated persons were unemployed compared with college-educated persons?   *LU 6-2(2)*

6–76. Due to increased mailing costs, the new rate will cost publishers $50 million; this is 12.5% more than they paid the previous year. How much did it cost publishers last year? Round to the nearest hundreds.   *LU 6-2(2)*

6–77. Jim Goodman, an employee at Walgreens, earned $45,900, an increase of 17.5% over the previous year. What were Jim's earnings the previous year? Round to the nearest cent.   *LU 6-2(2)*

6–78. If the number of personal loan applications declined by 7% to 1,625,415, what had been the previous year's number of applications?   *LU 6-2(2)*

6–79. If the price of a business math text rose to $150 and this was 8% more than the original price, what was the original selling price? Round to the nearest cent.   *LU 6-2(2)*

End-of-Chapter Problems (Continued)

6–80. Web Consultants, Inc., pays Alice Rose an annual salary of $48,000. Today, Alice's boss informs her that she will receive a $6,400 raise. What percent of Alice's old salary is the $6,400 raise? Round to the nearest tenth percent. *LU 6-2(2)*

6–81. Earl Miller, a lawyer, charges Lee's Plumbing, his client, 25% of what he can collect for Lee from customers whose accounts are past due. The attorney also charges, in addition to the 25%, a flat fee of $50 per customer. This month, Earl collected $7,000 from three of Lee's past-due customers. What is the total fee due to Earl? *LU 6-2(2)*

6–82. A local Petco ordered 100 dog calendars but received 60. What percent of the order was missing? *LU 6-2(2)*

6–83. Ray's Video uses MasterCard. MasterCard charges $2\frac{1}{2}\%$ on net deposits (credit slips less returns). Ray's made a net deposit of $4,100 for charge sales. How much did MasterCard charge Ray's? *LU 6-2(2)*

6–84. Globally the household savings rate varies significantly according to *Global Finance* magazine and Investopedia. The savings rate in Ireland is 22.8%, in Canada it is 15%, and in China it is 44.9%. In the United States, the savings rate had been around 7.5% for years until 2020, the year of the pandemic, when savings jumped to 16.1%. Finder.com states the median savings for American families is $4,830. If the average American couple should save at least $1.3 million for retirement, what percent has actually been saved on average to date? Round to the nearest hundredth percent. *LU 6-2(2)*

Word Problems (Fourth of Four Sets)

6–85. Chevrolet raised the base price of its Volt by $1,200 to $33,500. What was the percent increase? Round to the nearest tenth percent. *LU 6-2(2)*

6–86. The sales tax rate is 8%. If Jim bought a used Buick and paid a sales tax of $1,920, what was the cost of the Buick before the tax? *LU 6-2(2)*

6–87. Puthina Unge bought a new Dell computer system on sale for $1,800. It was advertised as 30% off the regular price. What was the original price of the computer? Round to the nearest dollar. *LU 6-2(2)*

6–88. John O'Sullivan has just completed his first year in business. His records show that he spent the following in advertising:

Website $600   Radio $650   Social Media $700   Local flyers $400

What percent of John's advertising was spent on the Social Media? Round to the nearest hundredth percent. *LU 6-2(2)*

6–89. Jay Miller sold his ski house at Attitash Mountain in New Hampshire for $35,000. This sale represented a loss of 15% off the original price. What was the original price Jay paid for the ski house? Round your answer to the nearest dollar. *LU 6-2(2)*

6–90. Out of 4,000 colleges surveyed, 60% reported that SAT scores were not used as a high consideration in viewing their applications. How many schools view the SAT as important in screening applicants? *LU 6-2(2)*

End-of-Chapter Problems (Continued)

6–91. If refinishing your basement at a cost of $45,404 would add $18,270 to the resale value of your home, what percent of your cost is recouped? Round to the nearest percent.   *LU 6-2(2)*

6–92. A major airline laid off 4,000 pilots and flight attendants. If this was a 12.5% reduction in the workforce, what was the size of the workforce after the layoffs?   *LU 6-2(2)*

6–93. Assume 450,000 people line up on the streets to see the Macy's Thanksgiving Parade. If attendance is expected to increase 30% next year, what will be the number of people lined up on the street to see the parade?   *LU 6-2(2)*

Challenge Problems

6–94. Each Tuesday, Ryan Airlines reduces its one-way ticket from Fort Wayne to Chicago from $125 to $40. To receive this special $40 price, the customer must buy a round-trip ticket. Ryan has a nonrefundable 25% penalty fare for cancellation; it estimates that about nine-tenths of 1% will cancel their reservations. The airline also estimates this special price will cause a passenger traffic increase from 400 to 900. Ryan expects revenue for the year to be 55.4% higher than the previous year. Last year, Ryan's sales were $482,000. To receive the special rate, Janice Miller bought two round-trip tickets. On other airlines, Janice has paid $100 round trip (with no cancellation penalty). Calculate the following: *LU 6-2(2)*

   a. Percent discount Ryan is offering.

   b. Percent passenger travel will increase.

   c. Sales for new year.

   d. Janice's loss if she cancels one round-trip flight.

   e. Approximately how many more cancellations can Ryan Airlines expect (after Janice's cancellation)?

6–95. A local Dunkin' Donuts shop reported that its sales have increased exactly 22% per year for the last 2 years. This year's sales were $82,500. What were Dunkin' Donuts' sales 2 years ago? Round each year's sales to the nearest dollar. *LU 6-2(2)*

Summary Practice Test

Do you need help? Connect videos have step-by-step worked-out solutions.

Convert the following decimals to percents.   *LU 6-1(1)*

1. .921   **2.** .4   **3.** 15.88   **4.** 8.00

Convert the following percents to decimals.   *LU 6-1(1)*

5. 42%   **6.** 7.98%   **7.** 400%   **8.** $\frac{1}{4}$%

Convert the following fractions to percents. Round to the nearest tenth percent.   *LU 6-1(1)*

9. $\frac{1}{6}$ **10.** $\frac{1}{3}$

Convert the following percents to fractions and reduce to the lowest terms as needed.   *LU 6-1(2)*

11. $19\frac{3}{8}$%   **12.** 6.2%

Solve the following problems for portion, base, or rate:

13. An Arby's franchise has a net income before taxes of $900,000. The company's treasurer estimates that 40% of the company's net income will go to federal and state taxes. How much will the Arby's franchise have left?   *LU 6-2(2)*

14. Domino's projects a year-end net income of $699,000. The net income represents 30% of its annual sales. What are Domino's projected annual sales?   *LU 6-2(2)*

15. Target ordered 400 iPhones. When Target received the order, 100 iPhones were missing. What percent of the order did Target receive?   *LU 6-2(2)*

16. Matthew Song, an employee at Putnam Investments, receives an annual salary of $120,000. Today his boss informed him that he would receive a $3,200 raise. What percent of his old salary is the $3,200 raise? Round to the nearest hundredth percent.   *LU 6-2(2)*

17. The price of a Delta airline ticket from Los Angeles to Boston increased to $440. This is a 15% increase. What was the old fare? Round to the nearest cent.   *LU 6-2(2)*

18. Scupper Grace earns a gross pay of $900 per week at Office Depot. Scupper's payroll deductions are 29%. What is Scupper's take-home pay?   *LU 6-2(2)*

19. Mia Wong is reviewing the total accounts receivable of Wong's department store. Credit customers paid $90,000 this month. This represents 60% of all receivables due. What is Mia's total accounts receivable?   *LU 6-2(2)*

MY MONEY

Q Pay Yourself First!

⤳ What I need to know

Establishing a savings plan and sticking to it is a very important part of healthy personal financial planning. It is important to get into the habit of paying yourself first. Learn to think of your monthly saving amount as an actual bill that you owe and are required to pay. Saving is not something we do once we have enough money; it is something to begin doing today. No matter the amount of saving we start with, the act of saving creates a healthy financial habit that will serve us well throughout our life.

⤳ What I need to do

Start saving right now, as there is no time like the present. As a student you may feel there is no way you can save due to your limited earnings and significant expenses. However, the act of saving begins with a focus on creating a good financial habit regardless of the amount of money being saved. Start small by determining an amount you will set aside from each paycheck to establish your savings account. Consider creating "no spending days" in which you forgo as many of your habitual daily expenditures as possible and instead place that amount into your savings.

Once you have established a savings plan you will want to modify the plan based on your income over time. Set a goal to increase your savings at a specified time in the future. It may be beneficial to set this plan to correspond with your annual performance review at work where your salary and potential raise is discussed. As your income increases each year, adjust your savings plan by increasing the amount based on the changes to your income. This will keep you focused on your long-term financial goals and make the increase in savings less noticeable as it corresponds to a time in which your income is increasing as well. Have a goal of saving 20% of all monies received.

Establish an emergency fund to protect yourself from surprise expenses. A car repair can sometimes be in the thousands of dollars and can have a significant negative impact on your budget if you are without an emergency fund. Being prepared is the best way to address these emergencies when they occur.

⤳ Steps I need to take

1. Create a plan to start saving today and establish a habit of saving.
2. Increase the amount you save per paycheck as your salary increases. Have a goal of saving 20% of all monies received.
3. Establish an emergency fund for life's unplanned events.

⤳ Resources I can use

- Digit: Save Money & Invest (mobile app)—establish your personal saving goals and receive regular feedback on your progress.
- Chime: Mobile Banking (mobile app)—allows you to save automatically by rounding up on your purchases.

MY MONEY ACTIVITY ✕

- Develop a plan to create an emergency fund of $500 this year.
- What are some specific actions you will take to achieve this goal?

PERSONAL FINANCE

A KIPLINGER APPROACH

"Retirement Planning During COVID." Kiplinger. February, 2021.

A KIPLINGER · PERSONAL CAPITAL POLL

Retirement Planning During COVID

Many Americans say they plan to delay retirement so they can work longer and boost savings.

The pandemic, a volatile stock market and an uncertain economy have rocked the confidence of Americans saving for retirement. In a new poll conducted by Kiplinger in partnership with wealth management firm Personal Capital, 43% of retirement savers said the pandemic made them less confident that they will have enough savings to retire comfortably. More than one-third said they plan to delay retirement and work longer. A similar percentage planned to save more.

A number of retirement savers fell even further behind by tapping their retirement accounts for living and other expenses: Nearly 60% of respondents took a withdrawal or loan from their retirement accounts in 2020. However, even as the stock market was touching new highs, investment mixes reported in the poll were very conservative. Stocks accounted for just 36% of the average allocation, and cash made up a whopping 24% of portfolios.

The poll, conducted in early November, surveyed a national sampling of 744 people ages 40 to 74, none of whom were fully retired, who had at least $50,000 in retirement savings. The median amount saved for retirement among all of the respondents was $188,800. The respondents were equally divided between men and women.

We've included highlights from the poll here. Figures are medians unless otherwise indicated.

How has the global COVID-19 pandemic changed your confidence about having enough income to retire comfortably?

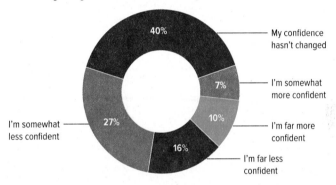

- 40% My confidence hasn't changed
- 7% I'm somewhat more confident
- 10% I'm far more confident
- 16% I'm far less confident
- 27% I'm somewhat less confident

How has the pandemic and its financial impact changed your retirement plan?*

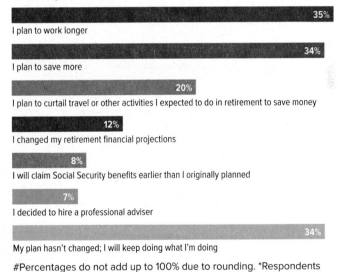

- 35% I plan to work longer
- 34% I plan to save more
- 20% I plan to curtail travel or other activities I expected to do in retirement to save money
- 12% I changed my retirement financial projections
- 8% I will claim Social Security benefits earlier than I originally planned
- 7% I decided to hire a professional adviser
- 34% My plan hasn't changed; I will keep doing what I'm doing

#Percentages do not add up to 100% due to rounding. *Respondents were asked to choose all applicable options.

Business Math Issue

Post COVID means all the percentages in the article are no longer valid today.

1. List the key points of the article and information to support your position.

2. Write a group defense of your position using math calculations to support your view. If you are in an online course, post to a discussion board.

Chapter 7

Discounts: Trade and Cash

Learning Unit Objectives

LU 7–1: Trade Discounts—Single and Chain (Includes Discussion of Freight)

1. Calculate single trade discounts with formulas and complements.
2. Explain the freight terms *FOB shipping point* and *FOB destination*.
3. Find list price when net price and trade discount rate are known.
4. Calculate chain discounts with the net price equivalent rate and single equivalent discount rate.

LU 7–2: Cash Discounts, Credit Terms, and Partial Payments

1. List and explain typical discount periods and credit periods that a business may offer.
2. Calculate outstanding balance for partial payments.

⊖ Essential Question

How can I use discounts to understand the business world and make my life easier?

UPS Adds Extra Fees As E-Commerce Soars

By Paul Ziobro

United Parcel Service Inc. is adding peak surcharges for companies that have been inundating its delivery network with many more packages and oversize items during the coronavirus pandemic, an unprecedented move to manage a summer flood of shipments and higher costs.

UPS typically imposes extra fees on merchants during the busy Christmas shopping season, but–for the first time in the e-commerce era–will add such surcharges starting May 31. The fees would apply to large online sellers like Amazon.com Inc. as well as traditional retailers like Target Corp. and Best Buy Co. that have Shifted heavily to e-commerce as many stores have closed temporarily.

Retailers will have to calculate whether to raise prices, absorb the added cost or a combination of the two. They can also try workarounds to avoid the fee by closely monitoring the amount and sizes of packages they ship with UPS, using another carrier or nudging customers to pick up online orders in stores.

Delivery companies like FedEx Corp., UPS and the U.S. Postal Service are struggling with an unexpected increase in online shopping over the past $2\frac{1}{2}$ months as consumers buy online everything from canned foods and toilet paper to office chairs and backyard pools. Digital sales at Target and Best Buy more than doubled in the most recent quarter.

Ziobro, Paul. "UPS Adds Extra Fees as E-Commerce Soars." *The Wall Street Journal* (May 29, 2020).

660389

Copyright © McGraw Hill   (tablet)Can Yesil/Shutterstock, Are Solution/Shutterstock

🌐 Math Around the World

The *Wall Street Journal* chapter opener clip reveals how UPS has added surcharges to companies that have shifted heavily to e-commerce during the pandemic resulting in a flood of shipments via UPS's delivery network. The *Wall Street Journal* clip to the left reports how Walmart plans to compete with Amazon Prime. We will look at various trade and shipping terms in this chapter but first we discuss two types of discounts taken by retailers— trade and cash. A **trade discount** is a reduction off the original selling price (list price) of an item and is not related to early payment. A **cash discount** is the result of an early payment based on the terms of the sale.

Wal-Mart Stores Inc. is testing a two-day shipping subscription service and building a regional delivery network, in the boldest attempt yet by a major traditional retailer to compete head-on with Amazon Prime.

Trade Discounts—Single and Chain (Includes Discussion of Freight)

The merchandise sold by retailers is bought from manufacturers and wholesalers who sell only to retailers and not to customers. These manufacturers and wholesalers offer retailer discounts so retailers can resell the merchandise at a profit. The discounts are off the manufacturers' and wholesalers' **list price** (suggested retail price), and the amount of discount that retailers receive off the list price is the **trade discount amount.** The following photo shows a Bed Bath and Beyond discount online coupon. Keep in mind that retailers can track customer purchases and preferences. The smartphone is a great tool customers can use to find discounts and retailers can use to gather marketing data.

When you make a purchase, the retailer (seller) gives you a purchase **invoice.** Invoices are important business documents that help sellers keep track of sales transactions and buyers keep track of purchase transactions. North Shore Community College Bookstore is a retail seller of textbooks to students. The bookstore usually purchases its textbooks directly from publishers. Figure 7.1 shows a sample of what a textbook invoice from McGraw Hill Education to the North Shore Community College Bookstore might look like. Note that the trade discount amount is given in percent. This is the **trade discount rate,** which is a percent off the list price that retailers can deduct. The following formula for calculating a trade discount amount gives the numbers from the Figure 7.1 invoice:

Trade Discount Amount Formula

Trade discount amount = List price × Trade discount rate

$2,887.50   = $11,550 ×   25%

The price that the retailer (bookstore) pays the manufacturer (publisher) or wholesaler is the **net price.** The following formula for calculating the net price gives the numbers from the Figure 7.1 invoice:

Net Price Formula

Net price = List price   –   Trade discount amount

$8,662.50 = $11,550   – $2,887.50

Frequently, manufacturers and wholesalers issue catalogs to retailers containing list prices of the seller's merchandise and the available trade discounts. To reduce printing costs when prices change, these sellers usually update the catalogs with new *discount sheets.* The discount sheet also gives the seller the flexibility of offering different trade discounts to different classes of retailers. For example, some retailers buy in quantity and service the products. They may receive a larger discount than the retailer who wants the manufacturer to service the products. Sellers may also give discounts to meet a competitor's price, to attract new retailers, and to reward the retailers who buy product-line products. Sometimes the ability of the retailer to negotiate with the seller determines the trade discount amount.

FIGURE 7.1 Bookstore invoice showing a trade discount

Invoice No.: 5582

McGraw Hill Education

Date: July 8, 2022
Ship: Two-day UPS
Terms: 2/10, n/30

Sold to: North Shore Community College Bookstore
1 Ferncroft Road
Danvers, MA 01923

Description	Unit list price	Total amount
50 Financial Management—Block/Hirt	$195	$9,750.00
10 Introduction to Business—Nichols	180	1,800.00
	Total List Price	11,550.00
	Less: Trade Discount 25%	2,887.50
	Net Price	8,662.50
	Plus: Prepaid Shipping Charge	+125.00
	Total Invoice Amount	$8,787.50

Retailers cannot take trade discounts on freight, returned goods, sales tax, and so on. Trade discounts may be single discounts or a chain of discounts. Before we discuss single trade discounts, let's study freight terms.

Learn: Freight Terms

The most common **freight terms** are *FOB shipping point* and *FOB destination.* These terms determine how the freight will be paid. The key words in the terms are *shipping point* and *destination.*

FOB shipping point means free on board at shipping point; that is, the buyer pays the freight cost of getting the goods to the place of business.

For example, assume that IBM in San Diego bought goods from Argo Suppliers in Boston. Argo ships the goods FOB Boston by plane. IBM takes title to the goods when the aircraft in Boston receives the goods, so IBM pays the freight from Boston to San Diego. Frequently, the seller (Argo) prepays the freight and adds the amount to the buyer's (IBM) invoice. When paying the invoice, the buyer takes the cash discount off the net price and adds the freight cost. FOB shipping point can be illustrated as follows:

FOB shipping point (Boston)

Boston — San Diego

Argo Suppliers (IBM takes title here) → **IBM** (Buyer pays the freight costs)

Buyer pays the freight

FOB destination means the seller pays the freight cost until it reaches the buyer's place of business. If Argo ships its goods to IBM FOB destination or FOB San Diego, the title to the goods remains with Argo. Then it is Argo's responsibility to pay the freight from Boston to IBM's place of business in San Diego. FOB destination can be illustrated as follows:

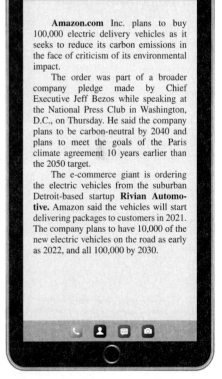

Amazon.com Inc. plans to buy 100,000 electric delivery vehicles as it seeks to reduce its carbon emissions in the face of criticism of its environmental impact.

The order was part of a broader company pledge made by Chief Executive Jeff Bezos while speaking at the National Press Club in Washington, D.C., on Thursday. He said the company plans to be carbon-neutral by 2040 and plans to meet the goals of the Paris climate agreement 10 years earlier than the 2050 target.

The e-commerce giant is ordering the electric vehicles from the suburban Detroit-based startup **Rivian Automotive.** Amazon said the vehicles will start delivering packages to customers in 2021. The company plans to have 10,000 of the new electric vehicles on the road as early as 2022, and all 100,000 by 2030.

Thomas, Patrick. "Amazon Goes Electric to Cut Emissions." *The Wall Street Journal* (September 20, 2019).

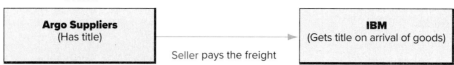

FOB destination (San Diego)

Boston

| Argo Suppliers (Has title) | → | IBM (Gets title on arrival of goods) |

Seller pays the freight

This *Wall Street Journal* clip reports that Amazon will be changing to electric delivery vehicles to cut carbon emissions.

Now you are ready for the discussion on single trade discounts.

Learn: Single Trade Discount

In the introduction to this unit, we showed how to use the trade discount amount formula and the net price formula to calculate the McGraw Hill Education textbook sale to the North Shore Community College Bookstore. Since McGraw Hill Education gave the bookstore only one trade discount, it is a **single trade discount.** In the following word problem, we use the formulas to solve another example of a single trade discount. Again, we will use a blueprint aid to help dissect and solve the word problem.

The Word Problem The list price of a MacBook Pro is $2,700. The manufacturer offers dealers a 40% trade discount. What are the trade discount amount and the net price?

	The facts	Solving for?	Steps to take	Key points
Blueprint	*List price:* $2,700. *Trade discount rate:* 40%.	Trade discount amount. Net price.	Trade discount amount = List price × Trade discount rate. Net price = List price − Trade discount amount.	Trade discount amount — Portion (?) — Base × Rate ($2,700) (.40) — List price / Trade discount rate

Steps to solving problem

1. Calculate the trade discount amount. $2,700 × .40 = $1,080
2. Calculate the net price. $2,700 − $1,080 = $1,620

Now let's learn how to check the dealers' net price of $1,620 with an alternate procedure using a complement.

How to Calculate the Net Price Using Complement of Trade Discount Rate
The **complement** of a trade discount rate is the difference between the discount rate and 100%. The following steps show you how to use the complement of a trade discount rate:

Calculating Net Price Using Complement of Trade Discount Rate
Step 1 To find the complement, subtract the single discount rate from 100%.
Step 2 Multiply the list price times the complement (from Step 1).

Think of a complement of any given percent (decimal) as the result of subtracting the percent from 100%.

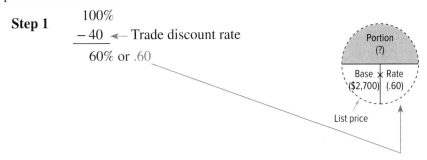

Step 1

100%
− 40 ← Trade discount rate
60% or .60

The complement means that we are spending 60 cents per dollar because we save 40 cents per dollar. Since we planned to spend $2,700, we multiply .60 by $2,700 to get a net price of $1,620.

Step 2 $1,620 = $2,700 × .60

Note how the portion ($1,620) and rate (.60) relate to the same piece of the base ($2,700). The portion ($1,620) is smaller than the base, since the rate is less than 100%.

Be aware that some people prefer to use the trade discount amount formula and the net price formula to find the net price. Other people prefer to use the complement of the trade discount rate to find the net price. The result is always the same.

Finding List Price When You Know Net Price and Trade Discount Rate
The following formula has many useful applications:

Calculating List Price When Net Price and Trade Discount Rate are Known

$$\text{List price} = \frac{\text{Net price}}{\text{Complement of trade discount rate}}$$

Next, let's see how to dissect and solve a word problem calculating list price.

The Word Problem A MacBook Pro has a $1,620 net price and a 40% trade discount. What is its list price?

	The facts	Solving for?	Steps to take	Key points
Blueprint	*Net price:* $1,620. *Trade discount rate:* 40%.	List price.	List price = $\dfrac{\text{Net price}}{\text{Complement of trade discount rate}}$	Net price Portion ($1,620) Base (?) × Rate (.60) List price 100% − 40%

Steps to solving problem

1. Calculate the complement of the trade discount

$$\begin{array}{r} 100\% \\ -\ 40 \\ \hline 60\% = .60 \end{array}$$

2. Calculate the list price.

$$\frac{\$1,620}{.60} = \boxed{\$2,700}$$

Note that the portion ($1,620) and rate (.60) relate to the same piece of the base.

Let's return to the McGraw Hill Education invoice in Figure 7.1 and calculate the list price using the formula for finding list price when the net price and trade discount rate are known. The net price of the textbooks is $8,662.50. The complement of the trade discount rate is 100% − 25% = 75% = .75. Dividing the net price $8,662.50 by the complement .75 equals $11,550, the list price shown in the McGraw Hill Education invoice. We can show this as follows:

$$\frac{\$8,662.50}{.75} = \$11,550,\ \text{the list price}$$

Learn: Chain Discounts

Frequently, manufacturers want greater flexibility in setting trade discounts for different classes of customers, seasonal trends, promotional activities, and so on. To gain this flexibility, some sellers give **chain** or **series discounts**—trade discounts in a series of two or more successive discounts.

Sellers list chain discounts as a group, for example, 20/15/10. Let's look at how Mick Company arrives at the net price of office equipment with a 20/15/10 chain discount.

Example: The list price of the office equipment is $15,000. The chain discount is 20/15/10. The long way to calculate the net price is as follows:

Step 1	Step 2	Step 3	Step 4
$15,000	$15,000	$12,000	$10,200
× .20	−3,000	− 1,800	− 1,020
$ 3,000	$12,000	$10,200	$ 9,180 net price
	× .15	× .10	
	$ 1,800	$ 1,020	

Note how we multiply the percent (in decimal) times the new balance after we subtract the previous trade discount amount. *Never add the 20/15/10 together.* For example, in Step 3, we change the last discount, 10%, to decimal form and multiply times $10,200. Remember that each percent is multiplied by a successively *smaller* base. You could write the 20/15/10 discount rate in any order and still arrive at the same net price. Thus, you would get the $9,180 net price if the discount were 10/15/20 or 15/20/10. However, sellers usually give larger discounts first. *Never try to shorten this step process by adding the discounts.* Your net price will be incorrect because, when done properly, each percent is calculated on a different base.

Aha!

Net Price Equivalent Rate   In the example above, you could also find the $9,180 net price with the **net price equivalent rate**—a shortcut method. Let's see how to use this rate to calculate net price.

Calculating Net Price Using Net Price Equivalent Rate

Step 1 Subtract each chain discount rate from 100% (find the complement) and convert each percent to a decimal.

Step 2 Multiply the decimals. Do not round off decimals, since this number is the net price equivalent rate.

Step 3 Multiply the list price times the net price equivalent rate (Step 2).

The following word problem with its blueprint aid illustrates how to use the net price equivalent rate method.

The Word Problem The list price of office equipment is $15,000. The chain discount is 20/15/10. What is the net price?

	The facts	Solving for?	Steps to take	Key points
Blueprint	*List price:* $15,000. *Chain discount:* 20/15/10.	Net price.	Net price equivalent rate. Net price = List price × Net price equivalent rate.	Do not round net price equivalent rate.

Steps to solving problem

1. Calculate the complement of each rate and convert each percent to a decimal.

$$\begin{array}{ccc} 100\% & 100\% & 100\% \\ -\ 20 & -\ 15 & -\ 10 \\ \hline 80\% & 85\% & 90\% \\ \downarrow & \downarrow & \downarrow \\ .8 & .85 & .9 \end{array}$$

2. Calculate the net price equivalent rate. (Do not round.)

$.8 \times .85 \times .9 = .612$

Net price equivalent rate. For each $1, you are spending about 61 cents.

3. Calculate the net price (actual cost to buyer).

$15,000 × .612 = \boxed{$9,180}$

Next we see how to calculate the trade discount amount with a simpler method. In the previous word problem, we could calculate the trade discount amount as follows:

$15,000 ← List price
− 9,180 ← Net price
$ 5,820 ← Trade discount amount

Single Equivalent Discount Rate You can use another method to find the trade discount by using the **single equivalent discount rate.**

Calculating Trade Discount Amount Using Single Equivalent Discount Rate

Step 1 Subtract the net price equivalent rate from 1. This is the single equivalent discount rate.

Step 2 Multiply the list price times the single equivalent discount rate. This is the trade discount amount.

Money Tip
Double-check invoices. On average 9 out of 10 invoices contain an error.

Let's now do the calculations.

Step 1   1.000 ←If you are using a calculator, just press 1.
        − .612
        ─────
         .388 ←This is the single equivalent discount rate.

Step 2 $15,000 × .388 = $5,820 → This is the trade discount amount.

Remember that when we use the net price equivalent rate, the buyer of the office equipment pays $.612 on each $1 of list price. Now with the single equivalent discount rate, we can say that the buyer saves $.388 on each $1 of list price. The .388 is the single equivalent discount rate for the 20/15/10 chain discount. Note how we use the .388 single equivalent discount rate as if it were the only discount.

 Aha! *Knowing the terminology for what you pay and what you save is an important step in understanding how to calculate net price and trade discount amounts. The pie charts show the terminology relating to each.*

It's time to try the Practice Quiz.

Practice Quiz

Complete this Practice Quiz to see how you are doing.[1]

1. The list price of a dining room set with a 40% trade discount is $12,000. What are the trade discount amount and net price? (Use the complement method for net price.)

2. The net price of a security camera system with a 30% trade discount is $1,400. What is the list price?

3. Lamps Outlet bought a shipment of lamps from a wholesaler. The total list price was $12,000 with a 5/10/25 chain discount. Calculate the net price and trade discount amount. (Use the net price equivalent rate and single equivalent discount rate in your calculation.)

✓ Solutions

1. Dining room set trade discount amount and net price:

The facts	Solving for?	Steps to take	Key points
Blueprint *List price:* $12,000. *Trade discount rate:* 40%.	Trade discount amount. Net price.	Trade discount amount = List price × Trade discount rate. Net price = List price × Complement of trade discount rate.	Trade discount amount — Portion (?) — Base × Rate ($12,000) (.40) — List price — Trade discount rate

[1]For all three problems we will show blueprint aids. You might want to draw them on scrap paper.

Steps to solving problem

1. Calculate the trade discount. $12,000 × .40 = $4,800 Trade discount amount

2. Calculate the net price. $12,000 × .60 = $7,200 (100% − 40% = 60%)

2. Security camera system list price:

The facts	Solving for?	Steps to take	Key points
Blueprint *Net price:* $1,400. *Trade discount rate:* 30%.	List price.	List price = $\dfrac{\text{Net price}}{\text{Complement of trade discount}}$	Net price — Portion ($1,400) — Base × Rate (?) (.70) — List price — 100% −30%

Practice Quiz *Continued*

Steps to solving problem

1. Calculate the complement of trade discount.

$$100\%$$
$$-\ 30$$
$$70\% = .70$$

2. Calculate the list price.

$$\frac{\$1,400}{.70} = \boxed{\$2,000}$$

3. Lamps Outlet's net price and trade discount amount:

	The facts	Solving for?	Steps to take	Key points
Blueprint	*List price:* $12,000. *Chain discount:* 5/10/25.	Net price. Trade discount amount.	Net price = List price × Net price equivalent rate. Trade discount amount = List price × Single equivalent discount rate.	Do not round off net price equivalent rate or single equivalent discount rate.

Steps to solving problem

1. Calculate the complement of each chain discount.

$$\begin{array}{ccc} 100\% & 100\% & 100\% \\ -\ 5 & -10 & -25 \\ \hline 95\% & 90\% & 75\% \end{array}$$
$$\downarrow \qquad \downarrow \qquad \downarrow$$

2. Calculate the net price equivalent rate.

$$.95 \ \times \ .90 \ \times \ .75 = .64125$$

3. Calculate the net price.

$$\$12,000 \times .64125 = \boxed{\$7,695}$$

4. Calculate the single equivalent discount rate.

$$\begin{array}{r} 1.00000 \\ -\ .64125 \\ \hline .35875 \end{array}$$

5. Calculate the trade discount amount.

$$\$12,000 \times .35875 = \boxed{\$4,305}$$

Learning Unit 7–2:

Cash Discounts, Credit Terms, and Partial Payments

To introduce this learning unit, we will use the New Hampshire Propane Company invoice that follows. The invoice shows that if you pay your bill early, you will receive a 19-cent discount. Every penny counts.

New Hampshire Propane Company				
Date	Description	Qty.	Price	Total
	Previous Balance			**$0.00**
06/24/22	PROPANE	3.60	$3.40	$12.24
	Invoice No. 004433L		**Totals this invoice:**	**$12.24**
			AMOUNT DUE:	**$12.24**
	Invoice Date 6/26/22		Prompt Pay Discount: $0.19	
			Net Amount Due if RECEIVED by 07/10/22:	$12.05
		Due Date 7/26/22		

Now let's study cash discounts.

Learn: Cash Discounts

In the New Hampshire Propane Company invoice, we receive a cash discount of 19 cents. This amount is determined by the **terms of the sale,** which can include the credit period, cash discount, discount period, and freight terms.

Buyers can often benefit from buying on credit. The time period that sellers give buyers to pay their invoices is the **credit period.** Frequently, buyers can sell the goods bought during this credit period. Then, at the end of the credit period, buyers can pay sellers with the funds from the sales of the goods. When buyers can do this, they can use the consumer's money to pay the invoice instead of their money.

Sellers can also offer a cash discount, or reduction from the invoice price, if buyers pay the invoice within a specified time. This time period is the **discount period,** which is part of the total credit period. Sellers offer this cash discount because they can use the dollars to better advantage sooner than later. Buyers who are not short of cash like cash discounts because the goods will cost them less and, as a result, provide an opportunity for larger profits.

A cash discount is for prompt payment. A trade discount is not.

Remember that buyers do not take cash discounts on freight, returned goods, sales tax, and trade discounts. Buyers take cash discounts on the *net price* of the invoice. Before we discuss how to calculate cash discounts, let's look at some aids that will help you calculate credit **due dates** and **end of credit periods.**

Trade discounts should be taken before cash discounts.

Aids in Calculating Credit Due Dates   Sellers usually give credit for 30, 60, or 90 days. Not all months of the year have 30 days. So you must count the credit days from the date of the invoice. The trick is to remember the number of days in each month. You can choose one of the following three options to help you do this.

Option 1: Days-in-a-Month Rule You may already know this rule. Remember that every 4 years is a leap year.

> Thirty days has September, April, June, and November; all the rest have 31, except February, which has 28, and 29 in leap years.

Option 2: Knuckle Months Some people like to use the knuckles on their hands to remember which months have 30 or 31 days. Note in the following diagram that each knuckle represents a month with 31 days. The short months are in between the knuckles.

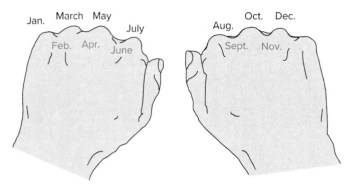

31 days: Jan., March, May, July, Aug., Oct., Dec.

 A financial calculator can calculate maturity date, the number of days between dates, and loan date .

Option 3: Days-in-a-Year Calendar The days-in-a-year calendar (excluding leap year) is another tool to help you calculate dates for discount and credit periods (Table 7.1). For example, let's use Table 7.1 to calculate 90 days from August 12.

Example: By Table 7.1:   August 12 =

$$\begin{array}{r} 224 \text{ days} \\ +\ \ 90 \\ \hline 314 \text{ days} \end{array}$$

Search for day 314 in Table 7.1. You will find that day 314 is November 10. In this example, we stayed within the same year. Now let's try an example in which we overlap from year to year.

 When using the days-in-a-year calendar, always put the number of days in the numerator and 365 (366 in leap years or 360 for ordinary interest) as the denominator.

Example: What date is 80 days after December 5?

Table 7.1 shows that December 5 is 339 days from the beginning of the year. Subtracting 339 from 365 (the end of the year) tells us that we have used up 26 days by the end of the year. This leaves 54 days in the new year. Go back in the table and start with the beginning of the year and search for 54 (80 − 26) days. The 54th day is February 23.

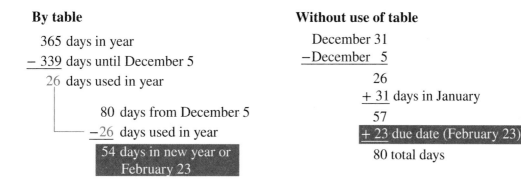

By table	Without use of table
365 days in year	December 31
− 339 days until December 5	−December 5
26 days used in year	26
80 days from December 5	+ 31 days in January
−26 days used in year	57
54 days in new year or February 23	+ 23 due date (February 23)
	80 total days

When you know how to calculate credit due dates, you can understand the common business terms sellers offer buyers involving discounts and credit periods. Remember that discount and credit terms vary from one seller to another.

Learn: Common Credit Terms Offered by Sellers

The common credit terms sellers offer buyers include *ordinary dating, receipt of goods (ROG),* and *end of month (EOM).* In this section we examine these credit terms. To determine the due dates, we use the exact days-in-a-year calendar (Table 7.1).

Ordinary Dating   Today, businesses frequently use the **ordinary dating** method. It gives the buyer a cash discount period that begins with the invoice date. The credit terms of two common ordinary dating methods are 2/10, n/30 and 2/10, 1/15, n/30.

2/10, n/30 Ordinary Dating Method The 2/10, n/30 is read as "two ten, net thirty." Buyers can take a 2% cash discount off the gross amount of the invoice if they pay the bill within 10 days from the invoice date. If buyers miss the discount period, the net amount—without a discount—is due between day 11 and day 30. *Freight, returned goods, sales tax, and trade discounts must be subtracted from the gross before calculating a cash discount.*

Example: $400 invoice dated July 5: terms 2/10, n/30; no freight; paid on July 11.

Step 1 Calculate end of 2% discount period:

July 5   date of invoice
+ 10 days
July 15 end of 2% discount period

Step 2 Calculate end of credit period:

July 5 by Table 7.1
186 days
+ 30
216 days

Search in Table 7.1 for 216 → August 4 → end of credit period

Table 7.1 Exact days-in-a-year calendar (excluding leap year)*

Day of month	31 Jan.	28 Feb.	31 Mar.	30 Apr.	31 May	30 June	31 July	31 Aug.	30 Sept.	31 Oct.	30 Nov.	31 Dec.
1	1	32	60	91	121	152	182	213	244	274	305	335
2	2	33	61	92	122	153	183	214	245	275	306	336
3	3	34	62	93	123	154	184	215	246	276	307	337
4	4	35	63	94	124	155	185	216	247	277	308	338
5	5	36	64	95	125	156	186	217	248	278	309	339
6	6	37	65	96	126	157	187	218	249	279	310	340
7	7	38	66	97	127	158	188	219	250	280	311	341
8	8	39	67	98	128	159	189	220	251	281	312	342
9	9	40	68	99	129	160	190	221	252	282	313	343
10	10	41	69	100	130	161	191	222	253	283	314	344
11	11	42	70	101	131	162	192	223	254	284	315	345
12	12	43	71	102	132	163	193	224	255	285	316	346
13	13	44	72	103	133	164	194	225	256	286	317	347
14	14	45	73	104	134	165	195	226	257	287	318	348
15	15	46	74	105	135	166	196	227	258	288	319	349
16	16	47	75	106	136	167	197	228	259	289	320	350
17	17	48	76	107	137	168	198	229	260	290	321	351
18	18	49	77	108	138	169	199	230	261	291	322	352
19	19	50	78	109	139	170	200	231	262	292	323	353
20	20	51	79	110	140	171	201	232	263	293	324	354
21	21	52	80	111	141	172	202	233	264	294	325	355
22	22	53	81	112	142	173	203	234	265	295	326	356
23	23	54	82	113	143	174	204	235	266	296	327	357
24	24	55	83	114	144	175	205	236	267	297	328	358
25	25	56	84	115	145	176	206	237	268	298	329	359
26	26	57	85	116	146	177	207	238	269	299	330	360
27	27	58	86	117	147	178	208	239	270	300	331	361
28	28	59	87	118	148	179	209	240	271	301	332	362
29	29	—	88	119	149	180	210	241	272	302	333	363
30	30	—	89	120	150	181	211	242	273	303	334	364
31	31	—	90	—	151	—	212	243	—	304	—	365

*Often referred to as a Julian calendar.

Step 3 Calculate payment on July 11:

$$.02 \times \$400 = \$8 \text{ cash discount}$$

$$\$400 - \$8 = \boxed{\$392} \text{ paid}$$

> *Note:* A 2% cash discount means that you save 2 cents on the dollar and pay 98 cents on the dollar. Thus, $.98 \times \$400 = \boxed{\$392}$.

The following time line illustrates the 2/10, n/30 ordinary dating method beginning and ending dates of the above example:

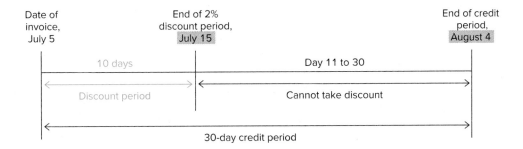

2/10, 1/15, n/30 Ordinary Dating Method The 2/10, 1/15, n/30 is read "two ten, one fifteen, net thirty." The seller will give buyers a 2% (2 cents on the dollar) cash discount if they pay within 10 days of the invoice date. If buyers pay between day 11 and day 15 from the date of the invoice, they can save 1 cent on the dollar. If buyers do not pay on day 15, the net or full amount is due 30 days from the invoice date.

Example: $600 invoice dated May 8; $100 of freight included in invoice price; paid on May 22. Terms 2/10, 1/15, n/30.

Step 1 Calculate the end of the 2% discount period:

May 8 date of invoice
+ 10 days
May 18 end of 2% discount period

Step 2 Calculate end of 1% discount period:

May 18 end of 2% discount period
+ 5   days
May 23 end of 1% discount period

Step 3 Calculate end of credit period:

May 8 by Table 7.1
128 days
+ 30
158 days

Search in Table 7.1 for 158 → June 7 → end of credit period

Step 4 Calculate payment on May 22 (14 days after date of invoice):

$600 invoice
−100 freight
$500
× .01
$5.00
$500 − $5.00 + $100 freight = $595

> A 1% discount means we pay $.99 on the dollar or
> $500 × $.99 = $495 + $100 freight = $595.
>
> *Note:* Freight is added back since no cash discount is taken on freight.

The following time line illustrates the 2/10, 1/15, n/30 ordinary dating method beginning and ending dates of the above example:

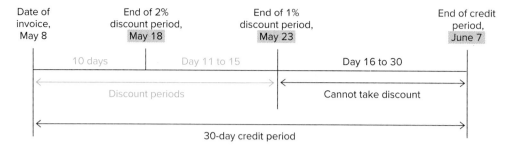

Receipt of Goods (ROG)

3/10, n/30 ROG With the **receipt of goods (ROG),** the cash discount period begins when the buyer receives the goods, *not* the invoice date. Industry often uses the ROG terms when buyers cannot expect delivery until a long time after they place the order. Buyers can take a 3% discount within 10 days *after* receipt of goods. The full amount is due between day 11 and day 30 if the cash discount period is missed.

Example: $900 invoice dated May 9; no freight or returned goods; the goods were received on July 8; terms 3/10, n/30 ROG; payment made on July 20.

Step 1 Calculate the end of the 3% discount period:

$$\begin{array}{l} \text{July 8 date goods arrive} \\ \underline{+\ 10 \text{ days}} \\ \boxed{\text{July 18}}\ \text{end of 3\% discount period} \end{array}$$

Step 2 Calculate the end of the credit period:

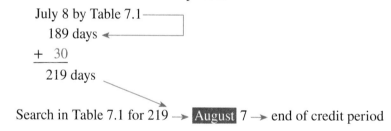

Step 3 Calculate payment on July 20:

Missed discount period and paid net or full amount of $900.

The following time line illustrates 3/10, n/30 ROG beginning and ending dates of the above example:

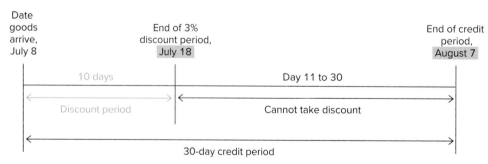

End of Month (EOM)[2]   In this section we look at invoices involving **end of month (EOM)** terms. If an invoice is dated the *25th or earlier* of a month, we follow one set of rules. If an invoice is dated after the 25th of the month, a new set of rules is followed. Let's look at each situation.

Invoice Dated 25th or Earlier in Month, 1/10 EOM If sellers date an invoice on the 25th or earlier in the month, buyers can take the cash discount if they pay the invoice by the first 10 days of the month following the sale (next month). If buyers miss the discount period, the full amount is due within 20 days after the end of the discount period.

Example: $600 invoice dated July 6; no freight or returns; terms 1/10 EOM; paid on August 8.

Step 1 Calculate the end of the 1% discount period:

August 10 ← First 10 days of month following sale

Step 2 Calculate the end of the credit period:

August 10
+ 20 days
August 30 → Credit period is 20 days after discount period.

Step 3 Calculate payment on August 8:

$.99 \times \$600 = \$594$

The following time line illustrates the beginning and ending dates of the EOM invoice of the previous example:

| Date of invoice, July 6 | Next month following sale, August* | End of 1% discount period, August 10 | End of credit period, August 30 |

| | ← 10 days → | ← 20 days → | |
| | Discount period | Cannot take discount | |

*Even though the discount period begins with the next month following the sale, if buyers wish, they can pay before the discount period (date of invoice until the discount period).

Invoice Dated after 25th of Month, 2/10 EOM When sellers sell goods *after* the 25th of the month, buyers gain an additional month. The cash discount period ends on the 10th day of the second month that follows the sale. Why? This occurs because the seller guarantees the 15 days' credit of the buyer. If a buyer bought goods on August 29, September 10 would be only 12 days. So the buyer gets the extra month.

---

[2] Sometimes the Latin term *proximo* is used. Other variations of EOM exist, but the key point is that the seller guarantees the buyer 15 days' credit. We assume a 30-day month.

Example: $800 invoice dated April 29; no freight or returned goods; terms 2/10 EOM; payment made on June 18.

Step 1 Calculate the end of the 2% discount period:

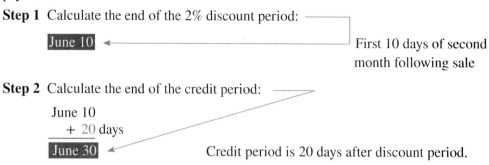

June 10 ← First 10 days of second month following sale

Step 2 Calculate the end of the credit period:

June 10
+ 20 days
June 30 ← Credit period is 20 days after discount period.

Step 3 Calculate the payment on June 18:

No discount; $800 paid.

The following time line illustrates the beginning and ending dates of the EOM invoice of the above example:

*Even though the discount period begins with the second month following the sale, if buyers wish, they can pay before the discount date (date of invoice until the discount period).

Learn: Solving a Word Problem with Trade and Cash Discount

Now that we have studied trade and cash discounts, let's look at a combination that involves both a trade and a cash discount.

The Word Problem Hardy Company sent Regan Corporation an invoice for office equipment with a $10,000 list price. Hardy dated the invoice July 29 with terms of 2/10 EOM (end of month). Regan receives a 30% trade discount and paid the invoice on September 6. Since terms were FOB destination, Regan paid no freight charge. What was the cost of office equipment for Regan?

	The facts	Solving for?	Steps to take	Key points
Blueprint	*List price:* $10,000. *Trade discount rate:* 30%. *Terms:* 2/10 EOM. *Invoice date:* 7/29. *Date paid:* 9/6.	Cost of office equipment.	Net price = List price × Complement of trade discount rate. After 25th of month for EOM. Discount period is first 10 days of second month that follows sale.	Trade discounts are deducted before cash discounts are taken. Cash discounts are not taken on freight or returns.

Steps to solving problem

1. Calculate the net price.

$$\$10,000 \times .70 = \$7,000 \qquad \begin{array}{l} 100\% \\ - \ 30\% \ \text{(trade discount)} \end{array}$$

2. Calculate the discount period.

Sale: 7/29 Month 1: Aug. Month 2: Sept 10 → Paid on Sept. 6—is entitled to 2% off.

3. Calculate the cost of office equipment.

$$\$7,000 \times .98 = \boxed{\$6,860} \qquad \begin{array}{l} 100\% \\ - \ 2\% \end{array}$$

If you save 2 cents on a dollar, you are spending 98 cents.

Learn: Partial Payments

Often buyers cannot pay the entire invoice before the end of the discount period. To calculate partial payments and outstanding balance, use the following steps:

Calculating Partial Payments and Outstanding Balance

Step 1 Calculate the complement of a discount rate.

Step 2 Divide partial payments by the complement of a discount rate (Step 1). This gives the amount credited.

Step 3 Subtract Step 2 from the total owed. This is the outstanding balance.

Example: Molly McGrady owed $400. Molly's terms were 2/10, n/30. Within 10 days, Molly sent a check for $80. The actual credit the buyer gave Molly is as follows:

Step 1 $100\% - 2\% = 98\% \rightarrow .98$

Step 2 $\dfrac{\$80}{.98} = \$81.63 \qquad\qquad \dfrac{\$80}{1 - .02} \ \leftarrow \text{Discount rate}$

Step 3
$$\begin{array}{r} \$400.00 \\ - \ 81.63 \\ \hline \boxed{\$ 318.37} \end{array}$$
    partial payment—although sent in $80

    outstanding balance

Note: We do not multiply .02 × $80 because the seller did not base the original discount on $80. When Molly makes a payment within the 10-day discount period, 98 cents pays each $1 she owes. Before buyers take discounts on partial payments, they must have permission from the seller. Not all states allow partial payments.

You have completed another unit. Let's check your progress.

Practice Quiz

Complete this Practice Quiz to see how you are doing.

Complete the following table:

	Date of invoice	Date goods received	Terms	Last day* of discount period	End of credit period
1.	July 6		2/10, n/30		
2.	February 19	June 9	3/10, n/30 ROG		
3.	May 9		4/10, 1/30, n/60		
4.	May 12		2/10 EOM		
5.	May 29		2/10 EOM		

*If more than one discount, assume date of last discount.

6. Metro Corporation sent Vasko Corporation an invoice for equipment with an $8,000 list price. Metro dated the invoice May 26. Terms were 2/10 EOM. Vasko receives a 20% trade discount and paid the invoice on July 3. What was the cost of equipment for Vasko? (A blueprint aid will be in the solution to help dissect this problem.)

7. Complete amount to be credited and balance outstanding:

   Amount of invoice: $600

   Terms: 2/10, 1/15, n/30

   Date of invoice: September 30

   Paid October 3: $400

✓ Solutions

1. End of discount period: July 6 + 10 days = July 16

   End of credit period: By Table 7.1, July 6 =

   $$\begin{array}{r} 187 \text{ days} \\ +\ 30 \text{ days} \\ \hline 217 \to \text{search} \longrightarrow \text{Aug. 5} \end{array}$$

2. End of discount period: June 9 + 10 days = July 19

   End of credit period: By Table 7.1, June 9 =

   $$\begin{array}{r} 160 \text{ days} \\ +\ 30 \text{ days} \\ \hline 190 \to \text{search} \longrightarrow \text{July 9} \end{array}$$

3. End of discount period: By Table 7.1, May 9 =

   $$\begin{array}{r} 129 \text{ days} \\ +30 \text{ days} \\ \hline 159 \to \text{search} \longrightarrow \text{June 8} \end{array}$$

   End of credit period: By Table 7.1, May 9 =

   $$\begin{array}{r} 129 \text{ days} \\ +60 \text{ days} \\ \hline 189 \to \text{search} \longrightarrow \text{July 8} \end{array}$$

4. End of discount period: June 10

   End of credit period: June 10 + 20 = June 30

Practice Quiz *Continued*

5. End of discount period: July 10

End of credit period: July 10 + 20 = July 30

6. Vasko Corporation's cost of equipment:

	The facts	Solving for?	Steps to take	Key points
Blueprint	*List price:* $8,000. *Trade discount rate:* 20%. *Terms:* 2/10 EOM. *Invoice date:* 5/26. *Date paid:* 7/3.	Cost of equipment.s	Net price = List price × Complement of trade discount rate. *EOM before 25th:* Discount period is first 10 days of month that follows sale.	Trade discounts are deducted before cash discounts are taken. Cash discounts are not taken on freight or returns.

Steps to solving problem

1. Calculate the net price.

$$\$8,000 \times .80 = \$6,400 \qquad \begin{matrix} 100\% \\ -\ 20\% \end{matrix}$$

2. Calculate the discount period.

Until July 10

3. Calculate the cost of office equipment.

$$\$6,400 \times .98 = \$6,272$$

$$\begin{pmatrix} 100\% \\ -\ 2\% \end{pmatrix}$$

7. $\dfrac{\$400}{.98} = \$408.16$, amount credited.

$600 − $408.16 = $191.84, balance outstanding.

Chapter 7 Review

Topic/Procedure/Formula	Example	You try it*
Trade discount amount $\dfrac{\text{Trade discount}}{\text{amount}} = \dfrac{\text{List}}{\text{price}} \times \dfrac{\text{Trade discount}}{\text{rate}}$	$600 list price 30% trade discount rate Trade discount amount = $600 × .30 = $180	**Calculate trade discount amount** $700 list price 20% trade discount
Calculating net price $\text{Net price} = \dfrac{\text{List}}{\text{price}} - \dfrac{\text{Trade discount}}{\text{amount}}$ or $\dfrac{\text{List}}{\text{price}} \times \dfrac{\text{Complement of trade}}{\text{discount price}}$	$600 list price 30% trade discount rate Net price = $600 × .70 = $420 $\begin{array}{r} 1.00 \\ -\ .30 \\ \hline .70 \end{array}$	**Calculate net price** $700 list price 20% trade discount
Freight FOB shipping point—buyer pays freight. FOB destination—seller pays freight.	Moose Company of New York sells equipment to Agee Company of Oregon. Terms of shipping are FOB New York. Agee pays cost of freight since terms are FOB shipping point.	**Calculate freight** If a buyer in Boston buys equipment with shipping terms of FOB destination, who will pay cost of freight?
Calculating list price when net price and trade discount rate are known $\text{List price} = \dfrac{\text{Net price}}{\text{Complement of trade discount rate}}$	40% trade discount rate Net price, $120 $\dfrac{\$120}{.60} = \$200$ list price $(1.00 - .40)$	**Calculate list price** 60% trade discount rate Net price, $240
Chain discounts Successively lower base.	5/10 on a $100 list item $\begin{array}{ll} \$\ \ 100 & \$\ \ 95 \\ \times\ \ .05 & \times\ \ .10 \\ \hline \$\ \ 5.00 & \$9.50 \ \text{(running balance)} \end{array}$ $\begin{array}{r} \$95.00 \\ -\ 9.50 \\ \hline \$85.50 \end{array}$ net price	**Calculate net price** 6/8 on $200 list item
Net price equivalent rate $\dfrac{\text{Actual cost}}{\text{to buyer}} = \dfrac{\text{List}}{\text{price}} \times \dfrac{\text{Net price}}{\text{equivalent rate}}$ Take complement of each chain discount and multiply—do not round. $\dfrac{\text{Trade discount}}{\text{amount}} = \dfrac{\text{List}}{\text{price}} - \dfrac{\text{Actual cost}}{\text{to buyer}}$	Given: 5/10 on $1,000 list price Take complement: .95 × .90 = .855 (net price equivalent) $1,000 × .855 = $855 (actual cost or net price) $\begin{array}{r} \$1,000 \\ -\ 855 \\ \hline \$\ \ 145 \end{array}$ trade discount amount	**Calculate net price equivalent rate, net price, and trade discount amount** 6/8 on $2,000 list

Chapter 7 Review (Continued)

Topic/Procedure/Formula	Example	You try it*
Single equivalent discount rate $$\text{Trade discount} \atop \text{amount} = {\text{List} \atop \text{price}} \times {1 - \text{Net price} \atop \text{equivalent rate}}$$	See preceding example for facts: $1 - .855 = .145$ $.145 \times \$1,000 = \boxed{\$145}$	**From the above You Try It, calculate single equivalent discount**
Cash discounts Cash discounts, due to prompt payment, are not taken on freight, returns, etc.	Gross $1,000 (includes freight) Freight $25 Terms 2/10, n/30 Returns $25 Purchased: Sept. 9; paid Sept. 15 Cash discount = $950 \times .02 = \boxed{\$19}$	**Calculate cash discount** Gross $2,000 (includes freight) Freight $40 Terms 2/10, n/30 Returns $40 Purchased: Sept. 2; paid Sept. 8
Calculating due dates *Option 1:* Thirty days has September, April, June, and November; all the rest have 31 except February has 28, and 29 in leap years. *Option 2:* Knuckles—31-day month; in between knuckles are short months. *Option 3:* Days-in-a-year table.	Invoice $500 on March 5; terms 2/10, n/30 March 5 *End of discount period:* $\underline{+ 10}$ → $\boxed{\text{March 15}}$ *End of credit* March 5 = 64 days *period by* $\underline{+ 30}$ *Table 7.1:* → 94 days Search in Table 7.1 $\boxed{\text{April 4}}$	**Calculate end of discount and end of credit periods** Invoice $600 on April 2; terms 2/10, n/30
Common terms of sale **a. Ordinary dating** Discount period begins from date of invoice. Credit period ends 20 days from the end of the discount period unless otherwise stipulated; example, 2/10, n/60—the credit period ends 50 days from end of discount period.	Invoice $600 (freight of $100 included in price) dated March 8; payment on March 16; 3/10, n/30. March 8 *End of discount* $\underline{+ 10}$ *period:* → $\boxed{\text{March 18}}$ *End of credit* March 8 = 67 days *period by* $\underline{+ 30}$ *Table 7.1:* → 97 days Search in Table 7.1 $\boxed{\text{April 7}}$ *If paid on March 16:* $.97 \times \$500 = \$\ \ 485$ $\underline{+\ 100}$ freight $\boxed{\$\ \ 585}$	**Calculate amount paid** Invoice $700 (freight of $100 included in price) dated May 7; payment on May 15; 2/10, n/30

Chapter 7 Review (Continued)

Topic/Procedure/Formula	Example	You try it*
b. Receipt of goods (ROG) Discount period begins when goods are received. Credit period ends 20 days from end of discount period.	4/10, n/30, ROG. $600 invoice; no freight; dated August 5; goods received October 2, payment made October 20. End of discount period: $\begin{array}{r} \text{October} \quad 2 \\ +\ 10 \\ \hline \boxed{\text{October} \ \ 12} \end{array}$ End of credit period by Table 7.1: $\begin{array}{r} \text{October } 2 = 275 \\ +\ 30 \\ \hline 305 \end{array}$ Search in Table 7.1 $\boxed{\text{November 1}}$ *Payment on October 20:* No discount, pay $\boxed{\$600}$	**Calculate amount paid** 3/10, n/30, ROG. $700 invoice; no freight; dated September 6; goods received September 20; payment made October 15.
c. End of month (EOM) On or before 25th of the month, discount period is 10 days after month following sale. After 25th of the month, an additional month is gained.	$1,000 invoice dated May 12; no freight or returns; terms 2/10 EOM. End of discount period → $\boxed{\text{June 10}}$ End of credit period → $\boxed{\text{June 30}}$	**Calculate end of discount and end of credit periods** $2,000 invoice dated October 11; terms 2/10 EOM
Partial payments $\text{Amount credited} = \dfrac{\text{Partial payment}}{1 - \text{Discount rate}}$	$200 invoice; terms 2/10, n/30; dated March 2; paid $100 on March 5. $\dfrac{\$100}{1-.02} = \dfrac{\$100}{.98} = \boxed{\$102.04}$	**Calculate amount credited** $400 invoice; terms 2/10, n/30; dated May 4; paid $300 on May 7.

Key Terms

Cash discount	FOB shipping point	Series discounts
Chain discounts	Freight terms	Single equivalent discount rate
Complement	Invoice	
Credit period	List price	Single trade discount
Discount period	Net price	Terms of the sale
Due dates	Net price equivalent rate	Trade discount
End of credit period		Trade discount amount
End of month (EOM)	Ordinary dating	Trade discount rate
FOB destination	Receipt of goods (ROG)	

*Worked-out solutions are in Appendix A.

Chapter 7 Review (Continued)

Critical Thinking Discussion Questions with Chapter Concept Check

1. What is the net price? June Long bought a jacket from a catalog company. She took her trade discount off the original price plus freight. What is wrong with June's approach? Who would benefit from June's approach—the buyer or the seller?

2. How do you calculate the list price when the net price and trade discount rate are known? A publisher tells the bookstore its net price of a book along with a suggested trade discount of 20%. The bookstore uses a 25% discount rate. Is this ethical when textbook prices are rising?

3. If Jordan Furniture ships furniture FOB shipping point, what does that mean? Does this mean you get a cash discount?

4. What are the steps to calculate the net price equivalent rate? Why is the net price equivalent rate not rounded?

5. What are the steps to calculate the single equivalent discount rate? Is this rate off the list or net price? Explain why this calculation of a single equivalent discount rate may not always be needed.

6. What is the difference between a discount and credit period? Are all cash discounts taken before trade discounts? Do you agree or disagree? Why?

7. Explain the following credit terms of sale:

   a. 2/10, n/30.

   b. 3/10, n/30 ROG.

   c. 1/10 EOM (on or before 25th of month).

   d. 1/10 EOM (after 25th of month).

8. Explain how to calculate a partial payment. Whom does a partial payment favor—the buyer or the seller?

9. **Chapter Concept Check.** Explain how online shopping during the pandemic affected Amazon. Visit Amazon's website to gather data.

End-of-Chapter Problems

Name _____   Date_____

Check figures for odd-numbered problems in Appendix A.

Drill Problems

For all problems, round your final answer to the nearest cent. Do not round net price equivalent rates or single equivalent discount rates.

Complete the following: *LU 7–1(4)*

Item	List price	Chain discount	Net price equivalent rate (in decimals)	Single equivalent discount rate (in decimals)	Trade discount	Net price
7–1. HP computer	$1,200	4/1				
7–2. LG Blu-Ray player	$199	8/4/3				
7–3. Canon document scanner	$269	7/3/1				

Complete the following: *LU 7–1(4)*

Item	List price	Chain discount	Net price	Trade discount
7–4. Trotter treadmill	$3,000	9/4		
7–5. Maytag dishwasher	$450	8/5/6		

7–6. Sony digital camera $320   3/5/9

7–7. Land Rover roofrack $1,850   12/9/6

e)(cel   **7–8.** Which of the following companies, A or B, gives a higher discount? Use the single equivalent discount rate to make your choice (convert your equivalent rate to the nearest hundredth percent).

Company A	Company B
8/10/15/3	10/6/16/5

Complete the following: *LU 7–2(1)*

Invoice	Date goods are received	Terms	Last day* of discount period	Final day bill is due (end of credit period)
7–9. June 18		1/10, n/30		
7–10. Nov. 27		2/10 EOM		
7–11. May 15	June 5	3/10, n/30, ROG		
7–12. April 10		2/10, 1/30, n/60		
7–13. June 12		3/10 EOM		
7–14. Jan. 10	Feb. 3 (no leap year)	4/10, n/30, ROG		

*If more than one discount, assume date of last discount.

End-of-Chapter Problems (Continued)

Complete the following by calculating the cash discount and net amount paid: *LU 7–2(1)*

	Gross amount of invoice (freight charge already included)	Freight charge	Date of invoice	Terms of invoice	Date of payment	Cash discount	Net amount paid
7–15.	$7,000	$100	4/8	2/10, n/60	4/15		
7–16.	$600	None	8/1	3/10, 2/15, n/30	8/13		
7–17.	$200	None	11/13	1/10 EOM	12/3		
7–18.	$500	$100	11/29	1/10 EOM	1/4		

Complete the following: *LU 7–2(2)*

	Amount of invoice	Terms	Invoice date	Actual partial payment made	Date of partial payment	Amount of payment to be credited	Balance outstanding
7–19.	$700	2/10, n/60	5/6	$400	5/15		
7–20.	$600	4/10, n/60	7/5	$400	7/14		

Word Problems (Round to Nearest Cent as Needed)

7–21. The list price of an iPhone 12 Pro Max is $1,099. Presume Amazon receives a trade discount of 20%. Find the trade discount amount and the net price. *LU 7–1(1)*

excel 7–22. A model NASCAR race car lists for $79.99 with a trade discount of 40%. What is the net price of the car? *LU 7–1(1)*

7–23. Lucky you! You went to couponcabin.com and found a 20% off coupon to your partner's favorite store. Armed with that coupon, you went to the store only to find a storewide sale offering 10% off everything in the store. In addition, your credit card has a special offer that allows you to save 10% if you use your credit card for all purchases that day. Using your credit card, what will you pay before tax for the $155 gift you found? Use the single equivalent discount to calculate how much you save and then calculate your final price. *LU 7–1(4)*

7–24. Levin Furniture buys a living room set with a $4,000 list price and a 55% trade discount. Freight (FOB shipping point) of $50 is not part of the list price. What is the delivered price (including freight) of the living room set, assuming a cash discount of 2/10, n/30, ROG? The invoice had an April 8 date. Levin received the goods on April 19 and paid the invoice on April 25. *LU 7–1(1, 2)*

7–25. DJI manufactures the Mavic Air 2 quadcopter and offered a 5/2/1 chain discount to primary customers. Drone Nerds is a primary customer and ordered 20 Mavic Air 2's for a total $17,000 list price. What was the net price of the quadcopters? What was the trade discount amount? *LU 7–1(4)*

excel 7–26. Home Depot wants to buy a new line of fertilizers. Manufacturer A offers a 21/13 chain discount. Manufacturer B offers a 26/8 chain discount. Both manufacturers have the same list price. What manufacturer should Home Depot buy from? *LU 7–1(4)*

7–27. Maplewood Supply received a $5,250 invoice dated 4/15/20. The $5,250 included $250 freight. Terms were 4/10, 3/30, n/60. **(a)** If Maplewood pays the invoice on April 27, 2020 what will it pay? **(b)** If Maplewood pays the invoice on May 21, what will it pay? *LU 7–2(1)*

excel 7–28. A local Dick's Sporting Goods ordered 50 pairs of tennis shoes from Nike Corporation. The shoes were priced at $85 for each pair with the following terms: 4/10, 2/30, n/60. The invoice was dated October 15. Dick's Sporting Goods sent in a payment on October 28. What should have been the amount of the check? *LU 7–2(1)*

End-of-Chapter Problems (Continued)

e)cel 7–29. Macy of New York sold LeeCo. of Chicago office equipment with a $6,000 list price. Sale terms were 3/10, n/30 FOB New York. Macy agreed to prepay the $30 freight. LeeCo. pays the invoice within the discount period. What does LeeCo. pay Macy? *LU 7–2(2)*

7–30. Royal Furniture bought a sofa for $800. The sofa had a $1,400 list price. What was the trade discount rate Royal received? Round to the nearest hundredth percent. *LU 7–2(1)*

7–31. After researching the cost of a shredder for your online business, you settle on a more expensive shredder because the credit terms are better and cash flow is a challenge for you during the summer months. If you buy a shredder for $1,399 with 3/15, net 30 terms on August 22, how much do you need to pay on September 5? *LU 7–2(1)*

7–32. Bally Manufacturing sent Intel Corp an invoice for machinery with a $14,000 list price. Bally dated the invoice July 23 with 2/10 EOM terms. Intel receives a 40% trade discount. Intel pays the invoice on August 5. What does Intel pay Bally? *LU 7–2(1)*

7–33. On August 1, Intel Corp (Problem 7–32) returns $100 of the machinery due to defects. What does Intel pay Bally on August 5? Round to nearest cent. *LU 7–2(1)*

7–34. Stacy's Dress Shop received a $1,050 invoice dated July 8 with 2/10, 1/15, n/60 terms. On July 22, Stacy's sent a $242 partial payment. What credit should Stacy's receive? What is Stacy's outstanding balance? *LU 7–2(2)*

7–35. On March 11, Jangles Corporation received a $20,000 invoice dated March 8. Cash discount terms were 4/10, n/30. On March 15, Jangles sent an $8,000 partial payment. What credit should Jangles receive? What is Jangles' outstanding balance? *LU 7–2(2)*

7–36. A used Porsche Macan Turbo starts at a consumer price of $72,300. If a dealership can purchase 10 with a 15/10/5 chain discount, what is the net price for the dealership? *LU 7–1(4)*

7–37. A local Barnes and Noble paid a $79.99 net price for each hardbound atlas. The publisher offered a 20% trade discount. What was the publisher's list price?   *LU 7–1(3)*

7–38. Rocky Mountain Chocolate Factory (RMCF) founder and president Frank Crail employs 220 people in 361 outlets in the United States, Canada, United Arab Emirates, Japan, and South Korea. If RMCF purchases 20 kilograms of premium dark chocolate at $16.25 per kilo, what is the net price with a 10/5 chain discount? Round to the nearest cent.   *LU 7–1(1)*

7–39. Vail Ski Shop received a $1,201 invoice dated July 8 with 2/10, 1/15, n/60 terms. On July 22, Vail sent a $485 partial payment. What credit should Vail receive? What is Vail's outstanding balance?   *LU 7–2(2)*

7–40. True Value received an invoice dated 4/15/20. The invoice had a $5,500 balance that included $300 freight. Terms were 4/10, 3/30, n/60. True Value pays the invoice on April 29. What amount does True Value pay?   *LU 7–1(1, 2)*

7–41. Marriott Hotels purchased seven iRobot Roomba i7+ for $850 each. It received a 15% discount for buying more than five and an additional 6% discount for immediate delivery. Terms of payment were 2/10, n/30. Marriott Hotels pays the bill within the cash discount period. How much should the check be written for? Round to the nearest cent.   *LU 7–1(4)*

7–42. On May 14, Talbots of Boston sold Forrest of Los Angeles $7,000 of fine clothes. Terms were 2/10 EOM FOB Boston. Talbots agreed to prepay the $80 freight. If Forrest pays the invoice on June 8, what will Forrest pay? If Forrest pays on June 20, what will Forrest pay?   *LU 7–1(2), LU 7–2(1)*

End-of-Chapter Problems (Continued)

7–43. Sam's Snowboards.com offers 5/4/1 chain discounts to many of its customers. The Ski Hut ordered 20 snowboards with a total list price of $1,200. What is the net price of the snowboards? What was the trade discount amount? Round to the nearest cent. *LU 7–1(4)*

7–44. Majestic Manufacturing sold Jordans Furniture a living room set for an $8,500 list price with 35% trade discount. The $100 freight (FOB shipping point) was not part of the list price. Terms were 3/10, n/30 ROG. The invoice date was May 30. Jordans received the goods on July 18 and paid the invoice on July 20. What was the final price (include cost of freight) of the living room set? *LU 7–1(1, 2), LU 7–2(1)*

7–45. Boeing Truck Company received an invoice showing 8 tires at $110 each, 12 tires at $160 each, and 15 tires at $180 each. Shipping terms are FOB shipping point. Freight is $400; trade discount is 10/5; and a cash discount of 2/10, n/30 is offered. Assuming Boeing paid within the discount period, what did Boeing pay? *LU 7–1(4)*

7–46. Verizon offers to sell cellular phones listing for $99.99 with a chain discount of 15/10/5. Cellular Company offers to sell its cellular phones that list at $102.99 with a chain discount of 25/5. If Irene is to buy six phones, how much could she save if she buys from the lower-priced company? *LU 7–1(4)*

7–47. Living Ornaments is offering a special for wedding planners. Wedding flower orders totaling over $500 receive a 10% discount, over $750 a 15% discount, over $1,000 a 20% discount. All orders $1,500 and above receive a 25% discount. The delivery charge is $75 on weekdays and $125 on weekends. Terms are 2/10 EOM. WeddingsRUs placed an order for Thursday, June 1, delivery. The list price on the chosen flowers totals $848.50. Calculate the trade discount and the net price for the flowers to be delivered. How much does WeddingsRUs owe if it pays the invoice on July 10? (Round to the nearest cent.) *LU 7–2(1)*

Challenge Problems

7–48. The original price of a Honda Shadow to the dealer was $17,995, but the dealer will pay only $16,495 after rebate. If the dealer pays Honda within 15 days, there is a 1% cash discount. **(a)** How much is the rebate? **(b)** What percent is the rebate? Round to nearest hundredth percent. **(c)** What is the amount of the cash discount if the dealer pays within 15 days? **(d)** What is the dealer's final price? **(e)** What is the dealer's total savings? Round answer to the nearest hundredth.   *LU 7–1(1), LU 7–2(1)*

7–49. On March 30, Century Link received an invoice dated March 28 from ACME Manufacturing for 50 televisions at a cost of $125 each. Century received a 10/4/2 chain discount. Shipping terms were FOB shipping point. ACME prepaid the $70 freight. Terms were 2/10 EOM. When Century received the goods, 3 sets were defective. Century returned these sets to ACME. On April 8, Century sent a $150 partial payment. Century will pay the balance on May 6. What is Century's final payment on May 6? Assume no taxes.   *LU 7–1(1, 2, 4), LU 7–2(1)*

Summary Practice Test
(Round to the Nearest Cent as Needed)

Do you need help? Connect videos have step-by-step worked-out solutions.

Complete the following: *LU 7–1(1)*

	Item	List price	Single trade discount	Net price
1.	Apple iPod	$350	5%	
2.	Palm Pilot		10%	$190

Calculate the net price and trade discount (use net price equivalent rate and single equivalent discount rate) for the following: *LU 7–1(4)*

	Item	List price	Chain discount	Net price	Trade discount
3.	Sony HD flat-screen TV	$899	5/4		

4. From the following, what is the last date for each discount period and credit period? *LU 7–1(1)*

	Date of invoice	Terms	End of discount period	End of credit period
a.	Nov. 4	2/10, n/30		
b.	Oct. 3, 2021	3/10, n/30 ROG (Goods received March 10, 2022)		
c.	May 2	2/10 EOM		
d.	Nov. 28	2/10 EOM		

excel **5.** Best Buy buys an iPad from a wholesaler with a $300 list price and a 5% trade discount. What is the trade discount amount? What is the net price of the iPad? *LU 7–1(1)*

6. Jordan's of Boston sold Lee Company of New York computer equipment with a $7,000 list price. Sale terms were 4/10, n/30 FOB Boston. Jordan's agreed to prepay the $400 freight. Lee pays the invoice within the discount period. What does Lee pay Jordan's? *LU 7–1(2), LU 7–2(1)*

7. Julie Ring wants to buy a new line of Tonka trucks for her shop. Manufacturer A offers a 14/8 chain discount. Manufacturer B offers a 15/7 chain discount. Both manufacturers have the same list price. Which manufacturer should Julie buy from? *LU 7–1(4)*

8. Office.com received an $8,000 invoice dated April 10. Terms were 2/10, 1/15, n/60. On April 14, Office.com sent a $1,900 partial payment. What credit should Office.com receive? What is Office.com's outstanding balance? Round to the nearest cent. *LU 7–2(2)*

9. Logan Company received from Furniture.com an invoice dated September 29. Terms were 1/10 EOM. List price on the invoice was $8,000 (freight not included). Logan receives an 8/7 chain discount. Freight charges are Logan's responsibility, but Furniture.com agreed to prepay the $300 freight. Logan pays the invoice on November 7. What does Logan Company pay Furniture.com? *LU 7–1(4)*

 **MY MONEY**

🔍 Buying Into It!

 What I need to know

As consumers we make a multitude of purchases throughout our lives. When making purchases we strive to gain the most benefit for our money spent. This is even more crucial when making large purchases (i.e., new car, house or even furniture). To ensure we are making wise decisions it is important to research our options. Are we getting the best price? Are there incentives available to lower the price in the form of coupons, rebates, etc.? Additionally, other indicators such as product warranties, company reputation, quality level, etc. could play a role in finding the best purchase for our money. Researching our purchase options creates value in the purchase and lessens the chance of experiencing regret from our purchase.

 What I need to do

Know where to look to find information to assist you in making larger purchases. The Internet makes a vast amount of information available. Use a simple Internet search to compare prices for the item you are considering for purchase. Comparisons should be of products possessing similar features and characteristics for you to make an accurate assessment of your options. Be sure to take note of all influences on the final price including tax, service fees, delivery charges, handling fees, etc. By understanding all these "extras" involved, you can make a more informed decision as to which is the least expensive way to go based on the overall price.

Consider any promotional offers available on your purchase. Some sellers may be offering a percentage/dollar off the ticket price for a limited time sale offer. Others may offer free shipping on the product, which could equate to substantial cost savings. Rebates from the manufacturer or retailer can also be valuable. Although rebates normally come in the form of a payment received later, they still provide you with a significant price reduction.

Research the quality and reputation of the company from which you purchase. Online reviews provide insight into the experiences of other consumers to help you determine whether your prospective purchase is a good fit for you. You can learn more about the quality of the item you are purchasing from users who have bought and used the item. This is a way to compare the seller's product description with user experiences to determine if the actual use of the product lives up to its description.

 Steps I need to take

1. Gather information about your potential purchase and compare available products.
2. Look for opportunities to save money on your purchase through promotional offers.
3. Investigate the product manufacturer and learn from the experience of other consumers.

 Resources I can use

- RetailMeNot (mobile app)—find deals and coupons on your purchases.
- ShopSavvy (mobile app)—compare prices before making your purchase.

MY MONEY ACTIVITY

- Conduct an Internet search for a new laptop computer.
- Note the price differences and incentives offered among retailers.
- Select the laptop you would most likely purchase and explain how you arrived at your purchase decision.

PERSONAL FINANCE

A KIPLINGER APPROACH

"AMERICANS ARE STOCKPILING CASH." Kiplinger's. November, 2020.

EPIDEMIC FRUGALITY

AMERICANS ARE STOCKPILING CASH

With no place to go and businesses closed, we are saving more than ever.

AMERICANS HAVE GRADUALLY begun to reopen their wallets, but we're still saving more than we have in decades. The personal savings rate, which measures the amount of money Americans have left over each month after spending and taxes, dropped to 17.8% in July after spiking to 33% in April, but that's still the highest rate since May 1975. Some 45% of Americans say they're saving more than usual, according to a survey by the Associated Press and the NORC Center for Public Affairs Research.

That means a lot of pent-up demand could be unleashed when the economy recovers.

But it could be a long time before Americans return to their previous spending habits, and that has serious implications for an economy that relies heavily on consumer spending to drive growth.

"There's a lot of uncertainty about how the pandemic is going to play out," says Mark Zandi, chief economist for Moody's Analytics. "Will I be working three months from now? Will my pay be cut? People are saving for a potentially stormy day."

The spike in April wasn't surprising because most of the country was in lockdown, bringing the economy to a halt. In addition, many Americans—particularly those who had jobs—saved the $1,200 stimulus checks they received through the CARES Act. Many jobless workers also stashed

the extra $600 a week they received in unemployment benefits through July, says Joao F. Gomes, professor of finance and economics at the University of Pennsylvania's Wharton School.

The unprecedented nature of this downturn makes it difficult to predict when Americans will feel confident enough to start spending again. "There's no real framework for this," says Tom Porcelli, chief economist of RBC Capital Markets. Porcelli doesn't expect spending to accelerate until the country returns to close to full employment, typically defined as an unemployment rate of 5%.

People could also be motivated to open their wallets if an effective vaccine

is introduced and widely distributed, Zandi says. But the rise in spending could come more slowly if the vaccine is rolled out over several months and only helps, say, half of the people who receive it, he says.

And even then, economists say, not everyone will be eager to whip out their credit cards. The Great Recession also led to a rise in the savings rate, and boomers continued to save after the downturn ended.

That could happen this time, too, Zandi says: "I think that group will be saving more, post-pandemic. They're the most vulnerable, and now they are even less prepared for retirement." **SANDRA BLOCK**

Saving Like It's 1975

The pandemic has led to a sharp increase in the amount of money Americans save each month.

Personal Savings Rate
January 1975 through July 2020

30%
25
20
15
10
5
0

1975 1980 1985 1990 1995 2000 2005 2010 2015 2020

SOURCE: U.S. Bureau of Economic Analysis. Personal savings rate retrieved from Federal Reserve Bank of St. Louis.

Business Math Issue

Stockpiling cash means more trade discounts.

1. List the key points of the article and information to support your position.

2. Write a group defense of your position using math calculations to support your view. If you are in an online course, post to a discussion board.

Chapter 8

Markups and Markdowns: Perishables and Breakeven Analysis

Learning Unit Objectives

LU 8–1: Markups1 Based on Cost (100%)

1. Calculate dollar markup and percent markup on cost.
2. Calculate selling price when you know the cost and percent markup on cost.
3. Calculate cost when you know the selling price and percent markup on cost.

LU 8–2: Markups Based on Selling Price (100%)

1. Calculate dollar markup and percent markup on selling price.
2. Calculate selling price when cost and percent markup on selling price are known.
3. Calculate cost when selling price and percent markup on selling price are known.
4. Convert from percent markup on cost to percent markup on selling price and vice versa.

LU 8–3: Markdowns and Perishables

1. Calculate markdowns; compare markdowns and markups.
2. Price perishable items to cover spoilage loss.

LU 8–4: Breakeven Analysis

1. Calculate contribution margin.
2. Calculate breakeven point.

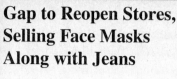

Gap to Reopen Stores, Selling Face Masks Along with Jeans

By Suzanne Kapner

When **Gap** Inc. begins reopening stores this weekend, shoppers will notice a few differences. Fitting rooms and restrooms will be closed, and it wants to turn face masks into a fashion statement.

The company, which owns the Gap, Old Navy and Banana Republic chains among others, plans to reopen roughly 800 North American stores this month, beginning this weekend with a handful of locations in Texas. Along with T-shirts and jeans, it will be selling fabric face masks, like a pack of two for $10 that was recently available on Gap.com.

Kapner, Suzanne. "Gap to Reopen Stores, Selling Face Masks Along with Jeans." *The Wall Street Journal* (May 5, 2020).

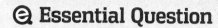

⊝ Essential Question

How can I use markups and markdowns to understand the business world and make my life easier?

🌐 Math Around the World

In the chapter opener *Wall Street Journal* clip, "Gap to Reopen Stores, Selling Face Along With Jeans", Gap is reopening after being shut down by the pandemic with plans to sell face masks as a fashion statement. During the pandemic many retailers have gone to curbside pickup.

We will look at some of Gap's pricing options for its fleece hoody jackets. Before we study the two pricing methods available to Gap (percent markup on cost and percent markup on selling price), we must know the following terms:

- **Selling price.** The price retailers charge consumers. The total selling price of all the goods sold by a retailer (like Gap) represents the retailer's total sales.
- **Cost.** The price retailers pay to a manufacturer or supplier to bring the goods into the store.
- **Markup, margin,** or **gross profit.** These three terms refer to the difference between the cost of bringing the goods into the store and the selling price of the goods.
- **Operating expenses** or **overhead.** The regular expenses of doing business such as wages, rent, utilities, insurance, and advertising.
- **Net profit** or **net income.** The profit remaining after subtracting the cost of bringing the goods into the store and the operating expenses from the sale of the goods (including any returns or adjustments). In Learning Unit 8–4 we will take a closer look at the point at which costs and expenses are covered. This is called the *breakeven* point.

From these definitions, we can conclude that *markup* represents the amount that retailers must add to the cost of the goods to cover their operating expenses and make a profit.[2]

Let's assume Gap plans to sell hooded fleece jackets for $23 that cost $18.[3]

Basic Selling Price Formula

Selling price (S) = Cost (C) + Markup (M)

$23 = $18 + $5
  (price paid to bring fleece jackets into store) (amount in dollars to cover operating expenses and make a profit)

In the Gap example, the markup is a dollar amount, or a **dollar markup.** Markup is also expressed in percent. When expressing markup in percent, retailers can choose a percent based on *cost* (Learning Unit 8–1) or a percent based on *selling price* (Learning Unit 8–2).

[1] Some texts use the term *markon* (selling price minus cost).

[2] In this chapter, we concentrate on the markup of retailers. Manufacturers and suppliers also use markup to determine selling price.

[3] These may not be actual store prices but we assume these prices in our examples.

Learning Unit 8–1:
Markups Based on Cost (100%)

In Chapter 6 you were introduced to the portion formula, which we used to solve percent problems. We also used the portion formula in Chapter 7 to solve problems involving trade and cash discounts. In this unit you will see how we use the basic selling price formula and the portion formula to solve percent markup situations based on cost. We will be using blueprint aids to show how to dissect and solve all word problems in this chapter.

Many manufacturers mark up goods on cost because manufacturers can get cost information more easily than sales information. Since retailers have the choice of using percent markup on cost or selling price, in this unit we assume Gap has chosen percent markup on cost. In Learning Unit 8–2 we show how Gap would determine markup if it decided to use percent markup on selling price.

Aha!

Businesses that use **percent markup on cost** recognize that cost is 100%. This 100% represents the base of the portion formula. All situations in this unit use cost as 100%.

For markups based on cost, the base is always cost (C).

To calculate percent markup on cost, we will use the hooded fleece jacket sold by Gap and begin with the basic selling price formula given in the chapter introduction. When we know the dollar markup, we can use the portion formula to find the percent markup on cost.

Markup expressed in dollars:

Selling price ($23) = Cost ($18) + Markup ($5)

Markup Expressed as a Percent Markup on Cost

Cost	100.00%
+ Markup	+ 27.78
= Selling price	127.78%

Cost is 100%—the base. Dollar markup is the portion, and percent markup on cost is the rate.

In Situation 1 (below) we show why Gap has a 27.78% markup (dollar markup [$5] divided by cost [$18]) based on cost by presenting the hooded fleece jacket as a word problem. We solve the problem with the blueprint aid used in earlier chapters. In the second column, however, you will see footnotes after two numbers. These refer to the steps we use below the blueprint aid to solve the problem. Throughout the chapter, the numbers that we are solving for are in red. Remember that cost is the base for this unit.

Situation 1: Calculating Dollar Markup and Percent Markup on Cost

Dollar markup is calculated with the basic selling price formula $S = C + M$. When you know the cost and selling price of goods, reverse the formula to $M = S - C$. Subtract the cost from the selling price, and you have the dollar markup.

The percent markup on cost is calculated with the portion formula. For Situation 1 the *portion* (P) is the dollar markup, which you know from the selling price formula. In this unit the *rate* (R) is always the percent markup on cost and the *base* (B) is always the cost (100%). To find the percent markup on cost (R), use the portion formula $R = \frac{P}{B}$ and divide the dollar markup (P) by the cost (B). Convert your answer to a percent and round if necessary.

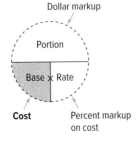

Dollar markup

Portion

Base × Rate

Cost   Percent markup on cost

Now we will look at the Gap example to see how to calculate the 27.78% markup on cost.

The Word Problem The Gap pays $18 for a hooded fleece jacket, which the store plans to sell for $23. What is Gap's dollar markup? What is the percent markup on cost (rounded to the nearest hundredth percent)?

	The facts	Solving for?			Steps to take	Key points
Blueprint	Hooded fleece jacket cost: $18. Hooded fleece jacket selling price: $23.		%	$	$\dfrac{\text{Dollar}}{\text{markup}} = \dfrac{\text{Selling}}{\text{price}} - \text{Cost}$	Dollar markup
		C	100.00%	$18		Portion ($5)
		$+M$	27.78[2]	5[1]	$\text{Percent markup on cost} = \dfrac{\text{Dollar markup}}{\text{Cost}}$	Base × Rate ($18) (?)
		$= S$	127.78%	$23		Cost

[1]Dollar markup. See Step 1, below.

[2]Percent markup on cost. See Step 2, below.

Steps to solving problem

1. Calculate the dollar markup.

$$\text{Dollar markup} = \text{Selling price} - \text{Cost}$$
$$\boxed{\$5} = \$23 - \$18$$

2. Calculate the percent markup on cost.

$$\text{Percent markup on cost} = \dfrac{\text{Dollar markup}}{\text{Cost}}$$

$$= \dfrac{\$5}{\$18} = \boxed{27.78\%}$$

To check the percent markup on cost, you can use the basic selling price formula $S = C + M$. Convert the percent markup on cost found with the portion formula to a decimal and multiply it by the cost. This gives the dollar markup. Then add the cost and the dollar markup to get the selling price of the goods.

You could also check the cost (B) by dividing the dollar markup (P) by the percent markup on cost (R).

Check

$\boxed{\text{Selling price} = \text{Cost} + \text{Markup}}$	or	$\boxed{\text{Cost }(B) = \dfrac{\text{Dollar markup }(P)}{\text{Percent markup on cost }(R)}}$

$$\$23 = \$18 + .2778(\$18)$$
$$\$23 = \$18 + \$5$$
$$\$23 = \$23$$

$$= \dfrac{\$5}{.2778} = \$18$$

Parentheses mean that you multiply the percent markup on cost in decimal by the cost.

Situation 2: Calculating Selling Price When You Know Cost and Percent Markup on Cost

When you know the cost and the percent markup on cost, you calculate the selling price with the basic selling formula $S = C + M$. Remember that when goods are marked up on cost, the cost is the base (100%). So you can say that the selling price is the cost plus the markup in dollars (percent markup on cost times cost).

Now let's look at Mel's Furniture where we calculate Mel's dollar markup and selling price.

The Word Problem Mel's Furniture bought a lamp that cost $100. To make Mel's desired profit, he needs a 65% markup on cost. What is Mel's dollar markup? What is his selling price?

	The facts	Solving for?		Steps to take	Key points
Blueprint	*Lamp cost:* $100. *Markup on cost:* 65%.	C $+M$ $=S$ % 100% 65 165% $ $100 65¹ $165²		Dollar markup: $S = C + M$ or $S = \text{Cost} \times \left(1 + \begin{array}{c}\text{Percent}\\ \text{markup}\\ \text{on cost}\end{array}\right)$	Selling price Portion (?) Base × Rate ($100) (1.65) Cost 100% +65%

¹ Dollar markup. See Step 1, below.

² Selling price. See Step 2, below.

Steps to solving problem

1. Calculate the dollar markup. $S = C + M$

      $S = \$100 + .65(\$100)$ ← Parentheses mean you multiply the percent markup in decimal by the cost

      $S = \$100 + \$65$ ← Dollar markup

2. Calculate the selling price. $S = \$165$

You can check the selling price with the formula $P = B \times R$. You are solving for the portion (P)—the selling price. Rate (R) represents the 100% cost plus the 65% markup on cost. Since in this unit the markup is on cost, the base is the cost. Convert 165% to a decimal and multiply the cost by 1.65 to get the selling price of $165.

Check

$$\begin{array}{ccc} \text{Selling price} = & \text{Cost} \times & (1 + \text{Percent markup on cost}) \\ (P) & (B) & (R) \end{array} = \$100 \times 1.65 = \$165$$

Situation 3: Calculating Cost When You Know Selling Price and Percent Markup on Cost

When you know the selling price and the percent markup on cost, you calculate the cost with the basic selling formula $S = C + M$. Since goods are marked up on cost, the percent markup on cost is added to the cost.

Let's see how this is done in the following Jill Sport example.

The Word Problem Jill Sport, owner of Sports, Inc., sells tennis rackets for $50. To make her desired profit, Jill needs a 40% markup on cost. What do the tennis rackets cost Jill? What is the dollar markup?

The facts	Solving for?			Steps to take	Key points
Selling price: $50. *Markup on cost:* 40%.		%	$	$S = C + M$ or $$\text{Cost} = \frac{\text{Selling price}}{\text{Percent}}$$ $$1 + \text{markup on cost}$$ $M = S - C$	
	C	100%	$ 35.71[1]		
	$+M$	40	14.29[2]		
	$= S$	140%	$ 50.00		

[1] Cost. See Step 1, below.

[2] Dollar markup. See Step 2, below.

Steps to solving problem

1. Calculate the cost.

$$S = C + M$$
$$\$50.00 = C + .40C \quad \leftarrow$$
$$\frac{\$50.00}{1.40} = \frac{1.40C}{1.40}$$
$$\boxed{\$35.71} = C$$

This means 40% times cost. C is the same as $1C$. Adding $.40C$ to $1C$ gives the percent markup on cost of $1.40C$ in decimal.

2. Calculate the dollar markup.

$$M = S - C$$
$$M = \$50.00 - \$35.71$$
$$M = \boxed{\$14.29}$$

You can check your cost answer with the portion formula $B = \frac{P}{R}$. Portion (P) is the selling price. Rate (R) represents the 100% cost plus the 40% markup on cost. Convert the percents to decimals and divide the portion by the rate to find the base, or cost.

Check

$$\text{Cost } (B) = \frac{\text{Selling price } (P)}{1 + \text{Percent markup on cost } (R)} = \frac{\$50.00}{1.40} = \boxed{\$35.71}$$

Now try the following Practice Quiz to check your understanding of this unit.

Money Tip
Automate your savings. Save more money from each paycheck starting NOW. If you have increments of 1% automatically taken out of your check, you will never miss it. Over time, it can add up to six figures and will help you start building a financially healthy retirement TODAY.

Practice Quiz

Complete this Practice Quiz to see how you are doing.

Solve the following situations (markups based on cost):

1. Irene Westing bought a desk for $400 from an office supply house. She plans to sell the desk for $600. What is Irene's dollar markup? What is her percent markup on cost? Check your answer.

2. Suki Komar bought dolls for her toy store that cost $12 each. To make her desired profit, Suki must mark up each doll 35% on cost. What is the dollar markup? What is the selling price of each doll? Check your answer.

3. Jay Lyman sells calculators. His competitor sells a new calculator line for $14 each. Jay needs a 40% markup on cost to make his desired profit, and he must meet price competition. At what cost can Jay afford to bring these calculators into the store? What is the dollar markup? Check your answer.

✓ Solutions

1. Irene's dollar markup and percent markup on cost:

	The facts	Solving for?		Steps to take	Key points
Blueprint	Desk cost: $400. Desk selling price: $600.	$\begin{array}{l} C \\ +M \\ = S \end{array}$ $\begin{array}{cc} \% & \$ \\ 100\% & \$100 \\ 50^2 & 200^1 \\ 150\% & \$600 \end{array}$		$\dfrac{\text{Dollar}}{\text{markup}} = \dfrac{\text{Selling}}{\text{price}} - \text{Cost}$ $\dfrac{\text{Percent}}{\text{markup}}_{\text{on cost}} = \dfrac{\text{Dollar markup}}{\text{Cost}}$	Dollar markup Portion ($200) Base × Rate ($400) (?) Cost

¹ Dollar markup. See Step 1, below.

² Percent markup on cost. See Step 2, below.

Steps to solving problem

1. Calculate the dollar markup.

$$\text{Dollar markup} = \text{Selling price} - \text{Cost}$$
$$\boxed{\$200} = \$600 - \$400$$

2. Calculate the percent markup on cost.

$$\text{Percent markup on cost} = \frac{\text{Dollar markup}}{\text{Cost}}$$
$$= \frac{\$200}{\$400} = \boxed{50\%}$$

Check

$$\begin{aligned} \text{Selling price} &= \text{Cost} + \text{Markup} \\ \$600 &= \$400 + .50(\$400) \\ \$600 &= \$400 + \$200 \\ \$600 &= \$600 \end{aligned}$$

or

$$\begin{aligned} \text{Cost (B)} &= \frac{\text{Dollar markup (P)}}{\text{Percent markup on cost (R)}} \\ &= \frac{\$200}{.50} = \$400 \end{aligned}$$

Practice Quiz *Continued*

2. Dollar markup and selling price of doll:

<table>
<tr><th>Blueprint</th><th>The facts</th><th colspan="3">Solving for?</th><th>Steps to take</th><th>Key points</th></tr>
<tr>
<td rowspan="2">Blueprint</td>
<td rowspan="2">Doll cost:
$12 each.

Markup
on cost:
35%.</td>
<td colspan="3">

	%	$
C	100%	$12.00
+M	35	4.20¹
= S	135%	$16.20²

</td>
<td>Dollar markup:
$S = C + M$
or

$S = \text{Cost} \times \left(1 + \dfrac{\text{Percent}}{\text{markup}}\right)$
on cost</td>
<td>

Selling price
Portion (?)
Base × Rate
($12) (1.35)
Cost   100% +35%

</td>
</tr>
</table>

¹ Dollar markup. See Step 1, below.

² Selling price. See Step 2, below.

Steps to solving problem

1. Calculate the dollar markup.

$$S = C + M$$
$$S = \$12.00 + .35(\$12.00)$$
$$S = \$12.00 + \boxed{\$4.20} \leftarrow \text{Dollar markup}$$

2. Calculate the selling price. $S = \boxed{\$16.20}$

Check

Selling price = Cost × (1 + Percent markup on cost) = $12.00 × 1.35 = $\boxed{\$16.20}$
   (P)   (B)   (R)

3. Cost and dollar markup:

<table>
<tr><th>Blueprint</th><th>The facts</th><th colspan="3">Solving for?</th><th>Steps to take</th><th>Key points</th></tr>
<tr>
<td rowspan="2">Blueprint</td>
<td rowspan="2">Selling
price: $14.

Markup on
cost: 40%.</td>
<td colspan="3">

	%	$
C	100%	$10¹
+M	40	4²
= S	140%	$14

</td>
<td>$S = C + M$
or
$\text{Cost} = \dfrac{\text{Selling price}}{\text{Percent}}$
$1 + \text{markup}$
on cost

$M = S - C$</td>
<td>

Selling price
Portion ($14)
Base × Rate
(?) (1.40)
Cost   100% +40%

</td>
</tr>
</table>

¹ Cost. See Step 1, below.

² Dollar markup. See Step 2, below.

Steps to solving problem

1. Calculate the cost.

$$S = C + M$$
$$\$14 = C + .40C$$
$$\frac{\$14}{1.40} = \frac{\cancel{1.40}C}{\cancel{1.40}}$$
$$\boxed{\$10} = C$$

2. Calculate the dollar markup.

$$M = S - C$$
$$M = \$14 - \$10$$
$$M = \boxed{\$4}$$

Check

$$\text{Cost } (B) = \frac{\text{Selling price } (P)}{1 + \text{Percent markup on cost } (R)} = \frac{\$14}{1.40} = \$10$$

Learning Unit 8–2:
Markups Based on Selling Price (100%)

Many retailers mark up their goods on the selling price since sales information is easier to get than cost information. These retailers use retail prices in their inventory and report their expenses as a percent of sales.

Aha!

For markups based on selling price, the base is always selling price (S).

Businesses that mark up their goods on selling price recognize that selling price is 100%. We begin this unit by assuming Gap has decided to use percent markup based on selling price. We repeat Gap's selling price formula expressed in dollars.

Markup expressed in dollars:

Selling price ($23) = Cost ($18) + Markup ($5)

Markup Expressed as Percent Markup on Selling Price

Cost	78.26%
+ Markup	+ 21.74
= Selling price	100.00%

> Selling price is 100%—the base. Dollar markup is the portion, and percent markup on selling price is the rate.

In Situation 1 (below) we show why Gap has a **21.74%** markup based on selling price. In the last unit, markups were based on *cost*. In this unit, markups are based on *selling price*.

Situation 1: Calculating Dollar Markup and Percent Markup on Selling Price

The dollar markup is calculated with the selling price formula used in Situation 1, Learning Unit 8–1: $M = S - C$. To find the percent markup on selling price, use the portion formula $R = \frac{P}{B}$, where rate (the percent markup on selling price) is found by dividing the portion (dollar markup) by the base (selling price). Note that when solving for percent markup on cost in Situation 1, Learning Unit 8–1, you divided the dollar markup by the cost.

The Word Problem The cost to Gap for a hooded fleece jacket is $18; the store then plans to sell the jacket for $23. What is Gap's dollar markup? What is its percent markup on selling price? (Round to the nearest hundredth percent.)

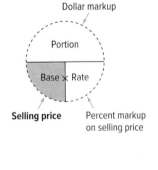

Dollar markup

Portion

Base × Rate

Selling price   Percent markup on selling price

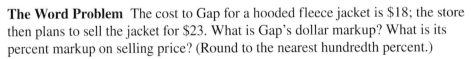

The facts	**Solving for?**		**Steps to take**	**Key points**
Hooded fleece jacket cost: $18. *Hooded fleece jacket price: $23.*	% $ C 78.26% $18 $+M$ 21.74² 5¹ $=S$ 100.00% $23		$\dfrac{\text{Dollar}}{\text{markup}} = \dfrac{\text{Selling}}{\text{price}} - \text{Cost}$ $\dfrac{\text{Percent}}{\text{markup on}} = \dfrac{\text{Dollar}}{\text{markup}}$ $\dfrac{\text{selling price}}{} \quad \dfrac{}{\text{Selling price}}$	Dollar markup Portion ($5) Base × Rate ($23) (?) Selling price

¹ Dollar markup. See Step 1, below.

² Percent markup on selling price. See Step 2, below.

Steps to solving problem

1. Calculate the dollar markup.

Dollar markup = Selling price − Cost

$$\boxed{\$5} = \$23 - \$18$$

2. Calculate the percent markup on selling price.

$$\frac{\text{Percent markup}}{\text{on selling price}} = \frac{\text{Dollar markup}}{\text{Selling price}}$$

$$= \frac{\$5}{\$23} = \boxed{21.74\%}$$

You can check the percent markup on selling price with the basic selling price formula $S = C + M$. You can also use the portion formula by dividing the dollar markup (P) by the percent markup on selling price (R).

Check

Selling price = Cost + Markup	or	Selling price $(B) = \dfrac{\text{Dollar markup } (P)}{\text{Percent markup on selling price } (R)}$

$\$23 \quad = \$18 + .2174(\$23)$ ←

$\$23 \quad = \$18 \ + \ \$5$

$\$23 \quad = \$23$

$$= \frac{\$5}{.2174} = \$23$$

Parentheses mean you multiply the percent markup on selling price in decimal by the selling price.

Situation 2: Calculating Selling Price When You Know Cost and Percent Markup on Selling Price

When you know the cost and percent markup on selling price, you calculate the selling price with the basic selling formula $S = C + M$. Remember that when goods are marked up on selling price, the selling price is the base (100%). Since you do not know the selling price, the percent markup is based on the unknown selling price. To find the dollar markup after you find the selling price, use the selling price formula $M = S - C$.

The Word Problem Mel's Furniture bought a lamp that cost $100. To make Mel's desired profit, he needs a 65% markup on selling price. What are Mel's selling price and his dollar markup?

	The facts	Solving for?		Steps to take	Key points
Blueprint	*Lamp cost:* $100. *Markup on selling price:* 65%.	C $+M$ $= S$ — % 35% 65 100% — $ $100.00 185.71[2] $285.71[1]		$S = C + M$ or $S = \dfrac{\text{Cost}}{1 - \dfrac{\text{Percent markup}}{\text{on selling price}}}$	Cost Portion ($100) Base × Rate (?) (.35) Selling price — 100% −65%

[1] Selling price. See Step 1, below.

[2] Dollar markup. See Step 2, below.

Steps to solving problem

1. Calculate the selling price.

$$S = C + M$$
$$S = \$100.00 + .65S$$

$$\begin{matrix} 1.00S \\ -.65S \\ \hline = .35S \end{matrix} \Bigg\} \longrightarrow \begin{matrix} -.65S && -.65S \\ \hline \dfrac{.35S}{.35} = \dfrac{\$100.00}{.35} \end{matrix}$$

$$S = \boxed{\$285.71}$$

Do not multiply the .65 times $100.00. The 65% is based on selling price not cost.

2. Calculate the dollar markup.

$$M = S - C$$
$$\boxed{\$185.71} = \$285.71 - \$100.00$$

You can check your selling price with the portion formula $B = \frac{P}{R}$. To find the selling price (B), divide the cost (P) by the rate (100% − Percent markup on selling price).

Check

$$\boxed{\text{Selling price } (B) = \dfrac{\text{Cost } (P)}{1 - \text{Percent markup on selling price } (R)}}$$

$$= \dfrac{\$100.00}{1 - .65} = \dfrac{\$100.00}{.35} = \boxed{\$285.71}$$

Situation 3: Calculating Cost When You Know Selling Price and Percent Markup on Selling Price

When you know the selling price and the percent markup on selling price, you calculate the cost with the basic formula $S = C + M$. To find the dollar markup, multiply the markup percent by the selling price. When you have the dollar markup, subtract it from the selling price to get the cost.

The Word Problem Jill Sport, owner of Sports, Inc., sells tennis rackets for $50. To make her desired profit, Jill needs a 40% markup on the selling price. What is the dollar markup? What do the tennis rackets cost Jill?

	The facts	Solving for?			Steps to take	Key points
Blueprint	*Selling price: $50.* *Markup on selling price: 40%.*	$\begin{matrix} C \\ +M \\ \hline = S \end{matrix}$	$\begin{matrix} \% \\ \boxed{60\%} \\ 40 \\ \hline 100\% \end{matrix}$	$\begin{matrix} \$ \\ \boxed{\$30}^2 \\ \underline{20}^1 \\ \$50 \end{matrix}$	$S = C + M$ or $\text{Cost} = \text{Selling price} \times \left(1 - \dfrac{\text{Percent markup}}{\text{on selling price}}\right)$	

[1] Dollar markup. See Step 1.
[2] Cost. See Step 2.

Steps to solving problem

1. Calculate the dollar markup.

$$\begin{matrix} S &=& C &+& M \\ \$50 &=& C &+& .40\,(\$50) \\ \$50 &=& C &+& \boxed{\$20} \quad \leftarrow \text{Dollar markup} \\ \underline{-20} &&&& \underline{-\ 20} \end{matrix}$$

2. Calculate the cost.

$$\boxed{\$30} = C$$

Money Tip
When analyzing a job offer, make sure you include the value of the benefits. Salary alone will not let you know the true value of the offer.

Copyright © McGraw Hill

To check your cost, use the portion formula Cost (P) = Selling price $(B) \times (100\%$ selling price − Percent markup on selling price) (R).

Check

$$\boxed{\begin{array}{l}\text{Cost} \\ (P)\end{array} = \begin{array}{c}\text{Selling} \\ \text{price} \\ (B)\end{array} \times \left(\begin{array}{c}\text{Percent markup} \\ 1 - \text{ on selling price} \\ (R)\end{array}\right)} = \$50 \times .60 = \boxed{\$30}$$

$$(1.00 - .40)$$

In Table 8.1, we compare percent markup on cost with percent markup on retail (selling price). This table is a summary of the answers we calculated from the word problems in Learning Units 8–1 and 8–2. The word problems in the units were the same except in Learning Unit 8–1, we assumed markups were on cost, while in Learning Unit 8–2, markups were on selling price. Note that in Situation 1, the dollar markup is the same $5, but the percent markup is different.

Let's now look at how to convert from percent markup on cost to percent markup on selling price and vice versa. We will use Situation 1 from Table 8.1.

Formula for Converting Percent Markup on Cost to Percent Markup on Selling Price

To convert percent markup on cost to percent markup on selling price:

$$\boxed{\dfrac{\text{Percent markup on cost}}{1 + \text{Percent markup on cost}}}$$

$$\dfrac{.2778}{1 + .2778} = \boxed{21.74\%}$$

TABLE 8.1
Comparison of
markup on cost
versus markup on
selling price

Markup based on cost—Learning Unit 8–1	Markup based on selling price—Learning Unit 8–2
Situation 1: Calculating dollar amount of markup and percent markup on cost. Hooded fleece jacket cost, $18. Hooded fleece jacket selling price, $23. $M = S - C$ $M = \$23 - \$18 = $ $5 markup $M \div C = \$5 \div \$18 = 27.78\%$	*Situation 1: Calculating dollar amount of markup and percent markup on selling price.* Hooded fleece jacket cost, $18. Hooded fleece jacket selling price, $23. $M = S - C$ $M = \$23 - \$18 = $ $5 markup $M \div S = \$5 \div \$23 = 21.74\%$
Situation 2: Calculating selling price on cost. Lamp cost, $100. 65% markup on cost $S = C \times (1 + \text{Percent markup on cost})$ $S = \$100 \times 1.65 = $ $165 $(100\% + 65\% = 165\% = 1.65)$	*Situation 2: Calculating selling price on selling price.* Lamp cost, $100. 65% markup on selling price $S = C \div (1 - \text{Percent markup on selling price})$ $S = \$100.00 \div .35$ $(100\% - 65\% = 35\% = .35)$ $S = $ $285.71
Situation 3: Calculating cost on cost. Tennis racket selling price, $50. 40% markup on cost $C = S \div (1 + \text{Percent markup on cost})$ $C = \$50.00 \div 1.40$ $(100\% + 40\% = 140\% = 1.40)$ $C = $ $35.71	*Situation 3: Calculating cost on selling price.* Tennis racket selling price, $50. 40% markup on selling price $C = S \times (1 - \text{Percent markup on selling price})$ $C = \$50 \times .60 = $ $30 $(100\% - 40\% = 60\% = .60)$

Formula for Converting Percent Markup on Selling Price to Percent Markup on Cost

To convert percent
markup on selling price
to percent markup on cost:

$$\frac{\text{Percent markup on selling price}}{1 - \text{Percent markup on selling price}}$$

$$\frac{.2174}{1 - .2174} = 27.78\%$$

Key point: A 21.74% markup on selling price or a 27.78% markup on cost results in the same dollar markup of $5.

Now let's test your knowledge of Learning Unit 8–2.

Practice Quiz

Complete this Practice Quiz to see how you are doing.

Solve the following situations (markups based on selling price). Note numbers 1, 2, and 3 are parallel problems to those in Practice Quiz LU 8–1.

1. Irene Westing bought a desk for $400 from an office supply house. She plans to sell the desk for $600. What is Irene's dollar markup? What is her percent markup on selling price (rounded to the nearest tenth percent)? Check your answer. Selling price will be slightly off due to rounding.

2. Suki Komar bought dolls for her toy store that cost $12 each. To make her desired profit, Suki must mark up each doll 35% on the selling price. What is the selling price of each doll? What is the dollar markup? Check your answer.

3. Jay Lyman sells calculators. His competitor sells a new calculator line for $14 each. Jay needs a 40% markup on the selling price to make his desired profit, and he must meet price competition. What is Jay's dollar markup? At what cost can Jay afford to bring these calculators into the store? Check your answer.

4. Dan Flow sells wrenches for $10 that cost $6. What is Dan's percent markup on cost? Round to the nearest tenth percent. What is Dan's percent markup on selling price? Check your answer.

✓ Solutions

1. Irene's dollar markup and percent markup on selling price:

	The facts	Solving for?		Steps to take	Key points
Blueprint	*Desk cost:* $400. *Desk selling price:* $600.		%	$	$\dfrac{\text{Dollar}}{\text{markup}} = \dfrac{\text{Selling}}{\text{price}} - \text{Cost}$ $\dfrac{\text{Percent}}{\text{markup on}} = \dfrac{\text{Dollar}}{\text{markup}}$ selling price $\quad$ Selling price	Markup Portion ($200) Base × Rate ($600) (?) Selling price
		C $+M$ $=S$	66.7% 33.3² 100%	$400 200¹ $600		

¹Dollar markup. See Step 1, below.

²Percent markup on selling price. See Step 2, below.

Steps to solving problem

1. Calculate the dollar markup.

$$\text{Dollar markup} = \text{Selling price} - \text{Cost}$$
$$\boxed{\$200} = \$600 - \$400$$

2. Calculate the percent markup on selling price.

$$\dfrac{\text{Percent markup}}{\text{on selling price}} = \dfrac{\text{Dollar markup}}{\text{Selling price}}$$
$$= \dfrac{\$200}{\$600} = \boxed{33.3\%}$$

Practice Quiz *Continued*

Check

$$\frac{\text{Selling}}{\text{price}} = \text{Cost} + \text{Markup} \quad \textbf{or} \quad \frac{\text{Selling}}{\text{price } (B)} = \frac{\text{Dollar markup } (P)}{\text{Percent markup on selling price } (R)}$$

$$\$600 = \$400 + .333(\$600)$$

$$\$600 = \$400 + \$199.80 \qquad\qquad = \frac{\$200}{.333} = \$600.60$$

$$\$600 = \$599.80 \text{ (off due to rounding)} \qquad \text{(not exactly \$600 due to rounding)}$$

2. Selling price of doll and dollar markup:

The facts	Solving for?			Steps to take	Key points
Blueprint *Doll cost:* $12 each. *Markup on selling price:* 35%.	C $+M$ $=S$	65% 35 100%	$12.00 6.46² $18.46¹	$S = C + M$ or $S = \dfrac{\text{Cost}}{1 - \dfrac{\text{Percent markup}}{\text{on selling price}}}$	Cost / Portion ($12) / Base × Rate (?) (.65) / Selling price 100% −35%

¹Selling price. See Step 1, below.

²Dollar markup. See Step 2, below.

Steps to solving problem

1. Calculate the selling price.

$$\begin{aligned}
S &= C &&+ M \\
S &= \$12.00 &&+ .35S \\
-.35S & &&- .35S \\[4pt]
\frac{.65S}{.65} &= \frac{\$12.00}{.65} \\[4pt]
S &= \boxed{\$18.46}
\end{aligned}$$

2. Calculate the dollar markup.

$$\begin{aligned}
M &= S &&- C \\
\boxed{\$6.46} &= \$18.46 &&- \$12.00
\end{aligned}$$

Check

$$\text{Selling price } (B) = \frac{\text{Cost } (P)}{1 - \text{Percent markup on selling price } (R)} = \frac{\$12.00}{.65} = \boxed{\$18.46}$$

3. Dollar markup and cost:

The facts	Solving for?			Steps to take	Key points
Blueprint *Selling price:* $14. *Markup on selling price:* 40%.	C $+M$ $=S$	60% 40 100%	$ 8.40² 5.60¹ $14.00	$S = C + M$ or $\text{Cost} = \text{Selling price} \times$ $\left(1 - \dfrac{\text{Percent markup}}{\text{on selling price}}\right)$	Cost / Portion (?) / Base × Rate ($14) (.60) / Selling price 100% −40%

¹Dollar markup. See Step 1, below.

²Cost. See Step 2, below.

Practice Quiz *Continued*

Steps to solving problem

1. Calculate the dollar markup.

$$S = C + M$$
$$\$14.00 = C + .40(\$14.00)$$
$$\$14.00 = C + \boxed{\$5.60} \leftarrow \text{Dollar markup}$$

2. Calculate the cost.

$$\begin{array}{cc} -5.60 & -5.60 \\ \hline \boxed{\$8.40} = C \end{array}$$

Check

$$\underset{(P)}{\text{Cost}} = \underset{(B)}{\text{Selling price}} \times \underset{(R)}{(1 - \text{Percent markup on selling price})} = \$14.00 \times .60 = \boxed{\$8.40}$$

$$(1.00 - .40)$$

4. $\text{Cost} = \dfrac{\$4}{\$6} = \boxed{66.7\%}$ $\dfrac{.40}{1-.40} = \dfrac{.40}{.60} = \dfrac{2}{3} = 66.7\%$

   $\text{Selling price} = \dfrac{\$4}{\$10} = \boxed{40\%}$ $\dfrac{.667}{1+.667} = \dfrac{.667}{1.667} = 40\% \text{ (due to rounding)}$

Learning Unit 8–3:
Markdowns and Perishables

The following *Wall Street Journal* clip shows that Lululemon has increased profit margins with fewer markdowns.

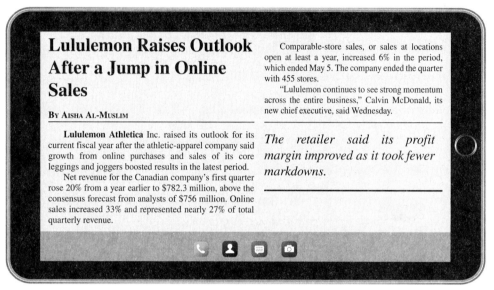

Al-Muslim, Aisha. "Lululemon Raises Outlook after a Jump in Online Sales." *The Wall Street Journal* (June 13, 2019).

This learning unit focuses your attention on how to calculate markdowns. Then you will learn how a business prices perishable items that may spoil before customers buy them.

Learn: Markdowns

Markdowns are reductions from the original selling price caused by seasonal changes, special promotions, style changes, and so on. We calculate the markdown percent as follows:

Markdown Percent Formula

$$\text{Markdown percent} = \frac{\text{Dollar markdown}}{\text{Selling price (original)}}$$

Let's look at the following Kmart example:

Example: Kmart marked down an $18 video to $10.80. Calculate the **dollar markdown** and the markdown percent.

$18.00 Original selling price
− 10.80 Sale price
$ 7.20 Markdown

$$\frac{\text{Dollar markdown, }\$7.20}{\text{Selling price (original), }\$18.00} = 40\%$$

Calculating a Series of Markdowns and Markups   Often the final selling price is the result of a series of markdowns (and possibly a markup in between markdowns). We calculate additional markdowns on the previous selling price. Note in the following example how we calculate markdown on selling price after we add a markup.

Example: Jones Department Store paid its supplier $400 for a TV. On January 10, Jones marked the TV up 60% on selling price. As a special promotion, Jones marked the TV down 30% on February 8 and another 20% on February 28. No one purchased the TV, so Jones marked it up 10% on March 11. What was the selling price of the TV on March 11?

January 10: Selling price = Cost   + Markup

$$S = \$400 + .60S$$
$$- .60S \qquad\qquad - .60S$$
$$\frac{.40S}{.40} = \frac{\$400}{.40}$$
$$S = \$1,000$$

Check

$$S = \frac{\text{Cost}}{1 - \text{Percent markup on selling price}}$$

$$S = \frac{\$400}{1 - .60} = \frac{\$400}{.40} = \$1,000$$

February 8 markdown:

100%
− 30
70% → .70 × $1,000 = $700 selling price

February 28 additional markdown:

100%
− 20
80% → .80 × $700 = $560

March 11 additional markup:

100%
− 10
110% → 1.10 × $560 = $616

Learn: Pricing Perishable Items

The following formula can be used to determine the price of goods that have a short shelf life such as fruit, flowers, and pastry. (We limit this discussion to obviously **perishable** items.)

Aha! *To calculate the selling price of a perishable item: (1) Calculate selling price based on cost or based on selling price (total dollar sales). (2) Divide total dollar sales by the number of units available for sale (after accounting for spoilage).*

Selling Price of Perishables Formula

$$\text{Selling price of perishables} = \frac{\text{Total dollar sales}}{\text{Number of units produced} - \text{Spoilage}}$$

Money Tip   Shopping for perishable items? Seek a quantity discount with the retailer especially towards the end of the business day.

The Word Problem Audrey's Bake Shop baked 20 dozen bagels. Audrey expects 10% of the bagels to become stale and not salable. The bagels cost Audrey $1.20 per dozen. Audrey wants a 60% markup on cost. What should Audrey charge for each dozen bagels so she will make her profit? Round to the nearest cent.

The facts	Solving for?	Steps to take	Key points
Bagels cost: $1.20 per dozen. *Not salable:* 10%. *Baked:* 20 dozen. *Markup on cost:* 60%.	Price of a dozen bagels.	Total cost. Total dollar markup. Total selling price. Bagel loss. $TS = TC + TM$	Markup is based on cost.

(Blueprint)

Steps to solving problem

1. Calculate the total cost.

$$TC = 20 \text{ dozen} \times \$1.20 = \$24.00$$

2. Calculate the total dollar markup.

$$\boxed{TS = TC + TM}$$

$$TS = \$24.00 + .60(\$24.00)$$

$$TS = \$24.00 + \$14.40 \leftarrow \text{Total dollar markup}$$

3. Calculate the total selling price.

$$TS = \$38.40 \leftarrow \text{Total selling price}$$

4. Calculate the bagel loss.

$$20 \text{ dozen} \times .10 = 2 \text{ dozen}$$

5. Calculate the selling price for a dozen bagels.

$$\frac{\$38.40}{18} = \boxed{\$2.13} \text{ per dozen} \quad \begin{matrix} 20 \\ -\ 2 \end{matrix}$$

It's time to try the Practice Quiz.

Practice Quiz

Complete this Practice Quiz to see how you are doing.

1. Sunshine Music Shop bought a stereo for $600 and marked it up 40% on selling price. To promote customer interest, Sunshine marked the stereo down 10% for 1 week. Since business was slow, Sunshine marked the stereo down an additional 5%. After a week, Sunshine marked the stereo up 2%. What is the new selling price of the stereo to the nearest cent? What is the markdown percent based on the original selling price to the nearest hundredth percent?

2. Alvin Rose owns a fruit and vegetable stand. He knows that he cannot sell all his produce at full price. Some of his produce will be markdowns, and he will throw out some produce. Alvin must put a high enough price on the produce to cover markdowns and rotted produce and still make his desired profit. Alvin bought 300 pounds of tomatoes at 14 cents per pound. He expects a 5% spoilage and marks up tomatoes 60% on cost. What price per pound should Alvin charge for the tomatoes?

✓ Solutions

1.

$$S = C \qquad + M$$
$$S = \$600 \quad + .40S$$
$$\underline{-.40S} \qquad \underline{-.40S}$$
$$\frac{.60S}{.60} = \frac{\$600}{.60}$$
$$S = \$1,000$$

Check

$$S = \frac{\text{Cost}}{1 - \text{Percent markup on selling price}}$$

$$S = \frac{\$600}{1 - .40} = \frac{\$600}{.60} = \$1,000$$

First markdown: $.90 \times \$1,000 = \$900$ selling price

Second markdown: $.95 \times \$900 = \$855$ selling price

Markup: $1.02 \times \$855 = \boxed{\$872.10}$ final selling price

$$\$1,000 - \$872.10 = \frac{\$127.90}{\$1,000} = \boxed{12.79\%}$$

2. Price of tomatoes per pound:

	The facts	Solving for?	Steps to take	Key points
Blueprint	300 lb. tomatoes at $.14 per pound. *Spoilage: 5%.* *Markup on cost: 60%.*	Price of tomatoes per pound.	Total cost. Total dollar markup. Total selling price. Spoilage amount. $TS = TC + TM$	Markup is based on cost.

Practice Quiz *Continued*

Steps to solving problem

1. Calculate the total cost. $TC = 300 \text{ lb.} \times \$.14 = \$42.00$

2. Calculate the total dollar markup.

    $TS = TC + TM$

    $TS = \$42.00 + .60(\$42.00)$

    $TS = \$42.00 + \$25.20 \leftarrow$ Total dollar markup

3. Calculate the total selling price. $TS = \$67.20 \leftarrow$ Total selling price

4. Calculate the tomato loss. $300 \text{ pounds} \times .05 = 15 \text{ pounds spoilage}$

5. Calculate the selling price per pound of tomatoes.

    $\dfrac{\$67.20}{285} = \boxed{\$.24}$ per pound (rounded to nearest hundredth)

    $(300 - 15)$

Learning Unit 8–4:
Breakeven Analysis

So far in this chapter, cost is the price retailers pay to a manufacturer or supplier to bring the goods into the store. In this unit, we view costs from the perspective of manufacturers or suppliers who produce goods to sell in units, such as polo shirts, pens, calculators, lamps, and so on. These manufacturers or suppliers deal with two costs—fixed costs (FC) and variable costs (VC).

To understand how the owners of manufacturers or suppliers that produce goods per unit operate their businesses, we must understand fixed costs (FC), variable costs (VC), contribution margin (CM), and breakeven point (BE). Carefully study the following definitions of these terms:

- **Fixed costs (FC).** Costs that *do not change* with increases or decreases in sales; they include payments for insurance, a business license, rent, a lease, utilities, labor, and so on.

- **Variable costs (VC).** Costs that *do change* in response to changes in the volume of sales; they include payments for material, some labor, and so on.

- **Selling price (S).** In this unit we focus on manufacturers and suppliers who produce goods to sell in units.

- **Contribution margin (CM).** The difference between selling price (S) and variable costs (VC). This difference goes *first* to pay off total fixed costs (FC); when they are covered, *profits (or losses)* start to accumulate.

- **Breakeven point (BE).** The point at which the seller has covered all expenses and costs of a unit and has not made any profit or suffered any loss. Every unit sold after the breakeven point (BE) will bring some profit or cause a loss.

Learning Unit 8–4 is divided into two sections: calculating a contribution margin (CM) and calculating a breakeven point (BE). You will learn the importance of these two concepts and the formulas that you can use to calculate them. Study the example given for each concept to help you understand why the success of business owners depends on knowing how to use these two concepts.

Learn: Calculating a Contribution Margin (CM)

Before we calculate the breakeven point, we must first calculate the contribution margin. The formula is as follows:

Contribution Margin (CM) formula

Contribution margin (CM) = Selling price (S) − Variable cost (VC)

Example: Assume Jones Company produces pens that have a selling price (S) of $2.00 and a variable cost (VC) of $.80. We calculate the contribution margin (CM) as follows:

Contribution margin (CM) = $2.00 ($S$) − $.80 ($VC$)

$$CM = \boxed{\$1.20}$$

This means that for each pen sold, $1.20 goes to cover fixed costs (FC) and results in a profit. It makes sense to cover fixed costs (FC) first because the nature of a fixed cost is that it does not change with increases or decreases in sales.

Now we are ready to see how Jones Company will reach a breakeven point (BE).

Learn: Calculating a Breakeven Point (BE)

Sellers like Jones Company can calculate their profit or loss by using a concept called the **breakeven point (BE)**. This important point results after sellers have paid all their expenses and costs. Study the following formula and the example:

Breakeven Point Formula

$$\text{Breakeven point } (BE) = \frac{\text{Fixed costs } (FC)}{\text{Contribution margin } (CM)}$$

Example: Jones Company produces pens. The company has a fixed cost (FC) of $60,000. Each pen sells for $2.00 with a variable cost (VC) of $.80 per pen.

Fixed cost (FC)	$60,000
Selling price (S) per pen	$2.00
Variable cost (VC) per pen	$.80

$$\text{Breakeven point } (BE) = \frac{\$60{,}000 \ (FC)}{\$2.00 \ (S) - \$.80 \ (VC)} = \frac{\$60{,}000 \ (FC)}{\$1.20 \ (CM)} = \boxed{50{,}000 \text{ units (pens)}}$$

At 50,000 units (pens), Jones Company is just covering its costs. Each unit after 50,000 brings in a profit of $1.20 ($CM$).

It is time to try the Practice Quiz.

Practice Quiz

Complete this Practice Quiz to see how you are doing.

Blue Company produces holiday gift boxes. Given the following, calculate (1) the contribution margin (*CM*) and (2) the breakeven point (*BE*) for Blue Company.

Fixed cost (*FC*)	$45,000
Selling price (*S*) per gift box	$20
Variable cost (*VC*) per gift box	$8

✓ Solutions

1. Contribution margin $(CM) = \$20\ (S) - \$8\ (VC) = \boxed{\$12}$

2. Breakeven point $(BE) = \dfrac{\$45,000\ (FC)}{\$20\ (S) - \$8\ (VC)} = \dfrac{\$45,000\ (FC)}{\$12\ (CM)} = \boxed{3,750\ \text{units (gift boxes)}}$

Chapter 8 Review

Topic/Procedure/Formula	Example	You try it*
Markups based on cost: **Cost is 100% (base)** Selling price (S) = Cost (C) + Markup (M)	$400 = \$300 + \$100$ $S \;\;=\;\; C \;+\; M$	**Calculate selling price** Cost, $400; Markup, $200
Percent markup on cost $\dfrac{\text{Dollar markup (portion)}}{\text{Cost (base)}} = \dfrac{\text{Percent markup}}{\text{on cost (rate)}}$ **Cost** $C = \dfrac{\text{Dollar markup}}{\text{Percent markup on cost}}$	$\dfrac{\$100}{\$300} = \dfrac{1}{3} = 33\dfrac{1}{3}\%$ $\dfrac{\$100}{.33} = \$303$ Off slightly due to rounding	**Calculate percent markup on cost** Dollar markup, $50; Cost, $200 **Calculate cost** Dollar markup, $50; Percent markup on cost, 25%
Calculating selling price $S = C + M$ **Check** $S = \text{Cost} \times (1 + \text{Percent markup on cost})$	Cost, $6; percent markup on cost, 20% $S = \$6 + .20(\$6)$ **Check** $S = \$6 + \$1.20$ $S = \$7.20$ $\boxed{\$6 \times 1.20 = \$7.20}$	**Calculate selling price** Cost, $8; Percent markup on cost, 10%
Calculating cost $S = C + M$ **Check** $\text{Cost} = \dfrac{\text{Selling price}}{1 + \text{Percent markup on cost}}$	$S = \$100;\; M = 70\%$ of cost $S = C + M$ $\left(\begin{array}{l}\textit{Remember,}\\ C = 1.00C\end{array}\right)$ $\$100 = C + .70C$ $\$100 = 1.7C$ $\dfrac{\$100}{1.7} = C$ **Check** $\$58.82 = C$ $\boxed{\dfrac{\$100}{1+.70} = \$58.82}$	**Calculate cost** Selling price, $200; Markup on cost, 60%
Markups based on selling price: **selling price is 100% (Base)** Dollar markup = Selling price − Cost	$M = S - C$ $\$600 = \$1{,}000 - \$400$	**Calculate dollar markup** Cost, $2,000; Selling price, $4,500
Percent markup on selling price $\dfrac{\text{Dollar markup (portion)}}{\text{Selling price (base)}} = \dfrac{\text{Percent markup}}{\text{on selling price (rate)}}$ **Selling price** $S = \dfrac{\text{Dollar markup}}{\text{Percent markup on selling price}}$	$\dfrac{\$600}{\$1{,}000} = 60\%$ $\dfrac{\$600}{.60} = \$1{,}000$	**Calculate percent markup on selling price** Dollar markup, $700; Selling price, $2,800 **Calculate selling price** Dollar markup, $700; Percent markup on selling price, 50%

Chapter 8 Review (Continued)

Topic/Procedure/Formula	Example	You try it*
Calculating selling price $S = C + M$ **Check** $\text{Selling price} = \dfrac{\text{Cost}}{1 - \begin{array}{c}\text{Percent markup}\\\text{on selling price}\end{array}}$	Cost, \$400; percent markup on S, 60% $S = C + M$ $S = \$400 + .60S$ $S - .60S = \$400 + .60S - .60S$ $\dfrac{.40S}{.40} = \dfrac{\$400}{.40} \quad S = \boxed{\$1,000}$ $\text{Check} \rightarrow \boxed{\dfrac{\$400}{1 - .60} = \dfrac{\$400}{.40} = \$1,000}$	**Calculate selling price** Cost, \$800; Markup on selling price, 40%
Calculating cost $S = C + M$ **Check** $\text{Cost} = \dfrac{\text{Selling}}{\text{price}} \times \left(1 - \begin{array}{c}\text{Percent markup}\\\text{on selling price}\end{array}\right)$	$\$1,000 = C + 60\%(\$1,000)$ $\$1,000 = C + \$600$ $\boxed{\$400} = C$ $\text{Check} \longrightarrow \boxed{\begin{array}{l}\$1,000 \times (1 - .60)\\\$1,000 \times .40 = \$400\end{array}}$	**Calculating cost** Selling price, \$2,000; 70% markup on selling price
Conversion of markup percent $\begin{array}{c}\text{Percent markup}\\\text{on cost}\end{array}$ to $\begin{array}{c}\text{Percent markup}\\\text{on selling price}\end{array}$ $\boxed{\dfrac{\text{Percent markup on cost}}{1 + \text{Percent markup on cost}}}$ $\begin{array}{c}\text{Percent markup}\\\text{on selling price}\end{array}$ to $\begin{array}{c}\text{Percent markup}\\\text{on cost}\end{array}$ $\boxed{\dfrac{\text{Percent markup on selling price}}{1 - \text{Percent markup on selling price}}}$	*Round to nearest percent:* 54% markup on cost → $\boxed{35\%}$ markup on selling price $\boxed{\dfrac{.54}{1 + .54} = \dfrac{.54}{1.54} = 35\%}$ 35% markup on selling price → $\boxed{54\%}$ markup on cost $\boxed{\dfrac{.35}{1 - .35} = \dfrac{.35}{.65} = 54\%}$	**Calculate percent markup on selling price** Convert 47% markup on cost to markup on selling price. Round to nearest percent.
Markdowns $\text{Markdown percent} = \dfrac{\text{Dollar markdown}}{\text{Selling price (original)}}$	\$40 selling price 10% markdown $\$40 \times .10 = \$4$ markdown $\dfrac{\$4}{\$40} = \boxed{10\%}$	**Calculate markdown percent** Selling price, \$50; Markdown, 20%
Pricing perishables 1. Calculate total cost and total selling price. 2. Calculate selling price per unit by dividing total sales in Step 1 by units expected to be sold after taking perishables into account.	50 pastries cost 20 cents each; 10 will spoil before being sold. Markup is 60% on cost. 1. $TC = 50 \times \$.20 = \$10$ $TS = TC + TM$ $TS = \$10 + .60(\$10)$ $TS = \$10 + \$6$ $TS = \boxed{\$16}$ 2. $\dfrac{\$16}{40 \text{ pastries}} = \boxed{\$.40}$ per pastry	**Calculate cost of each pastry** 30 pastries cost 30 cents each; 15 will spoil; markup is 30% on cost.

Chapter 8 Review (Continued)

Topic/Procedure/Formula	Example	You try it*
Breakeven point (*BE*) $$BE = \frac{\text{Fixed cost } (FC)}{\text{Contribution margin } (CM)} \\ \text{(Selling price, } S - \text{Variable cost, } VC\text{)}}$$	Fixed cost (*FC*) \$60,000 Selling price (*S*) \$90 Variable cost (*VC*) \$30 $$BE = \frac{\$60,000}{\$90 - \$30} = \frac{\$60,000}{\$60} = 1,000 \text{ units}$$	**Calculate *BE*** Fixed cost (*FC*) \$70,000 Selling price (*S*) \$80 Variable cost (*VC*) \$60

Key Terms

Breakeven point	Margin	Percent markup on selling price
Contribution margin	Markdowns	
Cost	Markup	Perishables
Dollar markdown	Net profit (net income)	Selling price
Dollar markup	Operating expenses (overhead)	Variable cost
Fixed cost		
Gross profit	Percent markup on cost	

*Worked-out solutions are in Appendix A.

Critical Thinking Discussion Questions with Chapter Concept Check

1. Assuming markups are based on cost, explain how the portion formula could be used to calculate cost, selling price, dollar markup, and percent markup on cost. Pick a company and explain why it would mark goods up on cost rather than on selling price.

2. Assuming markups are based on selling price, explain how the portion formula could be used to calculate cost, selling price, dollar markup, and percent markup on selling price. Pick a company and explain why it would mark up goods on selling price rather than on cost.

3. What is the formula to convert percent markup on selling price to percent markup on cost? How could you explain that a 40% markup on selling price, which is a 66.7% markup on cost, would result in the same dollar markup?

4. Explain how to calculate markdowns. Do you think stores should run 1-day-only markdown sales? Would it be better to offer the best price "all the time"?

5. Explain the five steps in calculating a selling price for perishable items. Recall a situation where you saw a store that did *not* follow the five steps. How did it sell its items?

6. Explain how Walmart uses breakeven analysis. Give an example.

7. **Chapter Concept Check.** Visit a retailer's website and find out how that retailer marks up goods and marks down specials. Present calculations based on this chapter to support your findings.

8. Research and discuss how many retailers went bankrupt due to the pandemic.

End-of-Chapter Problems

Name _____   Date_____

Check figures for odd-numbered problems in Appendix A.

Drill Problems

Assume markups in Problems 8–1 to 8–6 are based on cost. Find the dollar markup and selling price for the following problems. Round answers to the nearest cent.   *LU 8–1(1, 2)*

	Item	Cost	Markup percent	Dollar markup	Selling price
8–1.	Bell and Ross watch	$2,000	30%		
8–2.	Burberry men's watch	$425	200%		

Solve for cost (round to the nearest cent):   *LU 8–1(3)*

8–3. Selling price of office furniture at Staples, $6,000

Percent markup on cost, 40%

Actual cost?

8–4. Selling price of lumber at Home Depot, $4,000

Percent markup on cost, 30%

Actual cost?

Complete the following: *LU 8–1(1)*

	Cost	Selling price	Dollar markup	Percent markup on cost*
8–5.	$15.10	$22.00	?	?
8–6.	?	?	$4.70	102.17%

*Round to the nearest hundredth percent.

End-of-Chapter Problems (Continued)

Assume markups in Problems 8–7 to 8–12 are based on selling price. Find the dollar markup and cost (round answers to the nearest cent):   *Lu 8–2(1, 2)*

	Item	Selling price	Markup percent	Dollar markup	Cost
8–7.	Sony LCD TV	$1,000	45%		
8–8.	Canon scanner	$80	30%	$24.00	$56.00

Solve for the selling price (round to the nearest cent):   *LU 8–2(3)*

8–9. Selling price of a complete set of pots and pans at
Walmart 40% markup on selling price
Cost, actual, $66.50

8–10. Selling price of a dining room set at Macy's
55% markup on selling price
Cost, actual, $800

Complete the following: *LU 8–2(1)*

	Cost	Selling price	Dollar markup	Percent markup on selling price (round to nearest tenth percent)
8–11.	$14.80	$49.00	?	?
8–12.	?	?	$4.00	20%

By conversion of the markup formula, solve the following (round to the nearest whole percent as needed):   *LU 8–2(4)*

	Percent markup on cost	Percent markup on selling price
8–13.	12.4%	?
8–14.	?	13%

Complete the following: *LU 8–3(1, 2)*

excel 8–15. Calculate the final selling price to the nearest cent and markdown percent to the nearest hundredth percent:

Original selling price	First markdown	Second markdown	Markup	Final markdown
$5,000	20%	10%	12%	5%

	Item	Total quantity bought	Unit cost	Total cost	Percent markup on cost	Total selling price	Percent that will spoil	Selling price per brownie
8–16.	Brownies	20	$.79	?	60%	?	10%	?

Complete the following: *LU 8–4(1, 2)*

	Breakeven point	Fixed cost	Contribution margin	Selling price per unit	Variable cost per unit
8–17.		$65,000		$5.00	$1.00
8–18.		$90,000		$9.00	$4.00

Word Problems

8–19. Bari Jay, a gown manufacturer, received an order for 600 prom dresses from China. Her cost is $35 a gown. If her markup based on selling price is 79%, what is the selling price of each gown? Round to the nearest cent.   *LU 8–2(2)*

End-of-Chapter Problems (Continued)

8–20. Brian May, guitarist for Queen, does not know how to price his signature Antique Cherry Special that cost him £280 to make. He knows he wants 85% markup on cost. What price should Brian May ask for the guitar? *Note:* "£" is the pound sterling symbol.   *LU 8–1(2)*

8–21. You are buying and reselling items found at your local thrift shop. You found an antique pitcher for sale. If you need a 40% markup on cost and know most people will not pay more than $20 for it, what is the most you can pay for the pitcher? Round to the nearest cent.   *LU 8–1(4)*

8–22. Macy's was selling Calvin Klein jean shirts that were originally priced at $58.00 for $8.70. **(a)** What was the amount of the markdown? **(b)** Based on the selling price, what is the percent markdown?   *LU 8–3(1)*

8–23. Brownsville, Texas, boasts being the southernmost international seaport and the largest city in the lower Rio Grande Valley. Ben Supple, an importer in Brownsville, has just received a shipment of Peruvian opals that he is pricing for sale. He paid $150 for the shipment. If he wants a 75% markup, calculate the selling price based on selling price. Then calculate the selling price based on cost.   *LU 8–1(2), LU 8–2(2)*

8–24. Front Range Cabinet Distributors in Colorado Springs, Colorado, sells to its contractors with a 42% markup on cost. If the selling price for cabinets is $9,655, what is the cost to contractors based on cost? Round to the nearest tenth. Check your answer.   *LU 8–1(3)*

excel 8–25. Misu Sheet, owner of the Bedspread Shop, knows his customers will pay no more than $120 for a comforter. Misu wants a 30% markup on selling price. What is the most that Misu can pay for a comforter?   *LU 8–2(4)*

excel 8–26. Assume Misu Sheet (Problem 8–25) wants a 30% markup on cost instead of on selling price. What is Misu's cost? Round to the nearest cent.   *LU 8–1(4)*

8–27. Misu Sheet (Problem 8–25) wants to advertise the comforter as "percent markup on cost." What is the equivalent rate of percent markup on cost compared to the 30% markup on selling price? Check your answer. Is this a wise marketing decision? Round to the nearest hundredth percent.   *LU 8–2(4)*

excel 8–28. DeWitt Company sells a kitchen set for $475. To promote July 4, DeWitt ran the following advertisement:

Beginning each hour up to 4 hours we will mark down the kitchen set 10%. At the end of each hour, we will mark up the set 1%.

Assume Ingrid Swenson buys the set 1 hour 50 minutes into the sale. What will Ingrid pay? Round each calculation to the nearest cent. What is the markdown percent? Round to the nearest hundredth percent.   *LU 8–3(1)*

End-of-Chapter Problems (Continued)

excel 8–29. Angie's Bake Shop makes birthday chocolate chip cookies that cost $2 each. Angie expects that 10% of the cookies will crack and be discarded. Angie wants a 60% markup on cost and produces 100 cookies. What should Angie price each cookie? Round to the nearest cent.   *LU 8–3(2)*

8–30. Assume that Angie (Problem 8–29) can sell the cracked cookies for $1.10 each. What should Angie price each cookie?   *LU 8–3(2)*

8–31. Jane Corporation produces model toy cars. Each sells for $29.99. Its variable cost per unit is $14.25. What is the breakeven point for Jane Corporation assuming it has a fixed cost of $314,800?   *LU 8–4(2)*

8–32. Aunt Sally's "New Orleans Most Famous Pralines" sells pralines costing $1.10 each to make. If Aunt Sally's wants a 35% markup based on selling price and produces 45 pralines with an anticipated 15% spoilage (rounded to the nearest whole number), what should each praline be sold for?   *LU 8–3(2)*

8–33. On Black Friday, Amazon.com featured an Echo 4th Gen for $35 normally selling for $79. Calculate the dollar markdown and the markdown percent based on the selling price to the nearest whole percent.   *LU 8–3(1)*

8–34. The Food Co-op Club boasts that it has 5,000 members and a 200% increase in sales. Its markup is 36% based on cost. What would be its percent markup if selling price were the base? Round to the nearest hundredth percent.   *LU 8–2(4)*

8–35. At Bed Bath and Beyond, the manager, Jill Roe, knows her customers will pay no more than $300 for a bedspread. Jill wants a 35% markup on selling price. What is the most that Jill can pay for a bedspread?   *LU 8–2(4)*

8–36. Kathleen Osness purchased a $60,000 RV with a 40 percent markup on selling price. **(a)** What was the amount of the dealer's markup? **(b)** What was the dealer's original cost?   *LU 8–2(4)*

8–37. John's Smoothie Stand at Utah's Wasatch County's Demolition Derby sells bananas. If John bought 50 lbs. of bananas at $.23 per pound expecting 10% to spoil, how should he price his bananas to achieve 57% on selling price?   *LU 8–2(4)*

e**x**cel **8–38.** Arley's Bakery makes fat-free cookies that cost $1.50 each. Arley expects 15% of the cookies to fall apart and be discarded. Arley wants a 45% markup on cost and produces 200 cookies. What should Arley price each cookie? Round to the nearest cent. *LU 8–3(2)*

8–39. Assume that Arley (Problem 8–38) can sell the broken cookies for $1.40 each. What should Arley price each cookie?   *Lu 8–3(2)*

8–40. An Apple Computer store sells computers for $1,258.60. Assuming the computers cost $10,788 per dozen, find for each computer the **(a)** dollar markup, **(b)** percent markup on cost, and **(c)** percent markup on selling price, to the nearest hundredth percent.   *LU 8–1(1), Lu 8–2(1)*

End-of-Chapter Problems (Continued)

Prove (**b**) and (**c**) of the above problem using the equivalent formulas.

8–41. Pete Corporation produces bags of peanuts. Its fixed cost is $17,280. Each bag sells for $2.99 with a unit cost of $1.55. What is Pete's breakeven point? *LU 8–4(2)*

8–42. Yelp.com reported more than 97,966 businesses permanently shut down during the pandemic. Tuesday Morning is closing 230 out of 700 stores as part of its Chapter 11 bankruptcy filing. The store manager of one closing location had a going-out-of-business sale. They slashed the price of leather couches selling for $829 by 25%. They reduced them again by 15% the following week. For Black Friday, they increased the price 10%. All remaining leather couches had close-out prices of another 20% reduction. What was the final selling price of a leather couch? Round each answer to the nearest cent.

8–43. Friends who tasted your baking during the pandemic are encouraging you to open a bakery. To humor them you decide to price out your extra large snickerdoodle cupcake cookies. The cookies cost $2.55 each to make. You expect 15% to crack or not sell the first day and you want a 60% markup on selling price with the production of 48 cookies. How much should you sell each cookie for to achieve your goals? Round down to the whole cookie for the spoilage amount.

8–44. Lululemon needs help pricing a pair of leggings. The cost to produce one pair of leggings is $15.75. Compare the selling price of one pair of leggings with a 40% markup based on cost to the selling price of one pair of leggings with a 28.57% markup based on selling price. Compare the results.

Challenge Problems

8–45. Nissan Appliances bought two dozen camcorders at a cost of $4,788. The markup on the camcorders is 25% of the selling price. What was the original selling price of each camcorder?   *LU 8–2(3)*

8–46. On July 8, Leon's Kitchen Hut bought a set of pots with a $120 list price from Lambert Manufacturing. Leon's receives a 25% trade discount. Terms of the sale were 2/10, n/30. On July 14, Leon's sent a check to Lambert for the pots. Leon's expenses are 20% of the selling price. Leon's must also make a profit of 15% of the selling price. A competitor marked down the same set of pots 30%. Assume Leon's reduces its selling price by 30%. *LU 8–2(3)*

a. What is the sale price at Kitchen Hut?

b. What was the operating profit or loss?

Summary Practice Test

Do you need help? Connect videos have step-by-step worked-out solutions.

1. Sunset Co. marks up merchandise 40% on cost. A DVD player costs Sunset $90. What is Sunset's selling price? Round to the nearest cent. *LU 8–1(2)*

2. JCPenney sells jeans for $49.50 that cost $38.00. What is the percent markup on cost? Round to the nearest hundredth percent. Check the cost. *LU 8–1(1)*

3. Best Buy sells a flat-screen high-definition TV for $700. Best Buy marks up the TV 45% on cost. What is the cost and dollar markup of the TV? *LU 8–1(1, 3)*

4. REI marks up New Balance sneakers $30 and sells them for $109. Markup is on cost. What are the cost and percent markup to the nearest hundredth percent? *LU 8–1(1, 3)*

5. The Shoe Outlet bought boots for $60 and marks up the boots 55% on the selling price. What is the selling price of the boots? Round to the nearest cent. *LU 8–2(2)*

6. Office Max sells a desk for $450 and marks up the desk 35% on the selling price. What did the desk cost Office Max? Round to the nearest cent. *LU 8–2(4)*

7. Zales sells diamonds for $1,100 that cost $800. What is Zales's percent markup on selling price? Round to the nearest hundredth percent. Check the selling price.   *LU 8–2(1)*

8. Earl Miller, a customer of J. Crew, will pay $400 for a new jacket. J. Crew has a 60% markup on selling price. What is the most that J. Crew can pay for this jacket?   *LU 8–2(4)*

9. Home Liquidators marks up its merchandise 35% on cost. What is the company's equivalent markup on selling price? Round to the nearest tenth percent.   *LU 8–2(4)*

10. The Muffin Shop makes no-fat blueberry muffins that cost $.70 each. The Muffin Shop knows that 15% of the muffins will spoil. If The Muffin Shop wants 40% markup on cost and produces 800 muffins, what should The Muffin Shop price each muffin? Round to the nearest cent.   *LU 8–3(2)*

11. Angel Corporation produces calculators selling for $25.99. Its unit cost is $18.95. Assuming a fixed cost of $80,960, what is the breakeven point in units?   *LU 8–4(2)*

MY MONEY

🔍 Giving Yourself the Credit You Deserve

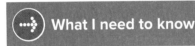
What I need to know

Your credit score stays with you for your entire adult life. It is an important component in many decisions that are made about you and your creditworthiness. Because of the vital role it plays in your financial life it is important you understand your credit score/report, determine where your score is today, and plan where you would like it to be in the future. What activities have an impact on your score? How and why does your credit score fluctuate over time? What are the benefits of having a good credit score and solid credit report? With this knowledge in hand, you will find your financial situation becomes a bit easier to understand and control.

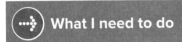
What I need to do

A credit score is like a photograph representing a point in time. Knowing your current credit score provides you with a baseline assessment of your creditworthiness now. Your current score helps you determine a course of action to improve your credit score. For example, when you finance a new car, your credit score is used to determine the interest rate of your loan. You can save a considerable amount of money by understanding your credit score and improving it prior to financing a large purchase. Access your free credit report from each of the three credit bureaus, Experian, TransUnion, and Equifax, once every 12 months through www.annualcreditreport.com so you are aware of your current score. Stagger your requests for timely credit information throughout the year.

Financial decisions you make today determine your credit score tomorrow. Understanding how common situations impact your credit are valuable life lessons to learn. Missing a scheduled payment (i.e., rent, utilities, etc.) and carrying high account balances compared to your credit limit can negatively impact your credit score. Conversely, paying bills on time, making frequent small payments, and obtaining higher credit limits positively boosts your credit rating. Knowledge truly is power when it comes to your credit score and knowing what impacts your credit is valuable information.

If it seems maintaining a good credit score takes some work, you are right! So, what do you gain from all this work? There are many benefits of a good credit score providing you with financial incentives to manage your money effectively. For instance, lower interest rates on loans will yield significant savings over the course of your loan. You will gain greater access and easier approval on credit accounts with larger purchase limits. A good credit score gives you the ability to negotiate purchases in your favor since you are now a more attractive customer to the seller and/or lender.

Steps I need to take

1. Know your current credit score. Request your free credit report from Experian through www.annualcreditreport.com in January, TransUnion in May, and Equifax in September.
2. Improve your credit score through sound financial decisions. Have a goal of a credit score of 740+.
3. Understand the benefits of establishing good credit and use it to your advantage.

Resources I can use

- Credit Karma (mobile app)—stay informed of your credit score and what impacts the score.

MY MONEY ACTIVITY ✕

- Retrieve your current credit report/score from www.annualcreditreport.com.
- Research ways to improve your credit score.
- Choose three of the suggestions for improving your score and discuss how you plan to implement each.

PERSONAL FINANCE

A KIPLINGER APPROACH

"$3,000-Plus UPGRADE TO AN E-BIKE." Kiplinger's. November, 2020.

$3,000-PLUS

UPGRADE TO AN E-BIKE

The COVID-based impulses that have driven bicycle sales have also boosted e-bike sales. These devices, which sport a battery-and-motor combination powerful enough to flatten small hills but weak enough to ensure they're classified as bicycles rather than motorcycles, have huge appeal to people who'd like to get places with a bit less sweat.

Like bicycles, they're scarce (you will wait for your order). Unlike bicycles, $1,000 doesn't get you much. We'd recommend prioritizing quality rather than features, as the bottom end of the market has plenty of sketchy products that look as if the builder wired the batteries from an exploding hoverboard to a 50-pound beach cruiser bike.

Buying from a well-established bicycle brand (think Specialized, Giant, Cannondale) is one way to ensure quality and get warranty support, but there are also some e-bike specialists worth considering. The Pedego Boomerang Plus has a frame that you can literally step through—no crossbar to swing a leg over ($2,995–$3,695), along with a five-year warranty. Rad Power Bikes' cheapest model, the RadMission I, lists at our $1,000 threshold and includes 500 watts of power on a conventional bicycle frame (with disc brakes to control the whoa). But you'll need to get on a waiting list, with no shipping date specified.

Business Math Issue

Post COVID means bike sales will not increase.

1. List the key points of the article and information to support your position.

2. Write a group defense of your position using math calculations to support your view. If you are in an online course, post to a discussion board.

Cumulative Review

A Word Problem Approach—Chapters 6, 7, 8

1. Assume Kellogg's produced 715,000 boxes of Corn Flakes this year. This was 110% of the annual production last year. What was last year's annual production?   *LU 6-2(2)*

2. A new Sony camera has a list price of $420. The trade discount is 10/20 with terms of 2/10, n/30. If a retailer pays the invoice within the discount period, what is the amount the retailer must pay?   *LU 7-1(4), LU 7-2(1)*

3. JCPenney sells loafers with a markup of $40. If the markup is 30% on cost, what did the loafers cost JCPenney? Round to the nearest dollar.   *LU 8-1(3)*

4. Aster Computers received from Ring Manufacturers an invoice dated August 28 with terms 2/10 EOM. The list price of the invoice is $3,000 (freight not included). Ring offers Aster a 9/8/2 trade chain discount. Terms of freight are FOB shipping point, but Ring prepays the $150 freight. Assume Aster pays the invoice on October 9. How much will Ring receive?   *LU 7-1(4), LU 7-2(1)*

5. Runners World marks up its Nike jogging shoes 25% on selling price. The Nike shoes sell for $65. How much did the store pay for them?   *LU 8-2(4)*

6. Ivan Rone sells antique sleds. He knows that the most he can get for a sled is $350. Ivan needs a 35% markup on cost. Since Ivan is going to an antiques show, he wants to know the maximum he can offer a dealer for an antique sled.   *LU 8-1(3)*

7. Bonnie's Bakery bakes 60 loaves of bread for $1.10 each. Bonnie's estimates that 10% of the bread will spoil. Assume a 60% markup on cost. What is the selling price of each loaf? If Bonnie's can sell the old bread for one-half the cost, what is the selling price of each loaf? *LU 8-3(2)*

Chapter 9
Payroll

Learning Unit Objectives

LU 9–1: Calculating Various Types of Employees' Gross Pay

1. Define, compare, and contrast weekly, biweekly, semimonthly, and monthly pay periods.
2. Calculate gross pay with overtime on the basis of time.
3. Calculate gross pay for piecework, differential pay schedule, straight commission with draw, variable commission scale, and salary plus commission.

LU 9–2: Computing Payroll Deductions for Employees' Pay; Employers' Responsibilities

1. Prepare and explain the parts of a payroll register.
2. Explain and calculate federal and state unemployment taxes.

Hilton To Cut Workforce, Salaries

BY ALLISON PRANG AND CRAIG KARMIN

Hilton Worldwide Holdings Inc. said it is cutting nearly 22% of its corporate workforce globally, in what analysts said is one of the deepest job cuts so far by a major lodging company in response to the coronavirus pandemic.

The job reductions amount to 2,100 corporate employees, Hilton said Tuesday. The hospitality company said it is also extending its corporate pay cuts, reduced hours and furloughs for up to three more months.

"Never in Hilton's 101-years history has our industry faced global crisis that brings travel to a virtual standstill," Chief Executive and President Christopher Nassetta said in prepared remarks.

Hilton's move is the latest sign that the hospitality industry continues to face headwinds from the Covid-19 pandemic, which caused travel to collapse and occupancy levels to fall below 30% earlier this spring in the U.S.

Prang, Allison, and Craig Karmin. "Hilton Cutting About 22% of Global Corporate Workforce." *The Wall Street Journal*, June 16, 2020.

⊖ Essential Question

How can I use payroll to understand the business world and make my life easier?

🌐 Math Around the World

The *Wall Street Journal* chapter opener clip, "Hilton To Cut Workforce, Salaries", discusses Hilton cutting its workforce and salaries due to the COVID-19 pandemic. The *Wall Street Journal* clip to the left shows how Facebook plans to hire many more people of color to their workforce by 2023.

This chapter discusses the payroll process for employees and employers. Keep in mind there are many software packages and computerized services that do payroll. In this chapter we look at a manual system to better understand the payroll process.

Facebook Spells Out
Hiring, contracts and content creation to be boosted as company is criticized on race issues

BY MATT GROSSMAN

Facebook Inc. laid out steps it said it would take to support black-owned businesses, black content creators and employees of color as protests across the U.S. and around the world have brought renewed attention to racial disparities.

The social-media giant is committing to a 30% increase in the number of people of color in leadership positions over the next five years, Sheryl Sandberg, Facebook's chief operating officer, wrote in a note published Thursday on the company's website. The company had already planned to ensure that half of its workforce will consist of minorities by 2023, she added.

Grossman, Matt. "Facebook Outlines Plans to Improve Workforce Diversity." *The Wall Street Journal*, June 18, 2020.

Learning Unit 9–1:
Calculating Various Types of Employees' Gross Pay

Logan Company manufactures dolls of all shapes and sizes. These dolls are sold worldwide. We study Logan Company in this unit because of the variety of methods Logan uses to pay its employees.

Companies usually pay employees **weekly, biweekly, semimonthly,** or **monthly.** How often employers pay employees can affect how employees manage their money. Some employees prefer a weekly paycheck that spreads the inflow of money. Employees who have monthly bills may find the twice-a-month or monthly paycheck more convenient. All employees would like more money to manage.

Let's assume you earn $50,000 per year. The following table shows what you would earn each pay period. Remember that 13 weeks equals one quarter. Four quarters or 52 weeks equals a year.

Salary paid	Period (based on a year)	Earnings for period (dollars)
Weekly	52 times (once a week)	$961.54 ($50,000 ÷ 52)
Biweekly	26 times (every two weeks)	$1,923.08 ($50,000 ÷ 26)
Semimonthly	24 times (twice a month)	$2,083.33 ($50,000 ÷ 24)
Monthly	12 times (once a month)	$4,166.67 ($50,000 ÷ 12)

Aha! *You can estimate an annual salary by doubling the full-time hourly rate and then multiplying by 1,000. Example: $15 an hour, $15 × 2 × 1,000 = $30,000. You can estimate an hourly full-time rate by dividing an annual salary by 1,000 and then dividing by 2. Example: $30,000/1,000 = $30; 30/2 = $15.*

Now let's look at some pay schedule situations and examples of how Logan Company calculates its payroll for employees of different pay status.

Situation 1: Hourly Rate of Pay; Calculation of Overtime

The **Fair Labor Standards Act** sets minimum wage standards and overtime regulations for employees of companies covered by this federal law. The law provides that employees working for an hourly rate receive time-and-a-half pay for hours worked in excess of their regular 40-hour week. Many managerial people, however, are exempt from the time-and-a-half pay for all hours in excess of a 40-hour week. Other workers may also be exempt.

The current federal hourly minimum wage is $7.25. At time of writing, Congress is investigating a $15 minimum wage proposal. Various states have passed their own minimum wages. The *Wall Street Journal* clip below shows that Bank of America has a $20-an-hour minimum wage. Note in the chart below how seven states have raised their minimum wage for 2021.

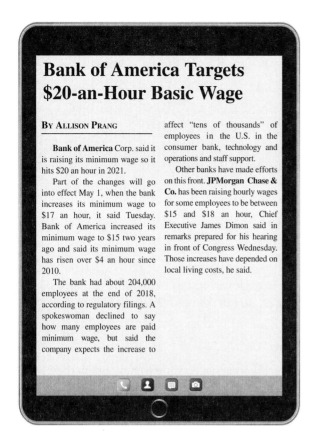

Bank of America Targets $20-an-Hour Basic Wage

BY ALLISON PRANG

Bank of America Corp. said it is raising its minimum wage so it hits $20 an hour in 2021.

Part of the changes will go into effect May 1, when the bank increases its minimum wage to $17 an hour, it said Tuesday. Bank of America increased its minimum wage to $15 two years ago and said its minimum wage has risen over $4 an hour since 2010.

The bank had about 204,000 employees at the end of 2018, according to regulatory filings. A spokeswoman declined to say how many employees are paid minimum wage, but said the company expects the increase to affect "tens of thousands" of employees in the U.S. in the consumer bank, technology and operations and staff support.

Other banks have made efforts on this front. **JPMorgan Chase & Co.** has been raising hourly wages for some employees to be between $15 and $18 an hour, Chief Executive James Dimon said in remarks prepared for his hearing in front of Congress Wednesday. Those increases have depended on local living costs, he said.

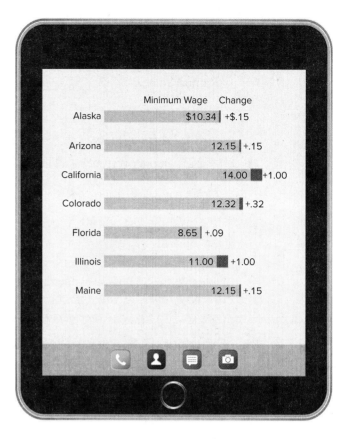

Prang, Allison. "Bank of America Aims to Raise Minimum Wage to $20 an Hour by 2021." *The Wall Street Journal,* April 9, 2019.

Now we return to our Logan Company example. Logan Company is calculating the weekly pay of Ramon Valdez, who works in its manufacturing division. For the first 40 hours Ramon works, Logan calculates his **gross pay** (earnings before **deductions**) as follows:

$$\text{Gross pay} = \text{Hours employee worked} \times \text{Rate per hour}$$

Ramon works more than 40 hours in a week. For every hour over his 40 hours, Ramon must be paid an **overtime** pay of at least 1.5 times his regular pay rate. The following formula is used to determine Ramon's overtime:

$$\text{Hourly overtime pay rate} = \text{Regular hourly pay rate} \times 1.5$$

Logan Company must include Ramon's overtime pay with his regular pay. To determine Ramon's gross pay, Logan uses the following formula:

$$\text{Gross pay} = \text{Earnings for 40 hours} + \text{Earnings at time-and-a-half rate (1.5)}$$

We are now ready to calculate Ramon's gross pay from the following data:

Example:

Employee	M	T	W	Th	F	S	Total hours	Rate per hour
Ramon Valdez	13	$8\frac{1}{2}$	10	8	$11\frac{1}{4}$	$10\frac{3}{4}$	$61\frac{1}{2}$	$9

$$
\begin{array}{ll}
61\frac{1}{2} & \text{total hours} \\
-40 & \text{regular hours} \\
\hline
21\frac{1}{2} & \text{hours overtime}^1 \quad \text{Time-and-a-half pay: } \$9 \times 1.5 = \$13.50
\end{array}
$$

$$
\begin{aligned}
\text{Gross pay} &= (40 \text{ hours} \times \$9) + (21\tfrac{1}{2} \text{ hours} \times \$13.50) \\
&= \quad \$360 \quad + \quad \$290.25 \\
&= \boxed{\$650.25}
\end{aligned}
$$

Note that the $13.50 overtime rate came out even. However, throughout the text, *if an overtime rate is greater than two decimal places, do not round it. Round only the final answer. This gives greater accuracy.*

Situation 2: Straight Piece Rate Pay

Some companies, especially manufacturers, pay workers according to how much they produce. Logan Company pays Ryan Foss for the number of dolls he produces in a week. This gives Ryan an incentive to make more money by producing more dolls. Ryan receives $.96 per doll, less any defective units. The following formula determines Ryan's gross pay:

> Gross pay = Number of units produced × Rate per unit

Companies may also pay a guaranteed hourly wage and use a piece rate as a bonus. However, Logan uses straight piece rate as wages for some of its employees.

Example: During the last week of April, Ryan Foss produced 900 dolls. Using the above formula, Logan Company paid Ryan $864.

$$
\begin{aligned}
\text{Gross pay} &= 900 \text{ dolls} \times \$.96 \\
&= \boxed{\$864}
\end{aligned}
$$

Situation 3: Differential Pay Schedule

Some of Logan's employees can earn more than the $.96 straight piece rate for every doll they produce. Logan Company has set up a **differential pay schedule** for these employees. The company determines the rate these employees make by the amount of units the employees produce at different levels of production.

---

[1] Some companies pay overtime for time over 8 hours in one day; Logan Company pays overtime for time over 40 hours per week.

Example: Logan Company pays Abby Rogers on the basis of the following schedule:

Units produced	Amount per unit
First 50 → 1–50	$ .50
Next 100 → 51–150	.62
Next 50 → 151–200	.75
Over 200	1.25

Last week Abby produced 300 dolls. What is Abby's gross pay?

Logan calculated Abby's gross pay as follows:

$(50 \times \$.50) + (100 \times \$.62) + (50 \times \$.75) + (100 \times \$1.25)$

$\$25 \quad + \quad \$62 \quad + \quad \$37.50 \quad + \quad \$125 \quad = \boxed{\$249.50}$

Now we will study some of the other types of employee commission payment plans.

Situation 4: Straight Commission with Draw

Companies frequently use **straight commission** to determine the pay of salespersons. This commission is usually a certain percentage of the amount the salesperson sells. An example of one group of companies ceasing to pay commissions is the rental-car companies.

Companies such as Logan Company allow some of their salespersons to draw against their commission at the beginning of each month. A **draw** is an advance on the salesperson's commission. Logan subtracts this advance later from the employee's commission earned based on sales. When the commission does not equal the draw, the salesperson owes Logan the difference between the draw and the commission.

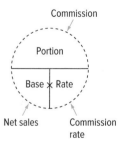

Example: Logan Company pays Jackie Okamoto a straight commission of 15% on her net sales (net sales are total sales less sales returns). In May, Jackie had net sales of $56,000. Logan gave Jackie a $600 draw in May. What is Jackie's gross pay?

Logan calculated Jackie's commission minus her draw as follows:

$\$56,000 \times .15 = \$8,400$

$\underline{ - 600}$

$\boxed{\$7,800}$

Logan Company pays some people in the sales department on a variable commission scale. Let's look at this, assuming the employee had no draw.

Situation 5: Variable Commission Scale

A company with a **variable commission scale** uses different commission rates for different levels of net sales.

Example: Last month, Jane Ring's net sales were $160,000. What is Jane's gross pay based on the following schedule?

Up to $35,000	4%
Excess of $35,000 to $45,000	6%
Over $45,000	8%

$$\text{Gross pay} = (\$35,000 \times .04) + (\$10,000 \times .06) + (\$115,000 \times .08)$$
$$= \quad \$1,400 \quad + \quad \$600 \quad + \quad \$9,200$$
$$= \boxed{\$11,200}$$

Situation 6: Salary Plus Commission

Logan Company pays Joe Roy a $3,000 monthly salary plus a 4% commission for sales over $20,000. Last month Joe's net sales were $50,000. Logan calculated Joe's gross monthly pay as follows:

$$\text{Gross pay} = \text{Salary} \; + \; (\text{Commission} \times \text{Sales over } \$20,000)$$
$$= \$3,000 + \qquad\qquad (.04 \times \$30,000)$$
$$= \$3,000 + \qquad\qquad \$1,200$$
$$= \boxed{\$4,200}$$

Before you take the Practice Quiz, you should know that many managers today receive **overrides.** These managers receive a commission based on the net sales of the people they supervise.

Practice Quiz

Complete this Practice Quiz to see how you are doing.

1. Jill Foster worked 52 hours in one week for Delta Airlines. Jill earns $10 per hour. What is Jill's gross pay, assuming overtime is at time-and-a-half for over 40 hours of work per week?

2. Matt Long had $180,000 in sales for the month. Matt's commission rate is 9%, and he had a $3,500 draw. What was Matt's end-of-month commission?

3. Bob Meyers receives a $1,000 monthly salary. He also receives a variable commission on net sales based on the following schedule (commission doesn't begin until Bob earns $8,000 in net sales):

$8,000–$12,000	1%	Excess of $20,000 to $40,000	5%
Excess of $12,000 to $20,000	3%	More than $40,000	8%

   Assume Bob earns $40,000 net sales for the month. What is his gross pay?

✓ Solutions

1. 40 hours × $10.00 = $400.00
   12 hours × $15.00 = $\underline{\ \ 180.00\ \ }$ ($10.00 × 1.5 = $15.00)
   $580.00

2. $180,000 × .09 = $16,200
   $\underline{-\ \ \ \ 3,500}$
   $12,700

3. Gross pay = $1,000 + ($4,000 × .01) + ($8,000 × .03) + ($20,000 × .05)
   = $1,000 + $40   + $240   + $1,000
   = $2,280

Computing Payroll Deductions for Employees' Pay; Employers' Responsibilities

This unit begins by dissecting a paycheck. Then we give you an insight into the tax responsibilities of employers.

Learn: Computing Payroll Deductions for Employees

Companies often record employee payroll information in a multicolumn form called a **payroll register.** The increased use of computers in business has made computerized registers a timesaver for many companies. The following payroll discussion is based on updates to the tax law effective 2021.

Glo Company uses a multicolumn payroll register. Below is Glo's partial payroll register showing the payroll information for Janet Wong during week 49. Let's check each column to see if Janet's take-home pay of $2,172.25 is correct. Note how the circled letters in the register correspond to the explanations that follow.

GLO COMPANY
Payroll Register
Week #49

Employee name	Status	Cum. earn.	Sal. per week	Earning Reg.	Earning Ovt.	Earning Gross	Cum. earn.	FICA Taxable Earning S.S.	FICA Taxable Earning Med.	FICA S.S.	FICA Med.	FIT	SIT	Health ins.	Net pay
Wong, Janet	M	$142,100	2,900	2,900	—	2,900	$145,000	700	2,900	43.40	42.05	368.30	174.00	100	$2,172.25
	Ⓐ	Ⓑ	Ⓒ		Ⓓ		Ⓔ	Ⓕ	Ⓖ	Ⓗ	Ⓘ	Ⓙ	Ⓚ	Ⓛ	Ⓜ

Payroll Register Explanations
Ⓐ—Status
Ⓑ, Ⓒ, Ⓓ—Cumulative earnings before payroll, salaries, earnings
Ⓔ—Cumulative earnings after payroll

When Janet was hired, she completed the **W-4 (Employee's Withholding Certificate)** form shown in Figure 9.1 stating that she is married. Glo Company will need this information to calculate the federal income tax (FIT) Ⓙ.

Before this pay period, Janet had earned $139,200. Since Janet receives no overtime, her $2,900 salary per week represents her gross pay (pay before any deductions).

After this pay period, Janet has earned $145,000($142,100 + $2,900).

The new W-4 for 2021 was developed to better estimate deductions for federal income tax. For our discussion of the W-4, we will skip Steps 2 to 4. We will assume all employees accept the standard deduction and only fill out Steps 1 and 5.

The **Federal Insurance Contribution Act (FICA)** funds the **Social Security** program. The program includes Old Age and Disability, Medicare, Survivor Benefits, and so on. The FICA tax requires separate reporting for Social Security and **Medicare.** We will use the following rates for Glo Company:

	Rate	Base
Social Security	6.20%	$142,800
Medicare	1.45	No base

These rates mean that Janet Wong will pay Social Security taxes on the first $142,800 she earns this year. That results in a year's maximum Social Security tax of $8,853.60 once the base is reached. After earning $142,800, Janet's wages will be exempt from Social Security. Note that Janet will be paying Medicare taxes on all wages since Medicare has no base cutoff.

To help keep Glo's record straight, the *taxable earnings column only shows what wages will be taxed. This amount is not the tax.* For example, in week 49, only $700 of Janet's salary will be taxable for Social Security.

$142,800 Social Security base
$- 142,100$ Ⓑ
$\$\quad 700$

Pepper ... And Salt

THE WALL STREET JOURNAL

"I don't like to brag, especially on a W2 form."

FIGURE 9.1
Employee's W-4 form

Form **W-4**	**Employee's Withholding Certificate**	OMB No. 1545-0074
(Rev. December 2020) Department of the Treasury Internal Revenue Service	▶ Complete Form W-4 so that your employer can withhold the correct federal income tax from your pay. ▶ Give Form W-4 to your employer. ▶ Your withholding is subject to review by the IRS.	**2021**

Step 1: Enter Personal Information

(a) First name and middle initial: Janet
Last name: Wong
(b) Social security number: 987-65-4321

Address: 1234 Rolling Hills Lane

City or town, state, and ZIP code: Reading, PA 19606

▶ Does your name match the name on your social security card? if not, to ensure you get credit for your earnings, contact SSA at 800-772-1213 or go to www.ssa.gov.

(c) ☐ Single or Married filing separately
☐ Married filing jointly or Qualifying widow(er)
☐ Head of household (Check only if you're unmarried and pay more than half the costs of keeping up a home for yourself and a qualifying individual.)

Step 5: Sign Here

Under penalties of perjury, I declare that this certificate, to the best of my knowledge and belief, is true, correct, and complete.

Janet Wong

Employee's signature (This form is not valid unless you sign it.) ▶
Date ▶ 1/1/20XX

Employers Only
Employer's name and address
First date of employment
Employer identification number (EIN)

For Privacy Act and Paperwork Reduction Act Notice, see page 3.
Cat. No. 10220Q
Form **W-4** (2021)

To calculate Janet's Social Security tax, we multiply $700 Ⓕ by 6.2%:

$$\$700 \times .062 = \boxed{\$43.40}$$

Since Medicare has no base, Janet's entire weekly salary is taxed 1.45%, which is multiplied by $2,900.

$$\$2,900 \times .0145 = \boxed{\$42.05}$$

Ⓕ, Ⓖ—Taxable earnings for Social Security and Medicare

Ⓗ—Social Security

Ⓘ—Medicare

Ⓙ—FIT

Using the W-4 form Janet completed, Glo deducts **federal income tax withholding (FIT).** Glo uses the percentage method to calculate FIT. The W-4 shows her status and how she is married filing jointly.

The Percentage Method   Since many companies do not want to store tax tables, they use computers for their payroll. We will look at how federal income tax would be calculated for Janet Wong using tables.

TABLE 9.1 Percentage method income tax withholding schedules

2021 Percentage Method Tables

STANDARD Withholding Rate Schedules
(Use these if the box in Step 2 of Form W-4 is **NOT** checked)

TABLE 1—WEEKLY Payroll Period

At least— (A)	But less than— (B)	The tentative amount to withhold is: (C)	Plus this percentage— (D)	of the amount that the Adjusted Wage exceeds— (E)
Married Filing Jointly				
$0	$483	$0.00	0%	$0
$483	$865	$0.00	10%	$483
$865	$2,041	$38.20	12%	$865
$2,041	$3,805	$179.32	22%	$2,041
$3,805	$6,826	$567.40	24%	$3,805
$6,826	$8,538	$1,292.44	32%	$6,826
$8,538	$12,565	$1,840.28	35%	$8,538
$12,565		$3,249.73	37%	$12,565
Single or Married Filing Separately				
$0	$241	$0.00	0%	$0
$241	$433	$0.00	10%	$241
$433	$1,021	$19.20	12%	$433
$1,021	$1,902	$89.76	22%	$1,021
$1,902	$3,413	$283.58	24%	$1,902
$3,413	$4,269	$646.22	32%	$3,413
$4,269	$10,311	$920.14	35%	$4,269
$10,311		$3,034.84	37%	$10,311
Head of Household				
$0	$362	$0.00	0%	$0
$362	$635	$0.00	10%	$362
$635	$1,404	$27.30	12%	$635
$1,404	$2,022	$119.58	22%	$1,404
$2,022	$3,533	$255.54	24%	$2,022
$3,533	$4,388	$618.18	32%	$3,533
$4,388	$10,431	$891.78	35%	$4,388
$10,431		$3,006.83	37%	$10,431

TABLE 2—BIWEEKLY Payroll Period

At least— (A)	But less than— (B)	The tentative amount to withhold is: (C)	Plus this percentage— (D)	of the amount that the Adjusted Wage exceeds— (E)
Married Filing Jointly				
$0	$965	$0.00	0%	$0
$965	$1,731	$0.00	10%	$965
$1,731	$4,083	$76.60	12%	$1,731
$4,083	$7,610	$358.84	22%	$4,083
$7,610	$13,652	$1,134.78	24%	$7,610
$13,652	$17,075	$2,584.86	32%	$13,652
$17,075	$25,131	$3,680.22	35%	$17,075
$25,131		$6,499.82	37%	$25,131
Single or Married Filing Separately				
$0	$483	$0.00	0%	$0
$483	$865	$0.00	10%	$483
$865	$2,041	$38.20	12%	$865
$2,041	$3,805	$179.32	22%	$2,041
$3,805	$6,826	$567.40	24%	$3,805
$6,826	$8,538	$1,292.44	32%	$6,826
$8,538	$20,621	$1,840.28	35%	$8,538
$20,621		$6,069.33	37%	$20,621
Head of Household				
$0	$723	$0.00	0%	$0
$723	$1,269	$0.00	10%	$723
$1,269	$2,808	$54.60	12%	$1,269
$2,808	$4,044	$239.28	22%	$2,808
$4,044	$7,065	$511.20	24%	$4,044
$7,065	$8,777	$1,236.24	32%	$7,065
$8,777	$20,862	$1,784.08	35%	$8,777
$20,862		$6,013.83	37%	$20,862

TABLE 3—MONTHLY Payroll Period

At least— (A)	But less than— (B)	The tentative amount to withhold is: (C)	Plus this percentage— (D)	of the amount that the Adjusted Wage exceeds— (E)
Married Filing Jointly				
$0	$2,092	$0.00	0%	$0
$2,092	$3,750	$0.00	10%	$2,092
$3,750	$8,846	$165.80	12%	$3,750
$8,846	$16,488	$777.32	22%	$8,846
$16,488	$29,579	$2,458.56	24%	$16,488
$29,579	$36,996	$5,600.40	32%	$29,579
$36,996	$54,450	$7,973.84	35%	$36,996
$54,450		$14,082.74	37%	$54,450
Single or Married Filing Separately				
$0	$1,046	$0.00	0%	$0
$1,046	$1,875	$0.00	10%	$1,046
$1,875	$4,423	$82.90	12%	$1,875
$4,423	$8,244	$388.66	22%	$4,423
$8,244	$14,790	$1,229.28	24%	$8,244
$14,790	$18,498	$2,800.32	32%	$14,790
$18,498	$44,679	$3,986.88	35%	$18,498
$44,679		$13,150.23	37%	$44,679
Head of Household				
$0	$1,567	$0.00	0%	$0
$1,567	$2,750	$0.00	10%	$1,567
$2,750	$6,083	$118.30	12%	$2,750
$6,083	$8,763	$518.26	22%	$6,083
$8,763	$15,308	$1,107.86	24%	$8,763
$15,308	$19,017	$2,678.66	32%	$15,308
$19,017	$45,200	$3,865.54	35%	$19,017
$45,200		$13,029.59	37%	$45,200

Source: Internal Revenue Service

Remember Janet is married filing jointly. (See Figure 9.1: Employee's W-4 form.)
Table 9.1 shows her salary falls between $2,041 and $3,805. This means her FIT tax is
$179.32 plus 22% of the excess over $2,041:

$$\begin{array}{r} \$2,900 \\ -2,041 \\ \hline \$\ \ \ 859 \text{ excess} \end{array}$$

FIT = $179.32 + ($859 × .22)

FIT = $179.32 + $188.98

FIT = $368.30

We assume a 6% **state income tax (SIT)**.

—SIT

$2,900 × .06 = $174

Janet contributes $100 per week for health insurance.
Janet's **net pay** is her gross pay less all deductions.

(L)—Health insurance
(M)—Net pay

$2,900.00	Gross
− 43.40	Social Security
− 42.05	Medicare
− 368.30	FIT
− 174.00	SIT
− 100.00	Health insurance
= $2,172.25	Net pay

Learn: Employers' Responsibilities

In the first section of this unit, we saw that Janet contributed to Social
Security and Medicare. Glo Company has the legal responsibility to match
her contributions. Besides matching Social Security and Medicare, Glo must
pay two important taxes that employees do not have to pay—federal and state
unemployment taxes.

Federal Unemployment Tax Act (FUTA)   The federal government participates in a
joint federal-state unemployment program to help unemployed workers. At this writing,
employers pay the government a 6% **FUTA** tax on the first $7,000 paid to employees as
wages during the calendar year. Any wages in excess of $7,000 per worker are exempt
wages and are not taxed for FUTA. If the total cumulative amount the employer owes the
government is less than $100, the employer can pay the liability yearly (end of January
in the following calendar year). If the tax is greater than $100, the employer must pay it
within a month after the quarter ends.

Companies involved in a state unemployment tax fund can usually take a 5.4% credit
against their FUTA tax. *In reality, then, companies are paying .6% (.006) to the federal
unemployment program.* In all our calculations, FUTA is .006.

Example: Assume a company had total wages of $19,000 in a calendar year. No employee earned more than $7,000 during the calendar year. The FUTA tax is .6% (6% minus the company's 5.4% credit for state unemployment tax). How much does the company pay in FUTA tax?

The company calculates its FUTA tax as follows:

$$
\begin{array}{rl}
& 6\% \text{ FUTA tax} \\
- & 5.4\% \text{ credit for SUTA tax} \\
= & .6\% \text{ tax for FUTA}
\end{array}
$$

.006 × $19,000 = **$114** FUTA tax due to federal government

State Unemployment Tax Act (SUTA) The current **SUTA** tax in many states is 5.4% on the first $7,000 the employer pays an employee. As an example, the SUTA rate base in Alabama is $8,000. Some states offer a merit rating system that results in a lower SUTA rate for companies with a stable employment period. The federal government still allows 5.4% credit on FUTA tax to companies entitled to the lower SUTA rate. Usually states also charge companies with a poor employment record a higher SUTA rate. However, these companies cannot take any more than the 5.4% credit against the 6% federal unemployment rate.

Example: Assume a company has total wages of $20,000 and $4,000 of the wages are exempt from SUTA. What are the company's SUTA and FUTA taxes if the company's SUTA rate is 5.8% due to a poor employment record?

The exempt wages (over $7,000 earnings per worker) are not taxed for SUTA or FUTA. So the company owes the following SUTA and FUTA taxes:

$$
\begin{array}{rl}
& \$20,000 \\
- & 4,000 \text{ (exempt wages)} \\
\hline
& \$16,000 \times .058 = \boxed{\$928} \text{ SUTA}
\end{array}
$$

Federal FUTA tax would then be:

$16,000 × .006 = **$96**

You can check your progress with the following Practice Quiz.

Practice Quiz

Complete this Practice Quiz to see how you are doing.

1. Calculate Social Security taxes, Medicare taxes, and FIT for Joy Royce. Joy's company pays her a monthly salary of $16,600. She is single. Before this payroll, Joy's cumulative earnings were $132,800. (Social Security maximum is 6.2% on $142,800, and Medicare is 1.45%.) Calculate FIT by the percentage method.

2. Jim Brewer, owner of Arrow Company, has three employees who earn $300, $700, and $900 a week. Assume a state SUTA rate of 5.1%. What will Jim pay for state and federal unemployment taxes for the first quarter?

✓ **Solutions**

1. **Social Security**   **Medicare**

   $142,800 $16,600 × .0145 = $240.70
   − 132,800
   $ 10,000 × .062 = $620

   FIT

   Percentage method: $16,600

   $14,790 to $18,498 → $2,800.32 plus 32% of excess over $14,790
   (Table 9.1)

   $16,600
   − 14,790
   1,810 × .32 = $ 579.20
                   + 2,800.32
                   $ 3,379.52

2. 13 weeks × $300 = $ 3,900
   13 weeks × $700 =   9,100 ($9,100 − $7,000) → $2,100 ⎫ Exempt wages
   13 weeks × $900 =   11,700 ($11,700 − $7,000) → 4,700 ⎬ (not taxed for
                      $24,700 $6,800 ⎭ FUTA or SUTA)

   $24,700 − $6,800 = $17,900 taxable wages   *Note:* FUTA remains at .006
   SUTA = .051 × $17,900 = $912.90   whether SUTA rate is higher or
   FUTA = .006 × $17,900 = $107.40   lower than standard.

Chapter 9 Review

Topic/Procedure/Formula	Example	You try it*
Gross pay $$\underset{\text{worked}}{\text{Hours employee}} \times \underset{\text{hour}}{\text{Rate per}}$$	$6.50 per hour at 36 hours Gross pay = 36 × $6.50 = $234	**Calculate gross pay** $9.25 per hour; 38 hours
Overtime $$\underset{\text{(pay)}}{\underset{\text{earnings} =}{\text{Gross}}} \; \underset{\text{pay}}{\text{Regular}} + \underset{(1\frac{1}{2})}{\underset{\text{overtime rate}}{\text{Earnings at}}}$$	$6 per hour; 42 hours Gross pay = (40 × $6) + (2 × $9) = $240 + $18 = $258	**Calculate gross pay** $7 per hour; 43 hours
Straight piece rate $$\underset{\text{pay}}{\text{Gross}} = \underset{\text{produced}}{\text{Number of units}} \times \underset{\text{unit}}{\text{Rate per}}$$	1,185 units; rate per unit, $.89 Gross pay = 1,185 × $.89 = $1,054.65	**Calculate gross pay** 2,250 units; $.79 per unit
Differential pay schedule Rate on each item is related to the number of items produced.	1–500 at $.84; 501–1,000 at $.96; 900 units produced. Gross pay = (500 × $.84) + (400 × $.96) = $420 + $384 = $804	**Calculate gross pay** 1–600 at $.79; 601–1,000 at $.88; 900 produced
Straight commission Total sales × Commission rate Any draw would be subtracted from earnings.	$155,000 sales; 6% commission $155,000 × .06 = $9,300	**Calculate straight commission** $175,000 sales; 7% commission
Variable commission scale Sales at different levels pay different rates of commission.	Up to $5,000, 5%; $5,001 to $10,000, 8%; over $10,000, 10% Sold: $6,500 Solution: ($5,000 × .05) + ($1,500 × .08) = $250 + $120 = $370	**Calculate commission** Up to $6,000, 5%; $6,001 to $8,000, 9%; Over $8,000, 12% Sold: $12,000
Salary plus commission $$\underset{\text{(fixed)}}{\text{Regular wages}} + \underset{\text{earned}}{\text{Commissions}}$$	Base $400 per week + 2% on sales over $14,000 Actual sales: $16,000 $400 (base) + (.02 × $2,000) = $440	**Calculate gross pay** Base $600 per week plus 4% on sales over $16,000. Actual sales $22,000.
Payroll register Multicolumn form to record payroll. Married filing jointly and paid weekly. (Table 9.1) FICA rates from chapter. Wage base has not been reached.	(see table below)	**Calculate net pay** Gross pay, $490; Married filing jointly, paid weekly. Use rates in text for Social Security, Medicare, and FIT.

Earnings	Deductions			Net
	FICA			pay
Gross	S.S.	Med.	FIT	
$865	53.63	12.54	38.20	$760.63

Topic/Procedure/Formula	Example	You try it*
FICA **Social Security Medicare** 6.2% on $142,800 (S.S.) 1.45% (Med.)	If John earns $150,000, what did he contribute for the year to Social Security and Medicare? S.S.: $142,800 × .062 = $8,853.60 Med.: $150,000 × .0145 = $2,175.00	**Calculate FICA** If John earns $160,000, what did he contribute to Social Security and Medicare?
FIT calculation (percentage method) *Facts:* Al Doe: Married Paid weekly: $1,600	$1,600.00 $865 to $2,041 → By Table 9.1 $ 1,600 − 865 $ 735 $38.20 + .12($735) $38.20 + $88.20 = $126.40	**Calculate FIT** Jim Smith, married, filing jointly, Paid weekly, $1,400
State and federal unemployment Employer pays these taxes. Rates are 6% on $7,000 for federal and 5.4% for state on $7,000 (6% − 5.4% = .6% federal rate after credit). If state unemployment rate is higher than 5.4%, no additional credit is taken. If state unemployment rate is less than 5.4%, the full 5.4% credit can be taken for federal unemployment.	Cumulative pay before payroll, $6,400; this week's pay, $800. What are state and federal unemployment taxes for employer, assuming a 5.2% state unemployment rate? State → .052 × $600 = $31.20 Federal → .006 × $600 = $3.60 ($6,400 + $600 = $7,000 maximum)	**Calculate SUTA and FUTA** Cumulative pay before payroll, $6,800. This week's payroll, $9,000. State rate is 5.4%.

Key Terms

Biweekly

Deductions

Differential pay schedule

Draw

Employee's Withholding Certificate (W-4)

Fair Labor Standards Act

Federal income tax withholding (FIT)

Federal Insurance Contribution Act (FICA)

Federal Unemployment Tax Act (FUTA)

Gross pay

Medicare

Monthly

Net pay

Overrides

Overtime

Payroll register

Percentage method

Semimonthly

Social Security

State income tax (SIT)

State Unemployment Tax Act (SUTA)

Straight commission

Variable commission scale

W-4

Weekly

*Worked-out solutions are in Appendix A.

Chapter 9 Review *(Continued)*

Critical Thinking Discussion Questions with Chapter Concept Check

1. Explain the difference between biweekly and semimonthly. Explain what problems may develop if a retail store hires someone on straight commission to sell cosmetics.

2. Explain what each column of a payroll register records and how each number is calculated. Social Security tax is based on a specific rate and base; Medicare tax is based on a rate but has no base. Do you think this is fair to all taxpayers?

3. What taxes are the responsibility of the employer? How can an employer benefit from a merit-rating system for state unemployment?

4. **Chapter Concept Check.** Visit Facebook's corporate website to see what benefits the company provides for its employees. Discuss the responsibilities of the employee and the employer.

End-of-Chapter Problems

Name _____ Date_____

Check figures for odd-numbered problems in Appendix A.

Drill Problems

Complete the following table: *LU 9–1(2)*

	Employee	M	T	W	Th	F	Hours	Rate per hour	Gross pay
9–1.	Bernie Roy	9	6	9	7	6		$8.95	
9–2.	Kristina Shaw	5	9	10	8	8		$8.10	

Complete the following table (assume the overtime for each employee is a time-and-a-half rate after 40 hours):   *LU 9–1(2)*

	Employee	M	T	W	Th	F	Sa	Total regular hours	Total overtime hours	Regular rate	Overtime rate	Gross earnings
9–3.	Blue	12	9	9	9	9	3			$8.00		
9–4.	Tagney	14	8	9	9	5	1			$7.60		

Calculate gross earnings: *LU 9–1(3)*

	Worker	Number of units produced	Rate per unit	Gross earnings
9–5.	Lang	480	$3.50	
9–6.	Swan	846	$ .58	

Calculate the gross earnings for each apple picker based on the following differential pay scale: *LU 9–1(3)*

1–1,000: $.03 each   1,001–1,600: $.05 each   Over 1,600: $.07 each

	Apple picker	Number of apples picked	Gross earnings
9–7.	Ryan	1,600	
9–8.	Rice	1,925	

Calculate the end-of-month commission.   *LU 9–1(3)*

	Employee	Total sales	Commission rate	Draw	End-of-month commission received
9–9.	Reese	$300,000	7%	$8,000	

End-of-Chapter Problems (Continued)

Ron Company has the following commission schedule:

Commission rate	Sales
2%	Up to $80,000
3.5%	Excess of $80,000 to $100,000
4%	More than $100,000

Calculate the gross earnings of Ron Company's two employees: *LU 9–1(3)*

	Employee	Total sales	Gross earnings
9–10.	Bill Moore	$ 70,000	
9–11.	Ron Ear	$155,000	

Complete the following table, given that A Publishing Company pays its salespeople a weekly salary plus a 2% commission on all net sales over $5,000 (no commission on returned goods):   *LU 9–1(3)*

	Employee	Gross sales	Return sales	Net quota	Given sales	Commission rates	Commission sales	Total commission	Regular wage	Total wage
excel 9–12.	Ring	$ 8,000	$ 25	$5,000		2%			$250	
excel 9–13.	Porter	$12,000	$100	$5,000		2%			$250	

Calculate the Social Security and Medicare deductions for the following employees (assume a tax rate of 6.2% on $142,800 for Social Security and 1.45% for Medicare):   *LU 9–2(1)*

	Employee	Cumulative earnings before this pay period	Pay amount this period	Social Security	Medicare
9–14.	Logan	$130,000	$4,000		
9–15.	Rouche	$142,000	$9,000		
9–16.	Cleaves	$175,000	$19,000		

Complete the following payroll register. Calculate FIT by the percentage method for this weekly period; Social Security and Medicare are the same rates as in the previous problems. No one will reach the maximum for FICA.   *LU 9–2(1)*

Employee	Marital status	Gross pay	FIT	FICA S.S.	FICA Med.	Net pay
9–17. Mike Rice	M, filing jointly	$2,000				
9–18. Pat Brown	S	$2,500				

9–19. Given the following, calculate the state (assume 5.3%) and federal unemployment taxes that the employer must pay for each of the first two quarters. The federal unemployment tax is .6% on the first $7,000. *LU 9–2(2)*

PAYROLL SUMMARY		
	Quarter 1	Quarter 2
Bill Adams	$4,000	$ 8,000
Rich Haines	8,000	14,000
Alice Smooth	3,200	3,800

Word Problems

9–20. Lai Xiaodong, a 22-year-old college-educated man, accepted a job at Foxconn Technology (where the iPad was being produced for Apple) in Chengdu, China, for $22 a day at 12 hours a day, 6 days a week. A company perk included company housing in dorms for the 70,000 employees. It was common for 20 people to be assigned to the same three-bedroom apartment. What were Lai's hourly (rounded to the nearest cent), weekly, and annual gross pay? *LU 9–1(1)*

9–21. Rhonda Brennan found her first job after graduating from college through the classifieds of the *Miami Herald.* She was delighted when the offer came through at $18.50 per hour. She completed her W-4 stating that she is married filing jointly. Her company will pay her biweekly for 80 hours. Calculate her take-home pay for her first check. Assume Social Security and Medicare rates from the chapter. *LU 9–2(1)*

End-of-Chapter Problems (Continued)

9–22. The Social Security Administration increased the taxable wage base from $137,700 to $142,800. The 6.2% tax rate is unchanged. Joe Burns earned over $130,000 each of the past two years. What is the percent increase in the base? Round to the nearest hundredth percent.   *LU 9–2(1)*

9–23. Calculate Social Security taxes, Medicare taxes, and FIT for Jordon Barrett. He earns a monthly salary of $12,000. He is single. Before this payroll, Barrett's cumulative earnings were $142,000. (Social Security maximum is 6.2% on $142,800 and Medicare is 1.45%.) Calculate FIT by the percentage method.   *LU 9–2(1)*

9–24. Maggie Vitteta, single, works 40 hours per week at $15.00 an hour. How much is taken out for federal income tax?   *LU 9–2(1)*

9–25. Robin Hartman earns $600 per week plus 3% of sales over $6,500. Robin's sales are $14,000. How much does Robin earn?   *LU 9–1(3)*

9–26. Pat Maninen earns a gross salary of $3,000 each week. What are Pat's first week's deductions for Social Security and Medicare? Will any of Pat's wages be exempt from Social Security and Medicare for the calendar year? Assume a rate of 6.2% on $142,800 for Social Security and 1.45% for Medicare.   *LU 9–2(1)*

9–27. Richard Gaziano is a manager for Health Care, Inc. Health Care deducts Social Security, Medicare, and FIT (by percentage method) from his earnings. Assume the same Social Security and Medicare rates as in Problem 9–26. Before this payroll, Richard is $1,000 below the maximum level for Social Security earnings. Richard is married, filing jointly and is paid weekly. What is Richard's net pay for the week if he earns $1,300?   *LU 9–2(1)*

9–28. Larren Buffett is concerned after receiving her weekly paycheck. She believes that her deductions for Social Security, Medicare, and federal income tax withholding (FIT) may be incorrect. Larren is paid a salary of $4,100 weekly. She is married, filing jointly and prior to this payroll check has total earnings of $140,000. What are the correct deductions for Social Security, Medicare, and FIT? *LU 9–2(2)*

e)(cel **9–29.** Westway Company pays Suzie Chan $3,000 per week. By the end of week 52, how much did Westway deduct for Suzie's Social Security and Medicare for the year? Assume Social Security is 6.2% on $142,800 and 1.45% for Medicare. What state and federal unemployment taxes does Westway pay on Suzie's yearly salary? The state unemployment rate is 5.1%. FUTA is .6%. *LU 9–2(1, 2)*

9–30. Sarah Jones earns $525 per week selling life insurance for Farmer's Insurance plus 5% of sales over $5,750. Sarah's sales this month (four weeks) are $20,000. How much does Sarah earn this month? *LU 9–2(2)*

e)(cel **9–31.** Tiffani Lind earned $1,200 during her biweekly pay period. She is married filing jointly. Her annual earnings to date are $52,800. Calculate her net pay. *LU 9–2(1)*

End-of-Chapter Problems (Continued)

Challenge Problems

9–32. The San Bernardino County Fair hires about 150 people during fair time. California has a state income tax of 9%. Sandy Denny earns $13.00 per hour; George Barney earns $14.00 per hour. They both worked 39 hours this week. Both are married; however, Sandy files jointly and George files separately. Assume a rate of 6.2% on $142,800 for Social Security and 1.45% for Medicare. **(a)** What is Sandy's net pay after FIT (use the tables in the text), Social Security tax, state income tax, and Medicare have been taken out? **(b)** What is George's net pay after the same deductions? **(c)** What is the difference between Sandy's net pay and George's net pay? Round to the nearest cent. *LU 9–2(1)*

9–33. Bill Rose is a salesperson for Boxes, Inc. He believes his $1,460.47 monthly paycheck is in error. Bill earns a $1,400 salary per month plus a 9.5% commission on sales over $1,500. Last month, Bill had $8,250 in sales. Bill believes his traveling expenses are 16% of his weekly gross earnings before commissions. Monthly deductions include Social Security, $126.56; Medicare, $29.60; FIT, $189.50; union dues, $25.00; and health insurance, $16.99. Calculate the following: **(a)** Bill's monthly take-home pay, and indicate the amount his check was under- or overstated, and **(b)** Bill's weekly traveling expenses. Round your final answer to the nearest dollar. *LU 9–2(1)*

Summary Practice Test

Do you need help? Connect videos have step-by-step worked-out solutions.

1. Calculate Sam's gross pay (he is entitled to time-and-a-half).   *LU 9–1(2)*

M	T	W	Th	F	Total hours	Rate per hour	Gross pay
$9\frac{1}{4}$	$9\frac{1}{4}$	$10\frac{1}{2}$	$8\frac{1}{2}$	$11\frac{1}{2}$		$8.00	

2. Mia Kaminsky sells shoes for Macy's. Macy's pays Mia $12 per hour plus a 5% commission on all sales. Assume Mia works 37 hours for the week and has $7,000 in sales. What is Mia's gross pay?   *LU 9–1(3)*

3. Lee Company pays its employees on a graduated commission scale: 6% on the first $40,000 sales, 7% on sales from $40,001 to $80,000, and 13% on sales of more than $80,000. May West, an employee of Lee, has $230,000 in sales. What commission did May earn? *LU 9–1(3)*

4. Matty Kim, an accountant for Vernitron, earned $135,000 from January to June. In July, Matty earned $20,000. Assume a tax rate of 6.2% for Social Security on $142,800 and 1.45% on Medicare. How much are the July taxes for Social Security and Medicare?   *LU 9–2(1)*

5. Grace Kelley earns $2,000 per week. She is married and files jointly. What is Grace's income tax? Use the percentage method.   *LU 9–2(1)*

6. Jean Michaud pays his two employees $900 and $1,200 per week. Assume a state unemployment tax rate of 5.7% and a federal unemployment tax rate of .6%. What state and federal unemployment taxes will Jean pay at the end of quarter 1 and quarter 2?   *LU 9–2(2)*

MY MONEY

My Paycheck: Where Did All My Money Go?

What I need to know

It is very important to understand how your pay is calculated and how deductions are made to arrive at your net pay. Knowing this allows you to verify your pay and budget your earnings more effectively. As your pay increases after college, these deductions become more significant in terms of their impact on your net pay. As your income changes, you may need to adjust your deductions to stay in line with your financial goals and tax responsibilities. FICA (the Federal Insurance Contributions Act) taxes, commonly called payroll tax, we see as deductions on our paychecks. FICA includes taxes to fund Social Security and Medicare. In addition to FICA, you may also see deductions for federal, state, and local taxes. Insurance costs and other voluntary deductions have a direct impact on net pay. Understanding your net pay allows you to make informed decisions about your money, letting you know how much you have available for saving and investing, critical parts of building personal financial health.

What I need to do

When considering your future career pay you will need to consider the tax implications of your earnings. By taking the time to determine the financial impact of your tax responsibility, you will be able to prepare yourself for future budgeting and stay on track with your financial goals. As you transition to higher earnings after college, the deductions from your paycheck will increase. Understanding these calculations is crucial to evaluating job offers when it comes to salary since you will gain a better understanding of the net pay you will ultimately receive. This allows you to determine if the salary amount under consideration will provide you the opportunity to meet your financial goals.

Ask for assistance from your employer to help you fully understand the impact of any mandatory or voluntary deductions on your net pay. There may be ancillary services offered by your employer to help with health costs you may incur. Some employers offer flexible spending accounts to help offset the out-of-pocket costs you incur when using health services. It is important for you to understand how these programs work and whether or not you could benefit from such services. Be sure to note if the flexible spending account is a "use it or lose it" program to ensure you are not missing out on the benefits of the program.

Steps I need to take

1. Know the deductions taken from your earnings and how they impact your net pay.
2. Familiarize yourself with the new (in 2020) W-4 form and its impact on your tax situation.
3. Review your tax situation yearly and adjust your withholding when needed.

Resources I can use

- Paycheck Calculator (mobile app)—estimate your paycheck and related deductions.
- https://www.irs.gov/newsroom/faqs-on-the-2020-form-w-4—frequently asked questions about the new W-4 form.

MY MONEY ACTIVITY

- Use your expected career pay from the Chapter 2 My Money feature.
- Identify your FIT, SS, and Medicare deductions to arrive at your estimated net pay.
- Discuss the pros and cons of adjusting your tax withholding via the (new in 2020) W-4 form.

PERSONAL FINANCE

A KIPLINGER APPROACH

"Keep Child Care Costs in Check." Kiplinger's. March 2021.

MILLENNIAL MONEY | Lisa Gerstner

Keep Child Care Costs in Check

The high price of child care is all too familiar to my family. The monthly cost of care for our toddler usually tops $1,000 and is second only to our mortgage payment. More than 70% of parents spend at least 10% of their income on child care, and more than half spend at least $10,000 per year, according to Care.com.

The coronavirus pandemic has added to the strain on working parents. Some have struggled to afford care following cuts in their pay or hours. Others have changed care providers or juggled job responsibilities and child care duties when schools and day care centers closed because of COVID-19. Parents have even left the workforce to care for their kids, removing care expenses from their budgets but losing income.

Choosing your care. Whether you're seeking child care for the first time or you're re-evaluating your options, be sure you understand the financial implications. A nanny, who comes to your home, is convenient. But the average weekly rate to have a nanny care for one infant is $565, according to Care.com—much higher than the $215-a-week average for a day care center and $201 for in-home day care. One way to reduce expenses is to share a nanny with another family, with whom you can split the cost.

Plus, hiring a nanny often comes with tax implications. If you pay a nanny at least $2,300 in 2021, the IRS requires you to treat him or her as a household employee. You must withhold Social Security and Medicare tax from the nanny's pay—and as the employer, you have to kick in Social Security and Medicare tax, too (you and the nanny each pay 7.65% of wages). You must also issue a Form W-2 each January and file other forms with the IRS. And you may be expected to cover transportation, meals and two weeks of vacation, says Dana Levin-Robinson, CEO of Upfront, a price-comparison website for child care services.

The expenses and tax complications that come with hiring a nanny were reason enough for me to take my son to day care instead. Consider other costs and savings, too. Day care centers and preschools may include snacks and meals in their rate. But they may also charge annual fees or penalties if you pick up your child late.

Don't overlook tax breaks. If you have earned

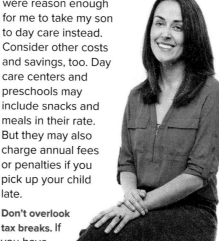

Poon Watchara-Amphaiwan

income from employment during the year and pay for care while you work or look for work, you can take a federal tax credit of 20% to 35% of care expenses (the percentage depends on your income) for up to $3,000 paid for one child or $6,000 for two or more children younger than 13. You can claim the credit whether the care is in or out of your home, and you must report the care provider's name, address and tax identification number.

Your employer may allow you to stash up to $5,000 of pretax money annually in a dependent care flexible spending account. You can use the funds to pay for a nanny or day care while you work, as well as for before- and after-school programs or summer day camp. The recently passed COVID relief law includes provisions through which employers may permit unlimited carryovers of unused FSA funds from the 2020 plan year to 2021 (and from 2021 to 2022), or extend the grace period to use 2020 or 2021 FSA funds from 2.5 months to 12 months.

You may have until the end of 2021, for example, to use money that you put in an FSA in 2020, depending on your employer's rules. The law also temporarily raises the limit of a child's age of eligibility for dependent FSA coverage from 12 to 13. ∎

TO SHARE THIS COLUMN, PLEASE GO TO KIPLINGER.COM/ LINKS/MILLENNIALS. YOU CAN CONTACT THE AUTHOR AT LISA_GERSTNER@KIPLINGER.COM.

Business Math Issue

Hiring a nanny has no tax implications.

1. List the key points of the article and information to support your position.

2. Write a group defense of your position using math calculations to support your view. If you are in an online course, post to a discussion board.

Chapter 10
Simple Interest

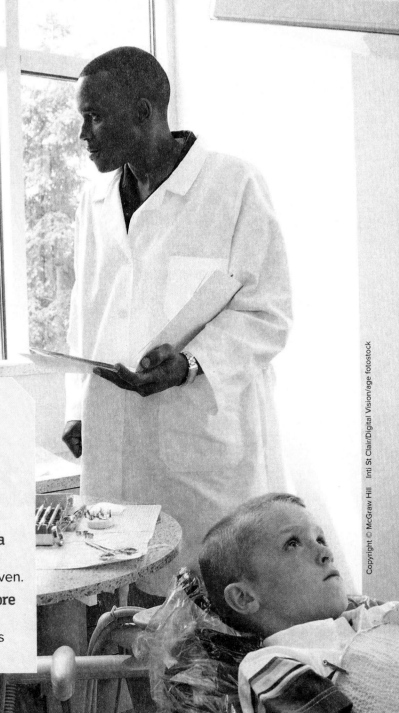

Learning Unit Objectives

LU 10–1: Calculation of Simple Interest and Maturity Value

1. Calculate simple interest and maturity value for months and years.
2. Calculate simple interest and maturity value by **(a)** exact interest and **(b)** ordinary interest.

LU 10–2: Finding Unknown in Simple Interest Formula

1. Using the interest formula, calculate the unknown when the other two (principal, rate, or time) are given.

LU 10–3: U.S. Rule—Making Partial Note Payments before Due Date

1. List the steps to complete the U.S. Rule as well as calculate proper interest credits.

Growing Interest

Monthly loan payment and total interest paid on a $20,000 loan, with a 4.5% interest rate

Legend:
- ■ Monthly payment
- ■ Total interest paid

LOAN PERIOD (in years): Three, Five, Seven, Ten

Source: Bankrate.com loan calculator

PETER OUMANSKI

""Growing Interest" Infographic." *The Wall Street Journal*, February 15, 2020.

Copyright © McGraw Hill   (tablet)Can Yesil/Shutterstock, PureSolution/Shutterstock

@ Essential Question

How can I use simple interest to understand the business world and make my life easier?

🌐 Math Around the World

The chapter opener *Wall Street Journal* clip, "The $1 Million Student Loan: 'Should I Be Doing This?'," shows that when student loan payments do not cover interest, the balance owed increases substantially. Since the pandemic, interest rates are at new lows. This should help many borrowers.

In this chapter, you will study simple interest. The principles discussed apply whether you are paying interest or receiving interest. Let's begin by learning how to calculate simple interest.

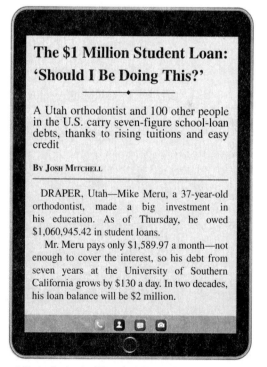

The $1 Million Student Loan: 'Should I Be Doing This?'

A Utah orthodontist and 100 other people in the U.S. carry seven-figure school-loan debts, thanks to rising tuitions and easy credit

BY JOSH MITCHELL

DRAPER, Utah—Mike Meru, a 37-year-old orthodontist, made a big investment in his education. As of Thursday, he owed $1,060,945.42 in student loans.

Mr. Meru pays only $1,589.97 a month—not enough to cover the interest, so his debt from seven years at the University of Southern California grows by $130 a day. In two decades, his loan balance will be $2 million.

Mitchell, Josh. "The $1 Million Student Loan: 'Should I Be Doing This?'" *The Wall Street Journal*, May 26, 2020.

Learning Unit 10–1:

Calculation of Simple Interest and Maturity Value

Hope Slater, a young attorney, rented an office in a professional building. Since Hope recently graduated from law school, she was short of cash. To purchase office furniture for her new office, Hope went to her bank and borrowed $40,000 for 6 months at a 4% annual interest rate. **Interest** expense is the cost of borrowing money.

The original amount Hope borrowed ($40,000) is the **principal** (face value) of the loan. Hope's price for using the $40,000 is the interest rate (4%) the bank charges on a yearly basis. Since Hope is borrowing the $40,000 for 6 months, Hope's loan will have a **maturity value** of $40,800—the principal plus the interest on the loan. Thus, Hope's price for using the furniture before she can pay for it is $800 interest, which is a percent of the principal for a specific time period. To make this calculation, we use the following formula:

Maturity Value (MV)

Maturity value (MV) = Principal (P) + Interest (I)

$40,800 = $40,000 + $800

Hope's furniture purchase introduces **simple interest**—the cost of a loan, usually for 1 year or less. Simple interest is only on the original principal or amount borrowed. Let's examine how the bank calculated Hope's $800 interest.

Learn: Simple Interest Formula

To calculate simple interest, we use the following **simple interest formula:**

Simple Interest Formula

Simple interest (I) = Principal (P) × Rate (R) × Time (T)

In this formula, rate is expressed as a decimal, fraction, or percent; and time is expressed in years or a fraction of a year.

Aha!

Do not round intermediate answers. Round only the final calculation.

Example: Hope Slater borrowed $40,000 for office furniture. The loan was for 6 months at an annual interest rate of 4%. What are Hope's interest and maturity value?

Using the simple interest formula, the bank determined Hope's interest as follows:

In your calculator, multiply $40,000 times .04 times 6. Divide your answer by 12. You could also use the % key—multiply $40,000 times 4% times 6 and then divide your answer by 12.

Step 1 Calculate the interest.

$$I = \$40,000 \times .04 \times \frac{6}{12}$$
$$\quad\quad (P)\quad\ (R)\quad (T)$$
$$= \$800$$

Step 2 Calculate the maturity value.

$$MV = \$40,000 + \$800$$
$$\quad\quad\quad (P)\quad\quad (I)$$
$$= \$40,800$$

Now let's use the same example and assume Hope borrowed $40,000 for 1 year. The bank would calculate Hope's interest and maturity value as follows:

Step 1 Calculate the interest.

$$I = \underset{(P)}{\$40,000} \times \underset{(R)}{.04} \times \underset{(T)}{1 \text{ year}}$$
$$= \$1,600$$

Step 2 Calculate the maturity value.

$$MV = \underset{(P)}{\$40,000} + \underset{(I)}{\$1,600}$$
$$= \boxed{\$41,600}$$

Let's use the same example again and assume Hope borrowed $40,000 for 18 months.[1] Then Hope's interest and maturity value would be calculated as follows:

Step 1 Calculate the interest.

$$I = \underset{(P)}{\$40,000} \times \underset{(R)}{.04} \times \underset{(T)}{\frac{18}{12}}$$
$$= \$2,400$$

Step 2 Calculate the maturity value.

$$MV = \underset{(P)}{\$40,000} + \underset{(I)}{\$2,400}$$
$$= \boxed{\$42,400}$$

Next we'll turn our attention to two common methods we can use to calculate simple interest when a loan specifies its beginning and ending dates.

Learn: Two Methods for Calculating Simple Interest and Maturity Value

Method 1: Exact Interest (365 Days) The Federal Reserve banks and the federal government use the **exact interest** method. The *exact interest* is calculated by using a 365-day year. For **time,** we count the exact number of days in the month that the borrower has the loan. The day the loan is made is not counted, but the day the money is returned is counted as a full day. This method calculates interest by using the following fraction to represent time in the formula:

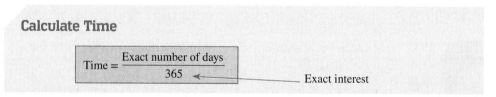

Calculate Time

$$\text{Time} = \frac{\text{Exact number of days}}{365} \longleftarrow \text{Exact interest}$$

For this calculation, we use the exact days-in-a-year calendar from the *Business Math Handbook.* You learned how to use this calendar in Chapter 7.

Example: On March 4, Joe Bench borrowed $50,000 at 5% interest. Interest and principal are due on July 6. What are the interest cost and the maturity value?

Step 1 Calculate the interest.

$$I = P \times R \times T$$
$$= \$50,000 \times .05 \times \frac{124}{365}$$
$$= \$849.32 \text{ (rounded to nearest cent)}$$

Step 2 Calculate the maturity value.

$$MV = P + I$$
$$= \$50,000 + \$849.32$$
$$= \boxed{\$50,849.32}$$

From the *Business Math Handbook*

July 6	187th day
March 4	− 63rd day
	124 days (exact time of loan)
March	31
	− 4
	27
April	30
May	31
June	30
July	+ 6
	124 days

Method 2: Ordinary Interest (360 Days) In the **ordinary interest** method, time in the formula $I = P \times R \times T$ is equal to the following:

[1] This is the same as 1.5 years.

Calculate Time

$$\text{Time} = \frac{\text{Exact number of days}}{360} \longleftarrow \text{Ordinary interest}$$

Since banks commonly use the ordinary interest method, it is known as the **Banker's Rule.** Banks charge a slightly higher rate of interest because they use 360 days instead of 365 in the denominator. (Here's a hint: The word *ordinary* starts with an "O" and "360" ends with a "0.") By using 360 instead of 365, the calculation is supposedly simplified. Consumer groups, however, are questioning why banks can use 360 days, since this benefits the bank and not the customer. The use of computers and calculators no longer makes the simplified calculation necessary. For example, after a court case in Oregon, banks began calculating interest on 365 days except in mortgages.

Now let's replay the Joe Bench example we used to illustrate Method 1 to see the difference in bank interest when we use Method 2.

Example: On March 4, Joe Bench borrowed $50,000 at 5% interest. Interest and principal are due on July 6. What are the interest cost and the maturity value?

Step 1 Calculate the interest.

$$I = \$50,000 \times .05 \times \frac{124}{360}$$

$$= \$861.11$$

Step 2 Calculate the maturity value.

$$MV = P + I$$
$$= \$50,000 + \$861.11$$
$$= \boxed{\$50,861.11}$$

Note: By using Method 2, the bank increases its interest by $11.79.

$$
\begin{array}{r}
\$861.11 \;\longleftarrow \text{Method 2}\\
-\;849.32 \;\longleftarrow \text{Method 1}\\
\hline
\$\;11.79
\end{array}
$$

 Use ordinary interest any time a problem does not specify to use exact interest.

The *Wall Street Journal* clip to the left looks at how to deal with loan defaults. Now you should be ready for your first Practice Quiz in this chapter.

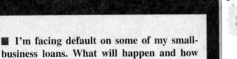

■ I'm facing default on some of my small-business loans. What will happen and how can I take steps to minimize the damage? Defaulting on a business loan can have business and personal consequences. Be proactive, says Ms. Kaufman, the New York City lawyer. Don't wait for the loan to default—contact your lender right away. Many lenders are offering relief in light of the pandemic.

Generally, when you miss a payment, your lender will reach out to see what can be done to get you back on track. If you don't respond—or don't start paying what you owe; the loan may be "accelerated." Instead of owing your monthly payments of principal and interest, the entire loan becomes due.

"Reader Question." *The Wall Street Journal,* August 1, 2020.

Practice Quiz

Complete this Practice Quiz to see how you are doing.

Calculate simple interest (rounded to the nearest cent):

1. $14,000 at 4% for 9 months

2. $25,000 at 7% for 5 years

3. $40,000 at $10\frac{1}{2}$% for 19 months

4. On May 4, Dawn Kristal borrowed $15,000 at 8%. Dawn must pay the principal and interest on August 10. What are Dawn's simple interest and maturity value if you use the exact interest method?

5. What are Dawn Kristal's (Problem 4) simple interest and maturity value if you use the ordinary interest method?

✓ Solutions

1. $\$14,000 \times .04 \times \dfrac{9}{12} = \$420$

2. $\$25,000 \times .07 \times 5 = \$8,750$

3. $\$40,000 \times .105 \times \dfrac{19}{12} = \$6,650$

4.
$$
\begin{array}{ll}
\text{August 10} \rightarrow & 222 \\
\text{May 4} \rightarrow & -\ 124 \\
\hline
& 98
\end{array}
$$

5. $\$15,000 \times .08 \times \dfrac{98}{360} = \$326.67$

   $\$15,000 \times .08 \times \dfrac{98}{365} = \$322.19$

   $MV = \$15,000 + \$322.19 = \$15,322.19$

   $MV = \$15,000 + \$326.67 = \$15,326.67$

Learning Unit 10–2:

Finding Unknown in Simple Interest Formula

This unit begins with the formula used to calculate the principal of a loan. Then it explains how to find the *principal, rate,* and *time* of a simple interest loan. In all the calculations, we use 360 days and round only final answers.

Learn: Finding the Principal

Example: Tim Jarvis paid the bank $19.48 interest at 9.5% for 90 days. How much did Tim borrow using the ordinary interest method?

The following formula is used to calculate the principal of a loan:

Calculate Principal

$$\text{Principal} = \frac{\text{Interest}}{\text{Rate} \times \text{Time}}$$

Note how we illustrated this in the margin. The shaded area is what we are solving for. When solving for principal, rate, or time, you are dividing. Interest will be in the numerator, and the denominator will be the other two elements multiplied by each other.

Step 1 When using a calculator, press

.095 $\times$ 90 $\div$ 360 M+ .

Step 1 Set up the formula.

$$P = \frac{\$19.48}{.095 \times \dfrac{90}{360}}$$

Step 2 Multiply the denominator.

.095 times 90 divided by 360 (do not round)

Step 2 When using a calculator, press

19.48 $\div$ MR $=$.

$$P = \frac{\$19.48}{.02375}$$

Step 3 Divide the numerator by the result of Step 2. $P = \$820.21$

Step 4 Check your answer.

$$\$19.48 = \$820.21 \times .095 \times \frac{90}{360}$$
$$(I) \qquad (P) \qquad (R) \qquad (T)$$

Learn: Finding the Rate

Example: Tim Jarvis borrowed $820.21 from a bank. Tim's interest is $19.48 for 90 days. What rate of interest did Tim pay using the ordinary interest method?

The following formula is used to calculate the rate of interest:

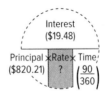

Calculate Rate Of Interest

$$\text{Rate} = \frac{\text{Interest}}{\text{Principal} \times \text{Time}}$$

Step 1 Set up the formula.

$$R = \dfrac{\$19.48}{\$820.21 \times \dfrac{90}{360}}$$

Step 2 Multiply the denominator.

$$R = \dfrac{\$19.48}{\$205.0525}$$

Do not round the answer.

Step 3 Divide the numerator by the result of Step 2. $R = .095 \times 100 = 9.5\%$

Step 4 Check your answer.

$$\underset{(I)}{\$19.48} = \underset{(P)}{\$820.21} \times \underset{(R)}{.095} \times \underset{(T)}{\dfrac{90}{360}}$$

Learn: Finding the Time

Example: Tim Jarvis borrowed $820.21 from a bank. Tim's interest is $19.48 at 9.5%. How much time does Tim have to repay the loan using the ordinary interest method?

The following formula is used to calculate time:

Calculate Time

$$\text{Time (in years)} = \dfrac{\text{Interest}}{\text{Principal} \times \text{Rate}}$$

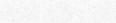 *Time is **always** over what makes up one year:* $\dfrac{\# \, of \, days}{360 \, or \, 365}, \dfrac{\# \, of \, weeks}{52},$

$$\dfrac{\# \, of \, months}{12}, \dfrac{\# \, of \, quarters}{4}.$$

Step 1 Set up the formula.

$$T = \dfrac{\$19.48}{\$820.21 \times .095}$$

Step 2 Multiply the denominator.
Do not round the answer.

$$T = \dfrac{\$19.48}{\$77.91995}$$

Step 3 Divide the numerator by the result of Step 2. $T = .25$ years

Step 4 Convert years to days (assume 360 days). $.25 \times 360 = \boxed{90 \, days}$

Step 5 Check your answer.

$$\underset{(I)}{\$19.48} = \underset{(P)}{\$820.21} \times \underset{(R)}{.095} \times \underset{(T)}{\dfrac{90}{360}}$$

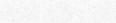 *Whole numbers in time represent full years. No adjustment to the time calculation is needed. However, when the calculation for number of days includes a decimal, multiply the decimal by 360 or 365 as indicated, to calculate the number of days the decimal represents in a year.*

Example: $T = .37$ $.37 \times 365 = 135.05 = 136$ days

When dealing with a fraction (or part) of a day, always round up to a full day even if the number being rounded is less than 5.

Before we go on to Learning Unit 10–3, let's check your understanding of this unit.

Money Tip Want to be offered the best interest rates by lending institutions? Make certain to work on building your credit score: Pay your bills on time. Make frequent micropayments throughout the month. Keep old accounts open. Become an authorized user of a responsible friend's or relative's credit card with a high credit limit. Use a secured credit card. Use less than 30% of your credit. No credit? Try Experian Boost or UltraFICO.

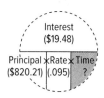

Interest
($19.48)

Principal × Rate × Time
($820.21) (.095) ?

Step 1 When using a calculator, press

[820.21] [×] [.095] [M+].

Step 2 When using a calculator, press

[19.48] [÷] [MR] [=].

Practice Quiz

Complete this Practice Quiz to see how you are doing.

Complete the following (assume 360 days):

	Principal	Interest rate	Time (days)	Simple interest
1.	?	5%	90 days	$8,000
2.	$7,000	?	220 days	$350
3.	$1,000	8%	?	$300

✓ **Solutions**

1. $\dfrac{\$8,000}{.05 \times \dfrac{90}{360}} = \dfrac{\$8,000}{.0125} = \boxed{\$640,000}$ $\qquad P = \dfrac{I}{R \times T}$

2. $\dfrac{\$350}{\$7,000 \times \dfrac{220}{360}} = \dfrac{\$350}{\$4,277.7777} = \boxed{8.18\%}$ $\qquad R = \dfrac{I}{P \times T}$
(do not round)

3. $\dfrac{\$300}{\$1,000 \times .08} = \dfrac{\$300}{\$80} = 3.75 \times 360 = \boxed{1,350 \text{ days}}$ $\qquad T = \dfrac{I}{P \times R}$

U.S. Rule—Making Partial Note Payments before Due Date

Often a person may want to pay off a debt in more than one payment before the maturity date. The **U.S. Rule** allows the borrower to receive proper interest credits. This rule states that any partial loan payment first covers any interest that has built up. The remainder of the partial payment reduces the loan principal. Courts or legal proceedings generally use the U.S. Rule. The Supreme Court originated the U.S. Rule in the case of *Story* v. *Livingston.*

> **Example:** Jeff Edsell owes $5,000 on a 4%, 90-day note. On day 50, Jeff pays $600 on the note. On day 80, Jeff makes an $800 additional payment. Assume a 360-day year. What is Jeff's adjusted balance after day 50 and after day 80? What is the ending balance due?

To calculate $600 payment on day 50:

Step 1 Calculate interest on principal from date of loan to date of first principal payment. Round to nearest cent.

$$I = P \times R \times T$$
$$I = \$5{,}000 \times .04 \times \frac{50}{360}$$
$$I = \$27.78$$

Step 2 Apply partial payment to interest due. Subtract remainder of payment from principal. This is the **adjusted balance** (principal).

$$\begin{array}{r} \$600.00 \text{ payment} \\ -\ 27.78 \text{ interest} \\ \hline \$572.22 \end{array}$$

$$\begin{array}{r} \$5{,}000.00 \text{ principal} \\ -\quad 572.22 \\ \hline \$4{,}427.78 \text{ adjusted} \\ \text{balance—} \\ \text{principal} \end{array}$$

To calculate $800 payment on day 80:

Step 3 Calculate interest on adjusted balance that starts from previous payment date and goes to new payment date. Then apply Step 2.

Compute interest on $4,427.78 for 30 days (80 − 50)

$$I = \$4{,}427.78 \times .04 \times \frac{30}{360}$$
$$I = \$14.76$$

$$\begin{array}{r} \$800.00 \text{ payment} \\ -\ 14.76 \text{ interest} \\ \hline \$785.24 \end{array}$$

$$\begin{array}{r} \$4{,}427.78 \\ -\quad 785.24 \\ \hline \$3{,}642.54 \text{ adjusted} \\ \text{balance} \end{array}$$

Step 4 At maturity, calculate interest from last partial payment. *Add* this interest to adjusted balance.

Ten days are left on note since last payment.

$$I = \$3{,}642.54 \times .04 \times \frac{10}{360}$$
$$I = \$4.05$$

$$\text{Balance owed} = \boxed{\$3{,}646.59} \left(\begin{array}{r} \$3{,}642.54 \\ +\quad 4.05 \end{array} \right)$$

> **Money Tip** Pay off debt instead of moving it around unless you have been offered 0% interest. Be wary of companies offering to consolidate your debt into a single loan. If you do, be certain to read and understand all the terms.

Note that when Jeff makes two partial payments, Jeff's total interest is $46.59 ($27.78 + $14.76 + $4.05). If Jeff had repaid the entire loan after 90 days, his interest payment would have been $50—a total savings of $3.41.

Let's check your understanding of the last unit in this chapter.

Practice Quiz

Complete this Practice Quiz to see how you are doing.

Polly Flin borrowed $5,000 for 60 days at 8%. On day 10, Polly made a $600 partial payment. On day 40, Polly made a $1,900 partial payment. What is Polly's ending balance due under the U.S. Rule (assuming a 360-day year)?

✓ **Solutions**

$$\$5{,}000 \times .08 \times \frac{10}{360} = \$11.11$$

$$
\begin{array}{r}
\$1{,}900.00 \\
- \quad 29.41 \\
\hline
\$1{,}870.59
\end{array}
$$

$$
\begin{array}{r}
\$4{,}411.11 \\
- \ 1{,}870.59 \\
\hline
\$2{,}540.52
\end{array}
$$

$$
\begin{array}{r}
\$600.00 \\
- \quad 11.11 \\
\hline
\$588.89
\end{array}
$$

$$
\begin{array}{r}
\$5{,}000.00 \\
- \quad 588.89 \\
\hline
\$4{,}411.11
\end{array}
$$

$$\$2{,}540.52 \times .08 \times \frac{20}{360} = \$11.29$$

$$\$4{,}411.11 \times .08 \times \frac{30}{360} = \$29.41$$

$$
\begin{array}{r}
\$ \quad 11.29 \ \leftarrow \\
+ \ 2{,}540.52 \\
\hline
\boxed{\$2{,}551.81}
\end{array}
$$

Chapter 10 Review

Topic/Procedure/Formula	Example	You try it*
Simple interest for months Interest = Principal × Rate × Time $\quad$(I)$\quad\quad$(P)$\quad\quad$(R)$\quad\quad$(T)	$2,000 at 9% for 17 months $I = \$2,000 \times .09 \times \dfrac{17}{12}$ $I =$ 255	**Calculate simple interest** $4,000 at 3% for 18 months
Exact interest $T = \dfrac{\text{Exact number of days}}{365}$ $I = P \times R \times T$	$1,000 at 10% from January 5 to February 20 $I = \$1,000 \times .10 \times \dfrac{46}{365}$ $\quad\quad\quad\quad$Feb. 20: $\quad$ 51 days $\quad\quad\quad\quad$Jan. 5: $\quad$ − 5 $\quad\quad\quad\quad\quad\quad\quad\quad$46 days $I =$ 12.60	**Calculate exact interest** $3,000 at 4% from January 8 to February 22
Ordinary interest (Banker's Rule) $T = \dfrac{\text{Exact number of days}}{360}$ $I = P \times R \times T$ $\quad$ Higher interest costs	$I = \$1,000 \times .10 \times \dfrac{46}{360} (51 - 5)$ $I =$ 12.78	**Calculate ordinary interest** $3,000 at 4% from January 8 to February 22
Finding unknown in simple interest formula (use 360 days) $I = P \times R \times T$	Use this example for illustrations of simple interest formula parts: $1,000 loan at 9%, 60 days $I = \$1,000 \times .09 \times \dfrac{60}{360} =$ 15	**Calculate interest (use 360 days)** $2,000 loan at 4%, 90 days
Finding the principal $P = \dfrac{I}{R \times T}$	$P = \dfrac{\$15}{.09 \times \dfrac{60}{360}} = \dfrac{\$15}{.015} =$ $1,000$	**Calculate principal** *Given:* interest, $20; rate, 4%; 90 days
Finding the rate $R = \dfrac{I}{P \times T}$	$R = \dfrac{\$15}{\$1,000 \times \dfrac{60}{360}} = \dfrac{\$15}{166.66666} = .09$ $\quad\quad\quad\quad\quad\quad\quad\quad\quad\quad = $ 9% *Note:* We did not round the denominator.	**Calculate rate** *Given:* interest, $20; principal, $2,000; 90 days
Finding the time $T = \dfrac{I}{P \times R}$ (in years) Multiply answer by 360 days to convert answer to days for ordinary interest.	$T = \dfrac{\$15}{\$1,000 \times .09} = \dfrac{\$15}{\$90} = .1666666$ $.1666666 \times 360 = 59.99 =$ 60 days	**Calculate number of days** *Given:* principal, $2,000; rate, 4%; interest, $20

Topic/Procedure/Formula	Example	You try it*
U.S. Rule (use 360 days) Calculate interest on principal from date of loan to date of first partial payment. Calculate adjusted balance by subtracting from principal the partial payment less interest cost. The process continues for future partial payments with the adjusted balance used to calculate cost of interest from last payment to present payment.	12%, 120 days, $2,000 *Partial payments:* On day 40: $250 On day 60: $200 *First payment:* $$I = \$2{,}000 \times .12 \times \frac{40}{360}$$ $I = \$26.67$ $\$250.00$ payment -26.67 interest $\overline{\$223.33}$ $\$2{,}000.00$ principal -223.33 $\overline{\$1{,}776.67}$ adjusted balance *Second payment:* $$I = \$1{,}776.67 \times .12 \times \frac{20}{360}$$ $I = \$11.84$ $\$200.00$ payment -11.84 interest $\overline{\$188.16}$ $\$1{,}776.67$ -188.16 $\overline{\$1{,}588.51}$ adjusted balance	**Calculate balance due and total interest** *Given:* $4,000; 4%; 90 days *Partial payments:* On day 30: $400 On day 70: $300
Balance owed equals last adjusted balance plus interest cost from last partial payment to final due date.	*60 days left:* $$\$1{,}588.51 \times .12 \times \frac{60}{360} = \$31.77$$ $\$1{,}588.51 + \$31.77 = \boxed{\$1{,}620.28 \text{ balance due}}$ Total interest = $\$26.67$ $\phantom{\text{Total interest} = +}11.84$ $\phantom{\text{Total interest} = }+\ 31.77$ $\phantom{\text{Total interest} = }\overline{\$70.28}$	

Key Terms

Adjusted balance	Maturity value	Simple interest formula
Banker's Rule	Ordinary interest	Time
Exact interest	Principal	U.S. Rule
Interest	Simple interest	

*Worked-out solutions are in Appendix A.

Critical Thinking Discussion Questions with Chapter Concept Check

1. What is the difference between exact interest and ordinary interest? With the increase of computers in banking, do you think that the ordinary interest method is a dinosaur in business today?

2. Explain how to use the portion formula to solve the unknowns in the simple interest formula. Why would rounding the answer of the denominator result in an inaccurate final answer?

3. Explain the U.S. Rule. Why in the last step of the U.S. Rule is the interest added, not subtracted?

4. Do you believe the government bailout of banks is in the best interest of the country? Defend your position.

5. **Chapter Concept Check.** How has the pandemic affected interest rates? Go to the Federal Reserve website for your data.

End-of-Chapter Problems

Name _____   Date _____

Check figures for odd-numbered problems in Appendix A.

Drill Problems

Calculate the simple interest and maturity value for the following problems. Round to the nearest cent as needed.   *LU 10–1(1)*

	Principal	Interest rate	Time	Simple interest	Maturity value
10–1.	$9,000	$2\frac{1}{4}\%$	18 mo.		
10–2.	$4,500	3%	6 mo.		
10–3.	$20,000	$6\frac{3}{4}\%$	9 mo.		

Complete the following, using ordinary interest: *LU 10–1(2)*

	Principal	Interest rate	Date borrowed	Date repaid	Exact time	Interest	Maturity value
10–4.	$1,000	8%	Mar. 8	June 9			
10–5.	$585	9%	June 5	Dec. 15			
10–6.	$1,200	12%	July 7	Jan. 10			

excel *(appears beside 10–4, 10–5, 10–6)*

Complete the following, using exact interest: *LU 10–1(2)*

	Principal	Interest rate	Date borrowed	Date repaid	Exact time	Interest	Maturity value
10–7.	$1,000	8%	Mar. 8	June 9			
10–8.	$585	9%	June 5	Dec. 15			
10–9.	$1,200	12%	July 7	Jan. 10			

Solve for the missing item in the following (round to the nearest hundredth as needed):   *LU 10–2(1)*

	Principal	Interest rate	Time (months or years)	Simple interest
10–10.	$400	5%	?	$100
10–11.	?	7%	$1\frac{1}{2}$ years	$200
10–12.	$5,000	?	6 months	$300

10–13. Use the U.S. Rule to solve for total interest costs, balances, and final payments (use ordinary interest).   *LU 10–3(1)*

> **Given**   Principal: $10,000, 8%, 240 days
> Partial payments: On 100th day, $4,000
> On 180th day, $2,000

Word Problems

10–14. Nolan Walker decided to buy a used snowmobile since his credit union was offering such low interest rates. He borrowed $2,700 at 3.5% on December 26, 2022, and paid it off February 21, 2023. How much did he pay in interest? (Assume ordinary interest and no leap year.)
LU 10–1(2)

End-of-Chapter Problems (Continued)

10–15. Harold Hill borrowed $15,000 to pay for his child's education at Riverside Community College. Harold must repay the loan at the end of 9 months in one payment with $5\frac{1}{2}\%$ interest. How much interest must Harold pay? What is the maturity value?   *LU 10–1(1)*

10–16. On September 12, Jody Jansen went to Sunshine Bank to borrow $2,300 at 9% interest. Jody plans to repay the loan on January 27. Assume the loan is on ordinary interest. What interest will Jody owe on January 27? What is the total amount Jody must repay at maturity?   *LU 10–1(2)*

10–17. Kelly O'Brien met Jody Jansen (Problem 10–16) at Sunshine Bank and suggested she consider the loan on exact interest. Recalculate the loan for Jody under this assumption. How much would she save in interest?   *LU 10–1(2)*

10–18. On May 3, 2022, Leven Corp. negotiated a short-term loan of $685,000. The loan is due October 1, 2022, and carries a 6.86% interest rate. Use ordinary interest to calculate the interest. What is the total amount Leven would pay on the maturity date?   *LU 10–1(2)*

10–19. Gordon Rosel went to his bank to find out how long it will take for $1,200 to amount to $1,650 at 8% simple interest. Please solve Gordon's problem. Round time in years to the nearest tenth.   *LU 10–2(1)*

10–20. Lucky Champ owes $191.25 interest on a 6% loan he took out on his March 17 birthday to upgrade an oven in his Irish restaurant, Lucky's Pub and Grub. The loan is due on August 17. What is the principal (assume ordinary interest)?   *LU 10–2(1)*

10–21. On April 5, 2022, Janeen Camoct took out an $8\frac{1}{2}$% loan for $20,000. The loan is due March 9, 2023. Use ordinary interest to calculate the interest. What total amount will Janeen pay on March 9, 2023? (Ignore leap year.)   *LU 10–1(2)*

10–22. Sabrina Bowers took out the same loan as Janeen (Problem 10–21). Sabrina's terms, however, are exact interest. What is Sabrina's difference in interest? What will she pay on March 9, 2023? (Ignore leap year.)   *LU 10–1(2)*

10–23. Max Wholesaler borrowed $2,000 on a 10%, 120-day note. After 45 days, Max paid $700 on the note. Thirty days later, Max paid an additional $630. What is the final balance due? Use the U.S. Rule to determine the total interest and ending balance due. Use ordinary interest.   *LU 10–3(1)*

10–24. Johnny Rockefeller had a bad credit rating and went to a local cash center. He took out a $100 loan payable in two weeks for $115. What is the percent of interest paid on this loan? Do not round denominator before dividing.   *LU 10–2(1)*

End-of-Chapter Problems (Continued)

10–25. You decided it is important to pay off some of your debt to help build your credit score. If you paid $1,307 interest on $45,000 at 4.0%, what was the time, using exact interest (rounded up to the nearest day)? *LU 10–2(1)*

10–26. On September 14, Jennifer Rick went to Park Bank to borrow $2,500 at $11\frac{3}{4}\%$ interest. Jennifer plans to repay the loan on January 27. Assume the loan is on ordinary interest. What interest will Jennifer owe on January 27? What is the total amount Jennifer must repay at maturity? *LU 10–1(2)*

10–27. Steven Linden met Jennifer Rick (Problem 10–26) at Park Bank and suggested she consider the loan on exact interest. Recalculate the loan for Jennifer under this assumption. *LU 10–1(2)*

excel **10–28.** Lance Lopes went to his bank to find out how long it will take for $1,000 to amount to $1,700 at 12% simple interest. Can you solve Lance's problem? Round time in years to the nearest tenth. *LU 10–2(1)*

10–29. Andres Michael bought a new boat. He took out a loan for $24,500 at 4.5% interest for 2 years. He made a $4,500 partial payment at 2 months and another partial payment of $3,000 at 6 months. How much is due at maturity? *LU 10–3(1)*

10–30. Shawn Bixby borrowed $17,000 on a 120-day, 12% note. After 65 days, Shawn paid $2,000 on the note. On day 89, Shawn paid an additional $4,000. What is the final balance due? Determine total interest and ending balance due by the U.S. Rule. Use ordinary interest.   *LU 10–3(1)*

10–31. Carol Miller went to Europe and forgot to pay her $740 mortgage payment on her New Hampshire ski house. For her 59 days overdue on her payment, the bank charged her a penalty of $15. What was the rate of interest charged by the bank? Round to the nearest hundredth percent. (Assume 360 days.)   *LU 10–2(1)*

10–32. Evander Holyfield (the champion boxer who had part of his ear bitten off by Mike Tyson) made $250 million during his boxing career but declared bankruptcy because of poor financial choices. His July interest at 15% was $155. What was Evander's principal at the beginning of July (assume 360 days)?   *LU 10–2(1)*

10–33. Kurt Busch won the 59th Daytona 500 in February. If he paid back a $6,800 loan with $20 interest at 7.5%, what was the time of the loan (Ignore leap year. Assume 360 days)?   *LU 10–2(1)*

excel **10–34.** Molly Ellen, bookkeeper for Keystone Company, forgot to send in the payroll taxes due on April 15. She sent the payment November 8. The IRS sent her a penalty charge of 8% simple interest on the unpaid taxes of $4,100. Calculate the penalty. (Remember that the government uses exact interest.)   *LU 10–1(2)*

End-of-Chapter Problems (Continued)

10–35. Oakwood Plowing Company purchased two new plows for the upcoming winter. In 200 days, Oakwood must make a single payment of $23,200 to pay for the plows. As of today, Oakwood has $22,500. If Oakwood puts the money in a bank today, what rate of interest will it need to pay off the plows in 200 days? (Assume 360 days.) *LU 10–2(1)*

10–36. IKEA has a living room set that will work perfectly in your new apartment. Chime, your online-only bank, is offering you 7% for 24 months. The interest cost will be $350. How much is the living room set selling for? *LU 10–2(1)*

10–37. You were furloughed for 8 months during the pandemic. The stimulus checks helped out but your emergency fund was still depleted. Your credit union loaned you $2,200 at 6.25% charging you $68 interest. How long did you take the loan out for? Round up to nearest day and assume ordinary interest. *LU 10–2(1)*

10–38. Hunter Kahn, an engineering student at Cornell University in New York, made $30,000 off his GameStop investment in early 2021. If he used some of the money to pay off $10,000 of his student loans paying $600 in interest for 550 days, what was the rate he paid? Round to the nearest percent. Assume ordinary interest. *LU 10–2(1)*

Challenge Problems

10–39. Debbie McAdams paid 8% interest on a $12,500 loan balance. Jan Burke paid $5,000 interest on a $62,500 loan. Based on 1 year: **(a)** What was the amount of interest paid by Debbie? **(b)** What was the interest rate paid by Jan? **(c)** Debbie and Jan are both in the 28% tax bracket. Since the interest is deductible, how much would Debbie and Jan each save in taxes? *LU 10–2(1)*

10–40. Janet Foster bought a computer and printer at Computerland. The printer had a $600 list price with a $100 trade discount and 2/10, n/30 terms. The computer had a $1,600 list price with a 25% trade discount but no cash discount. On the computer, Computerland offered Janet the choice of (1) paying $50 per month for 17 months with the 18th payment paying the remainder of the balance or (2) paying 8% interest for 18 months in equal payments.   *LU 10–1(2)*

    a. Assume Janet could borrow the money for the printer at 8% to take advantage of the cash discount. How much would Janet save? (Assume 360 days.)

    b. On the computer, what is the difference in the final payment between choices 1 and 2?

Summary Practice Test

Do you need help? Connect videos have step-by-step worked-out solutions.

1. Lorna Hall's real estate tax of $2,010.88 was due on December 14, 2022. Lorna lost her job and could not pay her tax bill until February 27, 2023. The penalty for late payment is $6\frac{1}{2}\%$ ordinary interest. *LU 10–1(1)*

   a. What is the penalty Lorna must pay?

   b. What is the total amount Lorna must pay on February 27?

2. Ann Hopkins borrowed $60,000 for her child's education. She must repay the loan at the end of 8 years in one payment with $5\frac{1}{2}\%$ interest. What is the maturity value Ann must repay? *LU 10–1(1)*

3. On May 6, Jim Ryan borrowed $14,000 from Lane Bank at $7\frac{1}{2}\%$ interest. Jim plans to repay the loan on March 11. Assume the loan is on ordinary interest. How much will Jim repay on March 11? *LU 10–1(2)*

4. Gail Ross met Jim Ryan (Problem 3) at Lane Bank. After talking with Jim, Gail decided she would like to consider the same loan on exact interest. Can you recalculate the loan for Gail under this assumption? *LU 10–1(2)*

5. Claire Russell is buying a car. Her November monthly interest was $210 at $7\frac{3}{4}\%$ interest. What is Claire's principal balance (to the nearest dollar) at the beginning of November? Use 360 days. Do not round the denominator in your calculation. *LU 10–2(1)*

6. Comet Lee borrowed $16,000 on a 6%, 90-day note. After 20 days, Comet paid $2,000 on the note. On day 50, Comet paid $4,000 on the note. What are the total interest and ending balance due by the U.S. Rule? Use ordinary interest. *LU 10–3(1)*

Q Control Your Debt, Don't Let Your Debt Control You!

 What I need to know

Debt is a common part of personal finances for North Americans. Because of this eventuality, it is important to plan debt effectively. Taking on debt is a serious decision and you need to fully understand how debt will play a role in your financial goals. Taking on too much debt will have a significant impact on your ability to meet your recurring expenses as well as to achieve your financial goals for the future. Before you take on any new debt, you should consider the plan you have to fully satisfy, or pay off, the debt. Being proactive in debt planning will make you better prepared to understand how much debt you can take on, help you avoid impulse purchases, and decide whether any new debt is a wise decision at the time.

 What I need to do

Consider your budget and stay realistic. You should have developed a financial budget coinciding with your current earnings and expenses. Don't take on too much risk with large debt that will not correspond with the budget you have established. As you enter your career after college many things are changing all at once. Don't let the excitement of these changes point you in the wrong financial direction. For example, you might be considering a vehicle purchase to reward yourself for successfully completing your degree and beginning your new career. Now is the time to consider if purchasing a new car for $40,000 is as financially sound as purchasing a used car for $15,000. If you allow emotion to overly influence your decision you may find yourself in a fancy new ride (depreciating 15% to 20% when you drive off the lot) you cannot afford.

Know how you will pay off your debt prior to taking it on. Let's assume you are considering a furniture purchase for your new residence at a cost of $2,400 and the furniture company is offering no interest on purchases for 12 months. Sounds like an incredible deal, right? To avoid paying interest on this purchase you need to pay off the entire balance within the 12-month window which equates to a monthly expense of $200 ($2,400/12). Therefore, you need to determine if your budget will allow for an additional $200 expense per month. If you have existing debt, consider plans for how to pay it off such as starting with the smallest debt and repaying it completely before moving to the next larger debt. Another option is to pay off the debt carrying the highest interest rate first. You could also research debt consolidation to combine multiple debts together, possibly with a lower interest rate. Make certain to read the fine print, however, by determining your payoff plan you will be more prepared to make purchases in line with your budget.

 Steps I need to take

1. Use your budget to determine if any new debt aligns with your current financial plan.
2. Decide how you intend to pay off your debt before you take it on.
3. Reduce current debt to improve your financial situation. Your debt-to-income ratio should be 30% or less.

 Resources I can use

- Debt Payoff Planner & Tracker (mobile app)—create a debt payoff plan and track your progress.
- https://www.thebalance.com/how-to-manage-your-debt-960856—tips and strategies for getting yourself out of debt.

MY MONEY ACTIVITY

- List the total cost of a purchase you would like to make (car, furniture, laptop, etc.).
- Describe your plan for payoff within 6 months, 1 year, and 18 months.

PERSONAL FINANCE

A KIPLINGER APPROACH

"Behind on Debts? Know Your Rights." Kiplinger's. May 2021.

FUNDAMENTALS

BASICS | Rivan Stinson

Behind on Debts? Know Your Rights

There are limits on what debt collectors can do to recoup what you owe. If you have medical debts, you have even more rights.

THE AVERAGE FICO CREDIT SCORE HIT
an all-time high last summer, but that doesn't mean debts aren't a problem. Nearly 30% of consumers with a credit report had some type of debt in collection last October, according to data from the Urban Institute, a policy think tank. And when debts go into collection, they end up in the hands of debt collectors—a common source of angst for consumers.

According to the Consumer Federation of America, a consumer advocacy group, credit and debit issues, including those related to debt collection, are among the top 10 complaints that consumers file with state and local consumer agencies. And thanks to a new debt-collection rule issued last year by the government's Consumer Financial Protection Bureau, complaints aren't likely to decline anytime soon.

The CFPB's two-part rule, set to take effect in November, will allow debt collectors to contact consumers via e-mail, text

and direct messages on Facebook and other social media sites—and there's no limit on how many times the collectors can reach out to you using these methods. On the plus side, you may not receive as many phone calls during dinner because collectors will be limited to seven calls a week for each debt. If you pick up the phone during one of these attempts, the collector can't call you again that week to discuss that particular debt.

That's still a lot of calls, and consumer advocates are concerned about the lack of a cap on electronic communications, says Linda Sherry, director of national priorities at Consumer Action, a consumer advocacy group. If you have, say, three debts in collection, you could receive up to 21 calls a week and mountains of social media messages. Consumer groups hope that the Biden administration's nominee to head the CFPB, Rohit Chopra, will support tweaking the rule to limit digital communications.

Know your rights. Under the federal Fair Debt Collection Practices Act, it's illegal for debt collectors to use abusive, unfair or deceptive tactics when trying to collect a debt. Collectors cannot threaten you or someone you know with violence or jail time or add interest or fees to the debt that weren't approved by the company you owe. And debt collectors usually can't call you before 8 A.M. or after 9 P.M.

You also have the right to write to the debt collector (or collectors) and instruct it to stop contacting you. After that, a debt collector may not contact you again unless it's to verify that it has received your request or to tell you that it plans to take some form of legal action.

It's also important to understand that a statute of limitations applies for debts in collection. The time period varies by state and the type of debt, but typically debt collectors have three to six years from when the debt goes into collection to file a lawsuit against you to collect payment. Once the statute of limitations has passed, the debt collector can't sue you. But that doesn't mean collection calls will stop, Sherry says. What's more, making a payment on an old debt may reset the clock on the statute of limitations, and the collector could sue you for the full amount. However, if you're able to repay the debt in full, it should improve your credit score.

Business Math Issue

The Fair Debt Collections Act is based on a statute of limitations.

1. List the key points of the article and information to support your position.

2. Write a group defense of your position using math calculations to support your view. If you are in an online course, post to a discussion board.

Chapter 11

Promissory Notes, Simple Discount Notes and the Discount Process

Learning Unit Objectives

LU 11–1: Structure of Promissory Notes; the Simple Discount Note

1. Differentiate between interest-bearing and non-interest-bearing notes.
2. Calculate bank discount and proceeds for simple discount notes.
3. Calculate and compare the interest, maturity value, proceeds, and effective rate of a simple interest note with a simple discount note.
4. Explain and calculate the effective rate for a Treasury bill.

LU 11–2: Discounting an Interest-Bearing Note before Maturity

1. Calculate the maturity value, bank discount, and proceeds of discounting an interest-bearing note before maturity.
2. Identify and complete the four steps of the discounting process.

⊖ Essential Question

How can I use promissory notes, simple discount notes, and the discount process to understand the business world and make my life easier?

🌐 Math Around the World

The *Wall Street Journal* chapter opener clip, "Banks Prepare For Wave Of Loan Defaults", reveals the expected losses to banks as a result of the pandemic.

This chapter begins with a discussion of the structure of promissory notes and simple discount notes. We also look at the application of discounting with Treasury bills. The chapter concludes with an explanation of how to calculate the discounting of promissory notes. Keep in mind that since the pandemic, interest rates are at new lows.

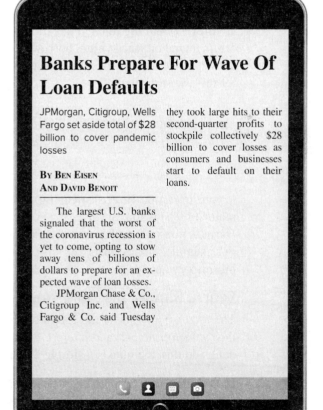

Banks Prepare For Wave Of Loan Defaults

JPMorgan, Citigroup, Wells Fargo set aside total of $28 billion to cover pandemic losses

BY BEN EISEN
AND DAVID BENOIT

The largest U.S. banks signaled that the worst of the coronavirus recession is yet to come, opting to stow away tens of billions of dollars to prepare for an expected wave of loan losses.

JPMorgan Chase & Co., Citigroup Inc. and Wells Fargo & Co. said Tuesday they took large hits to their second-quarter profits to stockpile collectively $28 billion to cover losses as consumers and businesses start to default on their loans.

Eisen, Ben, and David Benoit. "Banks Prepare for Wave of Loan Defaults." *The Wall Street Journal* (February 15, 2020).

Learning Unit 11–1:

Structure of Promissory Notes; the Simple Discount Note

Although businesses frequently sign promissory notes, customers also sign promissory notes. For example, some student loans may require the signing of promissory notes. Appliance stores often ask customers to sign a promissory note when they buy large appliances on credit. In this unit, promissory notes usually involve interest payments.

Learn: Structure of Promissory Notes

To borrow money, you must find a lender (a bank or a company selling goods on credit). You must also be willing to pay for the use of the money. In Chapter 10 you learned that interest is the cost of borrowing money for periods of time. Lenders charge interest as a rental fee on borrowing money.

Money lenders usually require that borrowers sign a **promissory note.** This note states that the borrower will repay a certain sum at a fixed time in the future. The note often includes the charge for the use of the money, or the rate of interest. Figure 11.1 shows a sample promissory note with its terms identified and defined. Take a moment to look at each term.

In this section you will learn the difference between interest-bearing notes and non-interest-bearing notes.

Interest-Bearing versus Non-Interest-Bearing Notes   A promissory note can be interest bearing or non–interest bearing. To be interest bearing, the note must state the rate of interest. Since the promissory note in Figure 11.1 states that its interest is 9%, it is an **interest-bearing note.** When the note matures, Regal Corporation will pay back the original amount **(face value)** borrowed plus interest. The simple interest formula (also known as the interest formula) and the maturity value formula from Chapter 10 are used for this transaction.

> **Interest-Bearing Note**
>
> Interest = Face value (principal) × Rate × Time
>
> Maturity value = Face value (principal) + Interest

If you sign a **non-interest-bearing** promissory note for $10,000, you pay back $10,000 at maturity. The maturity value of a non-interest-bearing note is the same as its face value. Usually, non-interest-bearing notes occur for short time periods under special conditions. For example, money borrowed from a relative could be secured by a non-interest-bearing promissory note.

Learn: Simple Discount Note

The total amount due at the end of the loan, or the **maturity value (MV),** is the sum of the face value (principal) and interest. Some banks deduct the loan interest in advance. When banks do this, the note is a **simple discount note.**

In the simple discount note, the **bank discount** is the interest that banks deduct in advance and the **bank discount rate** is the percent of interest. The amount that the borrower receives after the bank deducts its discount from the loan's maturity value is the note's **proceeds.** Sometimes we refer to simple discount notes as non-interest-bearing notes. Remember, however, that borrowers *do* pay interest on these notes.

In the example that follows, Pete Runnels has the choice of a note with a simple interest rate (Chapter 10) or a note with a simple discount rate (Chapter 11). Table 11.1 provides a summary of the calculations made in the example and gives the key points that you should remember. Now let's study the example, and then you can review Table 11.1.

Aha!

We will use 360 (not 365) days for all calculations in this chapter.

FIGURE 11.1
Interest-bearing promissory note

a. **Face value:** Amount of money borrowed—$10,000. The face value is also the principal of the note.
b. **Term:** Length of time that the money is borrowed—60 days.
c. **Date:** The date that the note is issued—October 2, 2022.
d. **Payee:** The company extending the credit—G.J. Equipment Company.
e. **Rate:** The annual rate for the cost of borrowing the money—4%.
f. **Maker:** The company issuing the note and borrowing the money—Regal Corporation.
g. **Maturity date:** The date the principal and interest are due—December 1, 2022.

Example: Pete Runnels has a choice of two different notes that both have a face value (principal) of $14,000 for 60 days. One note has a simple interest rate of 8%, while the other note has a simple discount rate of 8%. For each type of note, calculate (**a**) interest owed, (**b**) maturity value, (**c**) proceeds, and (**d**) effective rate.

Simple interest note—Chapter 10	Simple discount note—Chapter 11
Interest	**Interest**
a. $I = $ Face value (principal) $\times R \times T$	**a.** $I = $ Face value (principal) $\times R \times T$
$I = \$14{,}000 \times .08 \times \dfrac{60}{360}$	$I = \$14{,}000 \times .08 \times \dfrac{60}{360}$
$I = \$186.67$	$I = \$186.67$
Maturity value	**Maturity value**
b. $MV = $ Face value $+$ Interest	**b.** $MV = $ Face value
$MV = \$14{,}000 + \$186.67$	$MV = \$14{,}000$
$MV = \$14{,}186.67$	
Proceeds	**Proceeds**
c. Proceeds $=$ Face value	**c.** Proceeds $= MV - $ Bank discount
$= \$14{,}000$	$= \$14{,}000 - \$186.67$
	$= \$13{,}813.33$
Effective rate	**Effective rate**
d. Rate $= \dfrac{\text{Interest}}{\text{Proceeds} \times \text{Time}}$	**d.** Rate $= \dfrac{\text{Interest}}{\text{Proceeds} \times \text{Time}}$
$= \dfrac{\$186.67}{\$14{,}000 \times \dfrac{60}{360}}$	$= \dfrac{\$186.67}{\$13{,}813.33 \times \dfrac{60}{360}}$
$= 8\%$	$= 8.11\%$

Aha! *Do not round intermediate answers. Round only the final calculation.*

Simple interest note (Chapter 10)	Simple discount note (Chapter 11)
1. A promissory note for a loan with a term of usually less than 1 year. *Example:* 60 days.	**1.** A promissory note for a loan with a term of usually less than 1 year. *Example:* 60 days.
2. Paid back by one payment at maturity. Face value equals actual amount (or principal) of loan (this is not maturity value).	**2.** Paid back by one payment at maturity. Face value equals maturity value (what will be repaid).
3. Interest computed on face value or what is actually borrowed. *Example:* $186.67.	**3.** Interest computed on maturity value or what will be repaid and not on actual amount borrowed. *Example:* $186.67.
4. Maturity value = Face value + Interest. *Example:* $14,186.67.	**4.** Maturity value = Face value. *Example:* $14,000.
5. Borrower receives the face value. *Example:* $14,000.	**5.** Borrower receives proceeds = Face value − Bank discount. *Example:* $13,813.33.
6. Effective rate (true rate is same as rate stated on note). *Example:* 8%.	**6.** Effective rate is higher since interest was deducted in advance. *Example:* 8.11%.
7. Used frequently instead of the simple discount note.	**7.** Not used as much now because in 1969 congressional legislation required that the true rate of interest be revealed. Still used where legislation does not apply, such as personal loans.

Note that the interest of $186.67 is the same for the simple interest note and the simple discount note. The maturity value of the simple discount note is the same as the face value. In the simple discount note, interest is deducted in advance, so the proceeds are less than the face value. Note that the **effective rate** for a simple discount note is higher than the stated rate, since the bank calculated the rate on the face value of the note and not on what Pete received.

Application of Discounting—Treasury Bills   When the government needs money, it sells Treasury bills. A **Treasury bill** is a loan to the federal government for 28 days (4 weeks), 91 days (13 weeks), or 1 year. Check bankrate.com for the latest interest rates.

Treasury bills can be bought over the phone or on the government website. The purchase price (or proceeds) of a Treasury bill is the value of the Treasury bill less the discount. For example, if you buy a $10,000, 13-week Treasury bill at 4%, you pay $9,900 since you have not yet earned your interest $\left(\$10,000 \times .04 \times \frac{13}{52} = \$100\right)$. At maturity—13 weeks—the government pays you $10,000. You calculate your effective yield (4.04% rounded to the nearest hundredth percent) as follows:

$$(\$10,000 - \$100) \longrightarrow \frac{\$100}{\$9,900 \times \dfrac{13}{52}} = 4.04\% \text{ effective rate}$$

Now it's time to try the Practice Quiz and check your progress.

Money Tip

Paying your rent on time can help improve your credit rating. Leaving a lease before it is up and bouncing checks to a landlord will reduce it.

Practice Quiz

Complete this Practice Quiz to see how you are doing.

1. Warren Ford borrowed $12,000 on a non-interest-bearing, simple discount, $9\frac{1}{2}\%$, 60-day note. Assume ordinary interest. What are (a) the maturity value, (b) the bank's discount, (c) Warren's proceeds, and (d) the effective rate to the nearest hundredth percent?

2. Jane Long buys a $10,000, 13-week Treasury bill at 2%. What is her effective rate? Round to the nearest hundredth percent.

✓ Solutions

1. **a.** Maturity value = Face value = $12,000

    b. Bank discount = MV × Bank discount rate × Time

    $$= \$12,000 \times .095 \times \frac{60}{360}$$

    $$= \$190$$

    c. Proceeds = MV − Bank discount
    $$= \$12,000 - \$190$$
    $$= \$11,810$$

    d. Effective rate = $\dfrac{\text{Interest}}{\text{Proceeds} \times \text{Time}}$

    $$= \dfrac{\$190}{\$11,810 \times \dfrac{60}{360}}$$

    $$= 9.65\%$$

2. $\$10,000 \times .02 \times \dfrac{13}{52} = \$50$ interest

    $$\dfrac{\$50}{\$9,950 \times \dfrac{13}{52}} = .0201 = 2.01\%$$

Discounting an Interest-Bearing Note before Maturity

Manufacturers frequently deliver merchandise to retail companies and do not request payment for several months. For example, Roger Company manufactures outdoor furniture that it delivers to Home Depot in March. Payment for the furniture is not due until September. Roger will have its money tied up in this furniture until September. So Roger requests that Home Depot sign promissory notes.

If Roger Company needs cash sooner than September, what can it do? Roger Company can take one of its promissory notes to the bank, assuming the company that signed the note is reliable. The bank will buy the note from Roger. Now Roger has discounted the note and has cash instead of waiting until September when Home Depot would have paid Roger.

Remember that when Roger Company discounts the promissory note to the bank, the company agrees to pay the note at maturity if the maker of the promissory note fails to pay the bank. The potential liability that may or may not result from discounting a note is called a **contingent liability.**

Think of **discounting a note** as a three-party arrangement. Roger Company realizes that the bank will charge for this service. The bank's charge is a **bank discount.** The actual amount Roger receives is the **proceeds** of the note. The four steps below and the formulas in the example that follows will help you understand this discounting process.

Discounting A Note

Step 1 Calculate the interest and maturity value of the original simple interest note.

Step 2 Calculate the discount period (time the bank holds note).

Step 3 Calculate the bank discount.

Step 4 Calculate the proceeds.

Example: Roger Company sold the following promissory note to the bank:

Date of note	Face value of note	Length of note	Interest rate	Bank discount rate	Date of discount
March 8	$2,000	185 days	6%	5%	August 9

What are Roger's (1) interest and maturity value (*MV*)? What are the (2) discount period and (3) bank discount? (4) What are the proceeds?

1. *Calculate Roger's interest and maturity value (MV):*

$$MV = \text{Face value (principal)} + \text{Interest}$$

$$\text{Interest} = \$2,000 \times .06 \times \frac{185}{360} \quad \text{Actual number of days over 360}$$

$$= \$61.67$$
$$MV = \$2,000 + \$61.67$$
$$= \$2,061.67$$

2. *Calculate discount period*:
Determine the number of days that the bank will have to wait for the note to come due (discount period).

August 9   221 days
March 8   – 67
          154 days passed before note is discounted
          185 days
         – 154
           31 days bank waits for note to come due

Calculating days without table:

March	31
	– 8
	23
April	30
May	31
June	30
July	31
August	9
	154

  185 days—length of note
– 154 days Roger held note
   31 days bank waits

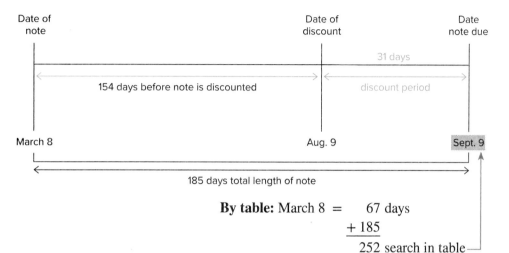

Date of note — Date of discount — Date note due

31 days

154 days before note is discounted — discount period

March 8 — Aug. 9 — Sept. 9

185 days total length of note

By table: March 8 = 67 days
               + 185
             252 search in table

3. *Calculate bank discount (bank charge):*

$$\$2,061.67 \times .05 \times \frac{31}{360} = \$8.88$$

$$\text{Bank discount} = MV \times \text{Bank discount rate} \times \frac{\text{Number of days bank waits for note to come due}}{360}$$

Step 1

↓

$$\text{Proceeds} = MV - \text{Bank discount (charge)}$$

Step 3

4. *Calculate proceeds:*

  $2,061.67
–   8.88
 $ 2,052.79

If Roger had waited until September 9, it would have received $2,061.67. Now, on August 9, Roger received $2,000 plus $52.79 interest.

Money Tip Protect your credit. At a minimum, review your credit report annually. Ideally, you can review your credit report for free three times per year if you stagger your free annual requests required by law from each of the three credit agencies: TransUnion, Equifax, and Experian. AnnualCreditReport.com has links to each.

Now let's assume Roger Company received a non-interest-bearing note. Then we follow the four steps for discounting a note except the maturity value is the amount of the loan. No interest accumulates on a non-interest-bearing note. Today, many banks use simple interest instead of discounting. Also, instead of discounting notes, many companies set up *lines of credit* so that additional financing is immediately available. The following *Wall Street Journal* clip shows that Gap Inc. has completely drawn down its line of credit.

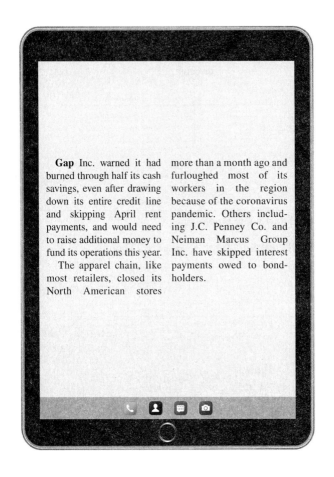

Gap Inc. warned it had burned through half its cash savings, even after drawing down its entire credit line and skipping April rent payments, and would need to raise additional money to fund its operations this year.

The apparel chain, like most retailers, closed its North American stores more than a month ago and furloughed most of its workers in the region because of the coronavirus pandemic. Others including J.C. Penney Co. and Neiman Marcus Group Inc. have skipped interest payments owed to bondholders.

Kapner, Suzanne, and Michael Dabaie. "Gap's Cash Pile is Dangerously Low." *The Wall Street Journal* (May 24, 2020).

The Practice Quiz that follows will test your understanding of this unit.

Practice Quiz

Complete this Practice Quiz to see how you are doing.

Date of note	Face value (principal) of note	Length of note	Interest rate	Bank discount rate	Date of discount
April 8	$35,000	160 days	11%	9%	June 8

From the above, calculate (**a**) interest and maturity value, (**b**) discount period, (**c**) bank discount, and (**d**) proceeds. Assume ordinary interest.

✓ Solutions

a. $I = \$35,000 \times .11 \times \dfrac{160}{360} = $ $\boxed{\$1,711.11}$

$MV = \$35,000 + \$1,711.11 = $ $\boxed{\$36,711.11}$

b. Discount period $= 160 - 61 = $ $\boxed{99 \text{ days}}$

April	30
	-8
	22
May	$+31$
	53
June	$+8$
	61

Or by table:

June 8	159
April 8	-98
	61

c. Bank discount $= \$36,711.11 \times .09 \times \dfrac{99}{360} = $ $\boxed{\$908.60}$

d. Proceeds $= \$36,711.11 - \$908.60 = $ $\boxed{\$35,802.51}$

Chapter 11 Review

Topic/Procedure/Formula	Example	You try it*
Simple discount note $\begin{array}{c}\text{Bank}\\\text{discount}\\(\text{interest})\end{array} = MV \times \begin{array}{c}\text{Bank}\\\text{discount}\\\text{rate}\end{array} \times \text{Time}$ Interest based on amount paid back and not on actual amount received.	$\$6{,}000 \times .09 \times \dfrac{60}{360} = \$90$ Borrower receives $\boxed{\$5{,}910}$ (the proceeds) and pays back \$6,000 at maturity after 60 days. A Treasury bill is a good example of a simple discount note.	**Calculate proceeds** \$4,000 note at 2% for 30 days
Effective rate $\dfrac{\text{Interest}}{\text{Proceeds} \times \text{Time}}$ $\uparrow$ What borrower receives (Face value − Discount)	*Example:* \$10,000 note, discount rate 12% for 60 days. $I = \$10{,}000 \times .12 \times \dfrac{60}{360} = \$200$ Effective rate: $\dfrac{\$200}{\$9{,}800 \times \dfrac{60}{360}} = \dfrac{\$200}{\$1{,}633.3333} = \boxed{12.24\%}$ $\uparrow$ Amount borrower received	**Calculate effective rate** \$15,000 note at 4% for 40 days
Discounting an interest-bearing note 1. Calculate interest and maturity value. $\quad I = \text{Face value} \times \text{Rate} \times \text{Time}$ $\quad MV = \text{Face value} + \text{Interest}$ 2. Calculate number of days bank will wait for note to come due (discount period). 3. Calculate bank discount (bank charge). $MV \times \begin{array}{c}\text{Bank}\\\text{discount}\\\text{rate}\end{array} \times \dfrac{\text{Number of days bank waits}}{360}$ 4. Calculate proceeds. $\quad MV - \text{Bank discount (charge)}$	*Example:* \$1,000 note, 6%, 60 days, dated November 1 and discounted on December 1 at 8%. 1. $\quad I = \$1{,}000 \times .06 \times \dfrac{60}{360} = \$10$ $\quad MV = \$1{,}000 + \$10 = \$1{,}010$ 2. 30 days 3. $\$1{,}010 \times .08 \times \dfrac{30}{360} = \$6.73$ 4. $1{,}010 - \$6.73 = \boxed{\$1{,}003.27}$	**Calculate proceeds** \$2,000 note, 3%, 60 days, dated November 5 and discounted on December 15 at 5%

Key Terms

Bank discount	Face value	note
Bank discount rate	Interest-bearing note	Payee
Contingent liability	Maker	Proceeds
Discount period	Maturity date	Promissory note
Discounting a note	Maturity value (MV)	Simple discount note
Effective rate	Non-interest-bearing	Treasury bill

* Worked-out solutions are in Appendix A.

Critical Thinking Discussion Questions with Chapter Concept Check

1. What are the differences between a simple interest note and a simple discount note? Which type of note would have a higher effective rate of interest? Why?

2. What are the four steps of the discounting process? Could the proceeds of a discounted note be less than the face value of the note?

3. What is a line of credit? What could be a disadvantage of having a large credit line?

4. Discuss the impact of a slow economy on small business borrowing.

5. **Chapter Concept Check.** Go to the Internet and determine the current status of business loans. Include concepts you learned in this chapter in your review.

6. How has the pandemic affected lines of credit?

End-of-Chapter Problems

Name _____   Date _____

Check figures for odd-numbered problems in Appendix A.

Drill Problems

Complete the following table for these simple discount notes. Use the ordinary interest method. *LU 11–1(2)*

	Amount due at maturity	Discount rate	Time	Bank discount	Proceeds
11–1.	$6,000	$3\frac{1}{2}\%$	160 days		
11–2.	$2,900	$6\frac{1}{4}\%$	180 days		

Calculate the discount period for the bank to wait to receive its money: *LU 11–2(1)*

	Date of note	Length of note	Date note discounted	Discount period
11–3.	April 12	45 days	May 2	
11–4.	March 7	120 days	June 8	

Solve for maturity value, discount period, bank discount, and proceeds (assume for Problems 11–5 and 11–6 a bank discount rate of 9%).   *LU 11–2(1, 2)*

	Face value (principal)	Rate of interest	Length of note	Maturity value	Date of note	Date note discounted	Discount period	Bank discount	Proceeds
11–5.	$50,000	11%	95 days		June 10	July 18			
11–6.	$25,000	9%	60 days		June 8	July 10			

excel **11–7.** Calculate the effective rate of interest (to the nearest hundredth percent) of the following Treasury bill. **Given:** $10,000 Treasury bill, 1% for 13 weeks. Round to the nearest thousandth percent. *LU 11–1(4)*

Word Problems

Use ordinary interest as needed.

11–8. Carl Sonntag wanted to compare what proceeds he would receive with a simple interest note versus a simple discount note. Both had the same terms: $19,500 at 8% for 2 years. Compare the proceeds. *LU 11–1(3)*

11–9. Paul and Sandy Moede signed an $8,000 note at Citizen's Bank. Citizen's charges a $6\frac{1}{2}\%$ discount rate. If the loan is for 300 days, find **(a)** the proceeds and **(b)** the effective rate charged by the bank (to the nearest tenth percent). *LU 11–1(3)*

11–10. You were offered either a simple interest note or a simple discount note with the following terms: $33,353 at 7% for 18 months. Based on the effective interest rate, which would you choose? *LU 11–1(3)*

11–11. On September 5, Sheffield Company discounted at Sunshine Bank a $9,000 (maturity value), 120-day note dated June 5. Sunshine's discount rate was 9%. What proceeds did Sheffield Company receive? *LU 11–2(1)*

11–12. The Treasury Department auctioned $21 billion in 3-month bills in denominations of $10,000 at a discount rate of 2.125%. What would be the effective rate of interest? Round only your final answer to the nearest hundredth percent. *LU 11–1(4)*

End-of-Chapter Problems (Continued)

11–13. There are some excellent free personal finance apps available: Mint.com, GoodBudget, Mvelopes, BillGuard, PocketExpense, HomeBudget, and Expensify. After using Mint.com, you realize you need to pay off one of your high interest loans to reduce your interest expense. You decide to discount a $5,250, 345-day note at 3% to your bank at a discount rate of 4.5% on day 210. What are your proceeds? Round each answer to the nearest cent.   *LU 11–2(1)*

11–14. Ron Prentice bought goods from Shelly Katz. On May 8, Shelly gave Ron a time extension on his bill by accepting a $3,000, 8%, 180-day note. On August 16, Shelly discounted the note at Roseville Bank at 9%. What proceeds does Shelly Katz receive?   *LU 11–2(1)*

excel 11–15. Rex Corporation accepted a $5,000, 8%, 120-day note dated August 8 from Regis Company in settlement of a past bill. On October 11, Rex discounted the note at Park Bank at 9%. What are the note's maturity value, discount period, and bank discount? What proceeds does Rex receive?   *LU 11–2(1)*

11–16. On May 12, Scott Rinse accepted an $8,000, 12%, 90-day note for a time extension of a bill for goods bought by Ron Prentice. On June 12, Scott discounted the note at Able Bank at 10%. What proceeds does Scott receive?   *LU 11–2(1)*

11–17. Robinson's, an electrical supply company, sold $4,800 of equipment to Jim Coates Wiring, Inc. Coates signed a promissory note May 12 with 4.5% interest. The due date was August 10. Short of funds, Robinson's contacted Capital One Bank on July 20; the bank agreed to take over the note at a 6.2% discount. What proceeds will Robinson's receive?   *LU 11–2(1)*

11–18. At www.daveramsey.com's Financial Peace University (FPU), Dave recommends Seven Baby Steps. One of these steps is "Pay off debt using the debt snowball." After graduating from FPU, Courtney Lopez-Munoz is trying to calculate the effective interest rate she is paying for a $1,789 simple discount note at $5\frac{1}{4}$% for 15 months. What rate has she been paying? Round to the nearest tenth percent. Do not round denominator calculation. *LU 11–2(1)*

11–19. Toyota Motor Company, headquartered in Nagoya, Japan, has faced operating challenges. Concerns about the strength of the steel used in its vehicle manufacturing; falling quarterly sales in North America, its largest market; and where the yen is trading against the dollar all affect Toyota's bottom line. If Toyota had a ¥20,000 note at 2.5% interest for 340 days, what would Toyota's proceeds be if it discounted the note on day 215 at 4%? (Round to the nearest yen for each answer.)   *LU 11–2(1)*

11–20. Wilson Montgomery was curious about the difference between a simple discount note and a simple interest note. He is considering taking out one or the other. Let him know which one he should take out based on proceeds and effective rate. Both have the same terms: $5,500 @ 6% for 18 months. Round to the nearest tenth.

End-of-Chapter Problems (Continued)

11–21. Elun Mosk wants to discount a $12,850, 225-day note at 6% at his bank at a discount rate of 4% on day 200. What are his proceeds? Round each answer to the nearest cent.

11–22. President Joe Biden signed the $1.9 trillion stimulus into law. Eligible individuals would receive $1,400 plus a $1,400 bonus for each dependent. Sammy Bailey needed funds prior to mid-March when the stimulus checks were to be funded and decided to look into discounting his 12-month $3,500 @ 5.75% simple discount note to his bank for an 8.5% discount. What was the maturity value for Sammy's note?

Challenge Problems

11–23. Assume that 3-month Treasury bills totaling $12 billion were sold in $10,000 denominations at a discount rate of 3.605%. In addition, the Treasury Department sold 6-month bills totaling $10 billion at a discount rate of 3.55%. **(a)** What is the discount amount for 3-month bills? **(b)** What is the discount amount for 6-month bills? **(c)** What is the effective rate for 3-month bills? **(d)** What is the effective rate for 6-month bills? Round to the nearest hundredth percent. *LU 11–1(4)*

11–24. Tina Mier must pay a $2,000 furniture bill. A finance company will loan Tina $2,000 for 8 months at a 9% discount rate. The finance company told Tina that if she wants to receive exactly $2,000, she must borrow more than $2,000. The finance company gave Tina the following formula:

$$\text{What to ask for} = \frac{\text{Amount in cash to be received}}{1 - (\text{Discount} \times \text{Time of loan})}$$

Calculate Tina's loan request and the effective rate of interest to the nearest hundredth percent. *LU 11–1(3)*

Summary Practice Test

Do you need help? Connect videos have step-by-step worked-out solutions.

1. On December 12, Lowell Corporation accepted a $160,000, 120-day, non-interest-bearing note from Able.com. What is the maturity value of the note? *LU 11–1(1)*

excel 2. The face value of a simple discount note is $17,000. The discount is 4% for 160 days. Calculate the following. *LU 11–1(3)*

    a. Amount of interest charged for each note.

    b. Amount borrower would receive.

    c. Amount payee would receive at maturity.

    d. Effective rate (to the nearest tenth percent).

3. On July 14, Gracie Paul accepted a $60,000, 6%, 160-day note from Mike Lang. On November 12, Gracie discounted the note at Lend Bank at 7%. What proceeds did Gracie receive? *LU 11–2(1)*

4. Lee.com accepted a $70,000, $6\frac{3}{4}$%, 120-day note on July 26. Lee discounts the note on October 28 at LB Bank at 6%. What proceeds did Lee receive? *LU 11–2(1)*

Summary Practice Test (Continued)

5. The owner of Lease.com signed a $60,000 note at Reese Bank. Reese charges a $7\frac{1}{4}\%$ discount rate. If the loan is for 210 days, find (a) the proceeds and (b) the effective rate charged by the bank (to the nearest tenth percent). *LU 11–2(1)*

6. Sam Slater buys a $10,000, 13-week Treasury bill at 1.875%. What is the effective rate? Round to the nearest hundredth percent. *LU 11–1(4)*

🔍 Protecting My Identity

 What I need to know

Personal and sensitive information about you exists in many forms across a variety of databases. It is imperative for you to be aware of the information about you and how to protect yourself from identity theft. Ensuring your information is accurate and protected is of increasing importance as we find ourselves conducting transactions electronically more frequently. Being proactive and taking the appropriate steps will help to ensure your information remains accurate and protected. Consider how you use your information online and how much information you are providing to various organizations. Using strong passwords online that are difficult to guess is a great way to protect your personal information. Ultimately, you are the best protector of your personal information. So, it is vital to protect yourself by using strong passwords (consider adding an authenticator app, too) for your online accounts and being aware of the locations you use and the amount of information you share.

 What I need to do

A credit report can be a great source of information about you. Requesting a credit report on yourself not only helps to show how you are progressing on your credit rating but can also identify potential issues with your personal information. It is a good idea to obtain your credit report on a yearly basis to monitor for any inconsistencies and correct any information that is not represented accurately. You can obtain your credit report for free once every 12 months from any of the three credit reporting agencies: Equifax, Experian, or TransUnion.

Protect yourself from potential fraud when making purchases, especially online purchases. Because of the protections included with many major credit cards, it is in your best interest to use a credit card instead of your bank debit card for purchases, especially online, due to the credit card's fraud protection. Consider freezing your credit with the three credit bureaus thus restricting access to any unauthorized new accounts. Be aware of child identity theft. Reference the article from the Federal Trade Commission found below to learn how to enable credit freezes and fraud alerts.

Don't share your social security number. Be vigilant about scammers and stay up-to-date on the latest scams and hacks. Closely review your financial and medical statements. Safeguard your mobile devices. Use alerts from your bank and credit card companies to alert you to transactions. Don't use public WiFi. Use USPS informed delivery to ensure your mail is being delivered. Avoid keying in passwords in public. Never give personal information over the phone.

 Steps I need to take

1. Request your credit report, and your children's, yearly and correct any inaccuracies.
2. Be vigilant and selective in where you provide or access personal information.
3. Keep online personal information secure by using strong passwords and authenticator apps.

 Resources I can use

- https://www.consumer.ftc.gov/articles/0279-extended-fraud-alerts-and-credit-freezes—credit freezes and fraud alerts information from the Federal Trade Commission.
- LifeLock ID Theft Protection (mobile app)—alerts and protection for your personal information.

MY MONEY ACTIVITY

- Request your free credit report: https://www.ftc.gov/faq/consumer-protection/get-my-free-credit-report.
- What types of information are posted on your credit report?
- Research and understand current data breaches, phone scams, skimming, phishing or spoofing, mailbox theft, and malware. Know the warning signs of each.

PERSONAL FINANCE

A KIPLINGER APPROACH

"HOW TO KEEP TABS ON YOUR CREDIT REPORTS." Kiplinger's. April, 2021.

EXTRA CREDIT

HOW TO KEEP TABS ON YOUR CREDIT REPORTS

Free weekly access is ending, but several services let you view your credit files more than once a year.

LAST SPRING, IN RESPONSE TO THE coronavirus crisis, the three major credit bureaus—Equifax, Experian and TransUnion—began offering consumers a free credit report every week at www.annualcreditreport.com, the federally authorized source of free credit reports. But unless the bureaus provide a last-minute extension, the free weekly reports will last only through April.

You'll still be able to get a free report from each bureau through Annual CreditReport.com once every 12 months, but you can see your reports for free more frequently through other websites that pull report data with your permission. If you create an account at CreditKarma.com, for example, you can see updated information from your Equifax and TransUnion reports once a week. You can also have the site monitor your reports for significant changes, such as the presence of a new loan or credit card, and send you alerts through e-mail or the site's mobile app. And Credit Karma offers free updates of your VantageScore credit scores based on data from each of the two bureaus.

If you would rather get free Equifax and TransUnion reports directly from those bureaus, each offers services through its website. At www.equifax.com/personal/products/credit/free-credit-score, you can register for free monthly updates of your Equifax credit report and VantageScore credit score. And by signing up for TransUnion's TrueIdentity at www.trans union.com/product/trueidentity free-identity-protection, you get unlimited access to your TransUnion report and credit monitoring alerts.

To check your report from the third major bureau, Experian, you can enroll at FreeCreditScore.com, which Experian sponsors. The site provides a new free credit report and FICO credit score based on Experian data every 30 days, as well as credit monitoring alerts.

To ensure that your reports remain free at any of these sites, skip pitches to upgrade to three-bureau report access or other services, and don't enter your credit card number or other payment information.

In addition to the yearly credit reports at AnnualCreditReport.com, you're entitled to a free report from the bureaus in certain other situations, including if you place a fraud alert on your report (a move you may make if you suspect identity theft); your report contains inaccurate information because of fraud; an adverse action has been taken against you (such as your application for credit being denied) because of information in the report; you're unemployed and expect to apply for employment in the next 60 days; or you receive public assistance.

Reviewing your reports. Regularly checking your credit reports is important in case a lender or other provider furnishes erroneous information to the bureaus, the bureaus mix up your file with that of someone else, or an identity thief opens fraudulent accounts in your name.

Missed Opportunities

Despite the availability of free, weekly credit reports, only one-third of Americans checked their credit reports in the past year,* down from previous years:

2018	37%
2019	39%
2020	33%

Even fewer Americans used alerts to notify them when there were changes in their credit reports:

2018	29%
2019	30%
2020	29%

And fewer than 10% put a freeze on their credit reports, which is one of the most effective ways to prevent ID theft:

2018	8%
2019	9%
2020	9%

*Survey was conducted August 14–18, 2020.
SOURCE: CompareCards.com

On your reports, make sure that all the accounts listed are yours and that the details on each—such as history of on-time payments, balances, credit limits and dates the accounts were opened—are accurate. Check that your address is listed correctly, too.

If you find a problem, contact the lender or company that provided the faulty data and file a dispute with each credit bureau that is reporting it. (You can get more information at www.equifax.com/personal/credit-report-services/credit-dispute, at www.experian.com/ disputes and at www.transunion.com/ disputes.) Include an explanation of your dispute, the resolution you expect, details such as the account number and name of the lender or other furnisher, and any supporting documents, such as a bank statement showing that you paid a bill on time despite a lender reporting that you didn't.
LISA GERSTNER

Business Math Issue

Credit Reports have little impact on establishing lines of credits.

1. List the key points of the article and information to support your position.

2. Write a group defense of your position using math calculations to support your view. If you are in an online course, post to a discussion board.

Chapter 12

Compound Interest and Present Value

Learning Unit Objectives

LU 12–1: Compound Interest (Future Value)—The Big Picture

1. Compare simple interest with compound interest.
2. Calculate the compound amount and interest manually and by table lookup.
3. Explain and compute the effective rate (APY).

LU 12–2: Present Value—The Big Picture

1. Compare compound interest (FV) with present value (PV).
2. Compute present value by table lookup.
3. Check the present value answer by compounding.

℮ Essential Question

How can I use compound interest and present value to understand the business world and make my life easier?

🌐 Math Around the World

Would you think of using your retirement savings to buy a home or payoff medical bills? The *Wall Street Journal* chapter opener clip, "What Would Withdrawing Retirement Savings Cost You", shows that using your savings will result in increased losses due to the compounding of money.

In this chapter we look at the power of compounding—interest paid on earned interest. Let's begin by studying Learning Unit 12–1, which shows you how to calculate compound interest.

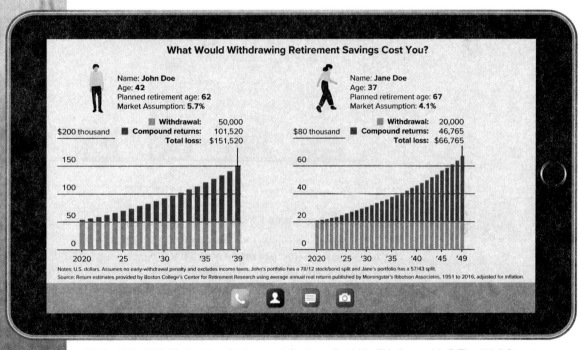

"'What Would Withdrawing Retirement Savings Cost You?' infographic." *The Wall Street Journal* (June 5, 2020).

Learning Unit 12–1:
Compound Interest (Future Value)— The Big Picture

So far we have discussed only simple interest, which is interest on the principal alone. Simple interest is either paid at the end of the loan period or deducted in advance. From the chapter introduction, you know that interest can also be compounded. Since the pandemic, interest rates have fallen to record lows.

Compounding involves the calculation of interest periodically over the life of the loan (or investment). After each calculation, the interest is added to the principal. Future calculations are on the adjusted principal (old principal plus interest). **Compound interest,** then, is the interest on the principal plus the interest of prior periods. **Future value (FV),** or the **compound amount,** is the final amount of the loan or investment at the end of the last period. In the beginning of this unit, do not be concerned with how to calculate compounding but try to understand the meaning of compounding.

FIGURE 12.1 Future value of $1 at 2% for four periods

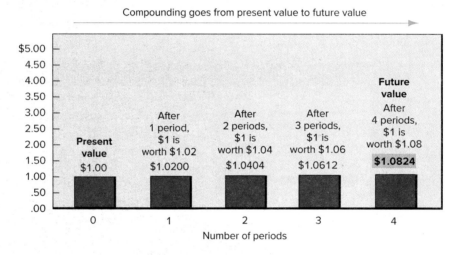

Figure 12.1 shows how $1 will grow if it is calculated for 4 years at 2% annually. This means that the interest is calculated on the balance once a year. In Figure 12.1, we start with $1, which is the **present value (PV).** After year 1, the dollar with interest is worth $1.02. At the end of year 2, the dollar is worth $1.04. By the end of year 4, the dollar is worth $1.08. Note how we start with the present and look to see what the dollar will be worth in the future. *Compounding goes from present value to future value.*

Before you learn how to calculate compound interest and compare it to simple interest, you must understand the terms that follow. These terms are also used in Chapter 13.

- **Compounded annually:** Interest calculated on the balance once a year.
- **Compounded semiannually:** Interest calculated on the balance every 6 months or every $\frac{1}{2}$ years.
- **Compounded quarterly:** Interest calculated on the balance every 3 months or every $\frac{1}{4}$ years.
- **Compounded monthly:** Interest calculated on the balance each month.
- **Compounded daily:** Interest calculated on the balance each day.

- **Number of periods:**[1] Number of years multiplied by the number of times the interest is compounded per year. For example, if you compound $1 for 4 years at 2% annually, semiannually, or quarterly, the following periods will result:

  Annually: 4 years × 1 = 4 periods
  Semiannually: 4 years × 2 = 8 periods
  Quarterly: 4 years × 4 = 16 periods

- **Rate for each period:**[2] Annual interest rate divided by the number of times the interest is compounded per year. Compounding changes the interest rate for annual, semiannual, and quarterly periods as follows:

  Annually: 2% ÷ 1 = 2%
  Semiannually: 2% ÷ 2 = 1%
  Quarterly: 2% ÷ 4 = $\frac{1}{2}$%

  Note that both the number of periods (4) and the rate (2%) for the annual example did not change. You will see later that rate and periods (not years) will always change unless interest is compounded yearly.

Now you are ready to learn the difference between simple interest and compound interest.

Learn: Simple versus Compound Interest

The following three situations involving Bill Smith will clarify the difference between simple interest and compound interest.

Situation 1: Calculating Simple Interest and Maturity Value

Example: Bill Smith deposited $80 in a savings account for 4 years at an annual interest rate of 8%. What is Bill's simple interest?

To calculate simple interest, we use the following simple interest formula:

$$\text{Interest } (I) = \text{Principal } (P) \times \text{Rate } (R) \times \text{Time } (T)$$

$$\$25.60 \quad = \quad \$80 \quad \times \quad .08 \quad \times \quad 4$$

In 4 years Bill receives a total of $105.60 ($80.00 + $25.60)—principal plus simple interest.

Now let's look at the interest Bill would earn if the bank compounded Bill's interest on his savings.

Situation 2: Calculating Compound Amount and Interest without Tables[3]

You can use the following steps to calculate the compound amount and the interest manually:

Calculating Compound Amount and Interest Manually

Step 1 Calculate the simple interest and add it to the principal. Use this total to figure next year's interest.

Step 2 Repeat for the total number of periods.

Step 3 Compound amount − Principal = Compound interest.

[1] Periods are often expressed with the letter *N* or *n* for number of periods.
[2] Rate is often expressed with the letter *i* for interest.
[3] For simplicity of presentation, round each calculation to the nearest cent before continuing the compounding process. The compound amount will be off by 1 cent.

Example: Bill Smith deposited $80 in a savings account for 4 years at an annual compounded rate of 8%. What are Bill's compound amount and interest?

The following shows how the compounded rate affects Bill's interest:

	Year 1	Year 2	Year 3	Year 4
	$80.00	$86.40	$ 93.31	$100.77
	× .08	× .08	× .08	× .08
Interest	$ 6.40	$ 6.91	$ 7.46	$ 8.06
Beginning balance	+ 80.00	+ 86.40	+ 93.31	+ 100.77
Amount at year-end	$86.40	$93.31	$100.77	$108.83

Note that the beginning year 2 interest is the result of the interest of year 1 added to the principal. At the end of each interest period, we add on the period's interest. This interest becomes part of the principal we use for the calculation of the next period's interest. We can determine Bill's compound interest as follows:

Compound amount	$108.83	
Principal	− 80.00	
Compound interest	$ 28.83	*Note:* In Situation 1 the interest was $25.60.

We could have used the following simplified process to calculate the compound amount and interest:

Year 1	Year 2	Year 3	Year 4
$80.00	$86.40	$ 93.31	$100.77
× 1.08	× 1.08	× 1.08	× 1.08
$86.40	$93.31	$100.77	$108.83 ← Future value

When using this simplification, you do not have to add the new interest to the previous balance. Remember that compounding results in higher interest than simple interest. Compounding is the *sum* of principal and interest multiplied by the interest rate we use to calculate interest for the next period. So, 1.08 above is 108%, with 100% as the base and 8% as the interest.

Situation 3: Calculating Compound Amount by Table Lookup   To calculate the compound amount with a future value table, use the following steps:

> **Calculating Compound Amount by Table Lookup**
>
> **Step 1** Find the periods: Years multiplied by number of times interest is compounded in 1 year.
>
> **Step 2** Find the rate: Annual rate divided by number of times interest is compounded in 1 year.
>
> **Step 3** Go down the Period column of the table to the number of periods desired; look across the row to find the rate. At the intersection of the two columns is the table factor for the compound amount of $1.
>
> **Step 4** Multiply the table factor by the amount of the loan. This gives the compound amount.

In Situation 2, Bill deposited $80 into a savings account for 4 years at an interest rate of 8% compounded annually. Bill heard that he could calculate the compound amount and interest by using tables. In Situation 3, Bill learns how to do this. Again, Bill wants to know the value of $80 in 4 years at 8%. He begins by using Table 12.1.

Looking at Table 12.1, Bill goes down the Period column to period 4 and then across the row to the 8% column. At the intersection, Bill sees the number 1.3605. The marginal notes show how Bill arrived at the periods and rate. The 1.3605 table number means that $1 compounded at this rate will increase in value in 4 years to about $1.36. Do you recognize the $1.36? Figure 12.1 showed how $1 grew to $1.36. Since Bill wants to know the value of $80, he multiplies the dollar amount by the table factor as follows:

Four Periods

1 × 4

No. of times compounded in 1 year

No. of years

$80.00 × 1.3605 = $108.84 *4

Principal × Table factor = Compound amount (future value)

*Off 1 cent due to rounding.

Aha!

Note all the table factors in the future value table are greater than 1. That is because investments increase in value over time due to interest being earned.

Figure 12.2 illustrates this compounding procedure. We can say that compounding is a future value (FV) since we are looking into the future. Thus,

$108.84 − $80.00 = $28.84 interest for 4 years at 8%
compounded annually on $80.00

8% Rate

$8\% \text{ rate} = \dfrac{8\%}{1}$ → Annual rate
→ No. of times compounded in 1 year

Now let's look at two examples that illustrate compounding more than once a year.

Example: Find the interest on $6,000 at 10% compounded semiannually for 5 years. We calculate the interest as follows:

Periods = 2 × 5 years = 10 $6,000 × 1.6289 = $9,773.40

Rate = 10% ÷ 2 = 5%   − 6,000.00

10 periods, 5%, in Table 12.1 = 1.6289 (table factor) $3,773.40
                                                                interest

FIGURE 12.2
Compounding (FV)

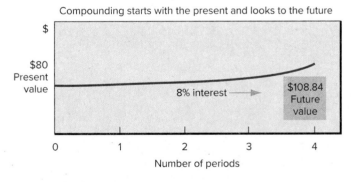

Compounding starts with the present and looks to the future

$

$80 Present value

8% interest → $108.84 Future value

0   1   2   3   4
Number of periods

Example: Pam Donahue deposits $8,000 in her savings account that pays 6% interest compounded quarterly. What will be the balance of her account at the end of 5 years?

Periods = 4 × 5 years = 20

Rate = 6% ÷ 4 = $1\frac{1}{2}\%$

20 periods, $1\frac{1}{2}\%$, in Table 12.1 = 1.3469 (table factor)

8,000 × 1.3469 = $10,775.20

Next, let's look at bank rates and how they affect interest.

---

[4] The formula for compounding is $A = P(1 + i)^N$, where *A* equals compound amount, *P* equals the principal, *i* equals interest per period, and *N* equals number of periods. The calculator sequence would be as follows for Bill Smith: 1 [+] .08 [y^x] 4 × 80 [=] 108.84. A Financial Calculator Guide booklet is available online that shows how to operate HP 10BII and TI BA II Plus.

TABLE 12.1 Future value of $1 at compound interest

Period	$\frac{1}{2}$%	1%	$1\frac{1}{2}$%	2%	3%	4%	5%	6%	7%	8%	9%	10%
1	1.0050	1.0100	1.0150	1.0200	1.0300	1.0400	1.0500	1.0600	1.0700	1.0800	1.0900	1.1000
2	1.0100	1.0201	1.0302	1.0404	1.0609	1.0816	1.1025	1.1236	1.1449	1.1664	1.1881	1.2100
3	1.0151	1.0303	1.0457	1.0612	1.0927	1.1249	1.1576	1.1910	1.2250	1.2597	1.2950	1.3310
4	1.0202	1.0406	1.0614	1.0824	1.1255	1.1699	1.2155	1.2625	1.3108	1.3605	1.4116	1.4641
5	1.0253	1.0510	1.0773	1.1041	1.1593	1.2167	1.2763	1.3382	1.4026	1.4693	1.5386	1.6105
6	1.0304	1.0615	1.0934	1.1262	1.1941	1.2653	1.3401	1.4185	1.5007	1.5869	1.6771	1.7716
7	1.0355	1.0721	1.1098	1.1487	1.2299	1.3159	1.4071	1.5036	1.6058	1.7138	1.8280	1.9487
8	1.0407	1.0829	1.1265	1.1717	1.2668	1.3686	1.4775	1.5938	1.7182	1.8509	1.9926	2.1436
9	1.0459	1.0937	1.1434	1.1951	1.3048	1.4233	1.5513	1.6895	1.8385	1.9990	2.1719	2.3579
10	1.0511	1.1046	1.1605	1.2190	1.3439	1.4802	1.6289	1.7908	1.9672	2.1589	2.3674	2.5937
11	1.0564	1.1157	1.1780	1.2434	1.3842	1.5395	1.7103	1.8983	2.1049	2.3316	2.5804	2.8531
12	1.0617	1.1268	1.1960	1.2682	1.4258	1.6010	1.7959	2.0122	2.2522	2.5182	2.8127	3.1384
13	1.0670	1.1381	1.2135	1.2936	1.4685	1.6651	1.8856	2.1329	2.4098	2.7196	3.0658	3.4523
14	1.0723	1.1495	1.2318	1.3195	1.5126	1.7317	1.9799	2.2609	2.5785	2.9372	3.3417	3.7975
15	1.0777	1.1610	1.2502	1.3459	1.5580	1.8009	2.0789	2.3966	2.7590	3.1722	3.6425	4.1772
16	1.0831	1.1726	1.2690	1.3728	1.6047	1.8730	2.1829	2.5404	2.9522	3.4259	3.9703	4.5950
17	1.0885	1.1843	1.2880	1.4002	1.6528	1.9479	2.2920	2.6928	3.1588	3.7000	4.3276	5.0545
18	1.0939	1.1961	1.3073	1.4282	1.7024	2.0258	2.4066	2.8543	3.3799	3.9960	4.7171	5.5599
19	1.0994	1.2081	1.3270	1.4568	1.7535	2.1068	2.5270	3.0256	3.6165	4.3157	5.1417	6.1159
20	1.1049	1.2202	1.3469	1.4859	1.8061	2.1911	2.6533	3.2071	3.8697	4.6610	5.6044	6.7275
21	1.1104	1.2324	1.3671	1.5157	1.8603	2.2788	2.7860	3.3996	4.1406	5.0338	6.1088	7.4002
22	1.1160	1.2447	1.3876	1.5460	1.9161	2.3699	2.9253	3.6035	4.4304	5.4365	6.6586	8.1403
23	1.1216	1.2572	1.4084	1.5769	1.9736	2.4647	3.0715	3.8197	4.7405	5.8715	7.2579	8.9543
24	1.1272	1.2697	1.4295	1.6084	2.0328	2.5633	3.2251	4.0489	5.0724	6.3412	7.9111	9.8497
25	1.1328	1.2824	1.4510	1.6406	2.0938	2.6658	3.3864	4.2919	5.4274	6.8485	8.6231	10.8347
26	1.1385	1.2953	1.4727	1.6734	2.1566	2.7725	3.5557	4.5494	5.8074	7.3964	9.3992	11.9182
27	1.1442	1.3082	1.4948	1.7069	2.2213	2.8834	3.7335	4.8223	6.2139	7.9881	10.2451	13.1100
28	1.1499	1.3213	1.5172	1.7410	2.2879	2.9987	3.9201	5.1117	6.6488	8.6271	11.1672	14.4210
29	1.1556	1.3345	1.5400	1.7758	2.3566	3.1187	4.1161	5.4184	7.1143	9.3173	12.1722	15.8631
30	1.1614	1.3478	1.5631	1.8114	2.4273	3.2434	4.3219	5.7435	7.6123	10.0627	13.2677	17.4494

Note: For more detailed tables, see your reference booklet, the *Business Math Handbook.*

Learn: Bank Rates—Nominal versus Effective Rates (Annual Percentage Yield, or APY)

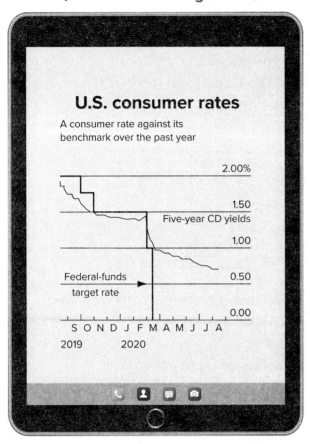

"'Consumer Rates and Returns to Investor' infographic."
The Wall Street Journal (August 16, 2020).

Banks often advertise their annual (nominal) interest rates and *not* their true or effective rate (annual percentage yield, or APY). This has made it difficult for investors and depositors to determine the actual rates of interest they were receiving. The Truth in Savings law forced savings institutions to reveal their actual rate of interest. The APY is defined in the Truth in Savings law as the percentage rate expressing the total amount of interest that would be received on a $100 deposit based on the annual rate and frequency of compounding for a 365-day period. As you can see in the *Wall Street Journal* clip, "U.S. consumer rates", to the left, due to the pandemic, CD rates have fallen from 2% to .5%.

Let's study the rates of two banks to see which bank has the better return for the investor. Blue Bank pays 8% interest compounded quarterly on $8,000. Sun Bank offers 8% interest compounded semiannually on $8,000. The 8% rate is the **nominal rate,** or stated rate, on which the bank calculates the interest. To calculate the **effective rate (annual percentage yield,** or **APY),** however, we can use the following formula:

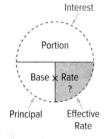

Effective Rate (Annual Percentage Yield, or APY)

$$\text{Effective rate (APY)} = \frac{\text{Interest for 1 year}}{\text{Principal}}$$

Now let's calculate the effective rate (APY)[5] for Blue Bank and Sun Bank.

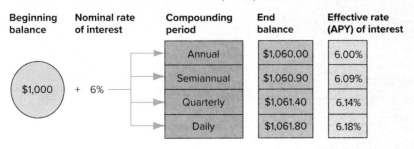

Beginning balance	Nominal rate of interest	Compounding period	End balance	Effective rate (APY) of interest
$1,000	+ 6%	Annual	$1,060.00	6.00%
		Semiannual	$1,060.90	6.09%
		Quarterly	$1,061.40	6.14%
		Daily	$1,061.80	6.18%

FIGURE 12.3
Nominal and effective rates (APY) of interest compared

[5] Round to the nearest hundredth percent as needed. In practice, the rate is often rounded to the nearest thousandth.

Blue, 8% compounded quarterly	Sun, 8% compounded semiannually
Periods = 4 (4 × 1)	Periods = 2 (2 × 1)
Percent = $\frac{8\%}{4}$ = 2%	Percent = $\frac{8\%}{2}$ = 4%
Principal = $8,000	Principal = $8,000
Table 12.1 lookup: 4 periods, 2%	Table 12.1 lookup: 2 periods, 4%

$$
\begin{array}{r}
1.0824 \\
\times\ \$8,000 \\
\hline
\$8,659.20
\end{array}
\qquad
\begin{array}{r}
1.0816 \\
\times\ \$8,000 \\
\hline
\$8,652.80
\end{array}
$$

Less principal
$$
\begin{array}{r}
\$8,659.20 \\
-\ 8,000.00 \\
\hline
\$\ \ 659.20
\end{array}
\qquad
\begin{array}{r}
\$8,652.80 \\
-\ 8,000.00 \\
\hline
\$\ \ 652.80
\end{array}
$$

$$
\text{Effective rate (APY)} = \frac{\$659.20}{\$8,000} = .0824 \qquad \frac{\$652.80}{\$8,000} = .0816
$$

$$= 8.24\% \qquad\qquad = 8.16\%$$

Figure 12.3 illustrates a comparison of nominal and effective rates (APY) of interest. This comparison should make you question any advertisement of interest rates before depositing your money.

Before concluding this unit, we briefly discuss compounding interest daily.

Learn: Compounding Interest Daily

Although many banks add interest to each account quarterly, some banks pay interest that is compounded daily, and other banks use *continuous compounding*. Remember that continuous compounding sounds great, but in fact, it yields only a fraction of a percent more interest over a year than daily compounding. Today, computers perform these calculations.

Table 12.2 is a partial table showing what $1 will grow to in the future by daily compounded interest, 360-day basis. For example, we can calculate interest compounded daily on $900 at 1% per year for 10 years as follows:

$900 × 1.1052 = $994.68 daily compounding

TABLE 12.2 Interest on $1 deposit compounded—daily-365-day year

Year	1%	2%	3%	4%	5%	6%	7%	8%
1	1.0101	1.0202	1.0305	1.0408	1.0513	1.0618	1.0725	1.0833
2	1.0202	1.0408	1.0618	1.0833	1.1052	1.1275	1.1503	1.1735
3	1.0305	1.0618	1.0942	1.1275	1.1618	1.1972	1.2337	1.2712
4	1.0408	1.0833	1.1275	1.1735	1.2214	1.2712	1.3231	1.3771
5	1.0513	1.1052	1.1618	1.2214	1.2840	1.3498	1.4190	1.4918
6	1.0618	1.1275	1.1972	1.2712	1.3498	1.4333	1.5219	1.6160
7	1.0725	1.1503	1.2337	1.3231	1.4190	1.5219	1.6322	1.7506
8	1.0833	1.1735	1.2712	1.3771	1.4918	1.6160	1.7506	1.8963
9	1.0942	1.1972	1.3099	1.4333	1.5683	1.7159	1.8775	2.0543
10	1.1052	1.2214	1.3498	1.4918	1.6487	1.8220	2.0136	2.2253

Now it's time to check your progress with the following Practice Quiz.

Practice Quiz

Complete this Practice Quiz to see how you are doing.

1. Complete the following without a table (round each calculation to the nearest cent as needed):

Principal	Time	Rate of compound interest	Compounded	Number of periods to be compounded	Total amount	Total interest
$200	1 year	4%	Quarterly	a.	b.	c.

2. Solve the previous problem by using compound value (FV) in Table 12.1.

3. Lionel Rodgers deposits $6,000 in Victory Bank, which pays 3% interest compounded semiannually. How much will Lionel have in his account at the end of 8 years?

4. Find the effective rate (APY) for the year: principal, $7,000; interest rate, 2%; and compounded quarterly.

5. Calculate by Table 12.2 what $1,500 compounded daily for 5 years will grow to at 7%.

✓ Solutions

1. **a.** 4 (4×1) **b.** $208.12 **c.** $8.12 ($208.12 − $200)

   $200 \times 1.01 = $202 \times 1.01 = $204.02 \times 1.01 = $206.06 \times 1.01 = $208.12

2. $200 \times 1.0406 = $208.12 (4 periods, 1%)

3. 16 periods, $1\frac{1}{2}\%$, $6,000 \times 1.2690 = $7,614

4. 4 periods, $\frac{1}{2}\%$

$$7,000 \times 1.0202 = \begin{array}{r} \$7,141.40 \\ -\ 7,000.00 \\ \hline \$\ \ 141.40 \end{array} \qquad \frac{\$141.40}{\$7,000.00} = 2.02\%$$

Check out the overlays that appear at the end of Chapter 13 to review these concepts.

5. $1,500 \times 1.4190 = $2,128.50

Learning Unit 12–2:
Present Value—The Big Picture

Figure 12.1 in Learning Unit 12–1 showed how by compounding, the *future value* of $1 became $1.08. This learning unit discusses *present value*. Before we look at specific calculations involving present value, let's look at the concept of present value.

Figure 12.4 shows that if we invested 92 cents today, compounding would cause the 92 cents to grow to $1 in the future. For example, let's assume you ask this question: "If I need $1 in 4 years in the future, how much must I put in the bank *today* (assume a 2% annual interest)?" To answer this question, you must know the present value of that $1 today. From Figure 12.4, you can see that the present value of $1 is .9238. Remember that the $1 is only worth 92 cents if you wait 4 periods to receive it. This is one reason why so many athletes get such big contracts—much of the money is paid in later years when it is not worth as much.

Figure 12.5 shows that if you want $108.84 in 4 years at 8% interest, you will need to invest $80 today.

FIGURE 12.4 Present value of $1 at 2% for four periods

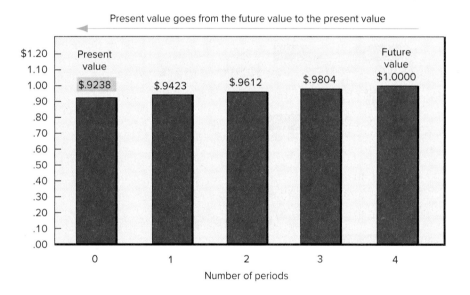

FIGURE 12.5 Present value

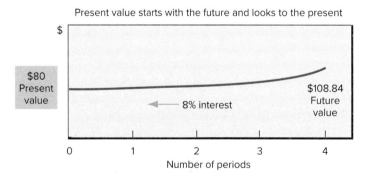

Learn: Relationship of Compounding (FV) to Present Value (PV)— The Bill Smith Example Continued

In Learning Unit 12–1, our consideration of compounding started in the *present* ($80) and looked to find the *future* amount of $108.84. Present value (PV) starts with the *future* and tries to calculate its worth in the *present* ($80). For example, in Figure 12.5, we assume

Bill Smith knew that in 4 years he wanted to buy a bike that cost $108.84 (future). Bill's bank pays 8% interest compounded annually. How much money must Bill put in the bank *today* (present) to have $108.84 in 4 years? To work from the future to the present, we can use a present value (PV) table. In the next section you will learn how to use this table.

Learn: How to Use a Present Value (PV) Table[6]

To calculate present value with a present value table, use the following steps:

Calculating Present Value by Table Lookup

Step 1 Find the periods: Years multiplied by number of times interest is compounded in 1 year.

Step 2 Find the rate: Annual rate divided by number of times interest is compounded in 1 year.

Step 3 Go down the Period column of the table to the number of periods desired; look across the row to find the rate. At the intersection of the two columns is the table factor for the compound value of $1.

Step 4 Multiply the table factor times the future value. This gives the present value.

Table 12.3 is a present value (PV) table that tells you what $1 is worth today at different interest rates. To continue our Bill Smith example, go down the Period column in Table 12.3 to 4. Then go across to the 8% column. At 8% for 4 periods, we see a table factor of .7350. This means that $1 in the future is worth approximately 74 cents today. If Bill invested 74 cents today at 8% for 4 periods, Bill would have $1 in 4 years.

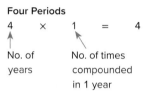

Four Periods

Since Bill knows the bike will cost $108.84 in the future, he completes the following calculation:

$$108.84 \times .7350 = \boxed{\$80.00}$$

This means that $108.84 in today's dollars is worth $80.00. Now let's check this.

Learn: Comparing Compound Interest (FV) Table 12.1 with Present Value (PV) Table 12.3

We know from our calculations that Bill needs to invest $80 for 4 years at 8% compound interest annually to buy his bike. We can check this by going back to Table 12.1 and comparing it with Table 12.3. Let's do this now.

Compound value Table 12.1			Present value Table 12.3		
Table 12.1	**Present value**	**Future value**	**Table 12.3**	**Future value**	**Present value**
1.3605	× $80.00	= $108.84	.7350	× $108.84	= $80.00
(4 per., 8%)			(4 per., 8%)		
We know the present dollar amount and find what the dollar amount is worth in the future.			We know the future dollar amount and find what the dollar amount is worth in the present.		

[6] The formula for present value is $PV = \dfrac{A}{(1+i)^N}$, where *A* equals future amount (compound amount), *N* equals number of compounding periods, and *i* equals interest rate per compounding period. The calculator sequence for Bill Smith would be as follows: 1 ⊞ 0.08 y^x 4 ⊟ M+ 108.84 ÷ MR ⊟ 80.00.

TABLE 12.1 Present value of $1 at end of period

Period	½%	1%	1½%	2%	3%	4%	5%	6%	7%	8%	9%	10%
1	.9950	.9901	.9852	.9804	.9709	.9615	.9524	.9434	.9346	.9259	.9174	.9091
2	.9901	.9803	.9707	.9612	.9426	.9246	.9070	.8900	.8734	.8573	.8417	.8264
3	.9851	.9706	.9563	.9423	.9151	.8890	.8638	.8396	.8163	.7938	.7722	.7513
4	.9802	.9610	.9422	.9238	.8885	.8548	.8227	.7921	.7629	.7350	.7084	.6830
5	.9754	.9515	.9283	.9057	.8626	.8219	.7835	.7473	.7130	.6806	.6499	.6209
6	.9705	.9420	.9145	.8880	.8375	.7903	.7462	.7050	.6663	.6302	.5963	.5645
7	.9657	.9327	.9010	.8706	.8131	.7599	.7107	.6651	.6227	.5835	.5470	.5132
8	.9609	.9235	.8877	.8535	.7894	.7307	.6768	.6274	.5820	.5403	.5019	.4665
9	.9561	.9143	.8746	.8368	.7664	.7026	.6446	.5919	.5439	.5002	.4604	.4241
10	.9513	.9053	.8617	.8203	.7441	.6756	.6139	.5584	.5083	.4632	.4224	.3855
11	.9466	.8963	.8489	.8043	.7224	.6496	.5847	.5268	.4751	.4289	.3875	.3505
12	.9419	.8874	.8364	.7885	.7014	.6246	.5568	.4970	.4440	.3971	.3555	.3186
13	.9372	.8787	.8240	.7730	.6810	.6006	.5303	.4688	.4150	.3677	.3262	.2897
14	.9326	.8700	.8119	.7579	.6611	.5775	.5051	.4423	.3878	.3405	.2992	.2633
15	.9279	.8613	.7999	.7430	.6419	.5553	.4810	.4173	.3624	.3152	.2745	.2394
16	.9233	.8528	.7880	.7284	.6232	.5339	.4581	.3936	.3387	.2919	.2519	.2176
17	.9187	.8444	.7764	.7142	.6050	.5134	.4363	.3714	.3166	.2703	.2311	.1978
18	.9141	.8360	.7649	.7002	.5874	.4936	.4155	.3503	.2959	.2502	.2120	.1799
19	.9096	.8277	.7536	.6864	.5703	.4746	.3957	.3305	.2765	.2317	.1945	.1635
20	.9051	.8195	.7425	.6730	.5537	.4564	.3769	.3118	.2584	.2145	.1784	.1486
21	.9006	.8114	.7315	.6598	.5375	.4388	.3589	.2942	.2415	.1987	.1637	.1351
22	.8961	.8034	.7207	.6468	.5219	.4220	.3418	.2775	.2257	.1839	.1502	.1228
23	.8916	.7954	.7100	.6342	.5067	.4057	.3256	.2618	.2109	.1703	.1378	.1117
24	.8872	.7876	.6995	.6217	.4919	.3901	.3101	.2470	.1971	.1577	.1264	.1015
25	.8828	.7798	.6892	.6095	.4776	.3751	.2953	.2330	.1842	.1460	.1160	.0923
26	.8784	.7720	.6790	.5976	.4637	.3607	.2812	.2198	.1722	.1352	.1064	.0839
27	.8740	.7644	.6690	.5859	.4502	.3468	.2678	.2074	.1609	.1252	.0976	.0763
28	.8697	.7568	.6591	.5744	.4371	.3335	.2551	.1956	.1504	.1159	.0895	.0693
29	.8653	.7493	.6494	.5631	.4243	.3207	.2429	.1846	.1406	.1073	.0822	.0630
30	.8610	.7419	.6398	.5521	.4120	.3083	.2314	.1741	.1314	.0994	.0754	.0573
35	.8398	.7059	.5939	.5000	.3554	.2534	.1813	.1301	.0937	.0676	.0490	.0356
40	.8191	.6717	.5513	.4529	.3066	.2083	.1420	.0972	.0668	.0460	.0318	.0221

Note: For more detailed tables, see your booklet, the *Business Math Handbook.*

Aha! *Note all the table factors in the present value table are less than 1. That is because money grows over time due to interest being earned so less money needs to be invested today to meet a future higher obligation.*

Note that the table factor for compounding at 2% for 4 periods is over 1 (1.0824) and the table factor for present value is less than 1 (.9238). The compound value table starts with the present and goes to the future. The present value table starts with the future and goes to the present.

Let's look at another example before trying the Practice Quiz.

Example: Rene Weaver needs $20,000 for college in 4 years. She can earn 4% compounded quarterly at her bank. How much must Rene deposit at the beginning of the year to have $20,000 in 4 years?

Remember that in this example the bank compounds the interest *quarterly*. Let's first determine the period and rate on a quarterly basis:

$$\text{Periods} = 4 \times 4 \text{ years} = 16 \text{ periods} \qquad \text{Rate} = \frac{4\%}{4} = 1\%$$

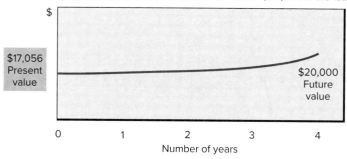

The present value is what we need **now** to have $20,000 in the future

$17,056 Present value

$20,000 Future value

Number of years

Now we go to Table 12.3 and find 16 under the Period column. We then move across to the 1% column and find the .8528 table factor.

$$\$20,000 \times .8528 = \boxed{\$17,056}$$

(future value)   (present value)

We illustrate this in Figure 12.6.

We can check the $17,056 present value by using the compound value Table 12.1:

16 periods, 1% column = 1.1726 × $17,056 = $19,999.87*

*Not quite $20,000 due to rounding of table factors.

Let's test your understanding of this unit with the Practice Quiz.

Money Tip
Almost 50% of full-time workers do **not** participate in their employer's retirement plan. Many companies have dollar matching programs allowing employees to receive "free" money in addition to helping reduce their taxable income. Check with your company to see what it offers.

FIGURE 12.6 Present value

Learning Unit 12–2

Practice Quiz

Complete this Practice Quiz to see how you are doing.

Use the present value Table 12.3 to complete:

	Future amount desired	Length of time	Rate compounded	Table period	Rate used	PV factor	PV amount
1.	$ 7,000	6 years	4% semiannually	_____	_____	_____	_____
2.	$15,000	20 years	1% annually	_____	_____	_____	_____

3. Bill Blum needs $20,000 6 years from today to attend V.P.R. Tech. How much must Bill put in the bank today (2% quarterly) to reach his goal?

4. Bob Fry wants to buy his grandson a Ford Taurus in 4 years. The cost of a car will be $24,000. Assuming a bank rate of 8% compounded quarterly, how much must Bob put in the bank today?

✓ Solutions

1. 12 periods (6 years × 2) 2% (4% ÷ 2) .7885 $5,519.50 ($7,000 × .7885)

2. 20 periods (20 years × 1) 1% (1% ÷ 1) .8195 $12,292.50 ($15,000 × .8195)

3. 6 years × 4 = 24 periods $\dfrac{2\%}{4} = \frac{1}{2}\%$ .8872 × $20,000 = $17,744

4. 4 × 4 years = 16 periods $\dfrac{8\%}{4} = 2\%$ .7284 × $24,000 = $17,481.60

Chapter 12 Review

Topic/Procedure/Formula	Example	You try it*
Calculating compound amount without tables (future value)† Determine new amount by multiplying rate times new balance (that includes interest added on). Start in present and look to future. $\dfrac{\text{Compound}}{\text{interest}} = \dfrac{\text{Compound}}{\text{amount}} - \text{Principal}$ $\longmapsto$Compounding$\longmapsto$ PV $\longrightarrow$ FV	$100 in savings account, compounded annually for 2 years at $\frac{1}{2}$%: $\begin{array}{ll} \$\quad 100 & 100.50 \\ \times\ 1.005 & \times\ 1.005 \\ \hline \$100.50 & \boxed{\$101}\ \text{(future value)} \end{array}$	**Calculate compound amount (future value)** $200 for 2 years at 4%, compounded annually
Calculating compound amount (future value) by table lookup $\text{Periods} = \begin{array}{c}\text{Number of times}\\ \text{compounded}\\ \text{per year}\end{array} \times \begin{array}{c}\text{Years of}\\ \text{loan}\end{array}$ $\text{Rate} = \dfrac{\text{Annual rate}}{\begin{array}{c}\text{Number of times compounded}\\ \text{per year}\end{array}}$ Multiply table factor (intersection of period and rate) times amount of principal.	*Example:* $2,000 at 2% for 5 years and compounded quarterly: Periods = 4×5 years = 20 Rate = $\dfrac{2\%}{4} = \frac{1}{2}\%$ 20 periods, $\frac{1}{2}$% = 1.1049 (table factor) $2,000 × 1.1049 = $\boxed{\$2,209.80}$ (future value)	**Calculate compound amount by table lookup** $4,000 at 6% for 6 years, compounded semiannually
Calculating effective rate (APY) $\text{Effective rate (APY)} = \dfrac{\text{Interest for 1 year}}{\text{Principal}}$ or Rate can be seen in Table 12.1 factor.	$1,000 at 3% compounded semiannually for 1 year. By Table 12.1: 2 periods, 1.5% 1.0302 means at end of year investor has earned 103.02% of original principal. Thus the interest is 3.02%. $\begin{array}{r} \$1,000 \times 1.0302 = \$1,030.20 \\ -\ 1,000.00 \\ \hline \$\quad 30.20 \end{array}$ $\dfrac{\$30.20}{\$1,000} = \boxed{3.02\%}$ effective rate (APY)	**Calculate effective rate** $4,000 at 6% for 1 year, compounded semiannually
Calculating present value (PV) by table lookup‡ Start with future and calculate worth in the present. Periods and rate computed like in compound interest. $\longmapsto$Present value$\longmapsto$ PV $\longleftarrow$ FV Find periods and rate. Multiply table factor (intersection of period and rate) times amount.	*Example:* Want $3,612.20 after 5 years with rate of 4% compounded quarterly: Periods = 4×5 = 20; 4%/4 = 1% By Table 12.3: 20 periods, 1% = .8195 $3,612.20 × .8195 = $\boxed{\$2,960.20}$ Invested today will yield desired amount in future	**Calculate present value by table lookup** Want $6,000 after 4 years with rate of 6%, compounded quarterly

Chapter 12 Review *(Continued)*

Key Terms

Annual percentage yield (APY)	Compounded monthly	Future value (FV)
Compound amount	Compounded quarterly	Nominal rate
Compound interest	Compounded semiannually	Number of periods
Compounded annually	Compounding	Present value (PV)
Compounded daily	Effective rate	Rate for each period

*Worked-out solutions are in Appendix A.

$A = P(1 + i)^N$.

$\dfrac{A}{(1 + i)^N}$ if table not used.

Critical Thinking Discussion Questions with Chapter Concept Check

1. Explain how periods and rates are calculated in compounding problems. Compare simple interest to compound interest.

2. What are the steps to calculate the compound amount by table? Why is the compound table factor greater than $1?

3. What is the effective rate (APY)? Why can the effective rate be seen directly from the table factor?

4. Explain the difference between compounding and present value. Why is the present value table factor less than $1?

5. **Chapter Concept Check.** Create a problem using present value and compounding to show the amount you would need to put away today in a bank to have enough money to pay for a child's costs through the age of 18. Assume your own rates and periods and that the amount you put in the bank is one lump sum that will grow through compounding (without new investments).

6. The pandemic has changed how people save. Agree or disagree and explain why.

End-of-Chapter Problems

Name _____ Date _____

Check figures for odd-numbered problems in Appendix A.

Drill Problems

Complete the following without using Table 12.1 (round to the nearest cent for each calculation) and then check your answer by Table 12.1 (check will be off due to rounding).   *LU 12–1(2)*

	Principal	Time (years)	Rate of compound interest	Compounded	Periods	Rate	Total amount	Total interest
12–1.	$575	1	4%	Quarterly				

Complete the following using compound future value Table 12.1 or the *Business Math Handbook*: *LU 12–1(2)*

	Time	Principal	Rate	Compounded	Amount	Interest
12–2.	12 years	$15,000	$3\frac{1}{2}\%$	Annually		
12–3.	6 months	$15,000	6%	Semiannually		
12–4.	2 years	$15,000	8%	Quarterly		

Calculate the effective rate (APY) of interest for 1 year.   *LU 12–1(3)*

12–5. Principal: $15,500

Interest rate: 6%
Compounded quarterly
Effective rate (APY):

12–6. Using Table 12.2, calculate what $700 would grow to at 6% per year compounded daily for 7 years.   *LU 12–1(3)*

End-of-Chapter Problems (Continued)

Complete the following using present value Table 12.3 or the present value table in the *Business Math Handbook*.   *LU 12–2(2)*

	Amount desired at end of period	Length of time	Rate	Compounded	On PV Table 12.3 Period used	Rate used	PV factor used	PV of amount desired at end of period
excel 12–7.	$6,000	8 years	3%	Semiannually				
excel 12–8.	$8,900	4 years	6%	Monthly				
excel 12–9.	$17,600	7 years	2%	Quarterly				
excel 12–10.	$20,000	20 years	8%	Annually				

12–11. Check your answer in Problem 12–9 by the compound value Table 12.1. The answer will be off due to rounding.   *LU 12–2(3)*

Word Problems

12–12. Sam Long anticipates he will need approximately $225,000 in 15 years to cover his 3-year-old daughter's college bills for a 4-year degree. How much would he have to invest today at an interest rate of 8% compounded semiannually?   *LU 12–2(2)*

12–13. Lynn Ally, owner of a local Subway shop, loaned $40,000 to Pete Hall to help him open a Subway franchise. Pete plans to repay Lynn at the end of 8 years with 6% interest compounded semiannually. How much will Lynn receive at the end of 8 years? *LU 12–1(2)*

12–14. Molly Hamilton deposited $50,000 at Bank of America at 8% interest compounded quarterly. What is the effective rate (APY) to the nearest hundredth percent? *LU 12–1(3)*

excel **12–15.** Melvin Indecision has difficulty deciding whether to put his savings in Mystic Bank or Four Rivers Bank. Mystic offers 10% interest compounded semiannually. Four Rivers offers 8% interest compounded quarterly. Melvin has $10,000 to invest. He expects to withdraw the money at the end of 4 years. Which bank gives Melvin the better deal? Check your answer.   *LU 12–1(3)*

12–16. Lee Holmes deposited $15,000 in a new savings account at 9% interest compounded semiannually. At the beginning of year 4, Lee deposits an additional $40,000 at 9% interest compounded semiannually. At the end of 6 years, what is the balance in Lee's account?   *LU 12–1(2)*

excel **12–17.** Lee Wills loaned Audrey Chin $16,000 to open Snip Its Hair Salon. After 6 years, Audrey will repay Lee with 8% interest compounded quarterly. How much will Lee receive at the end of 6 years?   *LU 12–1(2)*

12–18. Jazelle Momba wants to visit her family in Zimbabwe in 2028, which is 6 years from now. She knows that it will cost approximately $8,000 including flight costs, on-the-ground costs, and extra spending money to stay for 4 months. If she opens an account that compounds interest at 4% semiannually, how much does she need to deposit today to cover the total cost of her visit?   *LU 12–1(2)*

12–19. To protect her savings against further inflation and to help her prepare for a healthy financial future, Hanna Lind deposits $7,000 in an investment account earning 6% interest compounded quarterly. How much will Hanna have in her account in 10 years? (Use tables in the *Business Math Handbook*.)   *LU 12–1(2)*

12–20. The International Monetary Fund is trying to raise $500 billion in 5 years for new funds to lend to developing countries. At 6% interest compounded quarterly, how much must it invest today to reach $500 billion in 5 years?   *LU 12–2(2)*

12–21. You choose to invest your $2,985 income tax refund check (rather than spend it!) in an account earning 5% compounded annually. How much will the account be worth in 30 years?   *LU 12–1(2, 3)*

Imagine how much you would have in your account if you did this each year!

End-of-Chapter Problems (Continued)

12–22. Jim Ryan, an owner of a Burger King restaurant, assumes that his restaurant will need a new roof in 7 years. He estimates the roof will cost him $9,000 at that time. What amount should Jim invest today at 6% compounded quarterly to be able to pay for the roof? Check your answer. *LU 12–2(2)*

excel 12–23. Tony Ring wants to attend Northeast College. He will need $60,000 4 years from today. Assume Tony's bank pays 4% interest compounded semiannually. What must Tony deposit today so he will have $60,000 in 4 years? *LU 12–2(2)*

12–24. Check your answer (to the nearest dollar) in Problem 12–23 by using the compound value Table 12.1. The answer will be slightly off due to rounding. *LU 12–1(3)*

12–25. Pete Air wants to buy a used Jeep in 5 years. He estimates the Jeep will cost $15,000. Assume Pete invests $10,000 now at 6% interest compounded semiannually. Will Pete have enough money to buy his Jeep at the end of 5 years? *LU 12–1(2), LU 12–2(2)*

12–26. How much could you save for retirement if you chose to invest the money you spend on Starbucks coffee in one year? Assume you buy one venti cup of caffe latte for $4.15 each weekday for 50 weeks and can invest the total amount in a mutual fund earning 5% compounded annually for 30 years. *LU 12–1(2)*

12–27. Paul Havlik promised his grandson Jamie that he would give him $6,000 8 years from today for graduating from high school. Assume money is worth 6% interest compounded semiannually. What is the present value of this $6,000? *LU 12–2(2)*

12–28. Earl Ezekiel wants to retire in San Diego when he is 65 years old. Earl is now 50. He believes he will need $300,000 to retire comfortably. To date, Earl has set aside no retirement money. Assume Earl gets 6% interest compounded semiannually. How much must Earl invest today to meet his $300,000 goal? *LU 12–2(2)*

12–29. If you saved your tax refund (Problem 12–21), quit buying vendor coffee for one year (Problem 12–26), and decided to contribute $2,400 (you saved $200 per month) in your Roth IRA, how much would you have for retirement if you could invest these savings at 5% compounded annually for 30 years for this one year of savings?

12–30. Treasure Mountain International School in Park City, Utah, is a public middle school interested in raising money for next year's Sundance Film Festival. If the school raises $15,000 and invests it for 1 year at 3% interest compounded annually, what is the APY earned (round to nearest whole percent)? *LU 12–1(2, 3)*

12–31. David Bach, author of *Start Late, Finish Rich*, recommends paying yourself first, cutting small expenses such as by not going to Starbucks but brewing your own coffee, and saving 20% of your income. To get started saving, he recommends saving $2 per day. If you saved $2 per day for five years then invested that amount for an additional six years at 6% compounded quarterly, how much would you have saved? Round to the nearest cent.

12–32. A good idea is to invest any lump sum money you receive such as tax refund checks and bonus checks from work. Mallori Rouse invested her stimulus checks of $1,200, $600 and $1,400. How much will these checks be worth when she retires in 20 years if she can earn 7% interest compounded annually? Round to the nearest cent.

12–33. Steve and Ron want to begin a college fund for their daughter. They estimate she will need $80,000 for her education. How much do they need to invest today in order to reach their goal in 18 years if they can earn 8% compounded annually? Round to the nearest cent.

End-of-Chapter Problems (Continued)

Challenge Problems

12–34. Pete's Real Estate is currently valued at $65,000. Pete feels the value of his business will increase at a rate of 10% per year, compounded semiannually for the next 5 years. At a local fund-raiser, a competitor offered Pete $70,000 for the business. If he sells, Pete plans to invest the money at 6% compounded quarterly. What price should Pete ask? Verify your answer.   *LU 12–1(2), LU 12–2(2)*

12–35. You are the financial planner for Johnson Controls. Assume last year's profits were $700,000. The board of directors decided to forgo dividends to stockholders and retire high-interest outstanding bonds that were issued 5 years ago at a face value of $1,250,000. You have been asked to invest the profits in a bank. The board must know how much money you will need from the profits earned to retire the bonds in 10 years. Bank A pays 6% compounded quarterly, and Bank B pays $6\frac{1}{2}$% compounded annually. Which bank would you recommend, and how much of the company's profit should be placed in the bank? If you recommended that the remaining money not be distributed to stockholders but be placed in Bank B, how much would the remaining money be worth in 10 years? Use tables in the *Business Math Handbook.** Round final answer to nearest dollar.   *LU 12–1(2, 3), LU 12–2(2)*

*Check glossary for unfamiliar terms.

Copyright © McGraw Hill

Summary Practice Test

Do you need help? Connect videos have step-by-step worked-out solutions.

1. Lorna Ray, owner of a Starbucks franchise, loaned $40,000 to Lee Reese to help him open a new flower shop online. Lee plans to repay Lorna at the end of 5 years with 4% interest compounded semiannually. How much will Lorna receive at the end of 5 years? *LU 12–1(2)*

2. Joe Beary wants to attend Riverside College. Eight years from today he will need $50,000. If Joe's bank pays 6% interest compounded semiannually, what must Joe deposit today to have $50,000 in 8 years?   *LU 12–2(2)*

3. Shelley Katz deposited $30,000 in a savings account at 5% interest compounded semiannually. At the beginning of year 4, Shelley deposits an additional $80,000 at 5% interest compounded semiannually. At the end of 6 years, what is the balance in Shelley's account? Use the *Business Math Handbook*.   *LU 12–1(2)*

4. Earl Miller, owner of a Papa Gino's franchise, wants to buy a new delivery truck in 6 years. He estimates the truck will cost $30,000. If Earl invests $20,000 now at 5% interest compounded semiannually, will Earl have enough money to buy his delivery truck at the end of 6 years? Use the *Business Math Handbook*.   *LU 12–1(2), LU 12–2(2)*

Summary Practice Test (Continued)

5. Minnie Rose deposited $16,000 in Street Bank at 6% interest compounded quarterly. What was the effective rate (APY)? Round to the nearest hundredth percent.   *LU 12–1(2, 3)*

6. Lou Ling, owner of Lou's Lube, estimates that he will need $70,000 for new equipment in 7 years. Lou decided to put aside money today so it will be available in 7 years. Reel Bank offers Lou 6% interest compounded quarterly. How much must Lou invest to have $70,000 in 7 years?   *LU 12–2(2)*

excel 7. Bernie Long wants to retire to California when she is 60 years of age. Bernie is now 40. She believes that she will need $900,000 to retire comfortably. To date, Bernie has set aside no retirement money. If Bernie gets 8% compounded semiannually, how much must Bernie invest today to meet her $900,000 goal?   *LU 12–2(2)*

8. Jim Jones deposited $19,000 in a savings account at 7% interest compounded daily. At the end of 6 years, what is the balance in Jim's account?   *LU 12–1(2)*

MY MONEY

🔍 Step Up Your Retirement Game Plan

 What I need to know

As a student it may seem as though retirement is a lifetime away. Use that lifetime and the power of the time value of money. It will take proper planning today to achieve your retirement dreams in the future. Investing early provides you one of the most important advantages in reaching your financial goals—TIME! Time allows you the opportunity to grow your investments, recover from downturns in the economy, and achieve high financial goals. Beginning your retirement saving today will help you reach your retirement goals while allowing you flexibility along the way.

 What I need to do

Learn as much as you can about the retirement options offered by your employer to see how these can assist you in reaching your financial goals. Often these plans include an employer match up to a certain amount of what you are investing toward retirement. For instance, if your employer matches the first 6% you contribute, you are immediately doubling your investment by contributing 6% of your pay. While it may seem like a lot of money deducted from your paycheck initially, not taking full advantage of the employer match is like saying no to the offer of free money. In addition, your contribution may put you in a lower tax bracket reducing the negative affect on your paycheck.

Research private investment options such as Individual Retirement Accounts (IRAs). A Traditional IRA is funded up to a certain yearly limit with the possibility of receiving a tax break on your contributions. A Roth IRA is funded with after-tax dollars. When you withdraw money in retirement it will be taxed as ordinary income on a Traditional IRA or usually taken tax-free on a Roth IRA. Some IRA options are even coordinated via payroll deductions through your employer.

Diversify your retirement dollars through a variety of investment options to spread your risk across multiple opportunities. Investments can be made into stocks, bonds, CDs (certificate of deposit), Money Market accounts, real estate, to name a few of many options. By spreading your investments across multiple opportunities, you create a strategy to maintain your investments while providing the opportunity to weather the ups and downs of certain markets versus the stability of others.

 Steps I need to take

1. Begin investing TODAY while time is in your favor.
2. Research any employer-sponsored retirement options.
3. Spread your risk by diversifying, investing in a variety of options.

 Resources I can use

- Acorns (mobile app)—invests your spare change by rounding up to the next dollar on your everyday purchases.

MY MONEY ACTIVITY

- Interview three people about their retirement planning.
- Ask each person what they felt was done well and what they could have improved upon.
- Summarize your findings and discuss how this will impact what you do TODAY for your retirement planning.

PERSONAL FINANCE

A KIPLINGER APPROACH

"The Benefits of Being FIRE-ish." Kiplinger's. November, 2019.

RETHINKING RETIREMENT   Eileen Ambrose

The Benefits of Being FIRE-ish

You can adapt the Financial Independence, Retire Early movement to fit your lifestyle.

IT'S EASY TO DISMISS THE super savers who embrace the FIRE movement as starry-eyed dreamers or cultish extremists. Financial Independence, Retire Early followers—many of whom are millennials who consider working 9 to 5 to be drudgery—often save 50% to 70% of their annual income with the goal of retiring in 10 to 15 years. The aggressive "Lean FIRE" savers share tips online on how they manage on less than $40,000 a year—for example, by Dumpster diving, living in a van, not having children or living on a diet of rice and beans.

But FIRE is much more than that. FIRE followers track their money, invest in low-cost funds, avoid high-interest debt and focus their spending on what's important to them, rather than buying things because they can afford them. Adapting some of these FIRE principles to fit your less-austere lifestyle can go a long way toward helping you achieve your retirement goals.

If you're not ready to go full-blown FIRE, here's how to be FIRE-*ish*.

Supercharge your savings.

Boosting savings gives you more money to invest. But more important, "every time you increase your savings rate, you are decreasing your lifestyle," says Whitney Morrison, principal financial planner with Legal-Zoom. That means you won't need to accumulate as much to maintain your lifestyle in retirement.

FIRE folks typically watch every penny. You don't have to be that precise, but you should have an idea of your cash flow so you can find extra dollars to put toward savings. Some expenses, such as a car loan or kids' extracurricular activities, disappear over time, freeing up money that you can redirect into investments, says Melissa Sotudeh, a certified financial planner in Rockville, Md. "If you're at the point that you've got the kids launched, that's a big pay raise there," she says.

Or boost savings by cutting expenses. "There is a lot of low-hanging fruit that won't force you into depriving yourself," says Brad Barrett, cofounder of the ChooseFI website. You don't have to give up your lattes. Look to housing and transportation, the largest expenses for consumers. If the kids are grown and you no longer need a four-bedroom house, consider downsizing.

Many FIRE folks also move to areas with a lower cost of living. Barrett moved from a one-bedroom co-op in East Northport, N.Y., to a four-bedroom home in Richmond, Va., and slashed his housing expenses in half.

All FIREd Up

HOW SAVING MORE CAN BOOST YOUR NEST EGG

The table shows how much more you'll save after 15 years if you increase your savings rate from 15% to 20% or 25%. The final balance assumes that a 50-year-old starts with $500,000 in retirement assets, currently earns $100,000 and gets 3% annual raises, and earns an annual return of 7% on the savings until age 65.

$1,843,000	$1,998,000	$2,153,000
15% savings rate	20% savings rate	25% savings rate

SOURCE: T. Rowe Price

Some FIRE advocates also ditch cars in favor of bikes, or they move closer to work so they can walk to their jobs. Barrett says he and his wife each have a car for convenience, but they drive older vehicles (his is a 2003 model; hers is a 2013).

Also look for ways to increase income. FIRE practitioners often find creative ways to generate income, such as blogging about their path to financial independence. Sotudeh says a pair of her clients pull in $30,000 a year by leasing their basement through Airbnb. Consider consulting on the side or finding ways to turn hobbies into cash, such as selling your crafts. "It can potentially help you retire earlier, but it can also help you create more income in retirement," says Katrina Soelter, a CFP in Los Angeles.

Review the investment fees you pay, which can significantly erode your returns over time. FIRE acolytes favor low-cost index funds, such as those offered by Vanguard, Schwab and Fidelity. ■

CONTACT THE AUTHOR AT EAMBROSE@KIPLINGER.COM.

Business Math Issue

Only compounding affects being Fire-ish.

1. List the key points of the article and information to support your position.

2. Write a group defense of your position using math calculations to support your view. If you are in an online course, post to a discussion board.

Chapter 13
Annuities and Sinking Funds

Learning Unit Objectives

LU 13–1: Annuities: Ordinary Annuity and Annuity Due (Find Future Value)

1. Differentiate between contingent annuities and annuities certain.
2. Calculate the future value of an ordinary annuity and an annuity due manually and by table lookup.

LU 13–2: Present Value of an Ordinary Annuity (Find Present Value)

1. Calculate the present value of an ordinary annuity by table lookup and manually check the calculation.
2. Compare the calculation of the present value of one lump sum versus the present value of an ordinary annuity.

LU 13–3: Sinking Funds (Find Periodic Payments)

1. Calculate the payment made at the end of each period by table lookup.
2. Check table lookup by using ordinary annuity table.

℮ Essential Question

How can I use annuities and sinking funds to understand the business world and make my life easier?

🌐 Math Around the World

A *Boston Globe* article entitled "Cost of Living: A Cup a Day" began by explaining that each month the *Globe* runs a feature on an everyday expense to see how much it costs an average person. Since many people are coffee drinkers, the *Globe* assumed that a person drank 3 cups a day of Dunkin' Donuts coffee at the cost of $1.65 a cup. For a 5-day week, the person would spend $1,287 annually (52 weeks). If the person brewed the coffee at home, the cost of the beans per cup would be $0.10 with an annual expense of $78, saving $1,209 over the Dunkin' Donuts coffee. If a person gave up drinking coffee, the person would save $1,287.

The article continued with the discussion on "Investing Your Savings." Note how much you would have in 30 years if you invested your money in 0%, 6%, and 10% annual returns. Using the magic of compounding, if you saved $1,287 a year, your money could grow to a quarter of a million dollars.

This chapter shows how to compute compound interest that results from a *stream* of payments, or an annuity. Chapter 12 showed how to calculate compound interest on a lump-sum payment deposited at the beginning of a particular time. Knowing how to calculate interest compounding on a lump sum will make the calculation of interest compounding on annuities easier to understand.

We begin the chapter by explaining the difference between calculating the future value of an ordinary annuity and an annuity due. Then you learn how to find the present value of an ordinary annuity. The chapter ends with a discussion of sinking funds.

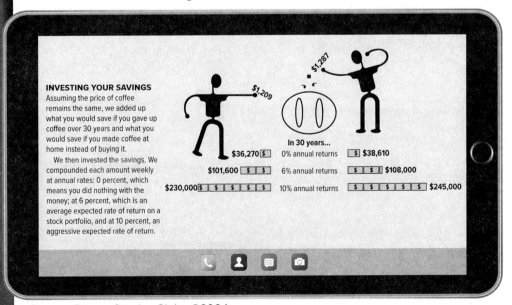

INVESTING YOUR SAVINGS
Assuming the price of coffee remains the same, we added up what you would save if you gave up coffee over 30 years and what you would save if you made coffee at home instead of buying it.
We then invested the savings. We compounded each amount weekly at annual rates: 0 percent, which means you did nothing with the money; at 6 percent, which is an average expected rate of return on a stock portfolio, and at 10 percent, an aggressive expected rate of return.

In 30 years...
$36,270 — 0% annual returns — $38,610
$101,600 — 6% annual returns — $108,000
$230,000 — 10% annual returns — $245,000

Source: Boston Sunday Globe ©2004

Copyright © McGraw Hill (tablet)Can Yesil/Shutterstock; PureSolution/Shutterstock

Learning Unit 13–1:

Annuities: Ordinary Annuity and Annuity Due (Find Future Value)

Many parents of small children are concerned about being able to afford to pay for their children's college educations. Some parents deposit a lump sum in a financial institution when the child is in diapers. The interest on this sum is compounded until the child is 18, when the parents withdraw the money for college expenses. Parents could also fund their children's educations with annuities by depositing a series of payments for a certain time. The concept of annuities is the first topic in this learning unit.

Learn: Concept of an Annuity—The Big Picture

All of us would probably like to win $1 million in a state lottery. What happens when you have the winning ticket? You take it to the lottery headquarters. When you turn in the ticket, do you immediately receive a check for $1 million? No. Lottery payoffs are not usually made in lump sums.

Lottery winners receive a series of payments over a period of time—usually years. This *stream* of payments is an **annuity.** By paying the winners an annuity, lotteries do not actually spend $1 million. The lottery deposits a sum of money in a financial institution. The continual growth of this sum through compound interest provides the lottery winner with a series of payments.

FIGURE 13.1 Future value of an annuity of $1 at 8%

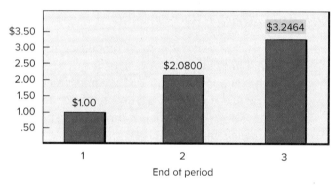

When we calculated the maturity value of a lump-sum payment in Chapter 12, the maturity value was the principal and its interest. Now we are looking not at lump-sum payments but at a series of payments (usually of equal amounts over regular **payment periods**) plus the interest that accumulates. So the **future value of an annuity** is the future *dollar amount* of a series of payments plus interest.[1] The **term of the annuity** is the time from the beginning of the first payment period to the end of the last payment period.

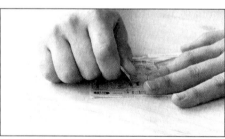

The concept of the future value of an annuity is illustrated in Figure 13.1. Do not be concerned about the calculations (we will do them soon). Let's first focus on the big picture of annuities. In Figure 13.1 we see the following:

At end of period 1: The $1 is still worth $1 because it was invested at the *end* of the period.

At end of period 2: An additional $1 is invested. The $2.00 is now worth $2.08. Note the $1 from period 1 earns interest but not the $1 invested at the end of period 2.

---

[1] The term *amount of an annuity* has the same meaning as *future value of an annuity.*

At end of period 3: An additional $1 is invested. The $3.00 is now worth $3.25. Remember that the last dollar invested earns no interest.

Before learning how to calculate annuities, you should understand the two classifications of annuities.

Learn: How Annuities Are Classified

Annuities have many uses in addition to lottery payoffs. Some of these uses are insurance companies' pension installments, Social Security payments, home mortgages, businesses paying off notes, bond interest, and savings for a vacation trip or college education.

Annuities are classified into two major groups: contingent annuities and annuities certain. **Contingent annuities** have no fixed number of payments but depend on an uncertain event (e.g., life insurance payments that cease when the insured dies). **Annuities certain** have a specific stated number of payments (e.g., mortgage payments on a home). Based on the time of the payment, we can divide each of these two major annuity groups into the following:

1. **Ordinary annuity**—regular deposits (payments) made at the *end* of the period. Periods could be months, quarters, years, and so on. An ordinary annuity could be salaries, stock dividends, and so on.

2. **Annuity due**—regular deposits (payments) made at the *beginning* of the period, such as rent or life insurance premiums.

The remainder of this unit shows you how to calculate and check ordinary annuities and annuities due. Remember that you are calculating the *dollar amount* of the annuity at the end of the annuity term or at the end of the last period.

Learn: Ordinary Annuities: Money Invested at End of Period (Find Future Value)

Before we explain how to use a table that simplifies calculating ordinary annuities, let's first determine how to calculate the future value of an ordinary annuity manually.

Calculating Future Value of Ordinary Annuities Manually   Remember that an ordinary annuity invests money at the *end* of each year (period). After we calculate ordinary annuities manually, you will see that the total value of the investment comes from the *stream* of yearly investments and the buildup of interest on the current balance.

Pepper ...
And Salt

THE WALL STREET JOURNAL

"Have you heard? There's talk about raising the retirement age to 170."

Used by permission of Cartoon Features Syndicate

Calculating Future Value of an Ordinary Annuity Manually

Step 1 For period 1, no interest calculation is necessary, since money is invested at the end of the period.

Step 2 For period 2, calculate interest on the balance and add the interest to the previous balance.

Step 3 Add the additional investment at the end of period 2 to the new balance.

Step 4 Repeat Steps 2 and 3 until the end of the desired period is reached.

Example: Find the value of an investment after 3 years for a $3,000 ordinary annuity at 8%. We calculate this manually as follows:

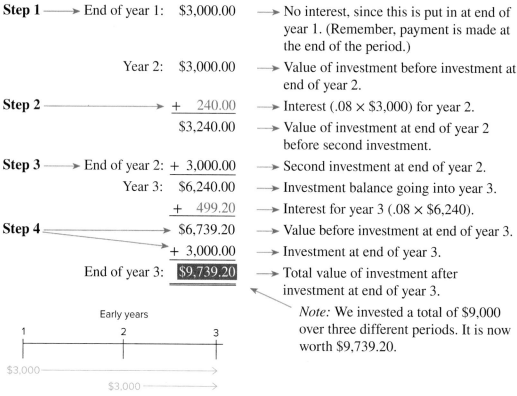

Step 1 ──→ End of year 1: $3,000.00 ──→ No interest, since this is put in at end of year 1. (Remember, payment is made at the end of the period.)

Year 2: $3,000.00 ──→ Value of investment before investment at end of year 2.

Step 2 ──────────→ + 240.00 ──→ Interest (.08 × $3,000) for year 2.

$3,240.00 ──→ Value of investment at end of year 2 before second investment.

Step 3 ──→ End of year 2: + 3,000.00 ──→ Second investment at end of year 2.

Year 3: $6,240.00 ──→ Investment balance going into year 3.

+ 499.20 ──→ Interest for year 3 (.08 × $6,240).

Step 4 ──────────→ $6,739.20 ──→ Value before investment at end of year 3.

+ 3,000.00 ──→ Investment at end of year 3.

End of year 3: $9,739.20 ──→ Total value of investment after investment at end of year 3.

Note: We invested a total of $9,000 over three different periods. It is now worth $9,739.20.

Early years

1   2   3

$3,000 ────────→
    $3,000 ────────→
        $3,000

When you deposit $3,000 at the end of each year at an annual rate of 8%, the total value of the annuity is $9,739.20. What we called *maturity value* in compounding is now called the *future value of the annuity*. Remember that Interest = Principal × Rate × Time, with the principal changing because of the interest payments and the additional deposits. We can make this calculation easier by using Table 13.1.

Calculating Future Value of Ordinary Annuities by Table Lookup Use the following steps to calculate the future value of an ordinary annuity by table lookup.[2]

Calculating Future Value of an Ordinary Annuity by Table Lookup

Step 1 Calculate the number of periods and rate per period.

Step 2 Look up the periods and rate in an ordinary annuity table. The intersection gives the table factor for the future value of $1.

Step 3 Multiply the payment each period by the table factor. This gives the future value of the annuity.

$$\text{Future value of ordinary annuity} = \text{Annuity payment each period} \times \text{Ordinary annuity table factor}$$

[2] The formula for an ordinary annuity is FV = PMT × [$\frac{(1+i)^n - 1}{i}$] where FV equals future value of an ordinary annuity, PMT equals annuity payment, i equals interest, and n equals number of periods. The calculator sequence for this example is: 1 ⊞ .08 = y^x 3 ⊟ 1 = ÷ .08 × 3,000 = 9,739.20. A *Financial Calculator Guide* is available online that shows how to operate HP 10BII and TI BA II Plus.

TABLE 13.1 Ordinary annuity table: Compound sum of an annuity of $1

Period	$\frac{1}{2}$%	1%	2%	3%	4%	5%	6%	7%	8%	9%	10%	11%	12%
1	1.0000	1.0000	1.0000	1.0000	1.0000	1.0000	1.0000	1.0000	1.0000	1.0000	1.0000	1.0000	1.0000
2	2.0050	2.0100	2.0200	2.0300	2.0400	2.0500	2.0600	2.0700	2.0800	2.0900	2.1000	2.1100	2.1200
3	3.0150	3.0301	3.0604	3.0909	3.1216	3.1525	3.1836	3.2149	3.2464	3.2781	3.3100	3.3421	3.3744
4	4.0301	4.0604	4.1216	4.1836	4.2465	4.3101	4.3746	4.4399	4.5061	4.5731	4.6410	4.7097	4.7793
5	5.0503	5.1010	5.2040	5.3091	5.4163	5.5256	5.6371	5.7507	5.8666	5.9847	6.1051	6.2278	6.3528
6	6.0755	6.1520	6.3081	6.4684	6.6330	6.8019	6.9753	7.1533	7.3359	7.5233	7.7156	7.9129	8.1152
7	7.1059	7.2135	7.4343	7.6625	7.8983	8.1420	8.3938	8.6540	8.9228	9.2004	9.4872	9.7833	10.0890
8	8.1414	8.2857	8.5830	8.8923	9.2142	9.5491	9.8975	10.2598	10.6366	11.0285	11.4359	11.8594	12.2997
9	9.1821	9.3685	9.7546	10.1591	10.5828	11.0266	11.4913	11.9780	12.4876	13.0210	13.5795	14.1640	14.7757
10	10.2280	10.4622	10.9497	11.4639	12.0061	12.5779	13.1808	13.8164	14.4866	15.1929	15.9374	16.7220	17.5487
11	11.2792	11.5668	12.1687	12.8078	13.4864	14.2068	14.9716	15.7836	16.6455	17.5603	18.5312	19.5614	20.6546
12	12.3356	12.6825	13.4121	14.1920	15.0258	15.9171	16.8699	17.8885	18.9771	20.1407	21.3843	22.7132	24.1331
13	13.3972	13.8093	14.6803	15.6178	16.6268	17.7130	18.8821	20.1406	21.4953	22.9534	24.5227	26.2116	28.0291
14	14.4642	14.9474	15.9739	17.0863	18.2919	19.5986	21.0151	22.5505	24.2149	26.0192	27.9750	30.0949	32.3926
15	15.5365	16.0969	17.2934	18.5989	20.0236	21.5786	23.2760	25.1290	27.1521	29.3609	31.7725	34.4054	37.2797
16	16.6142	17.2579	18.6393	20.1569	21.8245	23.6575	25.6725	27.8881	30.3243	33.0034	35.9497	39.1899	42.7533
17	17.6973	18.4304	20.0121	21.7616	23.6975	25.8404	28.2129	30.8402	33.7502	36.9737	40.5447	44.5008	48.8837
18	18.7858	19.6147	21.4122	23.4144	25.6454	28.1323	30.9057	33.9990	37.4502	41.3013	45.5992	50.3959	55.7497
19	19.8797	20.8109	22.8406	25.1169	27.6712	30.5390	33.7600	37.3790	41.4463	46.0185	51.1591	56.9395	63.4397
20	20.9791	22.0190	24.2974	26.8704	29.7781	33.0660	36.7856	40.9955	45.7620	51.1601	57.2750	64.2028	72.0524
25	26.5591	28.2432	32.0303	36.4593	41.6459	47.7271	54.8645	63.2490	73.1059	84.7009	98.3471	114.4133	133.3339
30	32.2800	34.7849	40.5681	47.5754	56.0849	66.4388	79.0582	94.4608	113.2832	136.3075	164.4940	199.0209	241.3327
40	44.1589	48.8864	60.4020	75.4013	95.0256	120.7998	154.7620	199.6351	259.0565	337.8824	442.5926	581.8261	767.0914
50	56.6452	64.4632	84.5790	112.7967	152.6671	209.3480	290.3359	406.5289	573.7702	815.0836	1,163.9085	1,668.7712	2,400.0182

Note: This is only a sampling of tables available. The *Business Math Handbook* shows tables from $\frac{1}{2}$% to 15%.

Example: Find the value of an investment after 3 years for a $3,000 ordinary annuity at 8%.

Step 1 Periods = 3 years × 1 = 3   Rate = $\dfrac{8\%}{\text{Annually}}$ = 8%

Step 2 Go to Table 13.1, an ordinary annuity table. Look for 3 under the Period column. Go across to 8%. At the intersection is the table factor, 3.2464. (This was the example we showed in Figure 13.1.)

Step 3 Multiply $3,000 × 3.2464 = $9,739.20 (the same figure we calculated manually).

Learn: Annuities Due: Money Invested at Beginning of Period (Find Future Value)

In this section we look at what the difference in the total investment would be for an annuity due. As in the previous section, we will first make the calculation manually and then use the table lookup.

Calculating Future Value of Annuities Due Manually   Use the steps that follow to calculate the future value of an annuity due manually.

Money Tip

Never cash out a retirement fund account until you have calculated the final cost to you. Consider having to pay a 10% early withdrawal penalty along with both federal and state income tax on the amount withdrawn. Determine if the withdrawal puts you in a higher tax bracket that may disqualify you from any aid (such as food stamps) you are currently receiving.

Calculating Future Value of an Annuity Due Manually

Step 1 Calculate the interest on the balance for the period and add it to the previous balance.

Step 2 Add additional investment at the *beginning* of the period to the new balance.

Step 3 Repeat Steps 1 and 2 until the end of the desired period is reached.

Remember that in an annuity due, we deposit the money at the *beginning* of the year and gain more interest. Common sense should tell us that the *annuity due* will give a higher final value. We will use the same example that we used before.

Example: Find the value of an investment after 3 years for a $3,000 annuity due at 8%. We calculate this manually as follows:

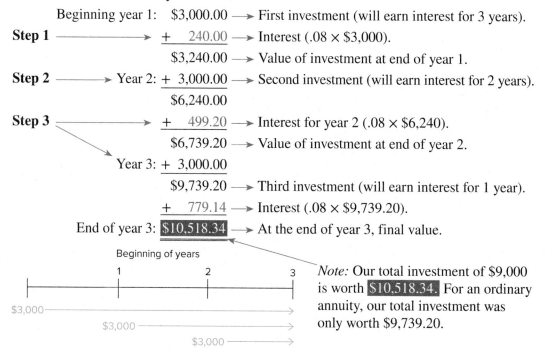

Beginning year 1: $3,000.00 ⟶ First investment (will earn interest for 3 years).

Step 1 ⟶ +   240.00 ⟶ Interest (.08 × $3,000).

$3,240.00 ⟶ Value of investment at end of year 1.

Step 2 ⟶ Year 2: + 3,000.00 ⟶ Second investment (will earn interest for 2 years).

$6,240.00

Step 3 ⟶ +   499.20 ⟶ Interest for year 2 (.08 × $6,240).

$6,739.20 ⟶ Value of investment at end of year 2.

Year 3: + 3,000.00

$9,739.20 ⟶ Third investment (will earn interest for 1 year).

+   779.14 ⟶ Interest (.08 × $9,739.20).

End of year 3: $10,518.34 ⟶ At the end of year 3, final value.

Beginning of years

1   2   3

$3,000 ⟶
$3,000 ⟶
$3,000 ⟶

Note: Our total investment of $9,000 is worth $10,518.34. For an ordinary annuity, our total investment was only worth $9,739.20.

Calculating Future Value of Annuities Due by Table Lookup   To calculate the future value of an annuity due with a table lookup, use the steps that follow.

Calculating Future Value of an Annuity Due by Table Lookup[3]

Step 1 Calculate the number of periods and the rate per period. Add one extra period.

Step 2 Look up in an ordinary annuity table the periods and rate. The intersection gives the table *factor* for future value of $1.

Step 3 Multiply payment each period by the table factor.

Step 4 Subtract 1 payment from Step 3.

$$\text{Future value of an annuity due} = \left(\begin{array}{c} \text{Annuity} \\ \text{payment} \\ \text{each period} \end{array} \times \begin{array}{c} \text{Ordinary}^* \\ \text{annuity} \\ \text{table factor} \end{array} \right) - 1 \text{ Payment}$$

*Add 1 period.

---

[3] The formula for an annuity due is $FV = PMT \times \frac{(1+i)^n - 1}{i} \times (1 + i)$, where FV equals future value of annuity due, PMT equals annuity payment, i equals interest, and n equals number of periods. This formula is the same as that in footnote 2 except we take one more step. Multiply the future value of annuity by $(1 + i)$ since payments are made at the beginning of the period. The calculator sequence for this step is: $1 \boxed{+} .08 \boxed{=} \boxed{\times} 9,739.20 \boxed{=} 10,518.34$.

Let's check the $10,518.34 by table lookup.

Step 1 Periods = 3 years × 1 = $\dfrac{\begin{array}{r} 3 \\ +\ 1 \text{ extra} \\ \hline 4 \end{array}}{}$ Rate = $\dfrac{8\%}{\text{Annually}} = 8\%$

Step 2 Table factor, 4.5061

Step 3 $3,000 × 4.5061 = $13,518.30

Step 4   − 3,000.00   ← Be sure to subtract 1 payment.

                            = **$10,518.30**   (off 4 cents due to rounding)

Note that the annuity due shows an ending value of $10,518.30, while the ending value of the ordinary annuity was $9,739.20. We had a higher ending value with the annuity due because the investment took place at the beginning of each period.

Annuity payments do not have to be made yearly. They could be made semiannually, monthly, quarterly, and so on. Let's look at one more example with a different number of periods and the same rate.

Different Number of Periods and Rates   By using a different number of periods and the same rate, we will contrast an ordinary annuity with an annuity due in the following example:

Example: Using Table 13.1, find the value of a $3,000 investment after 3 years made quarterly at 8%.

In the annuity due calculation, be sure to add one period and subtract one payment from the total value.

Ordinary annuity	Annuity due	
Step 1 Periods = 3 years × 4 = 12	Periods = 3 years × 4 = 12 + 1 = 13	**Step 1**
Rate = 8% ÷ 4 = 2%	Rate = 8% ÷ 4 = 2%	
Step 2 Table 13.1:	Table 13.1:	**Step 2**
12 periods, 2% = 13.4120	13 periods, 2% = 14.6803	
Step 3 $3,000 × 13.4120 = **$40,236**	$3,000 × 14.6803 = $44,040.90	**Step 3**
	− 3,000.00	**Step 4**
	$41,040.90	

Again, note that with the annuity due, the total value is greater since you invest the money at the beginning of each period.

Now check your progress with the Practice Quiz.

Practice Quiz

Complete this Practice Quiz to see how you are doing.

1. Using Table 13.1, **(a)** find the value of an investment after 4 years on an ordinary annuity of $4,000 made semiannually at 10%; and **(b)** recalculate, assuming an annuity due.

2. Wally Beaver won a lottery and will receive a check for $4,000 at the beginning of each 6 months for the next 5 years. If Wally deposits each check into an account that pays 6%, how much will he have at the end of the 5 years?

✓ **Solutions**

1. **a. Step 1.** Periods = 4 years × 2 = 8

   b. Periods = 4 years × 2   **Step 1**

   = 8 + 1 = 9

   $10\% \div 2 = 5\%$

   $10\% \div 2 = 5\%$

   Step 2. Factor = 9.5491

   Factor = 11.0265   **Step 2**

   Step 3. $4,000 × 9.5491

   $4,000 × 11.0265 = $44,106   **Step 3**

   = $38,196.40

   − 1 payment   − 4,000   **Step 4**

   $40,106

2. **Step 1.** 5 years × 2 = 10

   + 1

   11 periods

   $\dfrac{6\%}{2} = 3\%$

   Step 2. Table factor, 12.8078

   Step 3. $4,000 × 12.8078 = $51,231.20

   Step 4.   − 4,000.00

   $47,231.20

Present Value of an Ordinary Annuity (Find Present Value)[4]

This unit begins by presenting the concept of present value of an ordinary annuity. Then you will learn how to use a table to calculate the present value of an ordinary annuity.

Learn: Concept of Present Value of an Ordinary Annuity—The Big Picture

Let's assume that we want to know how much money we need to invest *today* to receive a stream of payments for a given number of years in the future. This is called the **present value of an ordinary annuity.**

In Figure 13.2 you can see that if you wanted to withdraw $1 at the end of one period, you would have to invest 93 cents *today*. If at the end of each period for three periods you wanted to withdraw $1, you would have to put $2.58 in the bank *today* at 8% interest. (Note that we go from the future back to the present.)

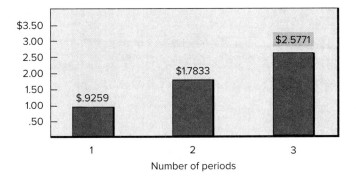

FIGURE 13.2 Present value of an annuity of $1 at 8%

Now let's look at how we could use tables to calculate the present value of annuities and then check our answer.

Learn: Calculating Present Value of an Ordinary Annuity by Table Lookup

Use the steps below to calculate the present value of an ordinary annuity by table lookup.[5]

Calculating Present Value of an Ordinary Annuity by Table Lookup

Step 1 Calculate the number of periods and rate per period.

Step 2 Look up the periods and rate in the present value of an annuity table. The intersection gives the table factor for the present value of $1.

Step 3 Multiply the withdrawal for each period by the table factor. This gives the present value of an ordinary annuity.

$$\frac{\text{Present value of}}{\text{ordinary annuity payment}} = \frac{\text{Annuity}}{\text{payment}} \times \frac{\text{Present value of ordinary}}{\text{annuity table factor}}$$

---

[4] For simplicity we omit a discussion of present value of annuity due that would require subtracting a period and adding a 1.

[5] The formula for the present value of an ordinary annuity is $\text{PV} = \text{PMT}\left[\frac{1-(1+i)^{-n}}{i}\right]$, where PV equals present value of annuity, PMT equals annuity payment, i equals interest, and n equals number of periods. The calculator sequence would be as follows for the John Fitch example that follows: $1\boxed{+}.08\boxed{y^x}3\boxed{+-}\boxed{-}1=\boxed{+-}\boxed{\div}.08\boxed{\times}8{,}000\boxed{=}20{,}616.78$.

TABLE 13.2 Present value of an annuity of $1

Period	$\frac{1}{2}$%	1%	2%	3%	4%	5%	6%	7%	8%	9%	10%	11%	12%
1	0.9950	0.9901	0.9804	0.9709	0.9615	0.9524	0.9434	0.9346	0.9259	0.9174	0.9091	0.9009	0.8929
2	1.9851	1.9704	1.9416	1.9135	1.8861	1.8594	1.8334	1.8080	1.7833	1.7591	1.7355	1.7125	1.6901
3	2.9702	2.9410	2.8839	2.8286	2.7751	2.7232	2.6730	2.6243	2.5771	2.5313	2.4869	2.4437	2.4018
4	3.9505	3.9020	3.8077	3.7171	3.6299	3.5459	3.4651	3.3872	3.3121	3.2397	3.1699	3.1024	3.0373
5	4.9259	4.8534	4.7134	4.5797	4.4518	4.3295	4.2124	4.1002	3.9927	3.8897	3.7908	3.6959	3.6048
6	5.8964	5.7955	5.6014	5.4172	5.2421	5.0757	4.9173	4.7665	4.6229	4.4859	4.3553	4.2305	4.1114
7	6.8621	6.7282	6.4720	6.2303	6.0021	5.7864	5.5824	5.3893	5.2064	5.0330	4.8684	4.7122	4.5638
8	7.8230	7.6517	7.3255	7.0197	6.7327	6.4632	6.2098	5.9713	5.7466	5.5348	5.3349	5.1461	4.9676
9	8.7791	8.5660	8.1622	7.7861	7.4353	7.1078	6.8017	6.5152	6.2469	5.9952	5.7590	5.5370	5.3282
10	9.7304	9.4713	8.9826	8.5302	8.1109	7.7217	7.3601	7.0236	6.7101	6.4177	6.1446	5.8892	5.6502
11	10.6770	10.3676	9.7868	9.2526	8.7605	8.3064	7.8869	7.4987	7.1390	6.8052	6.4951	6.2065	5.9377
12	11.6189	11.2551	10.5753	9.9540	9.3851	8.8632	8.3838	7.9427	7.5361	7.1607	6.8137	6.4924	6.1944
13	12.5562	12.1337	11.3483	10.6350	9.9856	9.3936	8.8527	8.3576	7.9038	7.4869	7.1034	6.7499	6.4235
14	13.4887	13.0037	12.1062	11.2961	10.5631	9.8986	9.2950	8.7455	8.2442	7.7862	7.3667	6.9819	6.6282
15	14.4166	13.8650	12.8492	11.9379	11.1184	10.3796	9.7122	9.1079	8.5595	8.0607	7.6061	7.1909	6.8109
16	15.3399	14.7179	13.5777	12.5611	11.6523	10.8378	10.1059	9.4466	8.8514	8.3126	7.8237	7.3792	6.9740
17	16.2586	15.5622	14.2918	13.1661	12.1657	11.2741	10.4773	9.7632	9.1216	8.5436	8.0216	7.5488	7.1196
18	17.1728	16.3983	14.9920	13.7535	12.6593	11.6896	10.8276	10.0591	9.3719	8.7556	8.2014	7.7016	7.2497
19	18.0824	17.2260	15.6784	14.3238	13.1339	12.0853	11.1581	10.3356	9.6036	8.9501	8.3649	7.8393	7.3658
20	18.9874	18.0455	16.3514	14.8775	13.5903	12.4622	11.4699	10.5940	9.8181	9.1285	8.5136	7.9633	7.4694
25	23.4457	22.0231	19.5234	17.4131	15.6221	14.0939	12.7834	11.6536	10.6748	9.8226	9.0770	8.4217	7.8431
30	27.7941	25.8077	22.3964	19.6004	17.2920	15.3724	13.7648	12.4090	11.2578	10.2737	9.4269	8.6938	8.0552
40	36.1723	32.8347	27.3554	23.1148	19.7928	17.1591	15.0463	13.3317	11.9246	10.7574	9.7790	8.9511	8.2438
50	44.1428	39.1961	31.4236	25.7298	21.4822	18.2559	15.7619	13.8007	12.2335	10.9617	9.9148	9.0417	8.3045

Example: John Fitch wants to receive an $8,000 annuity for 3 years. Interest on the annuity is 8% annually. John will make withdrawals at the end of each year. How much must John invest today to receive a stream of payments for 3 years? Use Table 13.2. Remember that interest could be earned semiannually, quarterly, and so on, as shown in the previous unit.

Step 1 3 years × 1 = 3 periods $\dfrac{8\%}{\text{Annually}} = 8\%$

Step 2 Table factor, 2.5771 (we saw this in Figure 13.2)

Step 3 8,000 × 2.5771 = $20,616.80

If John wants to withdraw $8,000 at the end of each period for 3 years, he will have to deposit $20,616.80 in the bank *today*.

$20,616.80
$+$ 1,649.34 → Interest at end of year 1 (.08 × $20,616.80)
$22,266.14
$-$ 8,000.00 → First payment to John
$14,266.14
$+$ 1,141.29 → Interest at end of year 2 (.08 × $14,266.14)
$15,407.43
$-$ 8,000.00 → Second payment to John
$ 7,407.43
$+$ 592.59 → Interest at end of year 3 (.08 × $7,407.43)
$ 8,000.02
$-$ 8,000.00 → After end of year 3 John receives his last $8,000
.02 (off 2 cents due to rounding)

Before we leave this unit, let's work out two examples that show the relationship of Chapter 13 to Chapter 12. Use the tables in your *Business Math Handbook*.

Learn: Lump Sum versus Annuities

Example: John Sands made deposits of $200 semiannually to Floor Bank, which pays 8% interest compounded semiannually. After 5 years, John makes no more deposits. What will be the balance in the account 6 years after the last deposit?

Step 1 Calculate amount of annuity: ⟋Table 13.1

  10 periods, 4% $200 × 12.0061 = $2,401.22

Step 2 Calculate how much the final value of the annuity will grow by the compound interest table. ⟋Table 12.1

  12 periods, 4% $2,401.22 × 1.6010 = $3,844.35

For John, the stream of payments grows to $2,401.22. Then this *lump sum* grows for 6 years to $3,844.35. Now let's look at a present value example.

Example: Mel Rich decided to retire in 8 years to New Mexico. What amount should Mel invest today so he will be able to withdraw $40,000 at the end of each year for 25 years after he retires? Assume Mel can invest money at 5% interest (compounded annually).

Step 1 Calculate the present value of the annuity: ⟋Table 13.2

  25 periods, 5% $40,000 × 14.0939 = $563,756

Step 2 Find the present value of $563,756 since Mel will not retire for 8 years:

  Table 12.3 ⟍

  8 periods, 5% (PV table) $563,756 × .6768 = $381,550.06

If Mel deposits $381,550 in year 1, it will grow to $563,756 after 8 years.

It's time to try the Practice Quiz and check your understanding of this unit.

Practice Quiz

Complete this Practice Quiz to see how you are doing.

1. What must you invest today to receive an $18,000 annuity for 5 years semiannually at a 10% annual rate? All withdrawals will be made at the end of each period.

2. Rase High School wants to set up a scholarship fund to provide five $2,000 scholarships for the next 10 years. If money can be invested at an annual rate of 9%, how much should the scholarship committee invest today?

3. Joe Wood decided to retire in 5 years in Arizona. What amount should Joe invest today so he can withdraw $60,000 at the end of each year for 30 years after he retires? Assume Joe can invest money at 6% compounded annually.

✓ Solutions

(Use tables in *Business Math Handbook*)

1. **Step 1.** Periods = 5 years × 2 = 10; Rate = 10% ÷ 2 = 5%

   Step 2. Factor, 7.7217

   Step 3. $18,000 × 7.7217 = $138,990.60

2. **Step 1.** Periods = 10; Rate = 9%

   Step 2. Factor, 6.4177

   Step 3. $10,000 × 6.4177 = $64,177

3. **Step 1.** Calculate present value of annuity: 30 periods, 6%.

   $60,000 × 13.7648 = $825,888

   Step 2. Find present value of $825,888 for 5 years: 5 periods, 6%.

   $825,888 × .7473 = $617,186.10

Sinking Funds
(Find Periodic Payments)

A **sinking fund** is a financial arrangement that sets aside regular periodic payments of a particular amount of money. Compound interest accumulates on these payments to a specific sum at a predetermined future date. Corporations use sinking funds to discharge bonded indebtedness, to replace worn-out equipment, to purchase plant expansion, and so on.

A sinking fund is a different type of an annuity. In a sinking fund, you determine the amount of periodic payments you need to achieve a given financial goal. In the annuity, you know the amount of each payment and must determine its future value. Let's work with the following formula:

> Sinking fund payment = Future value × Sinking fund table factor

Example: To retire a bond issue, Moore Company needs $60,000 in 18 years from today. The interest rate is 10% compounded annually. What payment must Moore make at the end of each year? Use Table 13.3.

We begin by looking down the Period column in Table 13.3 until we come to 18. Then we go across until we reach the 10% column. The table factor is .0219.

Now we multiply $60,000 by the factor as follows:

$60,000 × .0219 = **$1,314**

This states that if Moore Company pays $1,314 at the end of each period for 18 years, then $60,000 will be available to pay off the bond issue at maturity.

We can check this by using Table 13.1 (the ordinary annuity table):

$1,314 × 45.5992 = $59,917.35 (off due to rounding)

It's time to try the following Practice Quiz.

TABLE 13.3 Sinking fund table[6] based on $1

Period	2%	3%	4%	5%	6%	8%	10%
1	1.0000	1.0000	1.0000	1.0000	1.0000	1.0000	1.0000
2	0.4951	0.4926	0.4902	0.4878	0.4854	0.4808	0.4762
3	0.3268	0.3235	0.3203	0.3172	0.3141	0.3080	0.3021
4	0.2426	0.2390	0.2355	0.2320	0.2286	0.2219	0.2155
5	0.1922	0.1884	0.1846	0.1810	0.1774	0.1705	0.1638
6	0.1585	0.1546	0.1508	0.1470	0.1434	0.1363	0.1296
7	0.1345	0.1305	0.1266	0.1228	0.1191	0.1121	0.1054
8	0.1165	0.1125	0.1085	0.1047	0.1010	0.0940	0.0874
9	0.1025	0.0984	0.0945	0.0907	0.0870	0.0801	0.0736
10	0.0913	0.0872	0.0833	0.0795	0.0759	0.0690	0.0627
11	0.0822	0.0781	0.0741	0.0704	0.0668	0.0601	0.0540
12	0.0746	0.0705	0.0666	0.0628	0.0593	0.0527	0.0468
13	0.0681	0.0640	0.0601	0.0565	0.0530	0.0465	0.0408
14	0.0626	0.0585	0.0547	0.0510	0.0476	0.0413	0.0357
15	0.0578	0.0538	0.0499	0.0463	0.0430	0.0368	0.0315
16	0.0537	0.0496	0.0458	0.0423	0.0390	0.0330	0.0278
17	0.0500	0.0460	0.0422	0.0387	0.0354	0.0296	0.0247
18	0.0467	0.0427	0.0390	0.0355	0.0324	0.0267	0.0219
19	0.0438	0.0398	0.0361	0.0327	0.0296	0.0241	0.0195
20	0.0412	0.0372	0.0336	0.0302	0.0272	0.0219	0.0175
24	0.0329	0.0290	0.0256	0.0225	0.0197	0.0150	0.0113
25	0.0312	0.0274	0.0240	0.0210	0.0182	0.0137	0.0102
28	0.0270	0.0233	0.0200	0.0171	0.0146	0.0105	0.0075
32	0.0226	0.0190	0.0159	0.0133	0.0110	0.0075	0.0050
36	0.0192	0.0158	0.0129	0.0104	0.0084	0.0053	0.0033
40	0.0166	0.0133	0.0105	0.0083	0.0065	0.0039	0.0023

[6] Sinking fund table is the reciprocal of the ordinary annuity table.

Money Tip If you are trying to build credit by using a credit card, each time you make a purchase using the credit card, deduct that amount from your checking account. When your credit card bill is due, add up all your credit card deductions in your checking account. You will have enough to pay the credit card off in full.

Practice Quiz

Complete this Practice Quiz to see how you are doing.

Today, Arrow Company issued bonds that will mature to a value of $90,000 in 10 years. Arrow's controller is planning to set up a sinking fund. Interest rates are 12% compounded semiannually. What will Arrow Company have to set aside to meet its obligation in 10 years? Check your answer. Your answer will be off due to the rounding of Table 13.3.

✓ **Solutions**

10 years $\times$ 2 = 20 periods $\dfrac{12\%}{2} = 6\%$ $\$90,000 \times .0272 = \boxed{\$2,448}$

Check $\$2,448 \times 36.7855 = \$90,050.90$

Chapter 13 Review

Topic/Procedure/Formula	Examples	You try it*
Ordinary annuities (find future value) Invest money at end of each period. Find future value at maturity. Answers question of how much money accumulates. $\dfrac{\text{Future}}{\text{value of}}_{\text{ordinary}}_{\text{annuity}} = \dfrac{\text{Annuity}}{\text{payment}}_{\text{each}}_{\text{period}} \times \dfrac{\text{Ordinary}}{\text{annuity}}_{\text{table}}_{\text{factor}}$ $FV = PMT\left[\dfrac{(1+i)^n - 1}{i}\right]$	Use Table 13.1: 2 years, \$4,000 ordinary annuity at 8% annually. Value = \$4,000 × 2.0800 $\quad$ = \$8,320 (2 periods, 8%) $FV = 4{,}000\left[\dfrac{(1+0.8)^2 - 1}{0.8}\right] = \$8{,}320$	**Calculate value of ordinary annuity** \$6,000, 7% annually, 4 years
Annuities due (find future value) Invest money at beginning of each period. Find future value at maturity. Should be higher than ordinary annuity since it is invested at beginning of each period. Use Table 13.1, but add one period and subtract one payment from answer. $\dfrac{\text{Future}}{\text{value}}_{\text{of an}}_{\text{annuity}}_{\text{due}} = \left(\dfrac{\text{Annuity}}{\text{payment}}_{\text{each}}_{\text{period}} \times \dfrac{\text{Ordinary}^*}{\text{annuity}}_{\text{table}}_{\text{factor}}\right) - 1\ \text{Payment}$ *Add 1 period. $FV_{due} = PMT\left[\dfrac{(1+i)^n - 1}{i}\right](1+i)$	*Example:* Same example as above but invest money at beginning of period. $\begin{aligned}\$4{,}000 \times 3.2464 =\ & \$12{,}985.60\\ -\ & 4{,}000.00\\ \hline & \$\ 8{,}985.60\end{aligned}$ $\qquad\qquad$ (3 periods, 8%) $FV_{due} = 4{,}000\left(\dfrac{(1+.08)^2 - 1}{.08}\right)(1+.08)$ $\quad = \$8{,}985.60$	**Calculate value of annuity due** \$6,000, 7% annually, 4 years
Present value of an ordinary annuity (find present value) Calculate number of periods and rate per period. Use Table 13.2 to find table factor for present value of \$1. Multiply withdrawal for each period by table factor to get present value of an ordinary annuity. $\dfrac{\text{Present}}{\text{value of an}}_{\text{ordinary}}_{\text{annuity}}_{\text{payment}} = \dfrac{\text{Annuity}}{\text{payment}} \times \dfrac{\text{Present}}{\text{value of}}_{\text{ordinary}}_{\text{annuity}}_{\text{table factor}}$ $PV = PMT\left[\dfrac{1-(1+i)^{-n}}{i}\right]$	*Example:* Receive \$10,000 for 5 years. Interest is 10% compounded annually. Table 13.2: 5 periods, 10% $\qquad\qquad\qquad$ 3.7908 What you put in today = × \$10,000 $\qquad\qquad\qquad$ \$37,908 $PV = 10{,}000\left[\dfrac{1-(1+.1)^{-5}}{.1}\right] = \$37{,}907.88$	**Calculate present value of ordinary annuity** \$20,000, 6 years, 4% interest compounded annually

Chapter 13 Review (Continued)

Topic/Procedure/Formula	Examples	You try it*
Sinking funds (find periodic payment) Paying a particular amount of money for a set number of periodic payments to accumulate a specific sum. We know the future value and must calculate the periodic payments needed. Answer can be proved by ordinary annuity table. $$\begin{array}{l}\text{Sinking} \\ \text{fund} \\ \text{payment}\end{array} = \begin{array}{l}\text{Future} \\ \text{value}\end{array} \times \begin{array}{l}\text{Sinking} \\ \text{fund table} \\ \text{factor}\end{array}$$	*Example:* $200,000 bond to retire 15 years from now. Interest is 6% compounded annually. By Table 13.3: $200,000 × .0430 = $8,600 Check by Table 13.1: $8,600 × 23.2759 = $200,172.74	**Calculate periodic payment** $400,000 bond to retire 20 years from now. Interest is 5% compounded annually.

Key Terms

Annuities certain

Annuity

Annuity due

Contingent annuities

Future value of an annuity

Ordinary annuity

Payment periods

Present value of an ordinary annuity

Sinking fund

Term of the annuity

* Worked-out solutions are in Appendix A.

Critical Thinking Discussion Questions with Chapter Concept Check

1. What is the difference between an ordinary annuity and an annuity due? If you were to save money in an annuity, which would you choose and why?

2. Explain how you would calculate ordinary annuities and annuities due by table lookup. Create an example to explain the meaning of a table factor from an ordinary annuity.

3. What is a present value of an ordinary annuity? Create an example showing how one of your relatives might plan for retirement by using the present value of an ordinary annuity. Would you ever have to use lump-sum payments in your calculation from Chapter 12?

4. What is a sinking fund? Why could an ordinary annuity table be used to check the sinking fund payment?

5. With the tight economy, more businesses are cutting back on matching the retirement contributions of their employees. Do you think this is ethical?

6. **Chapter Concept Check.** Create a retirement plan. Back up your retirement plan with calculations involving ordinary annuities as well as the present value of annuities.

7. How did the pandemic affect your retirement plans?

End-of-Chapter Problems

Name _____   Date _____

Check figures for odd-numbered problems in Appendix A.

Drill Problems

Complete the ordinary annuities for the following using tables in the *Business Math Handbook:*
LU 13–1(2)

	Amount of payment	Payment payable	Years	Interest rate	Value of annuity
13–1.	$5,000	Annually	11	4%	
13–2.	$12,000	Semiannually	8	7%	

Redo Problem 13–1 as an annuity due:

13–3.

Calculate the value of the following annuity due without a table. Check your results by Table 13.1 or
the *Business Math Handbook* (they will be slightly off due to rounding): *LU 13–1(2)*

	Amount of payment	Payment payable	Years	Interest rate
13–4.	$2,000	Annually	3	6%

Complete the following using Table 13.2 or the *Business Math Handbook* for the present value of an
ordinary annuity: *LU 13–2(1)*

	Amount of annuity expected	Payment	Time	Interest rate	Present value (amount needed now to invest to receive annuity)
13–5.	$900	Annually	4 years	6%	
13–6.	$15,000	Quarterly	4 years	8%	

13–7. Check Problem 13–5 without the use of Table 13.2.

Using the sinking fund table, Table 13.3, or the *Business Math Handbook,* complete the following: *LU 13–3(1)*

	Required amount	Frequency of payment	Length of time	Interest rate	Payment amount end of each period
13–8.	$25,000	Quarterly	6 years	8%	
13–9.	$15,000	Annually	8 years	8%	

13–10. Check the answer in Problem 13–9 by Table 13.1.   *LU 13–3(2)*

Word Problems (Use Tables in the *Business Math Handbook*)

e)cel **13–11.** John Regan, an employee at Home Depot, made deposits of $800 at the end of each year for 4 years. Interest is 4% compounded annually. What is the value of Regan's annuity at the end of 4 years?   *LU 13–1(2)*

13–12. Suze Orman wants to pay $1,500 semiannually to her granddaughter for 10 years for helping her around the house. If Suze can invest money at 6% compounded semiannually, how much must she invest today to meet this goal?   *LU 13–2(1)*

13–13. Financial analysts recommend investing 15% to 20% of your annual income in your retirement fund to reach a replacement rate of 70% of your income by age 65. This recommendation increases to almost 30% if you start investing at 45 years old. Mallori Rouse is 25 years old and has started investing $3,000 at the end of each year in her retirement account. How much will her account be worth in 20 years at 8% interest compounded annually? How much will it be worth in 30 years? What about at 40 years? How much will it be worth in 50 years? Round to the nearest dollar.   *LU 13–1(2)*

End-of-Chapter Problems (Continued)

13–14. After paying off a car loan or credit card, don't remove this amount from your budget. Instead, invest in your future by applying some of it to your retirement account. How much would $450 invested at the end of each quarter be worth in 10 years at 4% interest? (Use the *Business Math Handbook* tables.)   *LU 13–1(2)*

13–15. You decide to reduce the amount you spend eating out by $150 a month and invest the total saved at the end of each year in your retirement account. How much will the account be worth at 5% in 15 years?   *LU 13–1(2)*

13–16. Rob Herndon, an accountant with Southwest Airlines, wants to retire 50% of Southwest Airlines bonds by 2038. Calculate the payment Rob needs to make at the end of each year at 6% compounded annually to reach his goal of paying off $300,000 in 20 years. *LU 13–1(2)*

excel 13–17. Josef Company borrowed money that must be repaid in 20 years. The company wants to make sure the loan will be repaid at the end of year 20, so it invests $12,500 at the end of each year at 12% interest compounded annually. What was the amount of the original loan?   *LU 13–1(2)*

13–18. Bankrate.com reported on a shocking statistic: only 54% of workers participate in their company's retirement plan. This means that 46% do not. With such an uncertain future for Social Security, this can leave almost 1 in 2 individuals without proper income during retirement. Jill Collins, 20, decided she needs to have $250,000 in her retirement account upon retiring at 60. How much does she need to invest each year at 5% compounded annually to meet her goal? *Tip:* She is setting up a sinking fund.   *LU 13–3(1)*

13–19. If you saved an average of $2,900 each year from your income tax return, $1,050 for not buying vendor coffee, and $2,400 (saving $200 each paycheck), how much would you have in your retirement account if you were able to invest this annual savings at the end of each year for 30 years at 5% interest compounded annually?

excel 13–20. Alice Longtree has decided to invest $400 quarterly for 4 years in an ordinary annuity at 8%. As her financial adviser, calculate for Alice the total cash value of the annuity at the end of year 4.   *LU 13–1(2)*

13–21. At the beginning of each period for 10 years, Merl Agnes invests $500 semiannually at 6%. What is the cash value of this annuity due at the end of year 10?   *LU 13–1(2)*

13–22. Jeff Associates needs to repay $30,000. The company plans to set up a sinking fund that will repay the loan at the end of 8 years. Assume a 12% interest rate compounded semiannually. What must Jeff pay into the fund each period of time? Check your answer by Table 13.1.   *LU 13–3(1, 2)*

excel 13–23. On Joe Martin's graduation from college, Joe's uncle promised him a gift of $12,000 in cash or $900 every quarter for the next 4 years after graduation. If money could be invested at 8% compounded quarterly, which offer is better for Joe?   *LU 13–1(2), LU 13–2(1)*

End-of-Chapter Problems (Continued)

excel 13–24. You are earning an average of $46,500 and will retire in 10 years. If you put 20% of your gross average income in an ordinary annuity compounded at 7% annually, what will be the value of the annuity when you retire?   *LU 13–1(2)*

13–25. A local Dunkin' Donuts franchise must buy a new piece of equipment in 5 years that will cost $88,000. The company is setting up a sinking fund to finance the purchase. What will the quarterly deposit be if the fund earns 8% interest?   *LU 13–3(1)*

13–26. Mike Macaro is selling a piece of land. Two offers are on the table. Morton Company offered a $40,000 down payment and $35,000 a year for the next 5 years. Flynn Company offered $25,000 down and $38,000 a year for the next 5 years. If money can be invested at 8% compounded annually, which offer is better for Mike?   *LU 13–2(1)*

13–27. Al Vincent has decided to retire to Arizona in 10 years. What amount should Al invest today so that he will be able to withdraw $28,000 at the end of each year for 15 years *after* he retires? Assume he can invest the money at 8% interest compounded annually. *LU 13–2(1)*

13–28. Victor French made deposits of $5,000 at the end of each quarter to Book Bank, which pays 8% interest compounded quarterly. After 3 years, Victor made no more deposits. What will be the balance in the account 2 years after the last deposit? *LU 13–1(2)*

13–29. Janet Woo decided to retire to Florida in 6 years. What amount should Janet invest today so she can withdraw $50,000 at the end of each year for 20 years after she retires? Assume Janet can invest money at 6% compounded annually. *LU 13–2(1)*

13–30. Researching retirement savings online, you found an article from NewRetirement.com with recommendations from financial guru, Dave Ramsey and others. Dave Ramsey recommends investing 15% of every paycheck into a Roth IRA and pre-tax retirement accounts. If your household earns $75,000 annually and you adhere to Ramsey's recommendation, how much will your retirement account be worth if your 15% ordinary annuity earns 9% annually for 30 years? Round to the nearest dollar. *LU 13–1(2)*

13–31. Robert Kiyosaki, author of *Rich Dad Poor Dad,* states it is basically impossible to know how much you need to save for retirement. Instead he asserts, "If you change your definition of nest egg to mean a pile of cash-flowing assets, rather than cash, your problem is solved." He recommends rather than selling your start-up home, rent it out instead. If you are able to rent your start-up home for $1,100 per month and invest this amount annually in an ordinary annuity for 30 years at 6% interest annually, how much will you have? Round to the nearest cent. *LU 13–1(2)*

End-of-Chapter Problems (Continued)

Challenge Problems

13–32. Assume that you can buy a $6,000 computer system in monthly installments for 3 years. The seller charges you 12% interest compounded monthly. What is your monthly payment? Assume your first payment is due at the end of the month. Use tables in the *Business Math Handbook.* LU 13–2(1)

$$\text{Monthly payment} = \frac{\text{Amount owed}}{\begin{array}{c}\text{Table factor}\\\text{for PV of annuity}\end{array}}$$

13–33. Ajax Corporation has hired Brad O'Brien as its new president. Terms included the company's agreeing to pay retirement benefits of $18,000 at the end of each semiannual period for 10 years. This will begin in 3,285 days. If the money can be invested at 8% compounded semiannually, what must the company deposit today to fulfill its obligation to Brad? LU 13–2(1)

Summary Practice Test

(Use Tables in the *Business Math Handbook*)

Do you need help? Connect videos have step-by-step worked-out solutions.

1. Lin Lowe plans to deposit $1,800 at the end of every 6 months for the next 15 years at 8% interest compounded semiannually. What is the value of Lin's annuity at the end of 15 years?   *LU 13–1(2)*

2. On Abby Ellen's graduation from law school, Abby's uncle, Bull Brady, promised her a gift of $24,000 or $2,400 every quarter for the next 4 years after graduating from law school. If the money could be invested at 6% compounded quarterly, which offer should Abby choose?   *LU 13–2(1, 2)*

3. Sanka Blunck wants to receive $8,000 each year for 20 years. How much must Sanka invest today at 4% interest compounded annually?   *LU 13–2(1)*

4. In 9 years, Rollo Company will have to repay a $100,000 loan. Assume a 6% interest rate compounded quarterly. How much must Rollo Company pay each period to have $100,000 at the end of 9 years?   *LU 13–3(1)*

5. Lance Industries borrowed $130,000. The company plans to set up a sinking fund that will repay the loan at the end of 18 years. Assume a 6% interest rate compounded semiannually. What amount must Lance Industries pay into the fund each period? Check your answer by Table 13.1. *LU 13–3(1, 2)*

6. Joe Jan wants to receive $22,000 each year for the next 22 years. Assume a 6% interest rate compounded annually. How much must Joe invest today?   *LU 13–2(1)*

7. Twice a year for 15 years, Warren Ford invested $1,700 compounded semiannually at 6% interest. What is the value of this annuity due?   *LU 13–1(2)*

8. Scupper Molly invested $1,800 semiannually for 23 years at 8% interest compounded semiannually. What is the value of this annuity due?   *LU 13–1(2)*

Summary Practice Test (*Continued*)

9. Nelson Collins decided to retire to Canada in 10 years. What amount should Nelson deposit so that he will be able to withdraw $80,000 at the end of each year for 25 years after he retires? Assume Nelson can invest money at 7% interest compounded annually.   *LU 13–2(1)*

10. Bob Bryan made deposits of $10,000 at the end of each quarter to Lion Bank, which pays 8% interest compounded quarterly. After 9 years, Bob made no more deposits. What will be the account's balance 4 years after the last deposit?   *LU 13–1(2)*

MY MONEY

Keeping Stress in Check

 What I need to know

Many financial situations you face will cause you anxiety and stress over making the right decision. The long-term nature of financial decisions can add stress as you consider how to reach your financial goals many years into the future. It is important to find ways to control and alleviate this stress and not let it get out of hand. Use it as a motivator for you to plan advantageously. You may even realize stressors can yield positive results as you use them to your advantage giving you the needed boost to carry out your financial plans.

 What I need to do

Since stress is part of everyday life, we need to consider the ways in which we can effectively deal with the stressful situations we face. It is helpful to identify those areas of your life you feel produce anxiety and stress. In doing so, you will be able to prepare yourself when you encounter stressful situations. Like delivering your first speech in front of an audience, which produces stress for many, preparing ourselves in advance helps to reduce the stress we feel in these situations.

Prepare for your financial decisions by setting a plan and establishing goals for your financial future. Get yourself organized by documenting your current financial situation. Gathering important documents such as those for your banking, investments, and loans and organizing your financial information in a secure online location ensures you will have the details you need to make effective decisions in the future without undue stress.

Do not overlook the power of your physical and emotional health in helping you deal with financial stressors. Take care of yourself. Regular exercise may help to reduce stress and improve your physical health. Simplify your life. Additionally, helping others may allow you to help yourself. Donating your time to causes which you care about helps to provide a sense of emotional well-being as you serve others. Be grateful. Reflect on your current financial situation and learn to appreciate what you have. Finally, being prepared and aware helps you avoid worrying about your financial future.

 Steps I need to take

1. Know yourself by identifying what causes you stress both daily and in regard to your financial future. Take a breath, by being prepared.
2. Seek advice from others close to you who have faced these situations and learn from them.
3. Use stress as a motivator to help you plan now for your financial future.

 Resources I can use

- Calm (mobile app)—provides stretching exercises and relaxing music among other tools to assist in anxiety and stress reduction.
- https://www.citizensbank.com/learning/ways-to-reduce-financial-stress.aspx—helpful advice to ease your financial stress.

MY MONEY ACTIVITY

- Identify 3 ways that help you reduce stress.
- Use these 3 stress reducers 3 times in the coming week.
- Explain how they worked and what other measures you may need to take.

PERSONAL FINANCE

A KIPLINGER APPROACH

"Should You Borrow From Your 401(k)?." Kiplinger's. November 2019.

MILLENNIAL MONEY   Lisa Gerstner

Should You Borrow From Your 401(k)?

For many millennials, a workplace 401(k) plan is their first venture into building a significant savings stash. According to Fidelity Investments, 40l(k) savers age 20 to 29 whose plans are managed by the firm have an average balance of $12,200, and those 30 to 39 have an average of $43,400. But not all young employees leave the money untouched. One in four adults age 18 to 34 with a 40l(k) have already made a withdrawal or borrowed against the account, according to a study from Merrill Lynch and Age Wave. The primary reason: Paying credit card debt.

When debt looms or a surprise expense arises, your 40l(k) balance may look like the perfect solution. But you shouldn't tap your 401(k) until you've exhausted other sources of funds. The more you tuck away—and keep—in retirement accounts when you're young, the more you'll benefit from compounding investment returns over time.

Fidelity's research also reveals that among survey respondents who had a financial emergency within the past two years and did not have an emergency fund, 42% took a loan or withdrawal from their retirement plan. Work on socking away at least three to six months' worth of living expenses in a no-fee, high-yield savings account.

The 401(k) loan. If you can't come up with any other sources of cash, taking money from your 401(k) as a loan instead of a withdrawal will minimize the harm to your retirement security. You can generally borrow up to 50% of your vested account balance or $50,000, whichever is less. Or, if half of the vested balance is less than $10,000, you may still be able to borrow up to $10,000 of your total balance, if your employer allows it. Instead of forking over principal and interest to a lender, you pay it back to your own retirement account. Often, interest is the prime rate plus one percentage point, which recently added up to 6%. With the average credit card interest rate at about 17%, paying off card debt with a 40l(k) loan can make sense.

YOU SHOULDN'T TAP YOUR RETIREMENT PLAN UNTIL YOU'VE EXHAUSTED OTHER SOURCES OF FUNDS.

Taking a 40l(k) loan once and paying it back in full typically has little impact on retirement security, says Eliza Badeau, Fidelity's director of workplace investing thought leadership.

Poon Watchara-Amphaiwan

The trouble, says Badeau, comes when employees take out multiple loans and stop or decrease contributions. To keep your savings on track, try to contribute at least enough to your 40l(k) to capture any employer match, in addition to your loan payments. And take a hard look at why you needed to borrow in the first place. If you struggle to control your spending, you're at risk of continually relying on your 401(k) for backup.

Generally, you have five years to repay the loan, and you must make payments at least quarterly. If you don't, the outstanding balance is subject to income tax and a 10% early-distribution penalty. And if you change jobs, you have to pay off the loan by the tax-return deadline for the year you leave your job (including extensions) to avoid taxes and penalties on the balance.

If your situation is truly dire and you can't afford to repay a loan, your employer may allow a hardship distribution. These are typically permitted for specific circumstances, such as medical expenses. Once you take a hardship withdrawal, you can't put the money back, and you'll typically owe income taxes and the early-distribution penalty.

If you leave your job, you can cash out your 40l(k) for any reason, and a striking 40% of workers younger than 30 do just that, according to Fidelity. Such distributions trigger taxes and penalties, and pulling the money from the stock market diminishes its earning power.

When my status with Kiplinger went from employee to see employed contractor five years a could no longer tribute to my 4 and I decided to it sit. Since then balance has growth more than $16,000. ∎

TO SHARE THIS COLUMN, PLEASE GO TO KIPLINGER.COM/LINK S/MILLENNIALS. YOU CAN CONTACT THE AUTHOR AT LGERSTNER@KIPLINGER.COM.

Business Math Issue

Borrowing from your 401K is a good financial decision.

1. List the key points of the article and information to support your position.

2. Write a group defense of your position using math calculations to support your view. If you are in an online course, post to a discussion board.

Cumulative Review

A Word Problem Approach—Chapters 10, 11, 12, 13

1. Amy O'Mally graduated from high school. Her uncle promised her as a gift a check for $2,000 or $275 every quarter for 2 years. If money could be invested at 6% compounded quarterly, which offer is better for Amy? (Use the tables in the *Business Math Handbook.*)   *LU 13–2(1)*

2. Alan Angel made deposits of $400 semiannually to Sag Bank, which pays 1% interest compounded semiannually. After 4 years, Alan made no more deposits. What will be the balance in the account 3 years after the last deposit? (Use the tables in the *Business Math Handbook.*)   *LU 13–1(2)*

3. Roger Disney decides to retire to Florida in 12 years. What amount should Roger invest today so that he will be able to withdraw $30,000 at the end of each year for 20 years *after* he retires? Assume he can invest money at 8% interest compounded annually. (Use tables in the *Business Math Handbook.*)   *LU 13–2(2)*

4. On September 15, Arthur Westering borrowed $3,000 from Vermont Bank at $10\frac{1}{2}$% interest. Arthur plans to repay the loan on January 25. Assume the loan is based on exact interest. How much will Arthur totally repay?   *LU 10–1(2)*

5. Sue Cooper borrowed $6,000 on an $11\frac{3}{4}$%, 120-day note. Sue paid $300 toward the note on day 50. On day 90, Sue paid an additional $200. Using the U.S. Rule, what is Sue's adjusted balance after her first payment?   *LU 10–3(1)*

6. On November 18, Northwest Company discounted an $18,000, 12%, 120-day note dated September 8. Assume a 10% discount rate. What will be the proceeds? Use ordinary interest.   *LU 11–2(1)*

7. Alice Reed deposits $16,500 into Rye Bank, which pays 1% interest compounded semiannually. Using the appropriate table, what will Alice have in her account at the end of 6 years?   *LU 12–1(2)*

8. Peter Regan needs $90,000 5 years from today to retire in Arizona. Peter invests this at 10% interest compounded semiannually. What will Peter have to invest today to have $90,000 in 5 years?   *LU 12–2(2)*

Time-Value Relationship Appendix

One Lump Sum (Single Amount)
FIGURE 13.3   Compound (future value) of $.68 at 10% for 4 periods
FIGURE 13.4   Present value of $1.00 at 10% for 4 periods

Annuity (Stream of Payments)
FIGURE 13.5   Present value of a 4-year annuity of $1.00 at 10%
FIGURE 13.6   Future value of a 4-year annuity of $1.00 at 10%

Compare to Figure 13.4 to see relationship of compounding to present value.

FIGURE 13.3 Compound (future value) of $.68 at 10% for 4 periods

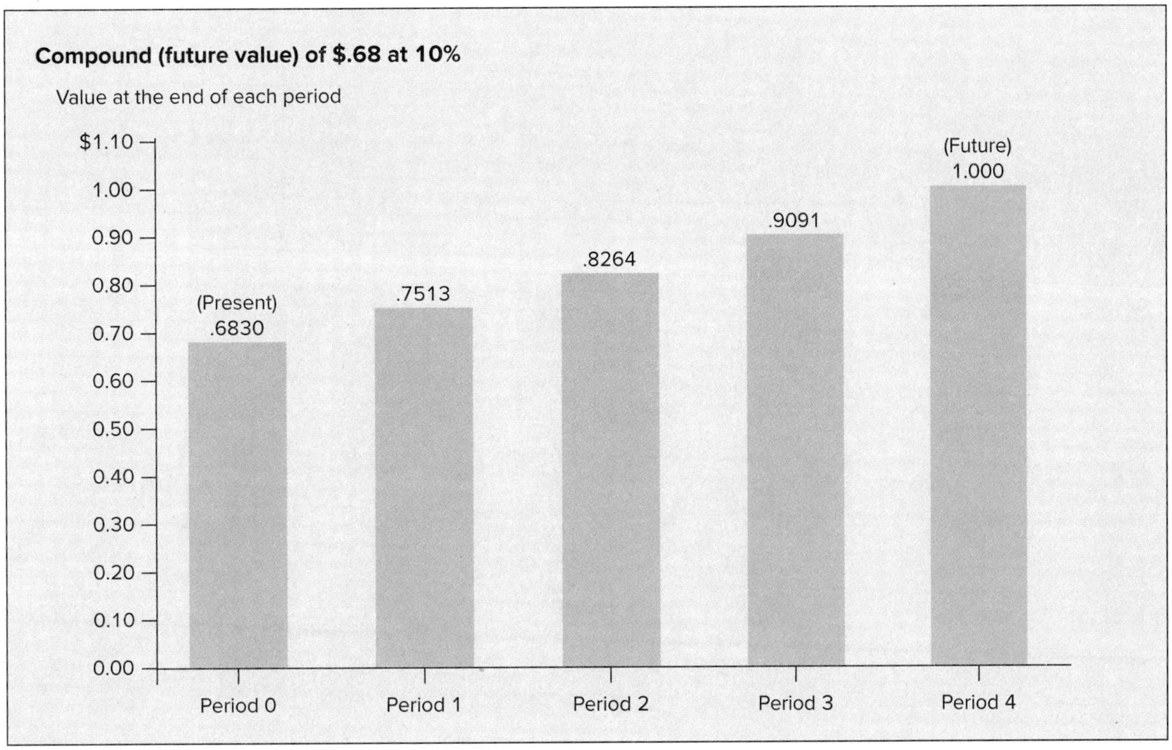

$.68 today will grow to $1.00 in the future.

What Figure 13.3 Means

If you invest $.68 with an average annual return of 10%, after 4 periods you will be able to get $1.00. The $.68 is the present value, and the $1.00 is the compound value or future value. Keep in mind that the $.68 is a one lump-sum investment.

FIGURE 13.4 Present value of $1.00 at 10% for 4 periods

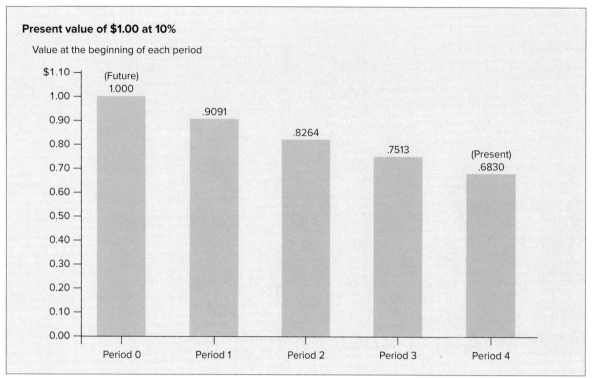

Present value of $1.00 at 10%

If I need $1 in four periods, I need to invest $0.68 today.

What Figure 13.4 Means

If you want to receive $1.00 at the end of 4 periods from an investment earning 10%, you will have to invest $.68 today. The longer you have to wait for your money, the less it is worth. The $1.00 is the compound or future amount, and the $.68 is the present value of a dollar that you will not receive for 4 periods.

FIGURE 13.5 Present value of a 4-year annuity of $1.00 at 10%

Present value of $1.00 at 10%

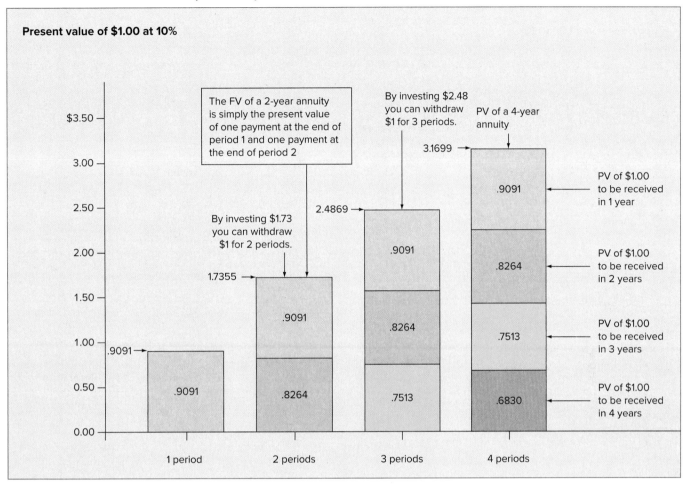

FIGURE 13.6 Future value of a 4-year annuity of $1.00 at 10%

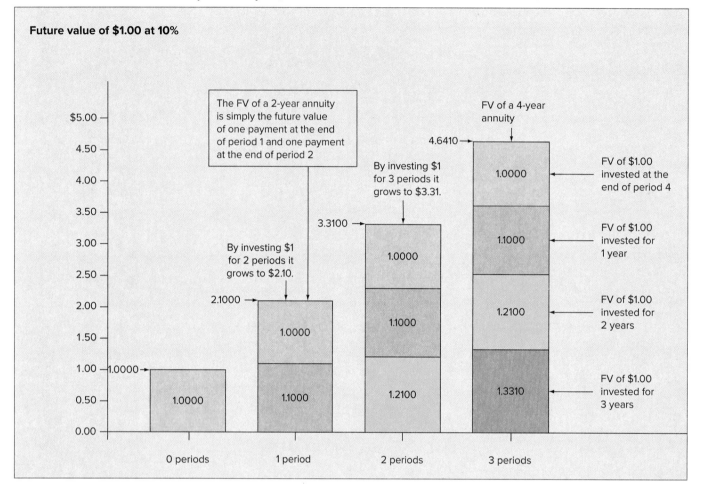

What Figure 13.5 Means*

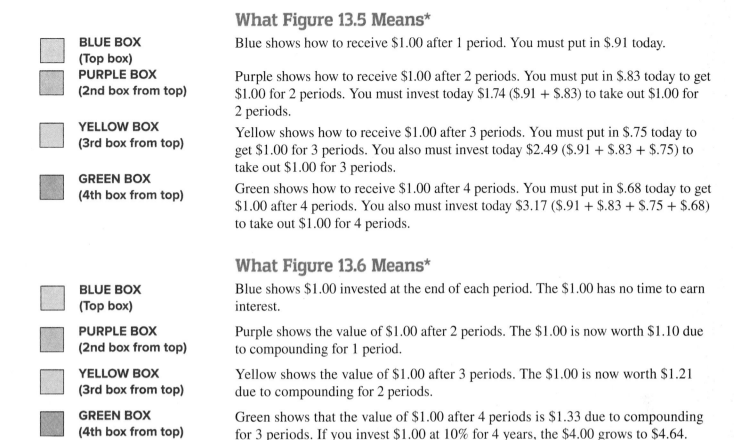

BLUE BOX
(Top box)

Blue shows how to receive $1.00 after 1 period. You must put in $.91 today.

PURPLE BOX
(2nd box from top)

Purple shows how to receive $1.00 after 2 periods. You must put in $.83 today to get $1.00 for 2 periods. You must invest today $1.74 ($.91 + $.83) to take out $1.00 for 2 periods.

YELLOW BOX
(3rd box from top)

Yellow shows how to receive $1.00 after 3 periods. You must put in $.75 today to get $1.00 for 3 periods. You also must invest today $2.49 ($.91 + $.83 + $.75) to take out $1.00 for 3 periods.

GREEN BOX
(4th box from top)

Green shows how to receive $1.00 after 4 periods. You must put in $.68 today to get $1.00 after 4 periods. You also must invest today $3.17 ($.91 + $.83 + $.75 + $.68) to take out $1.00 for 4 periods.

What Figure 13.6 Means*

BLUE BOX
(Top box)

Blue shows $1.00 invested at the end of each period. The $1.00 has no time to earn interest.

PURPLE BOX
(2nd box from top)

Purple shows the value of $1.00 after 2 periods. The $1.00 is now worth $1.10 due to compounding for 1 period.

YELLOW BOX
(3rd box from top)

Yellow shows the value of $1.00 after 3 periods. The $1.00 is now worth $1.21 due to compounding for 2 periods.

GREEN BOX
(4th box from top)

Green shows that the value of $1.00 after 4 periods is $1.33 due to compounding for 3 periods. If you invest $1.00 at 10% for 4 years, the $4.00 grows to $4.64.

*From table in Handbook for 10%.

Periods	Amount of annuity	Present value of an annuity
1	1.0000	.9091
2	2.1000	1.7355
3	3.3100	2.4869
4	4.6410	3.1699

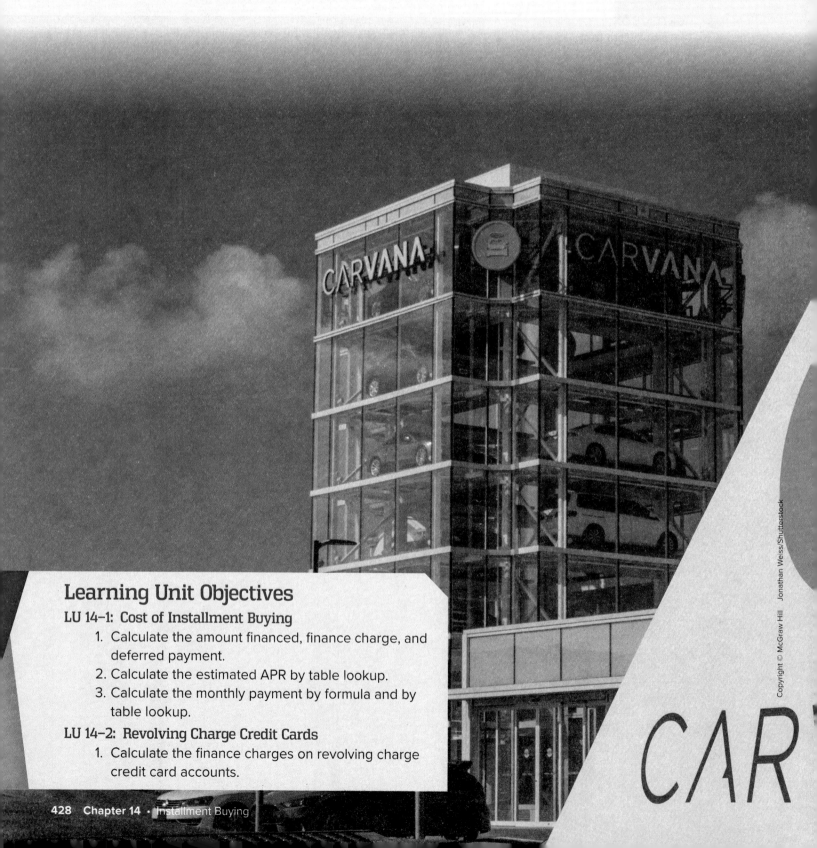

Chapter 14
Installment Buying

Learning Unit Objectives

LU 14–1: Cost of Installment Buying

1. Calculate the amount financed, finance charge, and deferred payment.
2. Calculate the estimated APR by table lookup.
3. Calculate the monthly payment by formula and by table lookup.

LU 14–2: Revolving Charge Credit Cards

1. Calculate the finance charges on revolving charge credit card accounts.

Online Shopping Is Coming To Cars

But slow-moving sector won't be changed quickly

You might buy your next car online, but don't expect that to upend the car industry.

Consultants have expected the disruptive power of e-commerce to spread to vehicles—the most valuable consumer products there are—ever since the dot-com boom more than 20 years ago. But the only significant change to the decades-old structure of U.S. vehicle retail has come from Tesla, which dispenses with independent dealerships in favor of directly controlled showrooms. Could the pandemic be the catalyst the wider industry previously lacked?

The coronavirus crisis will almost certainly push more vehicle sales online. Yet dealerships are such an entrenched part of the industry, particularly in the U.S., that the kind of radical transformation e-commerce has brought to other sectors won't follow quickly. For manufacturers, this slow pace of change is both frustrating and helpful.

"Online Shopping Is Coming to Cars." *The Wall Street Journal*, May 18, 2020.

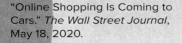

⊝ Essential Question

How can I use installment buying to understand the business world and make my life easier?

🌐 Math Around the World

The *Wall Street Journal* chapter opener clip, "Online Shopping Is Coming To Cars", suggests someday the way you buy a car may change. During and after the pandemic, Carvana, as well as traditional car dealers, is introducing new ways to buy cars without actually going to a dealer.

This chapter discusses the cost of buying products via installments (closed-end credit) and revolving credit card (open-end credit). You will see in Learning Unit 14–1 that to buy a 4×4 pickup truck a qualified buyer must have a credit score of 720 or higher (see the 4×4 pickup advertisement below).

Curtis ©2020 King Features Syndicate, Inc. World Rights Reserved

Learning Unit 14–1:
Cost of Installment Buying

Installment buying, a form of *closed-end credit,* can add a substantial amount to the cost of big-ticket purchases. To illustrate this, we follow the procedure of buying a pickup truck, including the amount financed, finance charge, and deferred payment price. Then we study the effect of the Truth in Lending Act.

Learn: Amount Financed, Finance Charge, and Deferred Payment

The advertisement for the sale of a pickup truck shown on the next page appeared in a local paper. As you can see from this advertisement, after customers make a **down payment,** they can buy the truck with an **installment loan.** This loan is paid off with a series of equal periodic payments. These payments include both interest and principal. The payment process is called **amortization.** In the promissory notes of earlier chapters, the loan was paid off in one ending payment. Now let's look at the calculations involved in buying a pickup truck.

Money Tip Consider the pros and cons of buying (new or used) versus leasing a vehicle. Negotiate the terms regardless of which option you choose. Note: you pay insurance and maintenance on either option.

Buy
Build equity
Substantial down
  payment
Higher monthly
  payments (but
  cheaper long term)
No worry about wear
  and tear (except for
  trade-in or resale
  value)
Unlimited miles

Lease
Don't own—can
  upgrade every few
  years
Lower down payment
Lower monthly
  payments
Responsible for
  above-average wear
  and tear
Limited miles

Checking Calculations in Pickup Advertisement

Calculating Amount Financed The **amount financed** is what you actually borrow. To calculate this amount, use the following formula:

$$\text{Amount financed} = \text{Cash price} - \text{Down payment}$$

$$\$9,045 = \$9,345 - \$300$$

Calculating Finance Charge The words **"finance charge"** in the advertisement represent the *interest* charge. The interest charge resulting in the finance charge includes the cost of credit reports, mandatory bank fees, and so on. You can use the following formula to calculate the total interest on the loan:

$$\begin{array}{ccc} \text{Total finance charge} & = & \text{Total of all} & - & \text{Amount} \\ \text{(interest charge)} & & \text{monthly payments} & & \text{financed} \end{array}$$

$$\$2,617.80 = \$11,662.80 - \$9,045$$
$$(\$194.38 \times 60 \text{ months})$$

Calculating Deferred Payment Price The **deferred payment price** represents the total of all monthly payments plus the down payment. The following formula is used to calculate the deferred payment price:

$$\boxed{\text{Deferred payment price}} = \frac{\text{Total of all}}{\text{monthly payments}} + \frac{\text{Down}}{\text{payment}}$$

$$\boxed{\$11{,}962.80} = \begin{array}{c}\$11{,}662.80 \\ (\$194.38 \times 60)\end{array} + \begin{array}{c}\$300\end{array}$$

Learn: Truth in Lending: APR Defined and Calculated

In 1969, the Federal Reserve Board established the **Truth in Lending Act** (Regulation Z). The law doesn't regulate interest charges; its purpose is to make the consumer aware of the true cost of credit.

The Truth in Lending Act requires that creditors provide certain basic information about the actual cost of buying on credit. Before buyers sign a credit agreement, creditors must inform them in writing of the amount of the finance charge and the **annual percentage rate (APR).** The APR represents the true or effective annual interest creditors charge. This is helpful to buyers who repay loans over different periods of time (1 month, 48 months, and so on).

To illustrate how the APR affects the interest rate, assume you borrow $100 for 1 year and pay a finance charge of $9. Your interest rate would be 9% if you waited until the end of the year to pay back the loan. Now let's say you pay off the loan and the finance charge in 12 monthly payments. Each month that you make a payment, you are losing some of the value or use of that money. So the true or effective APR is actually greater than 9%. The *Wall Street Journal* clip, " Generous Car-Loan Deals Lure Borrowers", shows that APRs have been lowered as a result of the pandemic.

The APR can be calculated by formula or by tables. We will use the table method since it is more exact.

Calculating APR Rate by Table 14.1 Note the following steps for using a table to calculate APR:

Calculating Apr by Table

Step 1 Divide the finance charge by amount financed and multiply by $100 to get the table lookup factor.

Step 2 Go to APR Table 14.1. At the left side of the table are listed the number of payments that will be made.

Step 3 When you find the number of payments you are looking for, move to the right and look for the two numbers closest to the table lookup number. This will indicate the APR.

Now let's determine the APR for the pickup truck advertisement given earlier in the chapter.

As stated in Step 1, we begin by dividing the finance charge by the amount financed and multiply by $100:

$$\frac{\text{Finance charge}}{\text{Amount financed}} \times \$100 = \begin{array}{c}\text{Table 14.1} \\ \text{lookup number}\end{array}$$

We multiply by $100, since the table is based on $100 of financing.

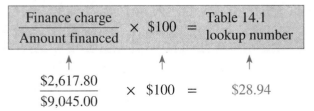

$$\frac{\$2{,}617.80}{\$9{,}045.00} \times \$100 = \$28.94$$

Generous Car-Loan Deals Lure Borrowers

By AnnaMaria Andriotes
And Ben Eisen

Auto lenders are dangling easy financing terms to try to revive halted car sales.

The catch: Those terms are typically just for people with strong credit histories. Those at the other end of the spectrum are finding it harder to get loans.

Lenders are offering loans for new and used cars that let borrowers delay making payments for up to four months. They are providing record amounts of financing at 0% interest rates, sometimes for loans as long as seven years. Some 26% of loans for new cars purchased in April had 0% interest, compared with about 5% of new-car loans in March, according to car-shopping website Edmunds.

Andriotis, AnnaMaria, and Ben Eisen. "Generous Car-Loan Deals Lure Borrowers." *The Wall Street Journal,* May 4, 2020.

To look up $28.94 in Table 14.1, we go down the left side of the table until we come to 60 payments (the advertisement states 60 months). Then, moving to the right, we look for $28.94 or the two numbers closest to it. The number $28.94 is between $28.22 and $28.96. So we look at the column headings and see a rate between 10.25% and 10.5%. The Truth in Lending Act requires that when creditors state the APR, it must be accurate to the nearest $\frac{1}{4}$ of 1%.[1]

Calculating the Monthly Payment by Formula and Table 14.2 The pickup truck advertisement showed a $194.38 monthly payment. We can check this by formula and by table lookup.

TABLE 14.1
Annual percentage rate table per $100

NUMBER OF PAYMENTS	2.00%	2.25%	2.50%	2.75%	3.00%	3.25%	3.50%	3.75%	4.00%	4.25%	4.50%	4.75%	5.00%	5.25%	5.50%	5.75%
					(FINANCE CHARGE PER $100 OF AMOUNT FINANCED)											
1	0.17	0.19	0.21	0.23	0.25	0.27	0.29	0.31	0.33	0.35	0.37	0.40	0.42	0.44	0.46	0.48
2	0.25	0.28	0.31	0.34	0.38	0.41	0.44	0.47	0.50	0.53	0.56	0.59	0.63	0.66	0.69	0.72
3	0.33	0.38	0.42	0.46	0.50	0.54	0.58	0.63	0.67	0.71	0.75	0.79	0.83	0.88	0.92	0.96
4	0.42	0.47	0.52	0.57	0.63	0.68	0.73	0.78	0.83	0.89	0.94	0.99	1.04	1.10	1.15	1.20
5	0.50	0.56	0.63	0.69	0.75	0.81	0.88	0.94	1.00	1.07	1.13	1.19	1.25	1.32	1.38	1.44
6	0.58	0.66	0.73	0.80	0.88	0.95	1.02	1.10	1.17	1.24	1.32	1.39	1.46	1.54	1.61	1.68
7	0.67	0.75	0.84	0.92	1.00	1.09	1.17	1.25	1.34	1.42	1.51	1.59	1.67	1.76	1.84	1.93
8	0.75	0.85	0.94	1.03	1.13	1.22	1.32	1.41	1.51	1.60	1.69	1.79	1.88	1.98	2.07	2.17
9	0.84	0.94	1.04	1.15	1.25	1.36	1.46	1.57	1.67	1.78	1.88	1.99	2.09	2.20	2.31	2.41
10	0.92	1.03	1.15	1.26	1.38	1.50	1.61	1.73	1.84	1.96	2.07	2.19	2.31	2.42	2.54	2.65
11	1.00	1.13	1.25	1.38	1.51	1.63	1.76	1.88	2.01	2.14	2.26	2.39	2.52	2.64	2.77	2.90
12	1.09	1.22	1.36	1.50	1.63	1.77	1.91	2.04	2.18	2.32	2.45	2.59	2.73	2.87	3.00	3.14
13	1.17	1.32	1.46	1.61	1.76	1.91	2.05	2.20	2.35	2.50	2.64	2.79	2.94	3.09	3.24	3.39
14	1.25	1.41	1.57	1.73	1.89	2.04	2.20	2.36	2.52	2.68	2.84	2.99	3.15	3.31	3.47	3.63
15	1.34	1.51	1.67	1.84	2.01	2.18	2.35	2.52	2.69	2.86	3.03	3.20	3.37	3.54	3.71	3.88
16	1.42	1.60	1.78	1.96	2.14	2.32	2.50	2.68	2.86	3.04	3.22	3.40	3.58	3.76	3.94	4.12
17	1.51	1.70	1.89	2.08	2.26	2.46	2.65	2.84	3.03	3.27	3.41	3.60	3.79	3.98	4.18	4.37
18	1.59	1.79	1.99	2.19	2.39	2.59	2.79	2.99	3.20	3.40	3.60	3.80	4.00	4.21	4.41	4.61
19	1.67	1.89	2.10	2.31	2.52	2.73	2.94	3.15	3.37	3.58	3.79	4.01	4.22	4.43	4.65	4.86
20	1.76	1.98	2.20	2.42	2.65	2.87	3.09	3.31	3.54	3.76	3.98	4.21	4.43	4.66	4.88	5.11
21	1.84	2.08	2.31	2.54	2.77	3.01	3.24	3.47	3.71	3.94	4.18	4.41	4.65	4.89	5.12	5.35
22	1.93	2.17	2.41	2.66	2.90	3.14	3.39	3.63	3.88	4.12	4.37	4.62	4.86	5.11	5.36	5.60
23	2.01	2.27	2.52	2.77	3.03	3.28	3.54	3.79	4.05	4.31	4.56	4.82	5.08	5.33	5.59	5.85
24	2.10	2.36	2.62	2.89	3.15	3.42	3.69	3.95	4.22	4.49	4.75	5.02	5.29	5.56	5.83	6.10
25	2.18	2.46	2.73	3.01	3.28	3.56	3.84	4.11	4.39	4.67	4.95	5.23	5.51	5.79	6.07	6.35
26	2.27	2.55	2.84	3.12	3.41	3.70	3.99	4.27	4.56	4.85	5.14	5.43	5.72	6.01	6.31	6.60
27	2.35	2.65	2.94	3.24	3.54	3.84	4.13	4.43	4.73	5.03	5.34	5.64	5.94	6.24	6.54	6.85
28	2.43	2.74	3.05	3.36	3.67	3.97	4.28	4.59	4.91	5.22	5.53	5.84	6.15	6.47	6.78	7.10
29	2.52	2.84	3.16	3.47	3.79	4.11	4.43	4.76	5.08	5.40	5.72	6.05	6.37	6.70	7.02	7.35
30	2.60	2.93	3.26	3.59	3.92	4.25	4.58	4.92	5.25	5.58	5.92	6.25	6.59	6.92	7.26	7.60
31	2.69	3.03	3.37	3.71	4.05	4.39	4.73	5.08	5.42	5.77	6.11	6.46	6.81	7.15	7.50	7.85
32	2.77	3.12	3.47	3.83	4.18	4.53	4.88	5.24	5.59	5.95	6.31	6.66	7.02	7.38	7.74	8.10
33	2.86	3.22	3.58	3.94	4.31	4.67	5.04	5.40	5.77	6.13	6.50	6.87	7.24	7.61	7.98	8.35
34	2.94	3.32	3.69	4.06	4.44	4.81	5.19	5.56	5.94	6.32	6.70	7.08	7.46	7.84	8.22	8.61
35	3.03	3.41	3.79	4.18	4.56	4.95	5.34	5.72	6.11	6.50	6.89	7.28	7.68	8.07	8.46	8.86
36	3.11	3.51	3.90	4.30	4.69	5.09	5.49	5.89	6.29	6.69	7.09	7.49	7.90	8.30	8.71	9.11
37	3.20	3.60	4.01	4.41	4.82	5.23	5.64	6.05	6.46	6.87	7.28	7.70	8.11	8.53	8.95	9.37
38	3.28	3.70	4.11	4.53	4.95	5.37	5.79	6.21	6.63	7.06	7.48	7.91	8.33	8.76	9.19	9.62
39	3.37	3.79	4.22	4.65	5.08	5.51	5.94	6.37	6.81	7.24	7.68	8.11	8.55	8.99	9.43	9.87
40	3.45	3.89	4.33	4.77	5.21	5.65	6.09	6.54	6.98	7.43	7.87	8.32	8.77	9.22	9.67	10.13
41	3.54	3.99	4.44	4.89	5.34	5.79	6.24	6.70	7.16	7.61	8.07	8.53	8.99	9.45	9.92	10.38
42	3.62	4.08	4.54	5.00	5.47	5.93	6.40	6.86	7.33	7.80	8.27	8.74	9.21	9.69	10.16	10.64
43	3.71	4.18	4.65	5.12	5.60	6.07	6.55	7.03	7.50	7.98	8.47	8.95	9.43	9.92	10.41	10.89
44	3.79	4.28	4.76	5.24	5.73	6.21	6.70	7.19	7.68	8.17	8.66	9.16	9.65	10.15	10.65	11.15
45	3.88	4.37	4.86	5.36	5.86	6.35	6.85	7.35	7.85	8.36	8.86	9.37	9.88	10.38	10.89	11.41
46	3.97	4.47	4.97	5.48	5.98	6.49	7.00	7.52	8.03	8.54	9.06	9.58	10.10	10.62	11.14	11.66
47	4.05	4.56	5.08	5.60	6.11	6.63	7.16	7.68	8.20	8.73	9.26	9.79	10.32	10.85	11.39	11.92
48	4.14	4.66	5.19	5.72	6.24	6.78	7.31	7.84	8.38	8.92	9.46	10.00	10.54	11.09	11.63	12.18
49	4.22	4.76	5.30	5.83	6.37	6.92	7.46	8.01	8.56	9.10	9.66	10.21	10.76	11.32	11.88	12.44
50	4.31	4.85	5.40	5.95	6.50	7.06	7.61	8.17	8.73	9.29	9.85	10.42	10.99	11.55	12.12	12.70
51	4.39	4.95	5.51	6.07	6.64	7.20	7.77	8.34	8.91	9.48	10.05	10.63	11.21	11.79	12.37	12.95
52	4.48	5.05	5.62	6.19	6.77	7.34	7.92	8.50	9.08	9.67	10.25	10.84	11.43	12.02	12.62	13.21
53	4.56	5.14	5.73	6.31	6.90	7.48	8.07	8.67	9.26	9.86	10.45	11.05	11.66	12.26	12.86	13.47
54	4.65	5.24	5.83	6.43	7.03	7.63	8.23	8.83	9.44	10.04	10.65	11.26	11.88	12.49	13.11	13.73
55	4.74	5.34	5.94	6.55	7.16	7.77	8.38	9.00	9.61	10.23	10.85	11.48	12.10	12.73	13.36	13.99
56	4.82	5.44	6.05	6.67	7.29	7.91	8.53	9.16	9.79	10.42	11.05	11.69	12.33	12.97	13.61	14.25
57	4.91	5.53	6.16	6.79	7.42	8.05	8.69	9.33	9.97	10.61	11.25	11.90	12.55	13.20	13.86	14.52
58	4.99	5.63	6.27	6.91	7.55	8.19	8.84	9.49	10.14	10.80	11.46	12.11	12.78	13.44	14.11	14.78
59	5.08	5.73	6.38	7.03	7.68	8.34	9.00	9.66	10.32	10.99	11.66	12.33	13.00	13.68	14.36	15.04
60	5.17	5.82	6.48	7.15	7.81	8.48	9.15	9.82	10.50	11.18	11.86	12.54	13.23	13.92	14.61	15.30

Note: For a more detailed set of tables from 2% to 21.75%, see the reference tables in the *Business Math Handbook*.

[1] If we wanted an exact reading of APR when the number is not exactly in the table, we would use the process of interpolating. We do not cover this method in this course.

By Formula

$$\boxed{\frac{\text{Finance charge} + \text{Amount financed}}{\text{Number of payments of loan}}} = \frac{\$2,617.80 + \$9,045}{60} = \boxed{\$194.38}$$

By Table 14.2 The **loan amortization table** (many variations of this table are available) in Table 14.2 can be used to calculate the monthly payment for the pickup truck. To calculate a monthly payment with a table, use the following steps:

TABLE 14.1 (continued)

NUMBER OF PAYMENTS	ANNUAL PERCENTAGE RATE															
	10.00%	10.25%	10.50%	10.75%	11.00%	11.25%	11.50%	11.75%	12.00%	12.25%	12.50%	12.75%	13.00%	13.25%	13.50%	13.75%
	(FINANCE CHARGE PER $100 OF AMOUNT FINANCED)															
1	0.83	0.85	0.87	0.90	0.92	0.94	0.96	0.98	1.00	1.02	1.04	1.06	1.08	1.10	1.12	1.15
2	1.25	1.28	1.31	1.35	1.38	1.41	1.44	1.47	1.50	1.53	1.57	1.60	1.63	1.66	1.69	1.72
3	1.67	1.71	1.76	1.80	1.84	1.88	1.92	1.96	2.01	2.05	2.09	2.13	2.17	2.22	2.26	2.30
4	2.09	2.14	2.20	2.25	2.30	2.35	2.41	2.46	2.51	2.57	2.62	2.67	2.72	2.78	2.83	2.88
5	2.51	2.58	2.64	2.70	2.77	2.83	2.89	2.96	3.02	3.08	3.15	3.21	3.27	3.34	3.40	3.46
6	2.94	3.01	3.08	3.16	3.23	3.31	3.38	3.45	3.53	3.60	3.68	3.75	3.83	3.90	3.97	4.05
7	3.36	3.45	3.53	3.62	3.70	3.78	3.87	3.95	4.04	4.12	4.21	4.29	4.38	4.47	4.55	4.64
8	3.79	3.88	3.98	4.07	4.17	4.26	4.36	4.46	4.55	4.65	4.74	4.84	4.94	5.03	5.13	5.22
9	4.21	4.32	4.43	4.53	4.64	4.75	4.85	4.96	5.07	5.17	5.28	5.39	5.49	5.60	5.71	5.82
10	4.64	4.76	4.88	4.99	5.11	5.23	5.35	5.46	5.58	5.70	5.82	5.94	6.05	6.17	6.29	6.41
11	5.07	5.20	5.33	5.45	5.58	5.71	5.84	5.97	6.10	6.23	6.36	6.49	6.62	6.75	6.88	7.01
12	5.50	5.64	5.78	5.92	6.06	6.20	6.34	6.48	6.62	6.76	6.90	7.04	7.18	7.32	7.46	7.60
13	5.93	6.08	6.23	6.38	6.53	6.68	6.84	6.99	7.14	7.29	7.44	7.59	7.75	7.90	8.05	8.20
14	6.36	6.52	6.69	6.85	7.01	7.17	7.34	7.50	7.66	7.82	7.99	8.15	8.31	8.48	8.64	8.81
15	6.80	6.97	7.14	7.32	7.49	7.66	7.84	8.01	8.19	8.36	8.53	8.71	8.88	9.06	9.23	9.41
16	7.23	7.41	7.60	7.78	7.97	8.15	8.34	8.53	8.71	8.90	9.08	9.27	9.46	9.64	9.83	10.02
17	7.67	7.86	8.06	8.25	8.45	8.65	8.84	9.04	9.24	9.44	9.63	9.83	10.03	10.23	10.43	10.63
18	8.10	8.31	8.52	8.73	8.93	9.14	9.35	9.56	9.77	9.98	10.19	10.40	10.61	10.82	11.03	11.24
19	8.54	8.76	8.98	9.20	9.42	9.64	9.86	10.08	10.30	10.52	10.74	10.96	11.18	11.41	11.63	11.85
20	8.98	9.21	9.44	9.67	9.90	10.13	10.37	10.60	10.83	11.06	11.30	11.53	11.76	12.00	12.23	12.46
21	9.42	9.66	9.90	10.15	10.39	10.63	10.88	11.12	11.36	11.61	11.85	12.10	12.34	12.59	12.84	13.08
22	9.86	10.12	10.37	10.62	10.88	11.13	11.39	11.64	11.90	12.16	12.41	12.67	12.93	13.19	13.44	13.70
23	10.30	10.57	10.84	11.10	11.37	11.63	11.90	12.17	12.44	12.71	12.97	13.24	13.51	13.78	14.05	14.32
24	10.75	11.02	11.30	11.58	11.86	12.14	12.42	12.70	12.98	13.26	13.54	13.82	14.10	14.38	14.66	14.95
25	11.19	11.48	11.77	12.06	12.35	12.64	12.93	13.22	13.52	13.81	14.10	14.40	14.69	14.98	15.28	15.57
26	11.64	11.94	12.24	12.54	12.85	13.15	13.45	13.75	14.06	14.36	14.67	14.97	15.28	15.59	15.89	16.20
27	12.09	12.40	12.71	13.03	13.34	13.66	13.97	14.29	14.60	14.92	15.24	15.56	15.87	16.19	16.51	16.83
28	12.53	12.86	13.18	13.51	13.84	14.16	14.49	14.82	15.15	15.48	15.81	16.14	16.47	16.80	17.13	17.46
29	12.98	13.32	13.66	14.00	14.33	14.67	15.01	15.35	15.70	16.04	16.38	16.72	17.07	17.41	17.75	18.10
30	13.43	13.78	14.13	14.48	14.83	15.19	15.54	15.89	16.24	16.60	16.95	17.31	17.66	18.02	18.38	18.74
31	13.89	14.25	14.61	14.97	15.33	15.70	16.06	16.43	16.79	17.16	17.53	17.90	18.27	18.63	19.00	19.38
32	14.34	14.71	15.09	15.46	15.84	16.21	16.59	16.97	17.35	17.73	18.11	18.49	18.87	19.25	19.63	20.02
33	14.79	15.18	15.57	15.95	16.34	16.73	17.12	17.51	17.90	18.29	18.69	19.08	19.47	19.87	20.26	20.66
34	15.25	15.65	16.05	16.44	16.85	17.25	17.65	18.05	18.46	18.86	19.27	19.67	20.08	20.49	20.90	21.31
35	15.70	16.11	16.53	16.94	17.35	17.77	18.18	18.60	19.01	19.43	19.85	20.27	20.69	21.11	21.53	21.95
36	16.16	16.58	17.01	17.43	17.86	18.29	18.71	19.14	19.57	20.00	20.43	20.87	21.30	21.73	22.17	22.60
37	16.62	17.06	17.49	17.93	18.37	18.81	19.25	19.69	20.13	20.58	21.02	21.46	21.91	22.36	22.81	23.25
38	17.08	17.53	17.98	18.43	18.88	19.33	19.78	20.24	20.69	21.15	21.61	22.07	22.52	22.99	23.45	23.91
39	17.54	18.00	18.46	18.93	19.39	19.86	20.32	20.79	21.26	21.73	22.20	22.67	23.14	23.61	24.09	24.56
40	18.00	18.48	18.95	19.43	19.90	20.38	20.86	21.34	21.82	22.30	22.79	23.27	23.76	24.25	24.73	25.22
41	18.47	18.95	19.44	19.93	20.42	20.91	21.40	21.89	22.39	22.88	23.38	23.88	24.38	24.88	25.38	25.88
42	18.93	19.43	19.93	20.43	20.93	21.44	21.94	22.45	22.96	23.47	23.98	24.49	25.00	25.51	26.03	26.55
43	19.40	19.91	20.42	20.94	21.45	21.97	22.49	23.01	23.53	24.05	24.57	25.10	25.62	26.15	26.68	27.21
44	19.86	20.39	20.91	21.44	21.97	22.50	23.03	23.57	24.10	24.64	25.17	25.71	26.25	26.79	27.33	27.88
45	20.33	20.87	21.41	21.95	22.49	23.03	23.58	24.12	24.67	25.22	25.77	26.32	26.88	27.43	27.99	28.55
46	20.80	21.35	21.90	22.46	23.01	23.57	24.13	24.69	25.25	25.81	26.37	26.94	27.51	28.08	28.65	29.22
47	21.27	21.83	22.40	22.97	23.53	24.10	24.68	25.25	25.82	26.40	26.98	27.56	28.14	28.72	29.31	29.89
48	21.74	22.32	22.90	23.48	24.06	24.64	25.23	25.81	26.40	26.99	27.58	28.18	28.77	29.37	29.97	30.57
49	22.21	22.80	23.39	23.99	24.58	25.18	25.78	26.38	26.98	27.59	28.19	28.80	29.41	30.02	30.63	31.24
50	22.69	23.29	23.89	24.50	25.11	25.72	26.33	26.95	27.56	28.18	28.80	29.42	30.04	30.67	31.29	31.92
51	23.16	23.78	24.40	25.02	25.64	26.26	26.89	27.52	28.15	28.78	29.41	30.05	30.68	31.32	31.96	32.60
52	23.64	24.27	24.90	25.53	26.17	26.81	27.45	28.09	28.73	29.38	30.02	30.67	31.32	31.98	32.63	33.29
53	24.11	24.76	25.40	26.05	26.70	27.35	28.00	28.66	29.32	29.98	30.64	31.30	31.97	32.63	33.30	33.97
54	24.59	25.25	25.91	26.57	27.23	27.90	28.56	29.23	29.91	30.58	31.25	31.93	32.61	33.29	33.98	34.66
55	25.07	25.74	26.41	27.09	27.77	28.44	29.13	29.81	30.50	31.18	31.87	32.56	33.26	33.95	34.65	35.35
56	25.55	26.23	26.92	27.61	28.30	28.99	29.69	30.39	31.09	31.79	32.49	33.20	33.91	34.62	35.33	36.04
57	26.03	26.73	27.43	28.13	28.84	29.54	30.25	30.97	31.68	32.39	33.11	33.83	34.56	35.28	36.01	36.74
58	26.51	27.23	27.94	28.66	29.37	30.10	30.82	31.55	32.27	33.00	33.74	34.47	35.21	35.95	36.69	37.43
59	27.00	27.72	28.45	29.18	29.91	30.65	31.39	32.13	32.87	33.61	34.36	35.11	35.86	36.62	37.37	38.13
60	27.48	28.22	28.96	29.71	30.45	31.20	31.96	32.71	33.47	34.23	34.99	35.75	36.52	37.29	38.06	38.83

(*continues*)

TABLE 14.1 (concluded)

NUMBER OF PAYMENTS	ANNUAL PERCENTAGE RATE															
	14.00%	14.25%	14.50%	14.75%	15.00%	15.25%	15.50%	15.75%	16.00%	16.25%	16.50%	16.75%	17.00%	17.25%	17.50%	17.75%
	(FINANCE CHARGE PER $100 OF AMOUNT FINANCED)															
1	1.17	1.19	1.21	1.23	1.25	1.27	1.29	1.31	1.33	1.35	1.37	1.40	1.42	1.44	1.46	1.48
2	1.75	1.78	1.82	1.85	1.88	1.91	1.94	1.97	2.00	2.04	2.07	2.10	2.13	2.16	2.19	2.22
3	2.34	2.38	2.43	2.47	2.51	2.55	2.59	2.64	2.68	2.72	2.76	2.80	2.85	2.89	2.93	2.97
4	2.93	2.99	3.04	3.09	3.14	3.20	3.25	3.30	3.36	3.41	3.46	3.51	3.57	3.62	3.67	3.73
5	3.53	3.59	3.65	3.72	3.78	3.84	3.91	3.97	4.04	4.10	4.16	4.23	4.29	4.35	4.42	4.48
6	4.12	4.20	4.27	4.35	4.42	4.49	4.57	4.64	4.72	4.79	4.87	4.94	5.02	5.09	5.17	5.24
7	4.72	4.81	4.89	4.98	5.06	5.15	5.23	5.32	5.40	5.49	5.58	5.66	5.75	5.83	5.92	6.00
8	5.32	5.42	5.51	5.61	5.71	5.80	5.90	6.00	6.09	6.19	6.29	6.38	6.48	6.58	6.67	6.77
9	5.92	6.03	6.14	6.25	6.35	6.46	6.57	6.68	6.78	6.89	7.00	7.11	7.22	7.32	7.43	7.54
10	6.53	6.65	6.77	6.88	7.00	7.12	7.24	7.36	7.48	7.60	7.72	7.84	7.96	8.08	8.19	8.31
11	7.14	7.27	7.40	7.53	7.66	7.79	7.92	8.05	8.18	8.31	8.44	8.57	8.70	8.83	8.96	9.09
12	7.74	7.89	8.03	8.17	8.31	8.45	8.59	8.74	8.88	9.02	9.16	9.30	9.45	9.59	9.73	9.87
13	8.36	8.51	8.66	8.81	8.97	9.12	9.27	9.43	9.58	9.73	9.89	10.04	10.20	10.35	10.50	10.66
14	8.97	9.13	9.30	9.46	9.63	9.79	9.96	10.12	10.79	10.45	10.62	10.78	10.95	11.11	11.28	11.45
15	9.59	9.76	9.94	10.11	10.29	10.47	10.64	10.82	11.00	11.17	11.35	11.53	11.71	11.88	12.06	12.24
16	10.20	10.39	10.58	10.77	10.95	11.14	11.33	11.52	11.71	11.90	12.09	12.28	12.46	12.65	12.84	13.03
17	10.82	11.02	11.22	11.42	11.62	11.82	12.02	12.22	12.42	12.62	12.83	13.03	13.23	13.43	13.63	13.83
18	11.45	11.66	11.87	12.08	12.29	12.50	12.72	12.93	13.14	13.35	13.57	13.78	13.99	14.21	14.42	14.64
19	12.07	12.30	12.52	12.74	12.97	13.19	13.41	13.64	13.86	14.09	14.31	14.54	14.76	14.99	15.22	15.44
20	12.70	12.93	13.17	13.41	13.64	13.88	14.11	14.35	14.59	14.82	15.06	15.30	15.54	15.77	16.01	16.25
21	13.33	13.58	13.82	14.07	14.32	14.57	14.82	15.06	15.31	15.56	15.81	16.06	16.31	16.56	16.81	17.07
22	13.96	14.22	14.48	14.74	15.00	15.26	15.52	15.57	15.78	16.04	16.30	16.57	16.83	17.09	17.36	17.62	17.88
23	14.59	14.87	15.14	15.41	15.68	15.96	16.23	16.50	16.78	17.05	17.32	17.60	17.88	18.15	18.43	18.70
24	15.23	15.51	15.80	16.08	16.37	16.65	16.94	17.22	17.51	17.80	18.09	18.37	18.66	18.95	19.24	19.53
25	15.87	16.17	16.46	16.76	17.06	17.35	17.65	17.95	18.25	18.55	18.85	19.15	19.45	19.75	20.05	20.36
26	16.51	16.82	17.13	17.44	17.75	18.06	18.37	18.68	18.99	19.30	19.62	19.93	20.24	20.56	20.87	21.19
27	17.15	17.47	17.80	18.12	18.44	18.76	19.09	19.41	19.74	20.06	20.39	20.71	21.04	21.37	21.69	22.02
28	17.80	18.13	18.47	18.80	19.14	19.47	19.81	20.15	20.48	20.82	21.16	21.50	21.84	22.18	22.52	22.86
29	18.45	18.79	19.14	19.49	19.83	20.18	20.53	20.88	21.23	21.58	21.94	22.29	22.64	22.99	23.35	23.70
30	19.10	19.45	19.81	20.17	20.54	20.90	21.26	21.62	21.99	22.35	22.72	23.08	23.45	23.81	24.18	24.55
31	19.75	20.12	20.49	20.87	21.24	21.61	21.99	22.37	22.74	23.12	23.50	23.88	24.26	24.64	25.02	25.40
32	20.40	20.79	21.17	21.56	21.95	22.33	22.72	23.11	23.50	23.89	24.28	24.68	25.07	25.46	25.86	26.25
33	21.06	21.46	21.85	22.25	22.65	23.06	23.46	23.86	24.26	24.67	25.07	25.48	25.88	26.29	26.70	27.11
34	21.72	22.13	22.54	22.95	23.37	23.78	24.19	24.61	25.03	25.44	25.86	26.28	26.70	27.12	27.54	27.97
35	22.38	22.80	23.23	23.65	24.08	24.51	24.94	25.36	25.79	26.23	26.66	27.09	27.52	27.96	28.39	28.83
36	23.04	23.48	23.92	24.35	24.80	25.24	25.68	26.12	26.57	27.01	27.46	27.90	28.35	28.80	29.25	29.70
37	23.70	24.16	24.61	25.06	25.51	25.97	26.42	26.88	27.34	27.80	28.26	28.72	29.18	29.64	30.10	30.57
38	24.37	24.84	25.30	25.77	26.24	26.70	27.17	27.64	28.11	28.59	29.06	29.53	30.01	30.49	30.96	31.44
39	25.04	25.52	26.00	26.48	26.96	27.44	27.92	28.41	28.89	29.38	29.87	30.36	30.85	31.34	31.83	32.32
40	25.71	26.20	26.70	27.19	27.69	28.18	28.68	29.18	29.68	30.18	30.68	31.19	31.68	32.19	32.69	33.20
41	26.39	26.89	27.40	27.91	28.41	28.92	29.44	29.95	30.46	30.97	31.49	32.01	32.52	33.04	33.56	34.08
42	27.06	27.58	28.10	28.62	29.15	29.67	30.19	30.72	31.25	31.78	32.31	32.84	33.37	33.90	34.44	34.97
43	27.74	28.27	28.81	29.34	29.88	30.42	30.96	31.50	32.04	32.58	33.13	33.67	34.22	34.76	35.31	35.86
44	28.42	28.97	29.52	30.07	30.62	31.17	31.72	32.28	32.83	33.39	33.95	34.51	35.07	35.63	36.19	36.76
45	29.11	29.67	30.23	30.79	31.36	31.92	32.49	33.06	33.63	34.20	34.77	35.35	35.92	36.50	37.08	37.66
46	29.79	30.36	30.94	31.52	32.10	32.68	33.26	33.84	34.43	35.01	35.60	36.19	36.78	37.37	37.96	38.56
47	30.48	31.07	31.66	32.25	32.84	33.44	34.03	34.63	35.23	35.83	36.43	37.04	37.64	38.25	38.86	39.46
48	31.17	31.77	32.37	32.98	33.59	34.20	34.81	35.42	36.03	36.65	37.27	37.88	38.50	39.13	39.75	40.37
49	31.86	32.48	33.09	33.71	34.34	34.96	35.59	36.21	36.84	37.47	38.10	38.74	39.37	40.01	40.65	41.29
50	32.55	33.18	33.82	34.45	35.09	35.73	36.37	37.01	37.65	38.30	38.94	39.59	40.24	40.89	41.55	42.20
51	33.25	33.89	34.54	35.19	35.84	36.49	37.15	37.81	38.46	39.12	39.79	40.45	41.11	41.78	42.45	43.12
52	33.95	34.61	35.27	35.93	36.60	37.27	37.94	38.61	39.28	39.96	40.63	41.31	41.99	42.67	43.36	44.04
53	34.65	35.32	36.00	36.68	37.36	38.04	38.72	39.41	40.10	40.79	41.48	42.17	42.87	43.57	44.27	44.97
54	35.35	36.04	36.73	37.42	38.12	38.82	39.52	40.22	40.92	41.63	42.33	43.04	43.75	44.47	45.18	45.90
55	36.05	36.76	37.47	38.17	38.88	39.60	40.31	41.03	41.74	42.47	43.19	43.91	44.64	45.37	46.10	46.83
56	36.76	37.48	38.20	38.92	39.65	40.38	41.11	41.84	42.57	43.31	44.05	44.79	45.53	46.27	47.02	47.77
57	37.47	38.20	38.94	39.68	40.42	41.16	41.91	42.65	43.40	44.15	44.91	45.66	46.42	47.18	47.94	48.71
58	38.18	38.93	39.68	40.43	41.19	41.95	42.71	43.47	44.23	45.00	45.77	46.54	47.32	48.09	48.87	49.65
59	38.89	39.66	40.42	41.19	41.96	42.74	43.51	44.29	45.07	45.85	46.64	47.42	48.21	49.01	49.80	50.60
60	39.61	40.39	41.17	41.95	42.74	43.53	44.32	45.11	45.91	46.71	47.51	48.31	49.12	49.92	50.73	51.55

TABLE 14.2 Loan amortization table (monthly payment per $1,000 to pay principal and interest on installment loan)

					Terms in months					
	6	12	18	24	30	36	42	48	54	60
7.50%	$170.34	$86.76	$58.92	$45.00	$36.66	$31.11	$27.15	$24.18	$21.88	$20.04
8.00%	170.58	86.99	59.15	45.23	36.89	31.34	27.38	24.42	22.12	20.28
8.50%	170.83	87.22	59.37	45.46	37.12	31.57	27.62	24.65	22.36	20.52
9.00%	171.20	87.46	59.60	45.69	37.35	31.80	27.85	24.77	22.59	20.76
10.00%	171.56	87.92	60.06	46.14	37.81	32.27	28.32	25.36	23.07	21.25
10.50%	171.81	88.15	60.29	46.38	38.04	32.50	28.55	25.60	23.32	21.49
11.00%	172.05	88.38	60.52	46.61	38.28	32.74	28.79	25.85	23.56	21.74
11.50%	172.30	88.62	60.75	46.84	38.51	32.98	29.03	26.09	23.81	21.99
12.00%	172.55	88.85	60.98	47.07	38.75	33.21	29.28	26.33	24.06	22.24
12.50%	172.80	89.08	61.21	47.31	38.98	33.45	29.52	26.58	24.31	22.50
13.00%	173.04	89.32	61.45	47.54	39.22	33.69	29.76	26.83	24.56	22.75
13.50%	173.29	89.55	61.68	47.78	39.46	33.94	30.01	27.08	24.81	23.01
14.00%	173.54	89.79	61.92	48.01	39.70	34.18	30.25	27.33	25.06	23.27
14.50%	173.79	90.02	62.15	48.25	39.94	34.42	30.50	27.58	25.32	23.53
15.00%	174.03	90.26	62.38	48.49	40.18	34.67	30.75	27.83	25.58	23.79
15.50%	174.28	90.49	62.62	48.72	40.42	34.91	31.00	28.08	25.84	24.05
16.00%	174.53	90.73	62.86	48.96	40.66	35.16	31.25	28.34	26.10	24.32

Calculating Monthly Payment by Table Lookup

Step 1 Divide the loan amount by $1,000 (since Table 14.2 is per $1,000):

$$\frac{\$9,045}{\$1,000} = 9.045$$

Step 2 Look up the rate (10.5%) and number of months (60). At the intersection is the table factor showing the monthly payment per $1,000.

Step 3 Multiply quotient in Step 1 by the table factor in Step 2:

9.045 × $21.49 = $194.38.

Remember that this $194.38 fixed payment includes interest and the reduction of the balance of the loan. As the number of payments increases, interest payments get smaller and the reduction of the principal gets larger.[2]

Now let's check your progress with the Practice Quiz.

Money Tip Go on a financial diet. Control your debt. Review your credit card's year-end summary to see where your money goes. Adjust your spending accordingly.

Copyright © McGraw Hill

[2] In Chapter 15 we give an amortization schedule for home mortgages that shows how much of each fixed payment goes to interest and how much reduces the principal. This repayment schedule also gives a running balance of the loan.

Installment Buying • **Chapter 14** 435

Learning Unit 14–1

Practice Quiz

Complete this Practice Quiz to see how you are doing.

From the partial advertisement at the right calculate the following:

1. **a.** Amount financed.

   b. Finance charge.

   c. Deferred payment price.

   d. APR by Table 14.1.

   e. Monthly payment by formula.

$288 per month	
Sale price	$14,150
Down payment	$ 1,450
Term/Number of payments	60 months

2. Jay Miller bought a New Brunswick boat for $7,500. Jay put down $1,000 and financed the balance at 10% for 60 months. What is his monthly payment? Use Table 14.2.

✓ Solutions

1. **a.** $14,150 − $1,450 = $12,700

   b. $17,280 ($288 × 60) − $12,700 = $4,580

   c. $17,280 ($288 × 60) + $1,450 = $18,730

   d. $\dfrac{\$4,580}{\$12,700} \times \$100 = \$36.06$; between 12.75% and 13%

   e. $\dfrac{\$4,580 + \$12,700}{60} = \$288$

2. $\dfrac{\$6,500}{\$1,000} = 6.5 \times \$21.25 = \$138.13$ (10%, 60 months)

Revolving Charge Credit Cards

Do you owe a balance on your credit card? Let's look at how long it will take to pay off your credit card balance by making payments for the minimum amount. Study the clipping "Pay Just the Minimum, and Get Nowhere Fast."

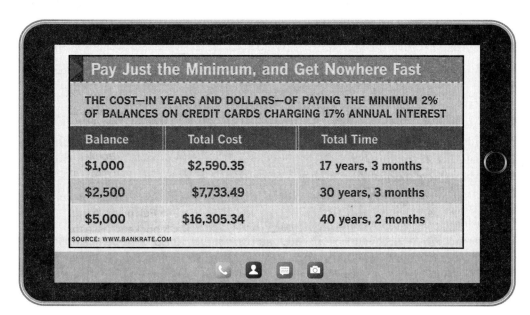

Pay Just the Minimum, and Get Nowhere Fast

THE COST—IN YEARS AND DOLLARS—OF PAYING THE MINIMUM 2% OF BALANCES ON CREDIT CARDS CHARGING 17% ANNUAL INTEREST

Balance	Total Cost	Total Time
$1,000	$2,590.35	17 years, 3 months
$2,500	$7,733.49	30 years, 3 months
$5,000	$16,305.34	40 years, 2 months

SOURCE: WWW.BANKRATE.COM

The *Wall Street Journal* clip above assumes that the minimum rate on the balance of a credit card is 2%. Note that if the annual interest cost is 17%, it will take 17 years, 3 months to pay off a balance of $1,000, and the total cost will be $2,590.35. If the balance on your revolving charge credit card is more than $1,000, you can see how fast the total cost rises. If you cannot afford the total cost of paying only the minimum, it is time for you to reconsider how you use your revolving credit card. This is why when you have financial difficulties, experts often advise you first to work on getting rid of your revolving credit card debt.

Do you know why revolving credit cards are so popular? Businesses encourage customers to use credit cards because consumers tend to buy more when they can use a credit card for their purchases. Consumers find credit cards convenient to use and valuable in establishing credit. The problem is that when consumers do not pay their balance in full each month, they do not realize how expensive it is to pay only the minimum of their balance.

To protect consumers, Congress passed the **Fair Credit and Charge Card Disclosure Act of 1988.** This act requires that for direct-mail application or solicitation, credit card companies must provide specific details involving all fees, grace period, calculation of finance charges, and so on.

We begin the unit by seeing how Moe's Furniture Store calculates the finance charge on Abby Jordan's previous month's credit card balance. Then we learn how to calculate the average daily balance on the partial bill of Joan Ring.

Pepper ... And Salt

THE WALL STREET JOURNAL

"...and the neighbors on both sides have credit scores over 800."

TABLE 14.3 Schedule of payments

Monthly payment number	Outstanding balance due	$1\frac{1}{2}\%$ interest payment	Amount of monthly payment	Reduction in balance due	Outstanding balance due
1	$8,000.00	$120.00	$500.00	$380.00	$7,620.00
		(.015 × $8,000.00)		($500.00 − $120.00)	($8,000.00 − $380.00)
2	$7,620.00	$114.30	$500.00	$385.70	$7,234.30
		(.015 × $7,620.00)		($500.00 − $114.30)	($7,620.00 − $385.70)
3	$7,234.30	$108.51	$500.00	$391.49	$6,842.81
		(.015 × $7,234.30)		($500.00 − $108.51)	($7,234.30 − $391.49)

Learn: Calculating Finance Charge on Previous Month's Balance

Abby Jordan bought a dining room set for $8,000 on credit. She has a **revolving charge account** at Moe's Furniture Store. A revolving charge account gives a buyer **open-end credit.** Abby can make as many purchases on credit as she wants until she reaches her maximum $10,000 credit limit.

Often customers do not completely pay their revolving charge accounts at the end of a billing period. When this occurs, stores add interest charges to the customers' bills. Moe's Furniture Store calculates its interest using the *unpaid balance method.* It charges $1\frac{1}{2}\%$ on the *previous month's balance,* or 18% per year. Moe's has no minimum monthly payment (many stores require $10 or $15, or a percent of the outstanding balance).

Abby has no other charges on her revolving charge account. She plans to pay $500 per month until she completely pays off her dining room set. Abby realizes that when she makes a payment, Moe's Furniture Store first applies the money toward the interest and then reduces the **outstanding balance** due. (This is the U.S. Rule we discussed in Chapter 10.) For her own information, Abby worked out the first 3-month schedule of payments, shown in Table 14.3. Note how the interest payment is the rate times the outstanding balance.

Today, most companies with credit card accounts calculate the finance charge, or interest, as a percentage of the average daily balance. Interest on credit cards can be very expensive for consumers; however, interest is a source of income for credit card companies. In the exhibit shown to the left, note the late payment warning issued by the credit card company. It states that a late payment could result in interest penalties close to 30%. The following is a letter I received from my credit card company when I questioned how my finance charge was calculated.

How Citibank Calculates My Finance Charge

Thank you for your recent inquiry regarding your Citi® / AAdvantage® MasterCard® account and how finance charges are calculated.

Finance charges for purchases, balance transfers, and cash advances will begin to accrue from the date the transaction is added to your balance. They will continue to accrue until payment in full is credited to your account. This means that when you make your final payment on these balances, you will be billed finance charges for the time between the date your last statement prints and the date your payment is received.

Paying your purchase balance in full each billing period by the payment due date saves you money because it allows you to take advantage of your grace period on purchases, which is not less than 20 days. You can avoid periodic finance charges on purchases (excluding balance transfers) that appear on your current billing statement if you paid the New Balance on the last statement by the payment due date on that statement and you pay your New Balance by the payment due date on your current statement. If you made a balance transfer, you may be unable to avoid periodic finance charges on new purchases, as described in the balance transfer offer.

Calculating Average Daily Balance   Let's look at the following steps for calculating the **average daily balance.** Remember that a **cash advance** is a cash loan from a credit card company.

Calculating Average Daily Balance and Finance Charge

Step 1 Calculate the daily balance or amount owed at the end of each day during the billing cycle:

$$\frac{\text{Daily}}{\text{balance}} = \frac{\text{Previous}}{\text{balance}} + \frac{\text{Cash}}{\text{advances}} + \text{Purchases} - \text{Payments} - \text{Credits}$$

Step 2 When the daily balance is the same for more than 1 day, multiply it by the number of days the daily balance remained the same, or the number of days of the current balance. This gives a cumulative daily balance.

Step 3 Add the cumulative daily balances.

Step 4 Divide the sum of the cumulative daily balances by the number of days in the billing cycle.

Step 5 Finance charge = Rate per month × Average daily balance.

Step 6* New balance = Previous balance + Cash advances + Purchases − Payments − Credits + Finance charge

* Note: There is a shortcut to this formula. Using data from the Joan Ring example (see below), take the last current daily balance and add the finance charge to it: $620 + $8.19 = $628.19.

Aha! *Always check the number of days in the billing cycle. The cycle may not be 30 or 31 days.*

Following is the partial bill of Joan Ring and an explanation of how Joan's average daily balance and finance charge were calculated. Note how we calculated each **daily balance** and then multiplied each daily balance by the number of days the balance remained the same. Take a moment to study how we arrived at 8 days. The total of the cumulative daily balances was

Pepper ...
And Salt

THE WALL STREET JOURNAL

"Yesterday my ID was stolen. This afternoon, after they checked my credit score, it was returned."

Used by permission of Cartoon Features Syndicate

$16,390. To get the average daily balance, we divided by the number of days in the billing cycle—30. Joan's finance charge is $1\frac{1}{2}\%$ per month on the average daily balance.

7 days had a balance of $450 30-day cycle − 22 (7 + 3 + 9 + 3) equals 8 days left with a balance of $620.

30-day billing cycle			
6/20	Billing date	Previous balance	$450
6/27	Payment		$ 50 cr.
6/30	Charge: JCPenney		200
7/9	Payment		40 cr.
7/12	Cash advance		60

	No. of days of current balance	Current daily balance	Extension	
Step 1 →	7	$450	$ 3,150	← Step 2
	3	400 ($450 − $50)	1,200	
	9	600 ($400 + $200)	5,400	
	3	560 ($600 − $40)	1,680	
	8	620 ($560 + $60)	4,960	
	30		$16,390	← Step 3

$$\text{Average daily balance} = \frac{\$16,390}{30} = \boxed{\$546.33} \leftarrow \textbf{Step 4}$$

Step 5 → Finance charge = $546.33 × .015 = $8.19

Step 6 → $450 + $60 + $200 − $40 − $50 + $8.19 = $628.19

Now try the following Practice Quiz to check your understanding of this unit.

Money Tip
Understand the costs of credit. Do not spend money you do not currently have—especially if it is for entertainment. Avoid the high cost of reaching or coming close to the maximum on your credit card(s). Be careful to fully understand the terms of any credit card you use. Know your debt-to-income ratio and keep it below 28% excluding mortgage or 36% including mortgage.

Practice Quiz

Complete this Practice Quiz to see how you are doing.

1. Calculate the balance outstanding at the end of month 2 (use U.S. Rule) given the following: purchased $600 desk at the beginning of month 1; pay back $40 per month; and charge of $2\frac{1}{2}\%$ interest on unpaid balance.

2. Calculate the average daily balance and finance charge from the information that follows.

31-day billing cycle			
8/20	Billing date	Previous balance	$210
8/27	Payment	$50 cr.	
8/31	Charge: Staples	30	
9/5	Payment	10 cr.	
9/10	Cash advance	60	

Rate = 2% per month on average daily balance.

✓ Solutions

1.

Month	Balance due	Interest	Monthly payment	Reduction in balance	Balance outstanding
1	$600	$15.00 (.025 × $600)	$40	$25.00 ($40 − $15)	$575.00
2	$575	$14.38 (.025 × $575)	$40	$25.62	$549.38

2. Average daily balance calculated as follows:

No. of days of current balance	Current balance	Extension
7	$210	$1,470
4	160 ($210 − $50)	640
5	190 ($160 + $30)	950
5	180 ($190 − $10)	900
10 ←	240 ($180 + $60)	2,400
31		$6,360

$$\text{Average daily balance} = \frac{\$6,360}{31} = \$205.16$$

$$\text{Finance charge} = \$4.10 \ (\$205.16 \times .02)$$

$$31 - 21(7 + 4 + 5 + 5)$$

Chapter 14 Review

Topic/Procedure/Formula	Examples	You try it*
Amount financed $\dfrac{\text{Amount}}{\text{financed}} = \dfrac{\text{Cash}}{\text{price}} - \dfrac{\text{Down}}{\text{payment}}$	60 payments of $125.67 per month; cash price $5,295 with a $95 down payment Cash price $5,295 − Down payment − 95 = Amount financed $5,200	**Calculate amount financed** 60 payments of $129.99 per month; Cash price $5,400 with a $100 down payment
Total finance charge (interest) $\dfrac{\text{Total}}{\text{finance}} = \dfrac{\text{Total of}}{\text{all monthly}} - \dfrac{\text{Amount}}{\text{financed}}$ charge payments	*(continued from above)* $\dfrac{\$125.67}{\text{per month}} \times \dfrac{60}{\text{months}} = \$7,540.20$ − Amount financed − 5,200.00 = Finance charge $2,340.20	**Calculate total finance charge** *(continued from above)*
Deferred payment price $\dfrac{\text{Deferred}}{\text{payment}} = \dfrac{\text{Total of}}{\text{all monthly}} + \dfrac{\text{Down}}{\text{payment}}$ price payments	*(continued from above)* $7,540.20 + $95 = $7,635.20	**Calculate deferred payment price** *(continued from above)*
Calculating APR by Table 14.1 $\dfrac{\text{Finance charge}}{\text{Amount financed}} \times \$100 = \dfrac{\text{Table 14.1}}{\text{lookup number}}$	*(continued from above)* $\dfrac{\$2,340.20}{\$5,200.00} \times \$100 = \$45.004$ Search in Table 14.1 between 15.50% and 15.75% for 60 payments.	**Calculate APR by table** *(continued from above)*
Monthly payment *By formula:* $\dfrac{\text{Finance charge} + \text{Amount financed}}{\text{Number of payments of loan}}$ *By table:* $\dfrac{\text{Loan}}{\$1,000} \times \dfrac{\text{Table}}{\text{factor}} \text{ (rate, months)}$	*(continued from above)* $\dfrac{\$2,340.20 + \$5,200.00}{60} = \$125.67$ Given: 15.5% 60 months $5,200 loan $\dfrac{\$5,200}{\$1,000} = 5.2 \times \$24.05 = \$125.06$ (off due to rounding of rate)	**Calculate monthly payment** *(continued from above; use 16%)*
Open-end credit Monthly payment applied to interest first before reducing balance outstanding	$4,000 purchase $250 a month payment $2\frac{1}{2}$% interest on unpaid balance $4,000 × .025 = $100 interest $250 − $100 = $150 to lower balance $4,000 − $150 = $3,850 Balance outstanding after month 1.	**Calculate balance outstanding after month 1** $5,000 purchase; $275 monthly payment; $3\frac{1}{2}$% interest on unpaid balance

Chapter 14 Review (Continued)

Topic/Procedure/Formula	Examples	You try it*

Average daily balance and finance charge
 	
$\dfrac{\text{Daily}}{\text{balance}} = \dfrac{\text{Previous}}{\text{balance}} + \dfrac{\text{Cash}}{\text{advances}}$
$+$ Purchases $-$ Payments
$-$ Credits
 	
$\dfrac{\text{Average}}{\text{daily}} = \dfrac{\text{Sum of cumulative daily balances}}{\text{Number of days in billing cycle}}$
balance
 	
$\dfrac{\text{Finance}}{\text{charge}} = \dfrac{\text{Monthly}}{\text{rate}} \times \dfrac{\text{Average daily balance}}{}$
	30-day billing cycle; $1\frac{1}{2}\%$ finance charge per month
	
	Example: 8/21 Balance $100
	              8/29 Payment $10
	              9/12 Charge 50
	30-day billing cycle less the 8 and 14.
	Average daily balance equals:
	8 days $\times$ \$100 $=$ \$ 800
	14 days $\times$ 90 $=$ 1,260
	8 days $\times$ 140 $=$ <u>1,120</u>
	                    \$3,180 $\div$ 30
	Average daily balance $=$ `$106`
	
	Finance charge $= \$106 \times .015 =$ `$1.59`
		Calculate daily balance and finance charge
		
		30-day billing cycle;
		$2\frac{1}{2}\%$ finance charge per month
		
		Given: 9/4 bal \$200
		           9/16 payment \$80
		           9/20 charge \$60

Key Terms

Amortization	Deferred payment price	Loan amortization table
Amount financed	Down payment	Open-end credit
Annual percentage rate (APR)	Fair Credit and Charge Card Disclosure Act of 1988	Outstanding balance
Average daily balance		Revolving charge account
Cash advance	Finance charge	Truth in Lending Act
Daily balance	Installment loan	

* Worked-out solutions are in Appendix A.

Critical Thinking Discussion Questions with Chapter Concept Check

1. Explain how to calculate the amount financed, finance charge, and APR by table lookup. Do you think the Truth in Lending Act should regulate interest charges?

2. Explain how to use the loan amortization table. Check with a person who owns a home and find out what part of each payment goes to pay interest versus the amount that reduces the loan principal.

3. What steps are used to calculate the average daily balance? Many credit card companies charge 18% annual interest. Do you think this is a justifiable rate? Defend your answer.

4. **Chapter Concept Check.** Visit the web and find information on how social networks like Facebook have had some influence on credit card companies' policies. Defend your position with the concepts learned in this chapter.

End-of-Chapter Problems

Name _____   Date _____

Check figures for odd-numbered problems in Appendix A.

Drill Problems

Complete the following table: *LU 14–1(1)*

	Purchase price of product	Down payment	Amount financed	Number of monthly payments	Amount of monthly payments	Total of monthly payments	Total finance charge
14–1.	Landcruiser $85,000	$60,000		72	$420		
14–2.	Pre-owned Specialized Mountain Bike $250	$100		12	$15.50		

Calculate **(a)** the amount financed, **(b)** the total finance charge, and **(c)** APR by table lookup. *LU 14–1(1, 2)*

	Purchase price of a used car	Down payment	Number of monthly payments	Amount finance	Total of monthly payments	Total finance charge	APR
14–3.	$5,673	$1,223	48		$5,729.76		
14–4.	$4,195	$95	60		$5,944.00		

Calculate the monthly payment for Problems 14–3 and 14–4 by table lookup and formula. (Answers will not be exact due to rounding of percents in table lookup.) *LU 14–1(3)*

14–5. **(14–3)** (Use 13% for table lookup.)

14–6. **(14–4)** (Use 15.5% for table lookup.)

14–7. Calculate the average daily balance and finance charge on the statement below.
LU 14–2(1)

30-day billing cycle			
9/16	Billing date	Previous balance	$2,000
9/19	Payment	$ 60 cr.	
9/30	Charge: Home Depot	1,500	
10/3	Payment	60 cr.	
10/7	Cash advance	70	
Finance charge is $1\frac{1}{2}$% on average daily balance			

Word Problems

14–8. Before purchasing a used car, Cody Lind checked www.kbb.com to learn what he should offer for the used car he wanted to buy. Then he conducted a carfax.com search on the car he found to see if the car had ever been in an accident. The Carfax was clean so he purchased the used car for $14,750. He put $2,000 down and financed the rest with a 48-month, 7.5% loan. What is his monthly car payment by table lookup?   *LU 14–1(1, 2)*

14–9. Troy Juth wants to purchase new dive equipment for Underwater Connection, his retail store in Colorado Springs. He was offered a $56,000 loan at 7.5% for 48 months. What is his monthly payment by table lookup?   *LU 14–1(3)*

14–10. Ramon Hernandez saw the following advertisement for a used Volkswagen Bug and decided to work out the numbers to be sure the ad had no errors. Please help Ramon by calculating **(a)** the amount financed, **(b)** the finance charge, **(c)** APR by table lookup, **(d)** the monthly payment by formula, and **(e)** the monthly payment by table lookup (will be off slightly).   *LU 14–1(1, 2, 3)*

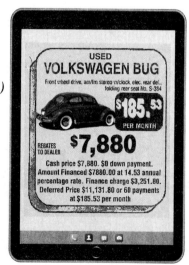

USED
VOLKSWAGEN BUG
Front wheel drive, am/fm stereo w/clock, elec. rear def., folding rear seat No. S-384

$185.53 PER MONTH

REBATES TO DEALER $7,880

Cash price $7,880. $0 down payment. Amount Financed $7880.00 at 14.53 annual percentage rate. Finance charge $3,251.80. Deferred Price $11,131.80 or 60 payments at $185.53 per month

a. Amount financed:

b. Finance charge:

c. APR by table lookup:

d. Monthly payment by formula:

e. Monthly payment by table lookup (use 14.50%):

End-of-Chapter Problems (Continued)

14–11. From this partial advertisement calculate: *LU 14–1(1, 2, 3)*

    a. Amount financed.

    b. Finance charge

    c. Deferred payment price.

    d. APR by Table 14.1.

    e. Monthly payment (by formula).

$95.10 per month
#43892 Used car. Cash price
$4,100. Down payment
$50. For 60 months.

e)cel 14–12. If you are trying to build credit by using a credit card, each time you make a purchase with the credit card, deduct that amount from your checking account. That way, when your credit card bill is due, you will have enough to pay the credit card off in full. Kathy Lehner is going to start doing this. She plans on paying her credit card bill in full this month. How much does she owe with a 12% APR and the following transactions? *LU 14–2(1)*

31-day billing cycle		
10/1	Previous balance	$1,168
10/3	Credit	$ 75 cr.
10/12	Charge: King Soopers	152
10/15	Payment	350 cr.
10/25	Charge: Delta	325
10/30	Charge: Holiday Fun	65

14–13. Dallas Pierce's most recent credit card statement follows. Her finance charge is 18% APR. Calculate Dallas's average daily balance, finance charge, and new balance. (Round final answers to the nearest cent.) *LU 14–2(1)*

30-day billing cycle		
9/2	Billing date	$1,200 previous balance
9/7	Payment	$ 100 cr.
9/13	Charge: Kohl's	350
9/17	Payment	200 cr.
9/28	Charge: Walmart	50

e)(cel **14–14.** First America Bank's monthly payment charge on a 48-month, $20,000 loan is $488.26. U.S. Bank's monthly payment fee is $497.70 for the same loan amount. What would be the APR for an auto loan for each of these banks? (Use the *Business Math Handbook*.) *LU 14–1(1, 2)*

e)(cel **14–15.** From the following facts, Molly Roe has requested you to calculate the average daily balance. The customer believes the average daily balance should be $877.67. Respond to the customer's concern. *LU 14–2(1)*

28-day billing cycle		
3/18	Billing date	Previous balance	$800
3/24	Payment	$ 60 cr.
3/29	Charge: Sears	250
4/5	Payment	20 cr.
4/9	Charge: Macy's	200

End-of-Chapter Problems (Continued)

14–16. Jill bought a $500 rocking chair. The terms of her revolving charge are $1\frac{1}{2}\%$ on the unpaid balance from the previous month. If she pays $100 per month, complete a schedule for the first 3 months like Table 14.3. Be sure to use the U.S. Rule.   *LU 14–2(1)*

Monthly payment number	Outstanding balance due	$1\frac{1}{2}\%$ interest payment	Amount of monthly payment	Reduction in balance due	Outstanding balance due

14–17. Dr. Dennis Natali plans to take advantage of a 0% interest balance transfer credit card offer to pay off a $7,250 loan he has. If his loan is at 7.5% interest for 12 months, what is his payment? How much will he save in interest?

14–18. Sonja Upton considered closing her travel agency, Opos Tours, during the pandemic, since travel had almost ceased. But she would prefer to keep her business open. If she didn't close the agency, she would need to take out a $60,000 loan to cover operating expenses. She found a business loan at 9% for 60 months. What would her monthly payment by table be if she decided to take out the loan?   *LU-14–1(3)*

14–19. Teri Silva's Whisker Watcher pet watching business shut down during the pandemic since people were no longer traveling and therefore didn't need her pet watching service. To help her until she found a new job, Teri's credit union offered her a $9,500 personal loan requiring no collateral with payments of $218.78 monthly for 4 years. Calculate her APR by table.   *LU-14–1(2)*

14–20. Cindi Cavalier was furloughed from her job at Red Robin Garden Store for six months at the start of the pandemic. She normally has paid her credit card off each month. Because of being furloughed, however, she has been unable to. Calculate Cindi's average daily balance, finance charge, and new balance. Her finance charge is 17% APR. (Round final answers to the nearest cent.)   *LU-14–2(1)*

30-day billing cycle		
June 1	Previous balance	$868
June 3	Charge, Safeway	102
June 10	Charge, CS Utilities	225
June 15	Credit, Amazon	75
June 20	Charge, Shell	68
June 29	Payment	50

Challenge Problems

14–21. Peg Gasperoni bought a $50,000 life insurance policy for $100 per year. Ryan Life Insurance Company sent her the following billing instructions along with a premium plan example:

"Your insurance premium notice will be mailed to you in a few days. You may pay the entire premium in full without a finance charge or you may pay the premium in installments after a down payment and the balance in monthly installments of $30. The finance charge will be added to the unpaid balance. The finance charge is based on an annual percentage rate of 15%."

If the total policy premium is:	And you put down:	The balance subject to finance charge will be:	The total number of monthly installments ($30 minimum) will be:	The monthly installment before adding the finance charge will be:	The total finance charge for all installments will be:	And the total deferred payment price will be:
$100	$30.00	$ 70.00	3	$30.00	$ 1.75	$101.75
200	50.00	150.00	5	30.00	5.67	205.67
300	75.00	225.00	8	30.00	12.84	312.84

End-of-Chapter Problems (*Continued*)

Peg feels that the finance charge of $1.75 is in error. Who is correct? Check your answer. *LU 14–2(1)*

14–22. You have a $1,100 balance on your 15% credit card. You have lost your job and been unemployed for 6 months. You have been unable to make any payments on your balance. However, you received a tax refund and want to pay off the credit card. How much will you owe on the credit card, and how much interest will have accrued? What will be the effective rate of interest after the 6 months (to the nearest hundredth percent)? *LU 14–2(1)*

Summary Practice Test

Do you need help? Connect videos have step-by-step worked-out solutions.

1. Walter Lantz buys a used Volvo SUV for $42,500. Walter made a down payment of $16,000 and paid $510 monthly for 60 months. What are the total amount financed and the total finance charge that Walter paid at the end of the 60 months?   *LU 14–1(1)*

2. Joyce Mesnic bought an HP laptop computer at Staples for $699. Joyce made a $100 down payment and financed the balance at 10% for 12 months. What is her monthly payment? (Use the loan amortization table.)   *LU 14–1(3)*

3. Lee Remick read the following partial advertisement: price, $22,500; down payment, $1,000 cash or trade; and $399.99 per month for 60 months. Calculate **(a)** the total finance charge and **(b)** the APR by Table 14.1 (or use the tables in *Business Math Handbook*) to the nearest hundredth percent. *LU 14–1(1, 2)*

4. Nancy Billows bought a $7,000 desk at Furniture.com. Based on her income, Nancy could only afford to pay back $700 per month. The charge on the unpaid balance is 3%. The U.S. Rule is used in the calculation. Calculate the balance outstanding at the end of month 2.   *LU 14–2(1)*

Month	Balance due	Interest	Monthly payment	Reduction in balance	Balance outstanding

5. Calculate the average daily balance and finance charge on the statement below.   *LU 14–2(1)*

30-day billing cycle		
7/3	Balance	$400
7/18	Payment	100 cr.
7/27	Charge Walmart	250

Assume 2% finance charge on average daily balance.

MY MONEY

Setting a Budget You Can Follow

 What I need to know

A budget is a necessary tool to effectively manage your finances. Initially you may view a budget as something restricting your ability to use your money in a manner you see fit. However, nothing could be further from the truth. Creating a budget will provide an accurate picture of your income and expenses, charting a course for you to achieve your financial goals. Good budgeting starts TODAY and is reviewed regularly as conditions change. Setting a budget for yourself during your college years will establish financial habits that will serve you well throughout your life. Your budget needs to be realistic, thorough, and flexible. When you begin the budgeting process look at the task as a learning opportunity to understand the big picture of your financial situation.

 What I need to do

Understand where your money is coming from and where it is going. This will involve taking the time to document your earnings as well as expenses you incur. Record all your financial transactions for one month to determine inflows and outflows of money. This process can be very eye opening as you learn how you are using the money you earn. Repeat this same process for another couple months to establish some consistency among your financial transactions and to account for earnings and expenses occurring less frequently.

Examining your expenses over a few months will identify patterns in how you spend your money. For instance, you may notice your expense for going out to eat is much higher than you would have guessed. Setting a limit to this spending is one step in establishing a budget in which you can be comfortable. You will also be able to identify regular monthly expenses such as rent/mortgage, utilities, and insurance. For each category of expense, set a realistic budget to cover the expense within the limit of your earnings. Don't forget to pay yourself first. Include a budget item for your saving and investing as well as for building your emergency fund. Create a reasonable entertainment or fun money category so you have funds set aside for yourself to enjoy your money as well. Add all budget categories and dollar amounts for each expense you incur during a typical month. Allow for some flexibility in categories that fluctuate such as groceries or transportation expense. Subtract this total from your earnings ensuring it balances. Make adjustments as needed. Keep in mind the goal of saving and investing 20% of all monies received.

 Steps I need to take

1. Know the amounts of money you earn and expenses you incur monthly.
2. Create budget categories for all income and expenses with realistic budget amounts.
3. Commit to following your budget and allowing for flexibility as conditions change.

 Resources I can use

*   EveryDollar: Budget Your Money (mobile app)—easily create a budget and track your spending.
*   https://www.daveramsey.com/blog/the-truth-about-budgeting—budgeting tips from a practical perspective.

MY MONEY ACTIVITY

*   Using your estimated salary from the Chapter 2 My Money feature, create a monthly budget for yourself.
*   Identify budget categories you are likely to encounter after college (car payment, mortgage, vacation fund, student loan payment, etc.) and set spending limits for each.
*   Discuss some practical actions you can take to help you to stick to your budget.

PERSONAL FINANCE

A KIPLINGER APPROACH

"Give Your Child Some Credit." Kiplinger's. February 2021.

BASICS

Give Your Child Some Credit

Naming your child as an authorized user on your credit card can be a great way to set them up with a healthy credit report.

AN AUTHORIZED USER IS
A secondary account holder on a credit card, which means the user has access to an existing credit card account but ultimately isn't responsible for making payments. Although someone can become an authorized user of a sibling's or even a friend's account, the most common arrangement is between a parent and a child. The arrangement benefits young adults who may not qualify for credit on their own because they have little or no credit history and limited income. Although most credit card companies won't issue a card to someone who is younger than 18, a child who is younger than that can be an authorized user.

Even though the card isn't in your child's name, your child can start to build credit. But if you miss payments on your credit card, that negative information could affect your child's credit report.

As long as you handle your credit responsibly, naming a child as an authorized user can help them establish a good credit record and learn how to use credit responsibly. Consider letting your child use it occasionally to pay for gas or food or other small expenses to gain an understanding of what it's like to use a credit card, says Ted Rossman, analyst at CreditCards.com.

You may also want to set spending limits for college-age children who are living away from home. You should be clear about how the card can be used—for textbooks, for example, but not for entertainment. Once your child is working and has established a credit history of their own, encourage them to apply for their own credit card, says Rossman.

The risks to you. Keep a close eye on your child's purchases, because you're legally liable for their spending. "It's really important that you talk about that with your kid when you're adding them as an authorized user," says Matt Schulz, chief industry analyst at CompareCards.com. Bear in mind, too, that your child's credit card charges can hurt your credit score even if you pay off the balance every month. One of the components of your credit score is the credit-utilization ratio, which is a ratio of debt you have relative to the amount of credit you have available. The general rule of thumb is to try to keep the ratio below 30%; below 10% is even better, Rossman says.

A good credit report is valuable for a young person who is starting out, and not just because it will make it easier for them to get their own credit card or a car loan. Even non-credit-related ventures often require a healthy credit record. For example, landlords may want to check your child's credit record when they apply to rent an apartment, as will utility companies and some cell-phone companies. And good credit is especially important for young people now. Nearly one-third of millennials age 24 to 29 were denied a financial product because of a low credit score in 2020, according to Bankrate.com. Making your child an authorized user may just be the leg up they need down the line. **EMMA PATCH**

Emma_Patch@kiplinger.com

Business Math Issue
Millennials do not have to establish credit.

1. List the key points of the article and information to support your position.

2. Write a group defense of your position using math calculations to support your view. If you are in an online course, post to a discussion board.

Chapter 15

The Cost of Home Ownership

Learning Unit Objectives

LU 15–1: Types of Mortgages and the Monthly Mortgage Payment

1. List the types of mortgages available.
2. Utilize an amortization chart to compute monthly mortgage payments.
3. Calculate the total cost of interest over the life of a mortgage.

LU 15–2: Amortization Schedule—Breaking Down the Monthly Payment

1. Calculate and identify the interest and principal portion of each monthly payment.
2. Prepare an amortization schedule.

@ Essential Question

How can I use my knowledge of the cost of home ownership to understand the business world and make my life easier?

🌐 Math Around the World

The *Wall Street Journal* chapter opener clip, "Rate for 30-Year Mortgage Falls to Lowest on Record", shows that during the pandemic mortgage rates fell below 3%.

Rate for 30-Year Mortgage Falls to Lowest on Record

BY ORLA McCAFFREY

In a year of financial firsts, this one stands out: Mortgage rates have fallen below the 3% mark.

The average rate on a 30-year fixed mortgage fell to 2.98%, mortgage-finance giant Freddie Mac said Thursday, its lowest level in almost 50 years of record-keeping. It is the third consecutive week and the seventh time this year that rates on America's most popular home loan have hit a fresh low.

The coronavirus pandemic has upended markets around the world, sending stocks on a wild ride and yields on U.S. government debt to record lows, but its effect on the 30-year mortgage is

Average rate on 30-year fixed mortgage

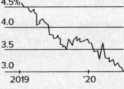

Spread between 10-year Treasury and 30-year fixed-rate mortgage

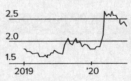

Sources: Freddie Mac (rate); Dow Jones Market Data (spread)

especially significant. In the early 1980s, it peaked above 18% after the Federal Reserve raised rates to fight runaway inflation.

Below 3% is a "tremendous benchmark," said Jeff Tucker, an economist at Zillow Group Inc. "It's also an indication that we remain in a crisis here."

The average rate on the 30-year mortgage stood at 3.72% at the beginning of the year and 3.81% a year ago, according to Freddie Mac. Mortgage rates tend to move in the same direction as the yield on the 10-year Treasury note. Yields fall as

Please turn to page A2

McCaffrey, Orla. "Rate For 30-Year Mortgage Falls to Lowest on Record." *The Wall Street Journal* (July 17, 2020).

Learning Unit 15–1:
Types of Mortgages and the Monthly Mortgage Payment

Figure 15.1 lists various loan types. A type of adjustable rate mortgage called a **subprime loan** was at the root of so many foreclosures during the subprime mortgage crisis from 2007–2010. This type of home loan allowed buyers to have a very low interest rate—sometimes even a zero rate. This helped customers qualify for expensive homes that they would not otherwise have qualified for. Lenders offering subprimes assumed prices of homes would rise and most buyers would convert to a fixed rate before the rate was substantially adjusted upward. As we now know, prices of homes fell. Since that time period, the Federal Reserve has been increasing interest rates.

Purchasing a home usually involves paying a large amount of interest. Note how your author was able to save $70,121.40. Over the life of a 30-year **fixed-rate mortgage** (see Figure 15.1) of $100,000, the interest would have cost $207,235. Monthly payments would have been $849.99. This would not include taxes, insurance, and so on. Your author chose a **biweekly mortgage** (see Figure 15.1). This meant that every 2 weeks (26 times a year) the bank would receive $425. By paying every 2 weeks instead of once a month, the mortgage would be paid off in 23 years instead of 30—a $70,121.40 *savings* on interest. Why? When a payment is made every 2 weeks, the principal is reduced more quickly, which substantially reduces the interest cost.

FIGURE 15.1 Types of mortgages available

The question facing prospective buyers concerns which type of mortgage will be best for them. Depending on how interest rates are moving when you purchase a home, you may find one type of **mortgage** to be the most advantageous for you (see Figure 15.1).

Loan types	Advantages	Disadvantages
30-year fixed-rate mortgage	A predictable monthly payment.	If interest rates fall, you are locked in to higher rate unless you refinance. (Application and appraisal fees along with other closing costs will result.)
15-year fixed-rate mortgage	Interest rate lower than 30-year fixed (usually $\frac{1}{4}$ to $\frac{1}{2}$ of a percent). Your equity builds up faster while interest costs are cut by more than one-half.	A larger down payment is needed. Monthly payment will be higher.
Graduated-payment mortgage (GPM)	Easier to qualify for than 30- or 15-year fixed rate. Monthly payments start low and increase over time.	May have higher APR than fixed or variable rates.
Biweekly mortgage	Shortens term loan; saves substantial amount of interest; 26 biweekly payments per year. Builds equity twice as fast.	Not good for those not seeking an early loan payoff. Extra payments per year.
Adjustable rate mortgage (ARM)	Lower rate than fixed. If rates fall, could be adjusted down without refinancing. Caps available that limit how high rate could go for each adjustment period over term of loan.	Monthly payment could rise if interest rates rise. Riskier than fixed-rate mortgage in which monthly payment is stable.
Home equity loan	Cheap and reliable accessible lines of credit backed by equity in your home. Tax-deductible. Rates can be locked in. Reverse mortgages may be available to those 62 or older.	Could lose home if not paid **(foreclosure)**. No annual or interest caps.
Interest-only mortgage	Borrowers pay interest but no principal in the early years (5 to 15) of the loan.	Early years build up no equity.
Jumbo mortgage	Borrowers can borrow more money to buy a home in an expensive area	Down payment of 10–20%, FICO score of 700+, debt-to-income ratio below 45%.

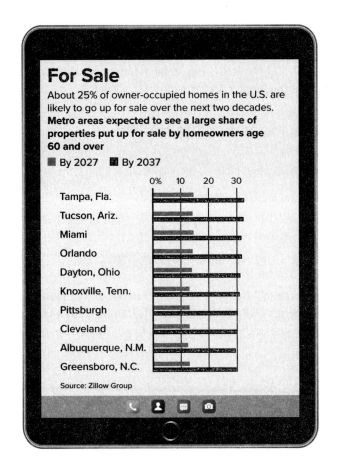

For Sale

About 25% of owner-occupied homes in the U.S. are likely to go up for sale over the next two decades. **Metro areas expected to see a large share of properties put up for sale by homeowners age 60 and over**

■ By 2027 ▨ By 2037

Tampa, Fla.
Tucson, Ariz.
Miami
Orlando
Dayton, Ohio
Knoxville, Tenn.
Pittsburgh
Cleveland
Albuquerque, N.M.
Greensboro, N.C.

Source: Zillow Group

"'For Sale' infographic."
The Wall Street Journal
(May 22, 2020).

The *Wall Street Journal* clip "For Sale" shows metro areas are expecting to see about 25% of owner-occupied homes being sold in the next 20 years.

Have you heard that elderly people who are house-rich and cash-poor can use their home to get cash or monthly income? The Federal Housing Administration makes it possible for older homeowners to take out a **reverse mortgage** on their homes. Under reverse mortgages, senior homeowners borrow against the equity in their property, often getting fixed monthly checks. The debt is repaid only when the homeowners or their estate sells the home.

If property values decrease, homeowners may have to sell their properties for less than is owed. When this happens, it is called a short sale.

Now let's learn how to calculate a monthly mortgage payment and the total cost of loan interest over the life of a mortgage. We will use the following example in our discussion:

Example: Gary bought a home for $200,000. He made a 20% down payment. The 3.5% mortgage is for 30 years (30 × 12 = 360 payments). What are Gary's monthly payment and total cost of interest?

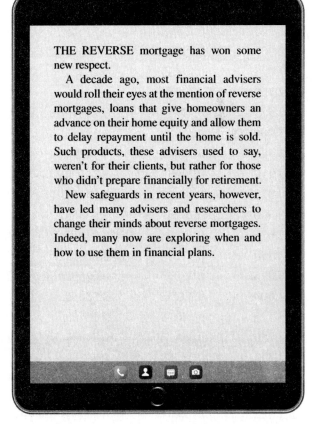

THE REVERSE mortgage has won some new respect.

A decade ago, most financial advisers would roll their eyes at the mention of reverse mortgages, loans that give homeowners an advance on their home equity and allow them to delay repayment until the home is sold. Such products, these advisers used to say, weren't for their clients, but rather for those who didn't prepare financially for retirement.

New safeguards in recent years, however, have led many advisers and researchers to change their minds about reverse mortgages. Indeed, many now are exploring when and how to use them in financial plans.

Reprinted by permission of *The Wall Street Journal*, © 2016 Dow Jones & Company, Inc. All rights reserved worldwide.

Money Tip Should you buy or rent your home? Buying actually saves you $1,743 more, on average, per year if you stay in your home for 6 years or longer. This savings results from the allowed deduction of property taxes and mortgage interest on your personal income taxes.

Learn: Computing the Monthly Payment for Principal and Interest

You can calculate the principal and interest of Gary's **monthly payment** using the **amortization table** shown in Table 15.1 and the following steps. (Remember that this is the same type of amortization table used in Chapter 14 for installment loans.)

Computing Monthly Payment by Using an Amortization Table

Step 1 Divide the amount of the mortgage by $1,000.

Step 2 Look up the rate and term in the amortization table. At the intersection is the table factor.

Step 3 Multiply Step 1 by Step 2.

For Gary, we calculate the following:

$$\frac{\$160,000 \text{ (amount of mortgage)}}{\$1,000} = 160 \times \$4.49045 \text{ (table rate)} = \boxed{\$718.47}$$

So $160,000 is the amount of the mortgage ($200,000 less 20%). The $4.49045 is the table factor of 3.5% for 30 years per $1,000. Since Gary is mortgaging 160 units of $1,000, the factor of $4.49045 is multiplied by 160. Remember that the $718.47 payment does not include taxes, insurance, and so on.

TABLE 15.1 Amortization table (mortgage principal and interest per $1,000)

Rate	Interest Only	10 Year	15 Year	20 Year	25 Year	30 Year	40 Year
2.000	**0.16667**	**9.20135**	**6.43509**	**5.05883**	**4.23854**	**3.69619**	**3.02826**
2.125	0.17708	9.25743	6.49281	5.11825	4.29966	3.75902	3.09444
2.250	0.18750	9.31374	6.55085	5.17808	4.36131	3.82246	3.16142
2.375	0.19792	9.37026	6.60921	5.23834	4.42348	3.88653	3.22921
2.500	0.20833	9.42699	6.66789	5.29903	4.48617	3.95121	3.29778
2.625	0.21875	9.48394	6.72689	5.36014	4.54938	4.01651	3.36714
2.750	0.22917	9.54110	6.78622	5.42166	4.61311	4.08241	3.43728
2.875	0.23958	9.59848	6.84586	5.48361	4.67735	4.14892	3.50818
3.000	**0.25000**	**9.65607**	**6.90582**	**5.54598**	**4.74211**	**4.21604**	**3.57984**
3.125	0.26042	9.71388	6.96609	5.60876	4.80738	4.28375	3.65226
3.250	0.27083	9.77190	7.02669	5.67196	4.87316	4.35206	3.72541
3.375	0.28125	9.83014	7.08760	5.73557	4.93945	4.42096	3.79930
3.500	0.29167	9.88859	7.14883	5.79960	5.00624	4.49045	3.87391
3.625	0.30208	9.94725	7.21037	5.86404	5.07352	4.56051	3.94923
3.750	0.31250	10.00612	7.27222	5.92888	5.14131	4.63116	4.02526
3.875	0.32292	10.06521	7.33440	5.99414	5.20959	4.70237	4.10198
4.000	**0.33333**	**10.12451**	**7.39688**	**6.05980**	**5.27837**	**4.77415**	**4.17938**
4.125	0.34375	10.18403	7.45968	6.12587	5.34763	4.84650	4.25746
4.250	0.35417	10.24375	7.52278	6.19234	5.41738	4.91940	4.33620
4.375	0.36458	10.30369	7.58620	6.25922	5.48761	4.99285	4.41559
4.500	0.37500	10.36384	7.64993	6.32649	5.55832	5.06685	4.49563
4.625	0.38542	10.42420	7.71397	6.39417	5.62951	5.14140	4.57629
4.750	0.39583	10.48477	7.77832	6.46224	5.70117	5.21647	4.65758
4.875	0.40625	10.54556	7.84297	6.53070	5.77330	5.29208	4.73947
5.000	**0.41667**	**10.60655**	**7.90794**	**6.59956**	**5.84590**	**5.36822**	**4.82197**

TABLE 15.1 (concluded)

Rate	Interest Only	10 Year	15 Year	20 Year	25 Year	30 Year	40 Year
5.125	0.42708	10.66776	7.97320	6.66881	5.91896	5.44487	4.90505
5.250	0.43750	10.72917	8.03878	6.73844	5.99248	5.52204	4.98870
5.375	0.44792	10.79079	8.10465	6.80847	6.06645	5.59971	5.07293
5.500	0.45833	10.85263	8.17083	6.87887	6.14087	5.67789	5.15770
5.625	0.46875	10.91467	8.23732	6.94966	6.21575	5.75656	5.24302
5.750	0.47917	10.97692	8.30410	7.02084	6.29106	5.83573	5.32888
5.875	0.48958	11.03938	8.37118	7.09238	6.36682	5.91538	5.41525
6.000	**0.50000**	**11.10205**	**8.43857**	**7.16431**	**6.44301**	**5.99551**	**5.50214**
6.125	0.51042	11.16493	8.50625	7.23661	6.51964	6.07611	5.58952
6.250	0.52083	11.22801	8.57423	7.30928	6.59669	6.15717	5.67740
6.375	0.53125	11.29130	8.64250	7.38232	6.67417	6.23870	5.76575
6.500	0.54167	11.35480	8.71107	7.45573	6.75207	6.32068	5.85457
6.625	0.55208	11.41850	8.77994	7.52950	6.83039	6.40311	5.94385
6.750	0.56250	11.48241	8.84909	7.60364	6.90912	6.48598	6.03357
6.875	0.57292	11.54653	8.91854	7.67814	6.98825	6.56929	6.12373
7.000	**0.58333**	**11.61085**	**8.98828**	**7.75299**	**7.06779**	**6.65302**	**6.21431**
7.125	0.59375	11.67537	9.05831	7.82820	7.14773	6.73719	6.30531
7.250	0.60417	11.74010	9.12863	7.90376	7.22807	6.82176	6.39672
7.375	0.61458	11.80504	9.19923	7.97967	7.30880	6.90675	6.48852
7.500	0.06250	11.87017	9.27101	8.05593	7.38991	6.99215	6.58071
7.625	0.63542	11.93552	9.34130	8.13254	7.47141	7.07794	6.67327
7.750	0.64583	12.00106	9.41276	8.20949	7.55329	7.16412	6.76620
7.875	0.65625	12.06681	9.48450	8.28677	7.63554	7.25069	6.85948
8.000	**0.66667**	**12.13276**	**9.55652**	**8.36440**	**7.71816**	**7.33765**	**6.95312**

Calculating Total Monthly Payment—Piti: Monthly Principal, Interest, Taxes, and Insurance

Step 1 Calculate principal and interest (see above).

Step 2 Determine 1/12 of the annual property tax.

Step 3 Determine 1/12 of annual homeowner's insurance.

Step 4 Add Step 1, Step 2, and Step 3 together to get monthly PITI.

Example: $718.47 + \dfrac{\$2,345}{12} + \dfrac{\$1,578}{12}$

$$\$718.47 + \$195.42 + \$131.50 = \$1,045.39 \text{ PITI}$$

 Aha! *Lenders typically require your PITI to not exceed 28% of your gross income. And your debt-to-income ratio should not be greater than 36%. Regardless of the loan amount you are offered, avoid being house poor. Don't buy more house than you can reasonably afford.*

Learn: What Is the Total Cost of Interest?

We can use the following formula to calculate Gary's total interest cost over the life of the mortgage:

$$\underset{\uparrow}{\text{Total cost of interest}} = \underset{\uparrow}{\text{Total of all monthly payments}} - \underset{\uparrow}{\text{Amount of mortgage}}$$

$$\underset{}{\boxed{\$98,649.20}} = \underset{(\$718.47 \times 360)}{\$258,649.20} - \$160,000$$

Learn: Effects of Interest Rates on Monthly Payment and Total Interest Cost

Table 15.2 shows the effect that an increase in interest rates would have on Gary's monthly payment and his total cost of interest. Note that if Gary's interest rate rises to 5.5%, the 2% increase will result in Gary paying an additional $85,248 in total interest.

For most people, purchasing a home is a major lifetime decision. Many factors must be considered before this decision is made. Being informed about related costs and the types of available mortgages can save you thousands of dollars.

TABLE 15.2 Effect of interest rates on monthly payments

	3.5%	5.5%	Difference
Monthly payment	$718.47	$908.46	$189.99 per month
	(160 × $4.49045)	(160 × $5.67789)	
Total cost of interest	$98,649.20	$167,045.60	$68,396.40
	($718.47 × 360) − $160,000	($908.46 × 360) − $160,000	(189.99 × 360)

In addition to the mortgage payment, buying a home can include the following costs:

- *Closing costs:* When property passes from seller to buyer, **closing costs** may include fees for credit reports, recording costs, lawyer's fees, points, title search, and so on. A **point** is a one-time charge that is a percent of the mortgage. Two points means 2% of the mortgage. The following *Wall Street Journal* clip shows how buyers typically pay between 2% and 5% in closing costs.

- *Escrow amount:* Usually, the lending institution, for its protection, requires that each month 1/12 of the insurance cost and 1/12 of the real estate taxes be kept in a special account called the **escrow account.** The monthly balance in this account will change depending on the cost of the insurance and taxes. Interest is paid on escrow accounts.

- *Repairs and maintenance:* This includes paint, wallpaper, landscaping, plumbing, electrical expenses, and so on.

- *PMI insurance:* When buying a house, if you do not have 20% in cash for a down payment, lenders will require you to purchase PMI (private mortgage insurance). This can be very expensive and only benefits the lender. It is important to know that as soon as 20% equity is reached in the home (determined by an appraisal), the borrower must petition to have the PMI removed. The mortgage lender will not be tracking this and you may continue to pay PMI after it is no longer required.

Note to house hunters on a budget: A home's sale price isn't really the sale price—there are lots of closing costs and expenses that jack up the final number.

According to online real-estate listings site Zillow, buyers typically pay between 2% and 5% of the purchase price in closing costs. So if a home costs $300,000, that buyer can expect to pay between $6,000 and $15,000.

As you can see, the cost of owning a home can be expensive. But remember that interest costs of your monthly payment and your real estate taxes are income tax deductible. And, your borrowing options increase if you have greater than 80% equity in your home. You can take out a **home equity loan** (a second mortgage for a fixed amount), a **home equity line of credit** (a revolving balance second mortgage), or a **cash-out refinance** (borrowing more than the amount owed) up to 80% loan to value of your home. For many, owning a home offers many advantages over renting.

Before you study Learning Unit 15–2, let's check your understanding of Learning Unit 15–1.

Practice Quiz

Complete this Practice Quiz to see how you are doing.

Given: Price of home, $225,000; 20% down payment; 4% interest rate; 25-year mortgage.
Solve for:

1. Monthly payment and total cost of interest over 25 years.

2. If rate fell to 3.25%, what would be the total decrease in interest cost over the life of the mortgage?

✓ **Solutions**

1. $225,000 − $45,000 = $180,000

$$\frac{\$180,000}{\$1,000} = 180 \times \$5.27837 = \boxed{\$950.11}$$

$$\boxed{\$105,033} = \underset{(\$950.11 \times 300)}{\$285,033} - \$180,000 \quad 25 \text{ years} \times 12 \text{ payments per year}$$

2. 3.25% = $877.17   monthly payment
         (180 × $4.87316)

    Total interest cost $83,151 = ($877.17 × 300) − $180,000

    Savings $\boxed{\$21,882}$ = ($285,033 − $263,151)

Learning Unit 15–2:
Amortization Schedule—Breaking Down the Monthly Payment

In Learning Unit 15–1, we saw that over the life of Gary's $160,000 loan, he would pay $98,649.20 in interest. Now let's use the following steps to determine what portion of Gary's first monthly payment reduces the principal and what portion is interest.

Calculating Interest, Principal, and New Balance of Monthly Payment

Step 1 Calculate the interest for a month (use current principal):

Interest = Principal × Rate × Time

Step 2 Calculate the amount used to reduce the principal:

Principal reduction = Monthly payment − Interest (Step 1)

Step 3 Calculate the new principal:

Current principal − Reduction of principal (Step 2) = New principal

Step 1   Interest $(I) = $ Principal $(P) \times$ Rate $(R) \times$ Time (T)

$$\$466.67 \ = \ \$160,000 \ \times \ .035 \ \times \ \frac{1}{12}$$

Step 2   The reduction of the $160,000 principal each month is equal to the payment less interest. So we can calculate Gary's new principal balance at the end of month 1 as follows:

Monthly payment at 3.5% (from Table 15.1)	$718.47 (160 × $4.49045)
− Interest for first month	− 466.67
= Principal reduction	$251.80

Step 3   As the years go by, the interest portion of the payment decreases and the principal portion increases.

Principal balance	$ 160,000
Principal reduction	− 251.80
Balance of principal	$159,748.20

Let's do month 2:

Step 1   Interest = Principal × Rate × Time

$$= \$159,748.20 \times .035 \times \frac{1}{12}$$

$$= \$465.93$$

Step 2
$718.47	monthly payment
− 465.93	interest for month 2
$252.54	principal reduction

Step 3
$159,748.20	principal balance
− 252.54	principal reduction
$159,495.66	balance of principal

Money Tip

Save money on interest by making 13 mortgage payments a year using one of the following methods:

1. Increase your monthly payment by 1/12.
2. Make one extra payment a year.
3. Pay half of your monthly payment every two weeks.

A $200,000 mortgage at 5% for 30 years will save you $32,699 on interest cost.

Note that in month 2, interest costs drop 74 cents ($466.67 − $465.93). So in 2 months, Gary has reduced his mortgage balance by $504.34 ($251.80 + $252.54). After 2 months, Gary has paid a total interest of $932.60 ($466.67 + $465.93).

Learn: Example of an Amortization Schedule

The partial **amortization schedule** given in Table 15.3 shows the breakdown of Gary's monthly payment. Note the amount that goes toward reducing the principal and toward payment of actual interest. Also note how the outstanding balance of the loan is reduced. After 10 months, Gary still owes $157,448.64. Often when you take out a mortgage loan, you receive an amortization schedule from the company that holds your mortgage.

It's time to test your knowledge of Learning Unit 15–2 with a Practice Quiz.

TABLE 15.3 Partial amortization schedule

Payment number	Principal (current)	MONTHLY PAYMENT, $718.47 Interest	Principal reduction	Balance of principal
1	$160,000	$466.67	$251.80	$159,748.20
2	$159,748.20	$465.93	$252.54	$159,495.66
3	$159,495.66	$465.20	$253.28	$159,242.38
4	$159,242.38	$464.46	$254.01	$158,988.37
5	$158,988.37	$463.72	$254.76	$158,733.61
6	$158,733.61	$462.97	$255.50	$158,478.11
7	$158,478.11	$462.23	$256.24	$158,221.87
8	$158,221.87	$461.48	$256.99	$157,964.88
9	$157,964.88	$460.73	$257.74	$157,707.14
10	$157,707.14	$459.98	$258.49	$157,448.64

Learning Unit 15-2

Practice Quiz

Complete this Practice Quiz to see how you are doing.

Prepare an amortization schedule for the first three periods for the following: mortgage, $100,000; 3.75%; 30 years.

✓ **Solutions**

Payment number	Principal (current)	PORTION TO— Interest	Principal reduction	Balance of principal
1	$100,000	$312.50	$150.62	$99,849.38
		$\left(\$100{,}000 \times .0375 \times \dfrac{1}{12}\right)$	($463.12 − $312.50)	($100,000 − $150.62)
2	$99,849.38	$312.03	$151.09	$99,698.29
		$\left(\$99{,}849.38 \times .0375 \times \dfrac{1}{12}\right)$	($463.12 − $312.03)	($99,849.38 − $151.09)
3	$99,698.29	$311.56	$151.56	$99,546.73
		$\left(\$99{,}698.29 \times .0375 \times \dfrac{1}{12}\right)$	($463.12 − $311.56)	($99,698.29 − $151.56)

Chapter 15 Review

Topic/Procedure/Formula	Examples	You try it*
Computing monthly mortgage payment Based on per $1,000 (Table 15.1): $\dfrac{\text{Amount of mortgage}}{\$1,000} \times$ Table rate	Use Table 15.1: 3.875% on $60,000 mortgage for 30 years. $\dfrac{\$60,000}{\$1,000} = 60 \times \$4.70237$ $= \boxed{\$282.14}$	**Calculate monthly payment** $70,000 mortgage at 3.5% for 30 years
Calculating total interest cost $\dfrac{\text{Total of all}}{\text{monthly payments}} - \dfrac{\text{Amount of}}{\text{mortgage}}$	Using example above: $\begin{aligned}30 \text{ years} = \quad & 360 \quad \text{(payments)}\\ \times\ & \$282.14\\ \hline & \$101,570.40\\ -\ & 60,000\\ \hline & \boxed{\$\ 41,570.40} \end{aligned}$ (mortgage interest over life of mortgage)	**Calculate total interest cost** Use the data from the problem above.
Amortization schedule $I = P \times R \times T$ $\left(I \text{ for month} = P \times R \times \dfrac{1}{12}\right)$ $\dfrac{\text{Principal}}{\text{reduction}} = \dfrac{\text{Monthly}}{\text{payment}} - \text{Interest}$ $\dfrac{\text{New}}{\text{principal}} = \dfrac{\text{Current}}{\text{principal}} - \dfrac{\text{Reduction of}}{\text{principal}}$	Using above example:	**Prepare amortization for first two payments** Use the data from the problem above.

Using above example:

Portion to—

Payment number	Interest	Principal reduction	Balance of principal
1	$193.75	$88.39	$59,911.61

$\left(\$60,000 \times .03875 \times \dfrac{1}{12}\right) \left(\begin{array}{c}\$282.14\\ -\$193.75\end{array}\right) \left(\begin{array}{c}\$60,000.00\\ -\$88.39\end{array}\right)$

2	$193.46	$88.68	$59,822.93

$\left(\$59,911.61 \times .03875 \times \dfrac{1}{12}\right) \left(\begin{array}{c}\$282.14\\ -\$193.46\end{array}\right) \left(\begin{array}{c}\$59,911.61\\ -\$88.68\end{array}\right)$

Key Terms

Adjustable rate mortgage (ARM)

Amortization schedule

Amortization table

Biweekly mortgage

Cash-out refinance

Closing costs

Escrow account

Fixed-rate mortgage

Foreclosure

Graduated-payment mortgages (GPM)

Home equity line of credit

Home equity loan

Interest-only mortgage

Monthly payment

Mortgage

Points

Reverse mortgage

Short sale

Subprime loan

* Worked-out solutions are in Appendix A.

Critical Thinking Discussion Questions with Chapter Concept Check

1. Explain the advantages and disadvantages of the following loan types: 30-year fixed rate, 15-year fixed-rate, graduated-payment mortgage, biweekly mortgage, adjustable rate mortgage, and home equity loan. Why might a bank require a home buyer to establish an escrow account?

2. How is an amortization schedule calculated? Is there a best time to refinance a mortgage?

3. What is a point? Is paying points worth the cost?

4. Explain how rising interest rates will affect the housing market.

5. Explain a short sale.

6. Explain subprime loans and how foreclosures result.

7. **Chapter Concept Check.** Locate three mortgage options for a house you would like to buy and calculate the payment and total interest for each. Which would you choose? Why? Use concepts in the chapter to support your case.

8. Explain how the pandemic affected people who rent.

End-of-Chapter Problems

Name _____ Date _____

Check figures for odd-numbered problems in Appendix A.

Drill Problems

Complete the following amortization chart by using Table 15.1. *LU 15–1(2)*

	Selling price of home	Down payment	Principal (loan)	Rate of interest	Years	Payment per $1,000	Monthly mortgage payment
e)cel **15–1.**	$160,000	$20,000		$3\frac{1}{2}\%$	30		
e)cel **15–2.**	$90,000	$5,000		$5\frac{1}{2}\%$	30		
e)cel **15–3.**	$190,000	$50,000		7%	25		

e)cel **15–4.** What is the total cost of interest in Problem 15–2? *LU 15–1(3)*

15–5. If the interest rate rises to 7% in Problem 15–2, what is the total cost of interest? *LU 15–1(3)*

Complete the following: *LU 15–2(1)*

	Selling price	Down payment	Amount mortgage	Rate	Years	Monthly payment	First Payment Broken Down Into— Interest	First Payment Broken Down Into— Principal	Balance at end of month
15–6.	$150,000	$30,000		7%	30				
15–7.	$225,000	$45,000		5%	15				

15–8. Bob Jones bought a new log cabin for $70,000 at 4% interest for 15 years. Prepare an amortization schedule for the first three periods. *LU 15–2(2)*

Payment number	Portion to— Interest	Portion to— Principal	Balance of loan outstanding

End-of-Chapter Problems (Continued)

Word Problems

15–9. CNBC.com reported mortgage applications increased 9.9% due to a decrease in the rate on 30-year fixed-rate mortgages. Joe Sisneros wants to purchase a vacation home for $235,000 with 20% down. Calculate his monthly payment for a 20-year mortgage at 3 3/8%. Calculate total interest.   *LU 15–1(3)*

15–10. If you buy a home with less than 20% down, you will pay an additional monthly fee, PMI (private mortgage insurance), until you reach 80% equity. Keep track of when you reach 80% equity so you can request to have your PMI removed. Ken Buckmiller's home recently appraised at $290,000. His mortgage was for $275,000 at 5% for 30 years with PMI of $229.17 per month. What is his monthly payment plus PMI? His mortgage balance is currently $222,990. Has he reached 80% equity?   *LU 15–1(2)*

excel 15–11. Joe Levi bought a home in Arlington, Texas, for $140,000. He put down 20% and obtained a mortgage for 30 years at $5\frac{1}{2}$%. What is Joe's monthly payment? What is the total interest cost of the loan?   *LU 15–1(2, 3)*

excel 15–12. If in Problem 15–11 the rate of interest is $6\frac{1}{8}$%, what is the difference in interest cost?   *LU 15–1(3)*

15–13. Mike Jones bought a new split-level home for $150,000 with 20% down. He decided to use Quicken Loans for his mortgage. Quicken was offering $3\frac{3}{4}$% for 25-year mortgages. Provide Mike with an amortization schedule for the first three periods.   *LU 15–2(1, 2)*

Payment number	Portion to—		Balance of loan outstanding
	Interest	Principal	

15–14. Harriet Marcus is concerned about the financing of a home. She saw a small cottage that sells for $50,000. If she puts 20% down, what will her monthly payment be at **(a)** 25 years, 3% **(b)** 25 years, 3.5%; **(c)** 25 years, 3.75%; and **(d)** 25 years, 4%? What is the total cost of interest over the cost of the loan for each assumption? **(e)** What is the savings in interest cost between 3% and 4%? **(f)** If Harriet uses 30 years instead of 25 for both 3% and 4%, what is the difference in interest?   *LU 15–1(2, 3)*

15–15. TheMortgageReports.com reported the median price of a home sold in the United States hit a record high in 2020 at just above $320,000. Juan Carlos Soto, Jr., wants to purchase a new home for $305,500. Juan puts 20% down and will finance the remainder of the purchase. Compare the following two mortgage options he has: 10 years at 3.5% or 15 years at 5%. Calculate Juan's monthly payment as well as his total cost of interest for both the 10- and 15-year mortgage. What is the difference in interest paid between the two options (round to nearest cent in calculations)?   *LU 15–1(3)*

End-of-Chapter Problems (Continued)

15–16. Daniel and Jan agreed to pay $560,000 for a four-bedroom colonial home in Waltham, Massachusetts, with a $60,000 down payment. They have a 25-year mortgage at a fixed rate of 6 3/8%. **(a)** How much is their monthly payment? **(b)** After the first payment, what would be the balance of the principal?   *LU 15–1(2), LU 15–2(1)*

15–17. Paying 13 mortgage payments instead of 12 per year can save you thousands in mortgage interest expense. If you had a $175,000 mortgage at 6% for 30 years, how much extra would you have to pay per year to make 13 instead of 12 mortgage payments per year? How much would you pay if you paid 1/12 of it per month?   *LU 15–1(1)*

15–18. Mortgage lenders base the mortgage interest rate they offer you on your credit rating. This makes it financially critical to maintain a credit score of 740 or higher. How much more interest would you pay on a $195,000 home if you put 20% down and financed the balance with a 30-year mortgage at 4 7/8% compared to a 30-year mortgage at 3 1/2%?   *LU 15–1(3)*

15–19. Should you refinance your mortgage? If you have an adjustable rate mortgage and plan on keeping your house, yes. If you can reduce your fixed interest rate by 1–2%, yes. If you have a high interest rate second mortgage and the balance is large, yes. Brett Gardner and Nellie Viner are remodeling their home. They took out a second mortgage for $120,000 at 4.5% for 10 years. Their first mortgage is $390,000 at 3% for 25 years. Should they refinance both mortgages into one 30-year mortgage at 2.875%? Why or why not?   *LU 15–1(3)*

15–20. Rebecca Johnson rents her home for $2,115 per month. She qualifies for a $475,000 mortgage for 30 years at 3.625%. Should she buy? Why or why not?   *LU 15–1(3)*

Challenge Problems

excel 15–21. Rick Rueta purchased a $90,000 home at 4 5/8% for 30 years with a down payment of $20,000. His annual real estate tax is $1,800 along with an annual insurance premium of $960. Rick's bank requires that his monthly payment include an escrow deposit for the tax and insurance. What is the total payment each month for Rick?   *LU 15–1(2)*

End-of-Chapter Problems (Continued)

15–22. Sharon Fox decided to buy a home in Marblehead, Massachusetts, for $275,000. Her bank requires a 30% down payment. Sue Willis, an attorney, has notified Sharon that besides the 30% down payment there will be the following additional costs:

Recording of the deed	$ 30.00
A credit and appraisal report	155.00
Preparation of appropriate documents	48.00

In addition, there will be a transfer tax of 1.8% of the purchase price and a loan origination fee of 2.5% of the mortgage amount.

Assume a 20-year mortgage at a rate of 3 5/8%.   *LU 15–1(2, 3)*

a. What is the initial amount of cash Sharon will need?

b. What is her monthly payment?

c. What is the total cost of interest over the life of the mortgage?

Summary Practice Test

Do you need help? Connect videos have step-by-step worked-out solutions.

excel 1. Pat Lavoie bought a home for $180,000 with a down payment of $10,000. Her rate of interest is 6% for 30 years. Calculate her **(a)** monthly payment; **(b)** first payment, broken down into interest and principal; and **(c)** balance of mortgage at the end of the month. *LU 15–1(2, 3)*

2. Jen Logan bought a home in Iowa for $110,000. She put down 20% and obtained a mortgage for 30 years at $5\frac{1}{2}$%. What are Jen's monthly payment and total interest cost of the loan? *LU 15–1(2, 3)*

3. Christina Sanders is concerned about the financing of a home. She saw a small Cape Cod–style house that sells for $90,000. If she puts 10% down, what will her monthly payment be at **(a)** 30 years, 5%; **(b)** 30 years, $5\frac{1}{2}$% **(c)** 30 years, 6%; and **(d)** 15 years, $3\frac{7}{8}$%? What is the total cost of interest over the cost of the loan for each assumption? *LU 15–1(2, 3)*

4. Loretta Scholten bought a home for $210,000 with a down payment of $30,000. Her rate of interest is $5\frac{5}{8}$% for 15 years. Calculate Loretta's payment per $1,000 and her monthly mortgage payment. *LU 15–1(2)*

5. Using Problem 4, calculate the total cost of interest for Loretta Scholten. *LU 15–1(3)*

🔍 Home Sweet Home!

 What I need to know

Buying your first home can be a very intimidating process. To lessen the anxiety it is best to learn all you can about buying your first home. This knowledge will give you the confidence you need to proceed positively into the process and understand your responsibilities as a new home buyer. A solid budget is an important component of making a home purchase within the limits of your income. Taking the time to understand how your home purchase will impact your budget equips you to find the home perfectly fit to your budget and lifestyle.

 What I need to do

Research the process of home ownership to understand how the process works as well as your responsibilities in this purchase. Knowledge truly is power and learning about how a home purchase works will help you confront this challenge with confidence. Work on increasing your credit score to 740+ and reducing your debt-to-income ratio to 36% or less. This means all your monthly debt payments, including the cost of your mortgage, must be 36% or less of your gross income. Seek out information from Internet sources on first-time homebuyer programs to assist you with understanding the down payment and other closing costs. Obtain information from others who have already been through this process. Family and friends may give you some valuable first-hand knowledge about the requirements involved in a home purchase.

Understand your mortgage loan. Go through the process of prequalification so you will know the price of home you can afford. Lenders will require documentation such as annual income, credit rating, employment verification, and value of assets to get you prequalified. Know the interest rate for your loan and do not hesitate to consider multiple lenders to get the best rate. Your interest rate will significantly impact the amount of home you can purchase and the total interest you will pay over the course of the loan. Also, be aware of the potential requirement to carry private mortgage insurance (PMI) on your loan. PMI is normally required on loans in which you are making a down payment of less than 20% of the home price. If your loan requires PMI at its origination, monitor your loan over time to identify when you have achieved 20% equity (reducing loan balance to 80% of appraised value) so you can eliminate the PMI expense from your loan. Loan details such as interest rates and PMI directly impact your monthly payment, so you want to be sure you fully understand how to use loan details to your advantage.

 Steps I need to take

1. Learn about the process of home buying to lessen your anxiety and increase your confidence.
2. Get prequalified for your loan to understand your buying power.
3. Understand all the details of your loan and their financial impact.

 Resources I can use

- Zillow (mobile app)—tools for buying, selling, or renting a home.
- https://www.nerdwallet.com/article/mortgages/tips-for-first-time-home-buyers—first-time home buyer tips.

MY MONEY ACTIVITY

- Work on increasing your credit rating to 740+. Discuss ways to accomplish this.
- Create a plan to reduce your debt-to-income ratio to 35% or less.
- Seek out first-time homebuyer's programs in your area and review the requirements.

PERSONAL FINANCE

A KIPLINGER APPROACH

"$10,000 CONTRIBUTE TO A MORTGAGE DOWN PAYMENT." Kiplinger's. November, 2020.

REWARDS

$10,000

CONTRIBUTE TO A MORTGAGE DOWN PAYMENT

With mortgage rates hovering at historic lows, now is an opportune time to buy a home for those who can swing the financing. If your child or grandchild has a strong enough income to qualify for a mortgage but is low on savings, consider kicking in all or a portion of the down payment. With a conventional mortgage backed by Fannie Mae or Freddie Mac, qualifying borrowers may put down as little as 3% of the purchase price. With $10,000, you could fund the full down payment for a home selling for as much as about $333,300. For those with a credit score of at least 580, Federal Housing Administration loans allow a down payment as low as 3.5% of the purchase price; $10,000 would cover the full down payment with a sale price of up to about $285,700.

Keep in mind that you'll usually need a down payment of at least 20% of the purchase price to avoid private mortgage insurance on a conventional loan. If the borrower already has some savings, a $10,000 contribution from you could push the down payment to the 20% threshold. PMI may cost about 0.5% to 2% of the loan amount per year. After the borrower pays off enough of the loan to gain 20% equity in the home, PMI is no longer required.

You'll have to write the mortgage lender a letter stating that the money is a gift and that you expect no repayment from the borrower. The lender will also want to confirm the gift's source, so you may have to provide a couple of recent statements for the bank account from which you withdraw the funds and document the source of any deposits during that period.

Business Math Issue
Private Mortgage Insurance is required when buying a house.

1. List the key points of the article and information to support your position.

2. Write a group defense of your position using math calculations to support your view. If you are in an online course, post to a discussion board.

Chapter 16
How to Read, Analyze, and Interpret Financial Reports

Learning Unit Objectives

LU 16–1: Balance Sheet—Report as of a Particular Date
1. Explain the purpose and the key items on the balance sheet.
2. Explain and complete vertical and horizontal analysis.

LU 16–2: Income Statement—Report for a Specific Period of Time
1. Explain the purpose and the key items on the income statement.
2. Explain and complete vertical and horizontal analysis.

LU 16–3: Trend and Ratio Analysis
1. Explain and complete a trend analysis.
2. List, explain, and calculate key financial ratios.

Kraft Heinz To Beef Up Accounting Controls

By HEATHER HADDON

Kraft Heinz Co. said it had concluded an internal investigation into accounting errors and changed financial practices that triggered a regulatory probe and steep decline in the food maker's stock price.

The company on Friday filed an overdue annual report to the Securities and Exchange Commission that it had withheld during an investigation into what it said were years of accounting errors. The misstatements, which understated the costs of goods sold by $208 million, involved the way Kraft Heinz booked rebates and costs tied to its contracts with suppliers.

"The company is taking actions to improve internal policies and procedures and to strengthen internal control over financial reporting," Kraft Heinz said in a statement.

Haddon, Heather. "Kraft Heinz to Beef Up Accounting Controls." *The Wall Street Journal*, June 8, 2019.

ⓔ Essential Question

How can I use my knowledge of financial reports to understand the business world and make my life easier?

🌐 Math Around the World

During the pandemic the government provided bailouts to large and small businesses. Since airline passenger travel was down by 95%, this industry received a huge bailout to keep them operational.

This chapter explains how to analyze two key financial reports: the *balance sheet* (shows a company's financial condition at a particular date) and the *income statement* (shows a company's profitability over a time period).[1] Business owners must understand their financial statements to avoid financial difficulties. This includes knowing how to read, analyze, and interpret financial reports.

The *Wall Street Journal* chapter opener clip, "Kraft Heinz To Beef Up Accounting Controls", highlights Heinz Co.'s report of accounting errors showing that some financial statements may not be as accurate as they are presented to be.

[1] The third key financial report is the statement of cash flows. We do not discuss this statement. For more information on the statement of cash flows, check your accounting text.

Learning Unit 16–1:
Balance Sheet—Report as of a Particular Date

The **balance sheet** gives a financial picture of what a company is worth as of a particular date, usually at the end of a month or year. This report lists (1) how much the company owns (assets), (2) how much the company owes (liabilities), and (3) how much the owner is worth (**owner's equity).**

Note that assets and liabilities are divided into two groups: current (*short term,* usually less than 1 year); and *long term,* usually more than 1 year. The basic formula for a balance sheet is as follows:

$$\text{Assets} - \text{Liabilities} = \text{Owner's equity}$$

Like all formulas, the items on both sides of the equals sign must balance.

By reversing the above formula, we have the following common balance sheet layout:

$$\boxed{\text{Assets} = \text{Liabilities} + \text{Owner's equity}}$$

To introduce you to the balance sheet, let's assume that you collect baseball cards and decide to open a baseball card shop. As the owner of The Card Shop, your investment, or owner's equity, is called **capital.** Since your business is small, your balance sheet is short. After the first year of operation, The Card Shop balance sheet is shown as follows:

> "Capital" does not mean "cash." It is the owner's investment in the company.

THE CARD SHOP
Balance Sheet
December 31, 2023

Report as of a particular date

Assets		Liabilities	
Cash	$ 3,000	Accounts payable	$ 2,500
Merchandise inventory (baseball cards)	4,000	**Owner's Equity**	
Equipment	3,000	E. Slott, capital	7,500
Total assets	$10,000	Total liabilities and owner's equity	$10,000

The heading gives the name of the company, title of the report, and date of the report. Note how the totals of both sides of the balance sheet are the same. This is true of all balance sheets.

We can take figures from the balance sheet of The Card Shop and use our first formula to determine how much the business is worth:

$$\boxed{\text{Assets} - \text{Liabilities} = \text{Owner's equity (capital)}}$$

$$\$10,000 - \$2,500 \quad = \quad \$7,500$$

Since you are the single owner of The Card Shop, your business is a **sole proprietorship.** If a business has two or more owners, it is a **partnership.** A **corporation** has many owners or stockholders,

Aha! and the equity of these owners is called **stockholders' equity.** *Anytime you create a balance sheet, check your accuracy by adding liabilities to owner's equity. If that total equals assets, you've got it right!*

Learn: Elements of the Balance Sheet

The format and contents of all corporation balance sheets are similar. Figure 16.1 shows the balance sheet of Mool Company. As you can see, the formula Assets = Liabilities + Stockholders' equity (we have a corporation in this example) is also the framework of this balance sheet.

To help you understand the three main balance sheet groups (assets, liabilities, and stockholders' equity) and their elements, we have labeled them in Figure 16.1. An explanation of these groups and their elements follows this paragraph. Do not try to memorize the elements. Just try to understand their meaning. Think of Figure 16.1 as a reference aid. You will find that the more you work with balance sheets, the easier it is for you to understand them.

FIGURE 16.1 Balance sheet

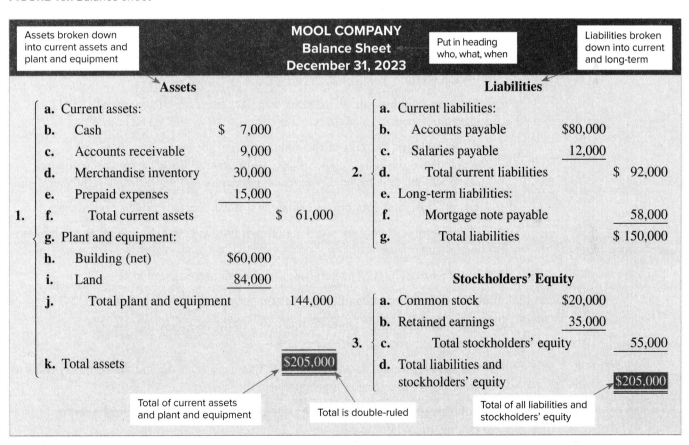

1. **Assets:** Things of value *owned* by a company (economic resources of the company) that can be measured and expressed in monetary terms.

   a. **Current assets:** Assets that companies consume or convert to cash *within 1 year* or a normal operating cycle.

   b. **Cash:** Total cash in checking accounts, savings accounts, and on hand.

   c. **Accounts receivable:** Money *owed* to a company by customers from sales on account (buy now, pay later).

Pepper ...
And Salt
THE WALL STREET JOURNAL
QUARTERLY REPORT WITH BACKUP SINGERS

d. **Merchandise inventory:** Cost of goods in stock for resale to customers.

e. **Prepaid expenses:** The purchases of a company are assets until they expire (insurance or rent) or are consumed (supplies).

f. **Total current assets:** Total of all assets that the company will consume or convert to cash within 1 year.

g. **Plant and equipment:** Assets that will last longer than 1 year. These assets are used in the operation of the company.

h. **Building (net):** The cost of the building minus the depreciation that has accumulated. Usually, balance sheets show this as "Building less accumulated depreciation." In Chapter 17 we discuss accumulated depreciation in greater detail.

i. **Land:** This asset does not depreciate, but it can increase or decrease in value.

j. **Total plant and equipment:** Total of building and land, including machinery and equipment.

k. **Total assets:** Total of current assets and plant and equipment.

2. **Liabilities:** Debts or obligations of the company.

a. **Current liabilities:** Debts or obligations of the company that are *due within 1 year.*

b. **Accounts payable:** A current liability that shows the amount the company owes to creditors for services or items purchased.

c. **Salaries payable:** Obligations that the company must pay within 1 year for salaries earned but unpaid.

d. **Total current liabilities:** Total obligations that the company must pay within 1 year.

e. **Long-term liabilities:** Debts or obligations that the company does not have to pay within 1 year.

f. **Mortgage note payable:** Debt owed on a building that is a long-term liability; often the building is the collateral.

g. **Total liabilities:** Total of current and long-term liabilities.

Copyright © McGraw Hill Cartoon Collections; (tablet)Can Yesil/Shutterstock, PureSolution/Shutterstock

3. **Stockholders' equity (owner's equity):** The rights or interest of the stockholders to assets of a corporation. If the company is not a corporation, the term *owner's equity* is used. The word *capital* follows the owner's name under the title *Owner's Equity.*

    a. **Common stock:** Amount of the initial and additional investment of corporation owners by the purchase of stock.

    b. **Retained earnings:** The amount of corporation earnings that the company retains, not necessarily in cash form.

    c. **Total stockholders' equity:** Total of stock plus retained earnings.

    d. **Total liabilities and stockholders' equity:** Total current liabilities, long-term liabilities, stock, and retained earnings. This total represents all the claims on assets—prior and present claims of creditors, owners' residual claims, and any other claims.

Aha! *Create a personal balance sheet by listing all personal assets and debts. Subtract debts from assets to calculate net worth. Do an Internet search to compare your net worth to what financial analysts recommend for your age group.*

Now that you are familiar with the common balance sheet items, you are ready to analyze a balance sheet.

Learn: Vertical and Horizontal Analyses and the Balance Sheet

Often financial statement readers want to analyze reports that contain data for two or more successive accounting periods. To make this possible, companies present a statement showing the data from these periods side by side. As you might expect, this statement is called a **comparative statement.**

Comparative reports help illustrate changes in data. Financial statement readers should compare the percents in the reports to industry percents and the percents of competitors.

Figure 16.2 shows the comparative balance sheet of Roger Company. Note that the statement analyzes each asset as a percent of total assets for a single period. The statement then analyzes each liability and equity as a percent of total liabilities and stockholders' equity. We call this type of analysis **vertical analysis.**

The following steps use the portion formula to prepare a vertical analysis of a balance sheet.

Preparing a Vertical Analysis of a Balance Sheet

Step 1 Divide each asset (the portion) as a percent of total assets (the base). Round as indicated.

Step 2 Round each liability and stockholders' equity (the portions) as a percent of total liabilities and stockholders' equity (the base). Round as indicated.

We can also analyze balance sheets for two or more periods by using **horizontal analysis.** Horizontal analysis compares each item in 1 year by amount, percent, or both with the same item of the previous year. Note the Abby Ellen Company horizontal analysis shown in Figure 16.3. To make a horizontal analysis, we use the portion formula and the steps that follow:

FIGURE 16.2 Comparative balance sheet: Vertical analysis

We divide each item by the total of assets.

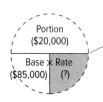

ROGER COMPANY				
Comparative Balance Sheet				
December 31, 2022 and 2023				
	2023		**2022**	
	Amount	**Percent**	**Amount**	**Percent**
Assets				
Current assets:				
Cash	$22,000	25.88	$18,000	22.22
Accounts receivable	8,000	9.41	9,000	11.11
Merchandise inventory	9,000	10.59	7,000	8.64
Prepaid rent	4,000	4.71	5,000	6.17
Total current assets	$43,000	50.59	$39,000	48.15*
Plant and equipment:				
Building (net)	$18,000	21.18	$18,000	22.22
Land	24,000	28.24	24,000	29.63
Total plant and equipment	$42,000	49.41*	$42,000	51.85
Total assets	$85,000	100.00	$81,000	100.00
Liabilities				
Current liabilities:				
Accounts payable	$14,000	16.47	$ 8,000	9.88
Salaries payable	18,000	21.18	17,000	20.99
Total current liabilities	$32,000	37.65	$25,000	30.86*
Long-term liabilities:				
Mortgage note payable	12,000	14.12	20,000	24.69
Total liabilities	$44,000	51.76*	$45,000	55.56*
Stockholders' Equity				
Common stock	$20,000	23.53	$20,000	24.69
Retained earnings	21,000	24.71	16,000	19.75
Total stockholders' equity	$41,000	48.24	$36,000	44.44
Total liabilities and stockholders' equity	$85,000	100.00	$81,000	100.00

We divide each item by the total of liabilities and stockholders' equity.

Note: All percents are rounded to the nearest hundredth percent.
* Due to rounding.

Money Tip

Are you financially "on track"? Here is a simple calculation to help estimate what your net worth should be:

(Age × Pretax income) ÷ 10

Preparing a Horizontal Analysis of a Comparative Balance Sheet

Step 1 Calculate the increase or decrease (portion) in each item from the base year.

Step 2 Divide the increase or decrease in Step 1 by the old or base year.

Step 3 Round as indicated.

You can see the difference between vertical analysis and horizontal analysis by looking at the example of vertical analysis in Figure 16.2. The percent calculations in Figure 16.2 are for each item of a particular year as a percent of that year's total assets or total liabilities and stockholders' equity.

Horizontal analysis needs comparative columns because we take the difference *between* periods. In Figure 16.3, for example, the accounts receivable decreased $1,000 from 2022 to 2023. Thus, by dividing $1,000 (amount of change) by $6,000 (base year), we see that Abby's receivables decreased 16.67%.

Copyright © McGraw Hill

FIGURE 16.3 Comparative balance sheet: Horizontal analysis

Difference between 2022 and 2023

Portion
−($1,000)

Base × Rate
($6,000) (?)

2022

ABBY ELLEN COMPANY
Comparative Balance Sheet
December 31, 2022 and 2023

	2023	2022	Increase (decrease) Amount	Increase (decrease) Percent
Assets				
Current assets:				
Cash	$ 6,000	$ 4,000	$2,000	50.00*
Accounts receivable	5,000	6,000	(1,000)	−16.67
Merchandise inventory	9,000	4,000	5,000	125.00
Prepaid rent	5,000	7,000	(2,000)	−28.57
Total current assets	$25,000	$21,000	$4,000	19.05
Plant and equipment:				
Building (net)	$12,000	$12,000	–0–	–0–
Land	18,000	18,000	–0–	–0–
Total plant and equipment	$30,000	$30,000	–0–	–0–
Total assets	$55,000	$51,000	$4,000	7.84
Liabilities				
Current liabilities:				
Accounts payable	$ 3,200	$ 1,800	$1,400	77.78
Salaries payable	2,900	3,200	(300)	− 9.38
Total current liabilities	$ 6,100	$ 5,000	$1,100	22.00
Long-term liabilities:				
Mortgage note payable	17,000	15,000	2,000	13.33
Total liabilities	$23,100	$20,000	$3,100	15.50
Owner's Equity				
Abby Ellen, capital	$31,900	$31,000	$ 900	2.90
Total liabilities and owner's equity	$55,000	$51,000	$4,000	7.84

* The percents are not summed vertically in horizontal analysis.

Let's now try the following Practice Quiz.

Learning Unit 16–1

Practice Quiz
Complete this Practice Quiz to see how you are doing.

1. Complete this partial comparative balance sheet by vertical analysis. Round percents to the nearest hundredth.

	2023		2022	
	Amount	**Percent**	**Amount**	**Percent**
Assets				
Current assets:				
a. Cash	$42,000		$40,000	
b. Accounts receivable	18,000		17,000	
c. Merchandise inventory	15,000		12,000	
d. Prepaid expenses	17,000		14,000	
•	•		•	
•	•		•	
•	•		•	
Total current assets	$160,000		$150,000	

2. What is the amount of change in merchandise inventory and the percent increase?

✓ Solutions

	2023	**2022**

1. a. Cash $\dfrac{\$42,000}{\$160,000} = $ 26.25% $\dfrac{\$40,000}{\$150,000} = $ 26.67%

b. Accounts receivable $\dfrac{\$18,000}{\$160,000} = $ 11.25% $\dfrac{\$17,000}{\$150,000} = $ 11.33%

c. Merchandise inventory $\dfrac{\$15,000}{\$160,000} = $ 9.38% $\dfrac{\$12,000}{\$150,000} = $ 8.00%

d. Prepaid expenses $\dfrac{\$17,000}{\$160,000} = $ 10.63% $\dfrac{\$14,000}{\$150,000} = $ 9.33%

2.

$$\begin{array}{r} \$15,000 \\ -\ 12,000 \\ \hline \end{array}$$

Amount = $3,000

$\text{Percent} = \dfrac{\$3,000}{\$12,000} = $ 25%

Income Statement—Report for a Specific Period of Time

One of the most important departments in a company is its accounting department. The job of the accounting department is to determine the financial results of the company's operations. Is the company making money or losing money?

In this learning unit we look at the **income statement**—a financial report that tells how well a company is performing (its profitability or net profit) during a specific period of time (month, year, etc.). In general, the income statement reveals the inward flow of revenues (sales) against the outward or potential outward flow of costs and expenses.

The form of income statements varies depending on the company's type of business. However, the basic formula of the income statement is the same:

$$\text{Revenues} - \text{Operating expenses} = \text{Net income}$$

In a merchandising business like The Card Shop, we can expand on this formula:

After any returns, allowances, or discounts

Revenues (sales)
– Cost of merchandise or goods ← Baseball cards
= Gross profit from sales
– Operating expenses
= Net income (profit)

THE CARD SHOP Income Statement For Month Ended December 31, 2023	
Revenues (sales)	$8,000
Cost of merchandise (goods) sold	3,000
Gross profit from sales	$5,000
Operating expenses	750
Net income	$4,250

Now let's look at The Card Shop's income statement to see how much profit The Card Shop made during its first year of operation. For simplicity, we assume The Card Shop sold all the cards it bought during the year. For its first year of business, The Card Shop made a profit of $4,250.

We can now go more deeply into the income statement elements as we study the income statement of a corporation.

Learn: Elements of the Corporation Income Statement

Figure 16.4 gives the format and content of the Mool Company income statement—a corporation. The five main items of an income statement are revenues, cost of merchandise (goods) sold, gross profit on sales, operating expenses, and net income. We will follow the same pattern we used in explaining the balance sheet and define the main items and the letter-coded subitems.

FIGURE 16.4 Income statement

MOOL COMPANY
Income Statement
For Month Ended December 31, 2023

Report for a specific period

Put in heading who, what, when

		Revenues:			
1.	a.	Gross sales		$22,080	
	b.	Less: Sales returns and allowances	$ 1,082		
	c.	Sales discounts	432	1,514	
	d.	Net sales			$20,566
		Cost of merchandise (goods) sold:			
2.	a.	Merchandise inventory, December 1, 2023		$ 1,248	
	b.	Purchases	$10,512		
	c.	Less: Purchase returns and allowances	$336		
	d.	Less: Purchase discounts	204	540	
	e.	Cost of net purchases		9,972	
	f.	Cost of merchandise (goods available for sale)		$11,220	
	g.	Less: Merchandise inventory, December 31, 2023		1,600	
	h.	Cost of merchandise (goods) sold			9,620
3. {		Gross profit from sales			$10,946
		Operating expenses:			
4.	a.	Salary	$ 2,200		
	b.	Insurance	1,300		
	c.	Utilities	400		
	d.	Plumbing	120		
	e.	Rent	410		
	f.	Depreciation	200		
	g.	Total operating expenses			4,630
5. {		Net income			$ 6,316

Actual sales after discounts and returns

Inventory not yet sold

Net sales – Cost of merchandise (goods) sold

Gross profit – Operating expenses

Note: Numbers are subtotaled from left to right.

1. **Revenues:** Total earned sales (cash or credit) less any sales returns and allowances or sales discounts.

   a. **Gross sales:** Total earned sales before sales returns and allowances or sales discounts.

   b. **Sales returns and allowances:** Reductions in price or reductions in revenue due to goods returned because of product defects, errors, and so on. When the buyer keeps the damaged goods, an allowance results.

   c. **Sales (not trade) discounts:** Reductions in the selling price of goods due to early customer payment. For example, a store may give a 2% discount to a customer who pays a bill within 10 days.

   d. **Net sales:** Gross sales less sales returns and allowances less sales discounts.

2. **Cost of merchandise (goods) sold:** All the costs of getting the merchandise that the company sold. The cost of all unsold merchandise (goods) will be subtracted from this item (ending inventory). See the chapter opener *WSJ* clip about Heinz Co.'s misstatement of cost of goods sold.

   a. **Merchandise inventory, December 1, 2023:** Cost of inventory in the store that was for sale to customers at the beginning of the month.

   b. **Purchases:** Cost of additional merchandise brought into the store for resale to customers.

   c. **Purchase returns and allowances:** Cost of merchandise returned to the store due to damage, defects, errors, and so on. Damaged goods kept by the buyer result in a cost reduction called an *allowance.*

   d. **Purchase discounts:** Savings received by the buyer for paying for merchandise before a certain date. These discounts can result in a substantial savings to a company.

   e. **Cost of net purchases:** Cost of purchases less purchase returns and allowances less purchase discounts.

   f. **Cost of merchandise (goods available for sale):** Sum of beginning inventory plus cost of net purchases.

   g. **Merchandise inventory, December 31, 2023:** Cost of inventory remaining in the store to be sold.

   h. **Cost of merchandise (goods) sold:** Beginning inventory plus net purchases less ending inventory.

3. **Gross profit (gross margin) from sales:** Net sales less cost of merchandise (goods) sold.

4. **Operating expenses:** Additional costs of operating the business beyond the actual cost of inventory sold.

   a.–f. **Expenses:** Individual expenses broken down.

   g. **Total operating expenses:** Total of all the individual expenses.

5. **Net income:** Gross profit less operating expenses.

Aha!

Create a personal income statement: Salaries − Expenses = Savings/Investment.

The *Wall Street Journal* clip "Companies Fix Errors in Accounting Quietly" shows that accounting mistakes may not always result in companies' reissuing their financial statements.

In the next section you will learn some formulas that companies use to calculate various items on the income statement.

Learn: Calculating Net Sales, Cost of Merchandise (Goods) Sold, Gross Profit, and Net Income of an Income Statement

It is time to look closely at Figure 16.4 and see how each section is built. Use the previous vocabulary as a reference. We will study Figure 16.4 step by step.

Companies Fix Errors In Accounting Quietly

BY JEAN EAGLESHAM

Papa John's International Inc. this spring discovered an accounting error that had caused years of misstated financial numbers. The pizza chain decided the mistake wasn't serious enough to require it to reissue its financial statements.

Eaglesham, Jean. "Companies Fix Errors in Accounting Quietly." *The Wall Street Journal,* December 6, 2019.

Step 1 Calculate the net sales—what Mool earned:

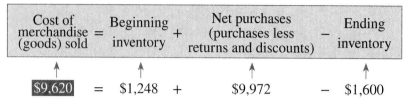

$$\text{Net sales} = \text{Gross sales} - \frac{\text{Sales returns}}{\text{and allowances}} - \text{Sales discounts}$$

$$\$20{,}566 = \$22{,}080 - \$1{,}082 - \$432$$

Money Tip Keep a record of how much you owe and to whom you owe it. Create a plan to pay each debt off, one by one. Pay off the highest interest rate balance first.

Step 2 Calculate the cost of merchandise (goods) sold:

$$\frac{\text{Cost of}}{\text{merchandise}} = \frac{\text{Beginning}}{\text{inventory}} + \frac{\text{Net purchases}}{\text{(purchases less}} - \frac{\text{Ending}}{\text{inventory}}$$
$$\text{(goods) sold} \qquad \text{returns and discounts)}$$

$$\$9{,}620 = \$1{,}248 + \$9{,}972 - \$1{,}600$$

Step 3 Calculate the gross profit (gross margin) from sales—profit before operating expenses:

$$\frac{\text{Gross profit}}{\text{from sales}} = \text{Net sales} - \frac{\text{Cost of merchandise}}{\text{(goods) sold}}$$

$$\$10{,}946 = \$20{,}566 - \$9{,}620$$

Step 4 Calculate the net income—profit after operating expenses:

$$\text{Net income} = \text{Gross profit} - \text{Operating expenses}$$

$$\$6{,}316 = \$10{,}946 - \$4{,}630$$

Learn: Analyzing Comparative Income Statements

We can apply the same procedures of vertical and horizontal analysis to the income statement that we used in analyzing the balance sheet. Let's first look at the vertical analysis for Royal Company, Figure 16.5. Then we will look at the horizontal analysis of Flint Company's 2022 and 2023 income statements shown in Figure 16.6. Note in the margin how numbers are calculated.

FIGURE 16.5 Vertical analysis

ROYAL COMPANY Comparative Income Statement For Years Ended December 31, 2022 and 2023				
	2023	**Percent of net**	**2022**	**Percent of net**
Net sales	$45,000	100.00	$29,000	100.00*
Cost of merchandise sold	19,000	42.22	12,000	41.38
Gross profit from sales	$26,000	57.78	$17,000	58.62
Operating expenses:				
Depreciation	$ 1,000	2.22	$ 500	1.72
Selling and advertising	4,200	9.33	1,600	5.52
Research	2,900	6.44	2,000	6.90
Miscellaneous	500	1.11	200	.69
Total operating expenses	$ 8,600	19.11†	$ 4,300	14.83

(*Continues*)

ROYAL COMPANY Comparative Income Statement For Years Ended December 31, 2022 and 2023				
	2023	Percent of net	2022	Percent of net
Income before interest and taxes	$17,400	38.67	$12,700	43.79
Interest expense	6,000	13.33	3,000	10.34
Income before taxes	$11,400	25.33†	$ 9,700	33.45
Provision for taxes	5,500	12.22	3,000	10.34
Net income	$ 5,900	13.11	$ 6,700	23.10†

* Net sales = 100%

† Off due to rounding.

 Aha! *When conducting a vertical analysis of an income statement, divide each line item by net sales.*

FLINT COMPANY Comparative Income Statement For Years Ended December 31, 2022 and 2023			INCREASE (DECREASE)	
	2023	2022	Amount	Percent
Sales	$90,000	$80,000	$10,000	
Sales returns and allowances	2,000	2,000	–0–	
Net sales	$88,000	$78,000	$10,000	+ 12.82
Cost of merchandise (goods) sold	45,000	40,000	5,000	+ 12.50
Gross profit from sales	$43,000	$38,000	$ 5,000	+ 13.16
Operating expenses:				
Depreciation	$ 6,000	$ 5,000	$ 1,000	+ 20.00
Selling and administrative	16,000	12,000	4,000	+ 33.33
Research	600	1,000	(400)	– 40.00
Miscellaneous	1,200	500	700	+140.00
Total operating expenses	$23,800	$18,500	$ 5,300	+ 28.65
Income before interest and taxes	$19,200	$19,500	$ (300)	– 1.54
Interest expense	4,000	4,000	–0–	
Income before taxes	$15,200	$15,500	$ (300)	– 1.94
Provision for taxes	3,800	4,000	(200)	– 5.00
Net income	$11,400	$11,500	$ (100)	– .87

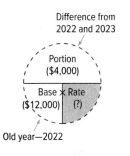

FIGURE 16.6
Horizontal analysis

The following Practice Quiz will test your understanding of this unit.

Learning Unit 16–2

Practice Quiz

Complete this Practice Quiz to see how you are doing.

From the following information, calculate:

a. Net sales. c. Gross profit from sales.

b. Cost of merchandise (goods) sold. d. Net income.

Given: Gross sales, $35,000; sales returns and allowances, $3,000; beginning inventory, $6,000; net purchases, $7,000; ending inventory, $5,500; operating expenses, $7,900.

✓ Solutions

a. $35,000 − $3,000 = $32,000 (Gross sales − Sales returns and allowances)

b. $6,000 + $7,000 − $5,500 = $7,500 (Beginning inventory + Net purchases − Ending inventory)

c. $32,000 − $7,500 = $24,500 (Net sales − Cost of merchandise sold)

d. $24,500 − $7,900 = $16,600 (Gross profit from sales − Operating expenses)

Trend and Ratio Analysis

Aha! *A balance sheet is like a snapshot. It reflects a company's or individual's financial position at a specific point in time. An income statement is like a video. It reflects a company's or individual's profitability over an interval of time.*

Now that you understand the purpose of balance sheets and income statements, you are ready to study how experts look for various trends as they analyze the financial reports of companies. This learning unit discusses trend analysis and ratio analysis. The study of these trends is valuable to businesses, financial institutions, and consumers.

Learn: Trend Analysis

Many tools are available to analyze financial reports. When data cover several years, we can analyze changes that occur by expressing each number as a percent of the base year. The base year is a past period of time that we use to compare sales, profits, and so on with other years. We call this **trend analysis.**

Using the data below, we complete a trend analysis with the following steps:

Completing a Trend Analysis

Step 1 Select the base year (100%).

Step 2 Express each amount as a percent of the base year amount (rounded to the nearest whole percent).

Money Tip Check out Financial Peace University at www.daveramsey.com for some excellent personal finance training and tips.

	Given (base year 2021)			
	2024	2023	2022	**2021**
Sales	$621,000	$460,000	$340,000	$420,000
Gross profit	182,000	141,000	112,000	124,000
Net income	48,000	41,000	22,000	38,000

	Trend analysis			
	2024	2023	2022	**2021**
Sales	148%	110%	81%	100%
Gross profit	147	114	90	100
Net income	126	108	58	100

How to Calculate Trend Analysis

$$\frac{\text{Each item}}{\text{Base amount}} = \frac{\$340,000}{\$420,000} = 80.95\% = 81\%$$

Sales for 2022 → $340,000
Sales for 2021 → $420,000

Portion ($340,000)

Base ($420,000) × Rate (?)

2021

What Trend Analysis Means Sales of 2022 were 81% of the sales of 2021. Note that you would follow the same process no matter which of the three areas you were analyzing. All categories are compared to the base year—sales, gross profit, or net income.

We now will examine **ratio analysis**—another tool companies use to analyze performance.

Learn: Ratio Analysis

A *ratio* is the relationship of one number to another. Many companies compare their ratios with those of previous years and with ratios of other companies in the industry. (See the *Wall Street Journal* clip, "Cash Burn", below.) Companies can get ratios of the performance of other companies from their bankers, accountants, local small business centers, libraries, and newspaper articles. It is important to choose companies from similar industries when comparing ratios. For example, ratios at McDonald's will be different from ratios at Best Buy. McDonald's sells more perishable products.

Percentage ratios are used by companies to determine the following:

1. How well the company manages its assets—*asset management ratios.*

2. The company's debt situation—*debt management ratios.*

3. The company's profitability picture—*profitability ratios.*

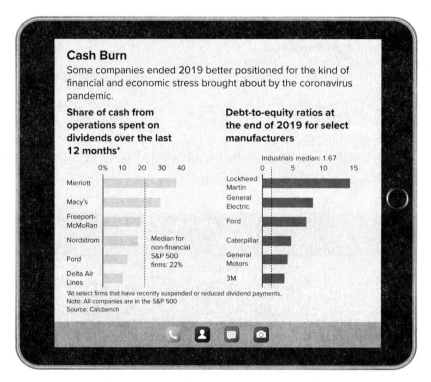

""Cash Burn" Infographic." *The Wall Street Journal,* March 26, 2020.

Each company must decide the true meaning of what the three types of ratios (asset management, debt management, and profitability) are saying. Table 16.1 gives a summary of the key ratios, their calculations (rounded to the nearest hundredth), and what they mean. All calculations are from Figures 16.1 and 16.4.

Now you can check your knowledge with the Practice Quiz that follows.

TABLE 16.1 Summary of key ratios: A reference guide*

Ratio	Formula	Actual calculations	What it says	Questions that could be raised
Current ratio[†]	Current assets / Current liabilities (Current assets include cash, accounts receivable, and marketable securities.)	$\dfrac{\$61,000}{\$92,000} = .66:1$ Industry average, 2 to 1	Business has 66¢ of current assets to meet each $1 of current debt.	Not enough current assets to pay off current liabilities. Industry standard is $2 for each $1 of current debt.
Acid test (quick ratio) Top of fraction often → referred to as *quick assets*	Current assets − Inventory − Prepaid expenses / Current liabilities (Inventory and prepaid expenses are excluded because it may not be easy to convert these to cash.)	$\dfrac{\$61,000 - \$30,000 - \$15,000}{\$92,000}$ $= .17:1$ Industry average, 1 to 1	Business has only 17¢ to cover each $1 of current debt. This calculation excludes inventory and prepaid expenses.	Same as above but more severe.
Average day's collection	Accounts receivable / $\left(\dfrac{\text{Net sales}}{360}\right)$	$\dfrac{\$9,000}{\left(\dfrac{\$20,566}{360}\right)} = 158\ \text{days}$ Industry average, 90–120 days	On the average, it takes 158 days to collect accounts receivable.	Could we speed up collection since industry average is 90–120 days?
Total debt to total assets	Total liabilities / Total assets	$\dfrac{\$150,000}{\$205,000} = 73.17\%$ Industry average, 50%–70%	For each $1 of assets, the company owes 73¢ in current and long-term debt.	73% is slightly higher than industry average.
Return on equity	Net income / Stockholders' equity	$\dfrac{\$6,316}{\$55,000} = 11.48\%$ Industry average, 15%–20%	For each $1 invested by the owner, a return of 11¢ results.	Could we get a higher return on money somewhere else?
Asset turnover	Net sales / Total assets	$\dfrac{\$20,566}{\$205,000} = 10¢$ Industry average, 3¢ to 8¢	For each $1 invested in assets, it returns 10¢ in sales.	Are assets being utilized efficiently?
Profit margin on net sales	Net income / Net sales	$\dfrac{\$6,316}{\$20,566} = 30.71\%$ Industry average, 25%–40%	For each $1 of sales, company produces 31¢ in profit.	Compared to competitors, are we showing enough profits versus our increased sales?

* Inventory turnover is discussed in Chapter 18.

[†] For example, Wal-Mart Stores, Inc., has a current ratio of .76:1.

Learning Unit 16–3

Practice Quiz

Complete this Practice Quiz to see how you are doing.

1. Prepare a trend analysis from the following sales, assuming a base year of 2021. Round to the nearest whole percent.

	2024	2023	2022	2021
Sales	$29,000	$44,000	$48,000	$60,000

2. **Given:** Total current assets (CA), $15,000; accounts receivable (AR), $6,000; total current liabilities (CL), $10,000; inventory (Inv), $4,000; net sales, $36,000; total assets, $30,000; net income (NI), $7,500.

   Calculate:

   a. Current ratio.

   b. Acid test.

   c. Average day's collection.

   d. Profit margin on sales (rounded to the nearest hundredth percent).

✓ **Solutions**

1.

	2024	2023	2022	2021
Sales	48%	73%	80%	100%

$$\left(\frac{\$29,000}{\$60,000}\right) \quad \left(\frac{\$44,000}{\$60,000}\right) \quad \left(\frac{\$48,000}{\$60,000}\right)$$

2. **a.** $\dfrac{CA}{CL} = \dfrac{\$15,000}{\$10,000} = 1.5{:}1$

b. $\dfrac{CA - Inv}{CL} = \dfrac{\$15,000 - \$4,000}{\$10,000} = 1.1{:}1$

c. $\dfrac{AR}{\left(\dfrac{Net\ sales}{360}\right)} = \dfrac{\$6,000}{\left(\dfrac{\$36,000}{360}\right)} = 60\ days$

d. $\dfrac{NI}{Net\ sales} = \dfrac{\$7,500}{\$36,000} = 20.83\%$

Chapter 16 Review

Topic/Procedure/Formula	Examples	You try it*
Balance sheet		
Vertical analysis Process of relating each figure on a financial report (down the column) to a total figure. The denominator for a balance sheet is total assets (or total liabilities + owner's equity); for an income statement it is net sales.	Current assets $520 52% Plant and equipment 480 48% Total assets $1,000 100%	Do vertical analysis CA $400 ? P + E 600 ? Total assets $1,000 ?

Horizontal analysis

Analyzing comparative financial reports shows rate and amount of change across columns item by item. (New line item amount − Old line item amount)/Old line item amount

	2023	2022	change	%
Cash, $5,000	$4,000	$1,000	25%	

$$\left(\frac{\$1,000}{\$4,000}\right)$$

Do horizontal analysis

	2023	2022	Change	%
Cash	$8,000	$2,000	?	?

Net sales

$$\text{Gross sales} - \text{Sales returns and allowances} - \text{Sales discounts}$$

$200 gross sales
− 10 sales returns and allowances
− 2 sales discounts
$188 net sales

Calculate net sales

Gross sales, $400
Sales returns and allowances, $20
Sales discount, $5

Cost of merchandise (goods) sold

$$\text{Beginning inventory} + \text{Net purchases} - \text{Ending inventory}$$

$50 + $100 − $20 = $130
Beginning inventory + Net purchases − Ending inventory = Cost of merchandise (goods) sold

Calculate cost of merchandise sold

Beginning inventory, $50
Net purchases, $200
Ending inventory, $20

Gross profit from sales

$$\text{Net sales} - \frac{\text{Cost of merchandise}}{\text{(goods) sold}}$$

$188 − $130 = $58 gross profit from sales

Net sales − Cost of merchandise (goods) sold = Gross profit from sales

Calculate gross profit

Net sales, $400
Cost of merchandise sold, $250

Net income

Gross profit − Operating expenses

$58 − $28 = $30

Gross profit from sales − Operating expenses = Net income

Calculate net income

Gross profit, $210
Operating expenses, $180

Trend analysis

Each number expressed as a percent of the base year.

$$\frac{\text{Each item}}{\text{Base amount}}$$

	2024	2023	2022
Sales	$200	$300	$400 ← Base year
	50%	75%	100%

$$\left(\frac{\$200}{\$400}\right)\left(\frac{\$300}{\$400}\right)$$

Prepare a trend analysis

2024	2023	2022
$1,200	$800	$1,000 ← Base year

Copyright © McGraw Hill

Chapter 16 Review (Continued)

Topic/Procedure/Formula	Examples	You try it*
Ratios Tools to interpret items on financial reports.	Use this example for calculating the following ratios: current assets, $30,000; accounts receivable, $12,000; total current liabilities, $20,000; inventory, $6,000; prepaid expenses, $2,000; net sales, $72,000; total assets, $60,000; net income, $15,000; total liabilities, $30,000.	**Use this example for calculating the following ratios:** Current assets, $40,000; Accounts receivable, $44,000; Total current liabilities, $160,000; Inventory, $2,000; Prepaid expenses, $3,000; Net sales, $60,000; Total assets, $70,000; Net income, $16,000; Total liabilities, $180,000.
Current ratio $\dfrac{\text{Current assets}}{\text{Current liabilities}}$	$\dfrac{\$30,000}{\$20,000} = \boxed{1.5:1}$	Use the information from the example above.
Acid test (quick ratio) Called quick assets $\dfrac{\text{Current assets} - \text{Inventory} - \text{Prepaid expenses}}{\text{Current liabilities}}$	$\dfrac{\$30,000 - \$6,000 - \$2,000}{\$20,000} = \boxed{1.1:1}$	Use the information from the example above.
Average day's collection $\dfrac{\text{Accounts receivable}}{\left(\dfrac{\text{Net sales}}{360}\right)}$	$\dfrac{\$12,000}{\left(\dfrac{\$72,000}{360}\right)} = \boxed{60\ \text{days}}$	Use the information from the example above.
Total debt to total assets $\dfrac{\text{Total liabilities}}{\text{Total assets}}$	$\dfrac{\$30,000}{\$60,000} = \boxed{50\%}$	Use the information from the example above.
Return on equity $\dfrac{\text{Net income}}{\text{Stockholders' equity (A} - \text{L)}}$	$\dfrac{\$15,000}{\$30,000} = \boxed{50\%}$	Use the information from the example above.
Asset turnover $\dfrac{\text{Net sales}}{\text{Total assets}}$	$\dfrac{\$72,000}{\$60,000} = \boxed{1.2}$	Use the information from the example above.
Profit margin on net sales $\dfrac{\text{Net income}}{\text{Net sales}}$	$\dfrac{\$15,000}{\$72,000} = .2083 = \boxed{20.83\%}$	Use the information from the example above.

Key Terms

Accounts payable
Accounts receivable
Acid test
Asset turnover
Assets
Average day's collection
Balance sheet
Capital
Common stock
Comparative statement
Corporation
Cost of merchandise (goods) sold
Current assets
Current liabilities
Current ratio
Expenses
Gross profit from sales

Gross sales
Horizontal analysis
Income statement
Liabilities
Long-term liabilities
Merchandise inventory
Mortgage note payable
Net income
Net purchases
Net sales
Operating expenses
Owner's equity
Partnership
Plant and equipment
Prepaid expenses
Profit margin on net sales
Purchase discounts

Purchase returns and allowances
Purchases
Quick assets
Quick ratio
Ratio analysis
Retained earnings
Return on equity
Revenues
Salaries payable
Sales (not trade) discounts
Sales returns and allowances
Sole proprietorship
Stockholders' equity
Total debt to total assets
Trend analysis
Vertical analysis

* Worked-out solutions are in Appendix A.

Critical Thinking Discussion Questions with Chapter Concept Check

1. What is the difference between current assets and plant and equipment? Do you think land should be allowed to depreciate?

2. What items make up stockholders' equity? Why might a person form a sole proprietorship instead of a corporation?

3. Explain the steps to complete a vertical or horizontal analysis relating to balance sheets. Why are the percents not summed vertically in horizontal analysis?

4. How do you calculate net sales, cost of merchandise (goods) sold, gross profit, and net income? Why do we need two separate figures for inventory in the cost of merchandise (goods) sold section?

5. Explain how to calculate the following: current ratios, acid test, average day's collection, total debt to assets, return on equity, asset turnover, and profit margin on net sales. How often do you think ratios should be calculated?

6. What is trend analysis? Explain how the portion formula assists in preparing a trend analysis.

7. In light of the pandemic, explain how companies such as GE are trying to gain market share and increase profit margins.

8. **Chapter Concept Check.** Visit the Delta Airlines website to see how the pandemic has affected passenger travel and profit projections.

End-of-Chapter Problems

Name _____   Date _____

Check figures for odd-numbered problems in Appendix A.

Drill Problems

16–1. Prepare a December 31, 2023, balance sheet for Long Print Shop like the one for The Card Shop from the following: cash, $50,000; accounts payable, $38,000; merchandise inventory, $14,000; Joe Ryan, capital, $46,000; and equipment, $20,000.   *LU 16–1(1)*

16–2. From the following, prepare a classified balance sheet for Bach Crawlers as of December 31, 2023. Ending merchandise inventory was $4,000 for the year.   *LU 16–1(1)*

Cash	$6,000	Accounts payable	$1,800
Prepaid rent	1,600	Salaries payable	1,600
Prepaid insurance	4,000	Note payable (long term)	8,000
Office equipment (net)	5,000	P. Bach, capital*	9,200

*What the owner supplies to the business. Replaces common stock and retained earnings section.

excel 16–3. Complete a horizontal analysis for Brown Company, rounding percents to the nearest hundredth: *LU 16–1(2)*

BROWN COMPANY Comparative Balance Sheet December 31, 2022 and 2023			INCREASE (DECREASE)	
	2023	2022	Amount	Percent
Assets				
Current assets:				
Cash	$ 15,750	$ 10,500		
Accounts receivable	18,000	13,500		
Merchandise inventory	18,750	22,500		
Prepaid advertising	54,000	45,000		
Total current assets	$106,500	$ 91,500		
Plant and equipment:				
Building (net)	$120,000	$126,000		
Land	90,000	90,000		
Total plant and equipment	$210,000	$216,000		
Total assets	$316,500	$307,500		
Liabilities				
Current liabilities:				
Accounts payable	$132,000	$120,000		
Salaries payable	22,500	18,000		
Total current liabilities	$154,500	$138,000		
Long-term liabilities:				
Mortgage note payable	99,000	87,000		
Total liabilities	$253,500	$ 225,000		
Owner's Equity				
J. Brown, capital	63,000	82,500		
Total liabilities and owner's equity	$316,500	$307,500		

End-of-Chapter Problems (Continued)

16–4. Prepare an income statement for Hansen Realty for the year ended December 31, 2023. Beginning inventory was $1,248. Ending inventory was $1,600.   *LU 16–2(1)*

Sales	$34,900
Sales returns and allowances	1,092
Sales discount	1,152
Purchases	10,512
Purchase discounts	540
Depreciation expense	115
Salary expense	5,200
Insurance expense	2,600
Utilities expense	210
Plumbing expense	250
Rent expense	180

16–5. Assume this is a partial list of financial highlights from a Best Buy annual report:

	2023	2022
	(dollars in millions)	
Net sales	$37,580	$33,075
Earnings before taxes	2,231	1,283
Net earnings	1,318	891

Complete a horizontal and vertical analysis from the above information. Round to the nearest hundredth percent.   *LU 16-2(2)*

16–6. From the French Instrument Corporation second-quarter report ended 2023, do a vertical analysis for the second quarter of 2023.   *LU 16–2(2)*

FRENCH INSTRUMENT CORPORATION AND SUBSIDIARIES Consolidated Statements of Operation (Unaudited) (In thousands of dollars, except share data)			
	SECOND QUARTER		
	2023	2022	Percent of net
Net sales	$6,698	$6,951	
Cost of sales	4,089	4,462	
Gross margin	2,609	2,489	
Expenses:			
Selling, general and administrative	1,845	1,783	
Product development	175	165	
Interest expense	98	123	
Other (income), net	(172)	(99)	
Total expenses	1,946	1,972	
Income before income taxes	663	517	
Provision for income taxes	265	209	
Net income	$398	$308	
Net income per common share*	$.05	$.03	
Weighted-average number of common shares and equivalents	6,673,673	6,624,184	

* Income per common share reflects the deduction of the preferred stock dividend from net income.

† Off due to rounding.

16–7. Complete the comparative income statement and balance sheet for Logic Company, rounding percents to the nearest hundredth: *LU 16–1(2), LU 16–2(2)*

LOGIC COMPANY Comparative Income Statement For Years Ended December 31, 2022 and 2023				
			INCREASE (DECREASE)	
	2023	2022	Amount	Percent
Gross sales	$19,000	$15,000		
Sales returns and allowances	1,000	100		
Net sales	$18,000	$14,900		
Cost of merchandise (goods) sold	12,000	9,000		
Gross profit	$ 6,000	$ 5,900		
Operating expenses:				
Depreciation	$ 700	$ 600		
Selling and administrative	2,200	2,000		

(Continues)

End-of-Chapter Problems (Continued)

LOGIC COMPANY Comparative Income Statement For Years Ended December 31, 2022 and 2023				
			INCREASE (DECREASE)	
	2023	2022	Amount	Percent
Research	550	500		
Miscellaneous	360	300		
Total operating expenses	$ 3,810	$ 3,400		
Income before interest and taxes	$ 2,190	$ 2,500		
Interest expense	560	500		
Income before taxes	$ 1,630	$ 2,000		
Provision for taxes	640	800		
Net income	$ 990	$ 1,200		

LOGIC COMPANY Comparative Balance Sheet December 31, 2022 and 2023				
	2023		2022	
	Amount	Percent	Amount	Percent
Assets				
Current assets:				
Cash	$12,000		$ 9,000	
Accounts receivable	16,500		12,500	
Merchandise inventory	8,500		14,000	
Prepaid expenses	24,000		10,000	
Total current assets*	$61,000		$45,500	
Plant and equipment:				
Building (net)	$14,500		$11,000	
Land	13,500		9,000	
Total plant and equipment	$28,000		$20,000	
Total assets	$89,000		$65,500	
Liabilities				
Current liabilities:				
Accounts payable	$13,000		$ 7,000	
Salaries payable	7,000		5,000	
Total current liabilities*	$20,000		$12,000	
Long-term liabilities:				
Mortgage note payable	22,000		20,500	
Total liabilities	$42,000		$32,500	

(Continues)

LOGIC COMPANY Comparative Balance Sheet December 31, 2022 and 2023				
	2023		**2022**	
	Amount	Percent	Amount	Percent
Stockholders' Equity				
Common stock	$21,000		$21,000	
Retained earnings	26,000		12,000	
Total stockholders' equity	$47,000		$33,000	
Total liabilities and stockholders' equity	$89,000		$65,500	

* Note that the percentages for total current assets and total current liabilities may be off due to rounding.

From Problem 16–7, your supervisor has requested that you calculate the following ratios, rounded to the nearest hundredth: *LU 16–3(2)*

	2023	**2022**

16–8. Current ratio.

16–9. Acid test.

16–10. Average day's collection.

16–11. Asset turnover.

16–12. Total debt to total assets.

16–13. Net income (after tax) to the net sales.

16–14. Return on equity (after tax).

16–8.

16–9.

16–10.

16–11.

16–12.

16–13.

16–14.

End-of-Chapter Problems (*Continued*)

Word Problems

excel 16–15. William Burris invested $100,000 in an Australian-based franchise, Rent Your Boxes, purchasing three territories in the Washington area. After finding out the company had gone bankrupt, he rallied 10 other franchisees to join him and created a new company, Rent Our Boxes. If Rent Our Boxes had net income of $38,902 with net sales of $286,585, what was its profit margin on net sales to the nearest hundredth percent? *LU 16–3(2)*

16–16. Assume General Motors announced a quarterly profit of $119 million for 4th quarter 2022. Below is a portion of its balance sheet. Conduct a horizontal analysis of the following line items (rounding percent to nearest hundredth): *LU 16–1(2)*

	2022 (dollars in millions)	2021 (dollars in millions)	Difference	% CHG
Cash and cash equivalents	$ 15,980	$ 15,499		
Marketable securities	9,222	16,148		
Inventories	13,642	14,324		
Goodwill	—	1,278		
Total liabilities and equity	$103,249	$144,603		

excel 16–17. Find the following ratios for Motorola Credit Corporation's annual report: **(a)** total debt to total assets, **(b)** return on equity, **(c)** asset turnover (to nearest cent), and **(d)** profit margin on net sales. Round to the nearest hundredth percent. *LU 16–3(2)*

	(dollars in millions)
Net revenue (sales)	$ 265
Net earnings	147
Total assets	2,015
Total liabilities	1,768
Total stockholders' equity	427

16–18. Assume figures were presented for the past 5 years on merchandise sold at Chicago department and discount stores ($ million). Sales in 2025 were $3,154; in 2024, $3,414; in 2023, $3,208; in 2022, $3,152; and in 2021, $3,216. Using 2021 as the base year, complete a trend analysis. Round each percent to the nearest whole percent. *LU 16–3(1)*

16–19. Don Williams received a memo requesting that he complete a trend analysis of the following numbers using 2021 as the base year and rounding each percent to the nearest whole percent. Could you help Don with the request?   *LU 16–3(1)*

	2024	2023	2022	2021
Sales	$340,000	$400,000	$420,000	$500,000
Gross profit	180,000	240,000	340,000	400,000
Net income	70,000	90,000	40,000	50,000

excel 16–20. If the French bank Société Générale reported its net income was 23,561 million euros and its operating expenses totaled 16,016 million euros, what was its gross profit? *LU 16–2(1)*

16–21. At age 32, you have assets of $275,658 and liabilities of $266,211. What is your net worth? Is this appropriate for your age if you have an annual household income of $69,200? See https://lifehacker.com/5859040/what-should-your-current-net-worth-be to determine if the net worth is adequate.   *LU 16–1(1)*

16–22. You did not have 20% to put down on the house you purchased two years ago so you are paying PMI (private mortgage insurance). If you bought your house for $375,000 and due to increasing prices it is now worth $435,000, what is your debt to asset ratio if your mortgage is $336,000? Do you have at least 20% equity to be able to request your PMI be removed? Round to the hundredth percent.   *LU 16–3(2)*

End-of-Chapter Problems (Continued)

16–23. Teri's Whisker Watcher had the following annual report information from the 12 months prior to the pandemic and the 12 months during the pandemic. Conduct a horizontal analysis for Teri's partial income statement. Round to the nearest hundredth percent. *LU 16–2(2)*

	During pandemic	**Pre-pandemic**
Sales	$625	$32,650
Operating expenses	21,751	28,549
Net income	($21,126)	$ 4,101

16–24. "United Airlines posts $1.9 billion loss in pandemic-laden 4Q" while "revenue plunged 69% in the fourth quarter compared with a year earlier," reported ABC news. Analysts predicted United would lose $6.62 per share but it lost $7 per share during the pandemic. Despite these statistics, United is optimistic about the future as it is cutting $2 billion in structural operating costs and is confident business travel will increase along with leisure travel resulting in higher profit margins in 2023. If United Airlines Holdings Inc. had fourth-quarter net income of negative $917 million with net sales of $3.412 billion, what was its fourth-quarter profit margin? Round to the nearest tenth percent.   *LU 16–3(2)*

Challenge Problems

16–25. On January 1, Pete Rowe bought a ski chalet for $51,000. Pete is renting the chalet for $55 per night. He estimates he can rent the chalet for 190 nights. Pete's mortgage for principal and interest is $448 per month. Real estate tax on the chalet is $500 per year.

Pete estimates that his heating bill will run $60 per month. He expects his monthly electrical bill to be $20 per month. He pays $12 per month for cable television.

What is Pete's return on the initial investment for this year? Assume rentals drop by 30% and monthly bills for heat and electricity drop by 10% each month. What would be Pete's return on initial investment? Round to the nearest tenth percent as needed.   *LU 16–3(2)*

excel 16–26. As the accountant for Tootsie Roll, you are asked to calculate the current ratio and the quick ratio for the following partial financial statement. Round to the nearest tenth. *LU 16–3(2)*

Assets		Liabilities	
Current assets:		Current liabilities:	
Cash and cash equivalents	$ 4,224,190	Notes payable to banks	$ 672,221
Investments	32,533,769	Accounts payable	7,004,075
Accounts receivable, less allowances of $748,000 and $744,000	16,206,648	Dividends payable	576,607
Inventories:		Accrued liabilities	9,826,534
Finished goods and work in progress	12,650,955	Income taxes payable	4,471,429
Raw materials and supplies	10,275,858		
Prepaid expenses	2,037,710		

Summary Practice Test

Do you need help? Connect videos have step-by-step worked-out solutions.

1. Given: Gross sales, $170,000; sales returns and allowances, $9,000; beginning inventory, $8,000; net purchases, $18,000; ending inventory, $5,000; and operating expenses, $56,000. Calculate (a) net sales, (b) cost of merchandise (goods) sold, (c) gross profit from sales, and (d) net income.   *LU 16–2(1)*

2. Complete the following partial comparative balance sheet by filling in the total current assets and percent column; assume no plant and equipment (round to the nearest hundredth percent as needed).   *LU 16–1(2)*

	Amount	Percent	Amount	Percent
Assets				
Current assets:				
Cash	$ 9,000		$ 8,000	
Accounts receivable	5,000		7,500	
Merchandise inventory	12,000		6,900	
Prepaid expenses	7,000		8,000	
Total current assets*				

* The percentages for total current assets will be off due to rounding.

3. Calculate the amount of increase or decrease and the percent change of each item, rounding to the nearest hundredth percent as needed.   *LU 16–1(2)*

	2023	2022	Amount	Percent
Cash	$19,000	$ 8,000		
Land	70,000	30,000		
Accounts payable	21,000	10,000		

4. Complete a trend analysis for sales, rounding to the nearest whole percent and using 2022 as the base year.   *LU 16–3(1)*

	2025	2024	2023	2022
Sales	$140,000	$350,000	$210,000	$190,000

5. From the following, prepare a balance sheet for True Corporation as of December 31, 2022. *LU 16–1(1)*

Building	$40,000	Mortgage note payable	$70,000
Merchandise inventory	12,000	Common stock	10,000
Cash	15,000	Retained earnings	37,000
Land	90,000	Accounts receivable	9,000
Accounts payable	50,000	Salaries payable	8,000
Prepaid rent	9,000		

6. Solve from the following facts, rounding to the nearest hundredth. *LU 16–3(2)*

Current assets	$14,000	Net sales	$40,000
Accounts receivable	$ 5,000	Total assets	$38,000
Current liabilities	$20,000	Net income	$10,100
Inventory	$ 4,000		

a. Current ratio

b. Acid test

c. Average day's collection

d. Asset turnover

e. Profit margin on sales

MY MONEY

Q For What It's Worth

What I need to know

Net worth is a snapshot of your current financial standing. At its basic level net worth subtracts what you owe from the value of what you own. Understanding your net worth is a valuable tool in assessing your financial situation and whether you are making progress toward your financial goals and a healthy financial future. Compare your net worth according to nerdwallet.com against the median and average, respectively, for your age: 35–44: $91,300; $436,200; 45–54: $168,600, $833,200; 55–64: $212,500, $1,175,900; 65–74: $266,400, $1,217,700.

Another helpful tool in assessing your financial standing is the personal income statement. This statement makes a direct comparison of the inflow of money (earnings) to the outflow of money (expenses). This is a quick way to assess how you are using the earnings generated from your employment. Make certain you are paying yourself first by saving and investing at least 20% of all monies received and your emergency fund contains at least three months worth of your monthly expenses.

What I need to do

Be aware of how taking on additional debt will impact your financial position. Since what you owe is compared directly to what you own in the assessment of your financial position, you need to seriously consider any new debt you incur. As you satisfy (pay off) debts over time your financial standing will improve. An example of this would be paying down the mortgage on your house, which reduces the amount you owe while increasing the value of the equity you have in the home, or what you own. Your debt-to-income ratio should be no more than 35% and less is better. Total your monthly debt payments (credit cards, auto loans, mortgage, etc.) and divide this amount by your gross income to determine your debt-to-income ratio.

Begin building your net worth in your twenties. The earlier you start investing, the less money you will need to invest. It is helpful to keep in mind why you are going to the effort of saving. Remind yourself the goal of saving and investing is to improve the quality of your life during your earning years by helping to reduce money stressors while striving to provide yourself with a quality retirement free from financial worries and with enough money to pursue your dreams. Financial advisors recommend having a net worth of $1,000,000 by age 60. If you invest $10,000 per year starting at age 25, with a 5% rate of return you will net almost $1 million by your 60th birthday. To estimate what your net worth should be by age use this formula: Your age × Your annual gross income ÷ 10.

Steps I need to take

1. Calculate what your net worth is and compare against what it should be at your age. Make a plan for how to increase your net worth.
2. Consider the impact of new debt on your net worth prior to incurring any new debt.
3. Pay off existing debts to yield a positive impact on your net worth.

Resources I can use

- Personal Capital (mobile app)—monitors your financial portfolio and shows your net worth.
- https://www.bankrate.com/calculators/smart-spending/personal-net-worth-calculator.aspx—net worth calculator.
- Why you should invest today: https://money.usnews.com/investing/investing-101/articles/2018-07-23/9-charts-showing-why-you-should-invest-today

MY MONEY ACTIVITY

- Use the net worth calculator provided above.
- Experiment with the Assets and Liabilities to see their impact on your net worth.

PERSONAL FINANCE

A KIPLINGER APPROACH

"Digesting Corporate Profits." Kiplinger's. October 2020.

FUNDAMENTALS

PRACTICAL PORTFOLIO   Ryan Ermey

Digesting Corporate Profits

You need to go beyond the headlines to interpret a company's earnings.

STOCK INVESTORS OF ALL STRIPES CARE about corporate earnings. Lovers of fast-growing firms prefer profits growing at a high, compounding rate, for example, and those looking for undervalued names may covet firms whose stock prices look cheap in comparison to earnings per share. Lately, the earnings picture for the broad stock market has been cloudier than ever, frustrating professional and armchair analysts alike.

Given the dire economic circumstances of the pandemic-induced recession, the expectations for corporate profits for the second quarter were beyond bleak. And yet, by early August, with close to 90% of S&P 500 companies having reported earnings for the quarter than ended in June, nearly 82% of them had exceeded Wall Street's dismal expectations, by an average of nearly 18%, according to investment research firm Refinitiv—the highest "earnings beat" percentages since Refinitiv began tracking earnings data in 1994.

But exceeding such a low bar isn't much to celebrate, and the market responded to the better-than-expected quarter with a collective shrug. When reports are all in, profits for the quarter still are expected to have declined by 33.9% from the same quarter a year ago, per Refinitiv. "That kind of collapse in S&P 500 earnings is not going to turn bears into bulls, especially at high stock valuations and given ongoing COVID-19 uncertainty," says

Jeff Buchbinder, a market strategist at investment research firm LPL Financial.

Where earnings are headed in the second half of 2020 is still difficult to divine given the uncertainty surrounding the reopening of the economy, Buchbinder says. "Until we get a vaccine, or dramatic leaps forward in treatments that make people comfortable resuming some semblance of normal life, earnings will have an extremely difficult time returning to pre-pandemic levels," he says.

Making things trickier for market prognosticators: About half of the companies in the S&P 500 have rescinded the guidance they usually supply regarding sales and profit expectations for 2020. So, with perhaps a little less confidence in their forecasts than usual, Wall Street analysts expect per-share earnings for S&P 500 companies overall to fall 23% for the calendar year, followed by a 31% bounce in 2021.

Business Math Issue

Post pandemic means all corporations will see rising profits.

1. List the key points of the article and information to support your position.

2. Write a group defense of your position using math calculations to support your view. If you are in an online course, post to a discussion board.

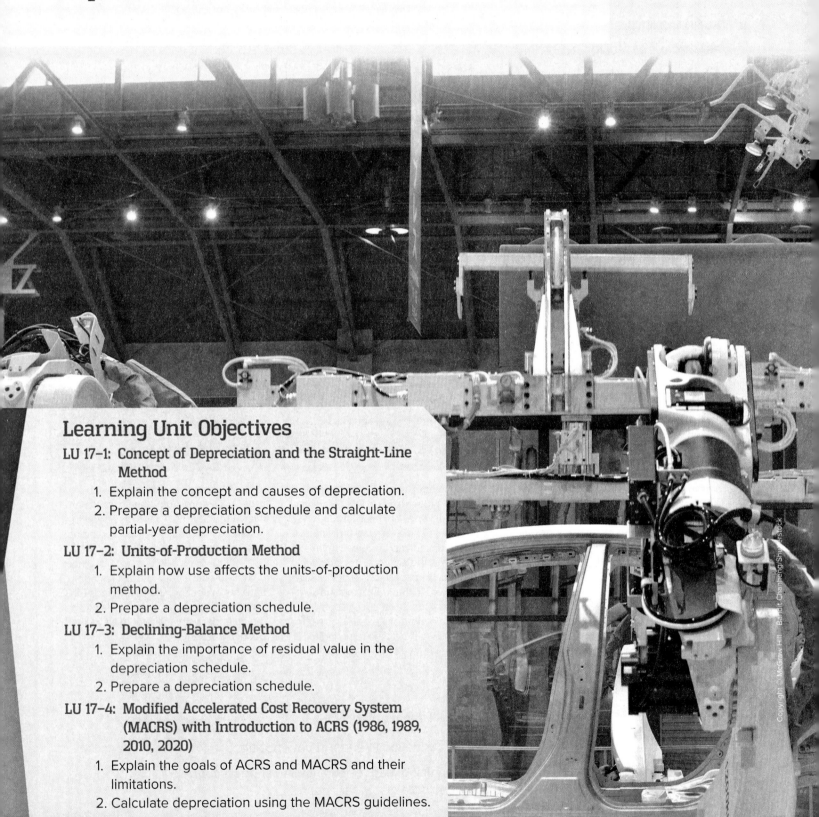

Chapter 17
Depreciation

Learning Unit Objectives

LU 17–1: Concept of Depreciation and the Straight-Line Method

1. Explain the concept and causes of depreciation.
2. Prepare a depreciation schedule and calculate partial-year depreciation.

LU 17–2: Units-of-Production Method

1. Explain how use affects the units-of-production method.
2. Prepare a depreciation schedule.

LU 17–3: Declining-Balance Method

1. Explain the importance of residual value in the depreciation schedule.
2. Prepare a depreciation schedule.

LU 17–4: Modified Accelerated Cost Recovery System (MACRS) with Introduction to ACRS (1986, 1989, 2010, 2020)

1. Explain the goals of ACRS and MACRS and their limitations.
2. Calculate depreciation using the MACRS guidelines.

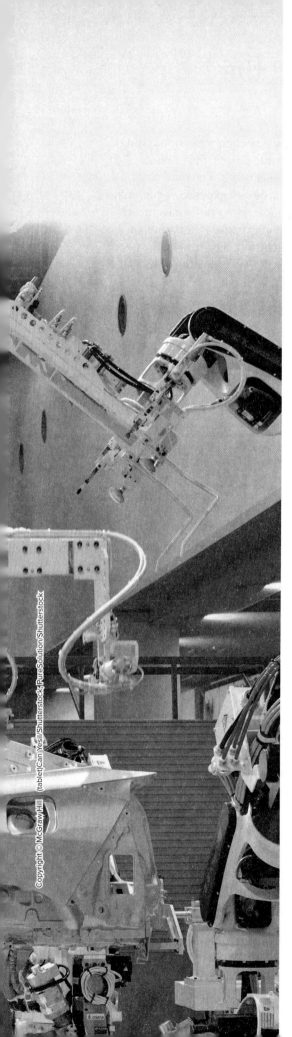

@ Essential Question

How can I use depreciation to understand the business world and make my life easier?

🌐 Math Around the World

The pandemic affected car sales dramatically. In the *Wall Street Journal* chapter opener clip "Overheard," which reports dealers offering risky long-term loans, Carfax estimates a 5-year-old car will have depreciated by 60% of its original value. However, any unsold cars also depreciate while sitting on the car lot.

In Learning Units 17–1 to 17–3, we discuss methods of calculating depreciation for financial reporting. In Learning Unit 17–4, we look at how tax laws force companies to report depreciation for tax purposes during the pandemic timeframe. Financial reporting methods and the tax-reporting methods are both legal.

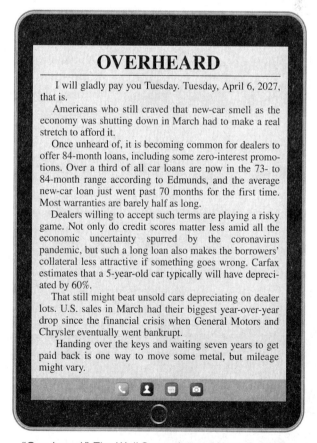

OVERHEARD

I will gladly pay you Tuesday. Tuesday, April 6, 2027, that is.

Americans who still craved that new-car smell as the economy was shutting down in March had to make a real stretch to afford it.

Once unheard of, it is becoming common for dealers to offer 84-month loans, including some zero-interest promotions. Over a third of all car loans are now in the 73- to 84-month range according to Edmunds, and the average new-car loan just went past 70 months for the first time. Most warranties are barely half as long.

Dealers willing to accept such terms are playing a risky game. Not only do credit scores matter less amid all the economic uncertainty spurred by the coronavirus pandemic, but such a long loan also makes the borrowers' collateral less attractive if something goes wrong. Carfax estimates that a 5-year-old car typically will have depreciated by 60%.

That still might beat unsold cars depreciating on dealer lots. U.S. sales in March had their biggest year-over-year drop since the financial crisis when General Motors and Chrysler eventually went bankrupt.

Handing over the keys and waiting seven years to get paid back is one way to move some metal, but mileage might vary.

"Overheard." *The Wall Street Journal* (April 5, 2020).

Concept of Depreciation and the Straight-Line Method

Pepper ...
And Salt

THE WALL STREET JOURNAL

CERTIFIED PUBLIC ACCOUNTANT

SCHWADRn.

"I'm the muse of tax loopholes."

Used by permission of
Cartoon Features Syndicate

Companies frequently buy assets such as equipment or buildings that will last longer than 1 year. As time passes, these assets depreciate, or lose some of their market value. The total cost of these assets cannot be shown in *1 year* as an expense of running the business. In a systematic and logical way, companies must estimate the asset cost they show as an expense of a particular period. This process is called **depreciation.** The next time you fly in a plane think how the airline will depreciate the cost of that plane over a number of years.

Remember that depreciation *does not* measure the amount of deterioration or decline in the market value of the asset. Depreciation is simply a means of recognizing that these assets are depreciating.

The depreciation process results in **depreciation expense** that involves three key factors: (1) **asset cost**—the amount the company paid for the asset including freight and charges relating to the asset; (2) **estimated useful life**—the number of years or time periods for which the company can use the asset; and (3) **residual value (salvage** or **trade-in value)**—the expected cash value at the end of the asset's useful life.

Depreciation expense is listed on the income statement. The **accumulated depreciation** title on the balance sheet gives the amount of the asset's depreciation taken to date. Asset cost less accumulated depreciation is the asset's book value. The **book value** shows the unused amount of the asset cost that the company may depreciate in future accounting periods. At the end of the asset's life, the asset's book value is the same as its residual value—book value cannot be less than residual value.

Aha!

Book value is important to a business owner because it provides a way to calculate the value of assets within a business.

Depending on the amount and timetable of an asset's depreciation, a company can increase or decrease its profit. If a company shows greater depreciation in earlier years, the company will have a lower reported profit and pay less in taxes. Thus, depreciation can be an indirect tax savings for the company.

Later in the chapter we will discuss the different methods of computing depreciation that spread the cost of an asset over specified periods of time. However, first let's look at some of the major causes of depreciation.

Learn: Causes of Depreciation

As assets, all machines have an estimated amount of usefulness simply because as companies use the assets, the assets gradually wear out. The cause of this depreciation is *physical deterioration.*

The growth of a company can also cause depreciation. Many companies begin on a small scale. As the companies grow, they often find their equipment and buildings inadequate. The use of depreciation enables

Hero Images/Getty Images; (tablet)Can Yesil/Shutterstock; PureSolution/Shutterstock; (puzzle)frender/Getty Images

Aha!

these businesses to "write off" their old, inadequate equipment and buildings. *In fact, the depreciation method chosen by a company can be determined by when tax write-offs are needed most.* Companies cannot depreciate land. For example, a garbage dump can be depreciated but not the land.

Another cause of depreciation is the result of advances in technology. The computers that companies bought a few years ago may be in perfect working condition but outdated. Companies may find it necessary to replace these old computers with more sophisticated, faster, and possibly more economical machines. Thus, *product obsolescence* is a key factor contributing to depreciation.

Now we are ready to begin our study of depreciation methods. The first method we will study is straight-line depreciation. It is also the most common of the three depreciation methods (straight line, units of production, and declining balance).

End of year	Depreciation cost of Equipment	Depreciation expense for year	Accumulated depreciation at end of year	Book value at end of year (Cost − Depreciation at end of year)
1	$2,500	$400	$ 400	$2,100
				($2,500 − $400)
2	2,500	400	800	1,700
3	2,500	400	1,200	1,300
4	2,500	400	1,600	900
5	2,500	400	2,000	500
	↑	↑	↑	↑
	Cost stays the same.	Depreciation expense is same each year.	Accumulated depreciation increases by $400 each year.	Book value is lowered by $400 until residual value of $500 is reached.

TABLE 17.1 Depreciation schedule for straight-line method

$$\frac{100\%}{\text{Number of years}} = \frac{100\%}{5} = 20\%$$

Thus, the company is depreciating the equipment at a 20% rate each year.

Learn: Straight-Line Method

The **straight-line method** of depreciation is used more than any other method. It tries to distribute the same amount of expense to each period of time. Germany and Japan depreciate equipment over 3 to 50 years. Most large companies, such as Gillette Corporation, Campbell's Soup, and General Mills, use the straight-line method. *Today, more than 90% of U.S. companies depreciate by straight line.* For example, let's assume Ajax Company bought equipment for $2,500. The company estimates that the equipment's period of "usefulness"—or *useful life*—will be 5 years. After 5 years the equipment will have a residual value (salvage value) of $500. The company decides to calculate its depreciation with the straight-line method and uses the following formula:

$$\frac{\text{Depreciation expense}}{\text{each year}} = \frac{\text{Cost} - \text{Residual value}}{\text{Estimated useful life in years}}$$

$$\frac{\$2,500 - \$500}{5 \text{ years}} = \$400 \text{ depreciation expense taken each year}$$

Money Tip Do not open several credit card accounts in a short period of time if you are trying to purchase or refinance a home. Doing so negatively affects your credit rating.

Table 17.1 gives a summary of the equipment depreciation that Ajax Company will take over the next 5 years. Companies call this summary a **depreciation schedule.** Buildings for BigLot are depreciated over 40 years while equipment is depreciated from 3 to 15 years.

You may be able to deduct allowable depreciation and other business expenses for a home office.

Learn: Depreciation for Partial Years

If a company buys an asset before the 15th of the month, the company calculates the asset's depreciation for a full month. Companies do not take the full month's depreciation for assets bought after the 15th of the month. For example, assume Ajax Company (Table 17.1) bought the equipment on May 6. The company would calculate the depreciation for the first year as follows:

$$\frac{\$2,500 - \$500}{5 \text{ years}} = \$400 \times \frac{8}{12} = \$266.67$$

Now let's check your progress with the Practice Quiz before we look at the next depreciation method.

Practice Quiz

Complete this Practice Quiz to see how you are doing.

1. Prepare a depreciation schedule using straight-line depreciation for the following:

   Cost of truck $16,000

   Residual value $ 1,000

   Life   5 years

2. If the truck were bought on February 3, what would the depreciation expense be in the first year?

✓ Solutions

1.

End of year	Cost of truck	Depreciation expense for year	Accumulated depreciation at end of year	Book value at end of year (Cost – Accumulated depreciation)
1	$16,000	$3,000	$ 3, 000	$13,000 ($16,000 – $3,000)
2	16,000	3,000	6,000	10,000
3	16,000	3,000	9,000	7,000
4	16,000	3,000	12,000	4,000
5	16,000	3,000	15,000	1,000 ← Note that we are down to residual value

2. $\dfrac{\$16,000 - \$1,000}{5} = \$3,000 \times \dfrac{11}{12} = \boxed{\$2,750}$

Learning Unit 17–2:
Units-of-Production Method

Unlike in the straight-line depreciation method, in the **units-of-production method** the passage of time is not used to determine an asset's depreciation amount. Instead, the company determines the asset's depreciation according to how much the company uses the asset. This use could be miles driven, tons hauled, or units that a machine produces. For example, when a company such as Ajax Company (in Learning Unit 17–1) buys equipment, the company estimates how many units the equipment can produce. Let's assume the equipment has a useful life of 4,000 units. The following formulas are used to calculate the equipment's depreciation for the units-of-production method.

$$\frac{\text{Depreciation}}{\text{per unit}} = \frac{\text{Cost} - \text{Residual value}}{\text{Total estimated units produced}} = \frac{\$2,500 - \$500}{4,000 \text{ units}} = \$.50 \text{ per unit}$$

$$\frac{\text{Depreciation}}{\text{amount}} = \frac{\text{Unit}}{\text{depreciation}} \times \frac{\text{Units}}{\text{produced}} = \$.50 \text{ times actual number of units}$$

Now we can complete Table 17.2. Note that the table gives the units produced each year.

TABLE 17.2 Depreciation schedule for units-of-production method

End of year	Cost of equipment	Units produced	Depreciation expense for year	Accumulated depreciation at end of year	Book value at end of year (Cost – Accumulated depreciation)
1	$2,500	300	$ 150 (300 × $.50)	$ 150	$2,350 ($2,500 – $150)
2	2,500	400	200	350	2,150
3	2,500	600	300	650	1,850
4	2,500	2,000	1,000	1,650	850
5	2,500	700 ↑	350 ↑	2,000	500 ↑

At the end of 5 years, the equipment produced 4,000 units. If in year 5 the equipment produced 1,500 units, only 700 could be used in the calculation, or it will go below the equipment's residual value.

Units produced per year times $.50 equals depreciation expense.

Residual value of $500 is reached. (Be sure depreciation is not taken below the residual value.)

Let's check your understanding of this unit with the Practice Quiz.

Practice Quiz

Complete this Practice Quiz to see how you are doing.

From the following facts prepare a depreciation schedule:

Machine cost $20,000

Residual value $ 4,000

Expected to produce 16,000 units over its expected life

$$\frac{\$20{,}000 - \$4{,}000}{16{,}000} = \$1$$

	2019	**2020**	**2021**	**2022**	**2023**
Units produced:	2,000	8,000	3,000	1,800	1,600

✓ Solutions

End of year	Cost of machine	Units produced	Depreciation expense for year	Accumulated depreciation at end of year	Book value at end of year (Cost − Accumulated depreciation)
1	$20,000	2,000	$2,000 (2,000 × $1)	$ 2,000	$18,000
2	20,000	8,000	8,000	10,000	10,000
3	20,000	3,000	3,000	13,000	7,000
4	20,000	1,800	1,800	14,800	5,200
5	20,000	1,600	1,200*	16,000	4,000

* Note that we can depreciate only 1,200 units since we cannot go below the residual value of $4,000.

Declining-Balance Method

In the declining-balance method, we cannot depreciate below the residual value.

The **declining-balance method** is another type of **accelerated depreciation** that takes larger amounts of depreciation expense in the earlier years of the asset. The straight-line method, you recall, estimates the life of the asset and distributes the same amount of depreciation expense to each period. To take larger amounts of depreciation expense in the asset's earlier years, the declining-balance method uses up to *twice* the **straight-line rate** in the first year of depreciation. A key point to remember is that the declining-balance method does not deduct the residual value in calculating the depreciation expense. Today, the declining-balance method is the basis of current tax depreciation.

For all problems, we will use double the straight-line rate unless we indicate otherwise. Today, the rate is often 1.5 or 1.25 times the straight-line rate. Again we use our $2,500 equipment with its estimated useful life of 5 years. As we build the depreciation schedule in Table 17.3, note the following steps:

TABLE 17.3 Depreciation schedule for declining-balance method

End of year	Cost of equipment	Accumulated depreciation at beginning of year	Book value at beginning of year (Cost – Accumulated depreciation)	Depreciation (Book value at beginning of year × Rate)	Accumulated depreciation at end of year	Book value at end of year (Cost – Accumulated depreciation)
1	$2,500	—	$2,500	$1,000 ($2,500 × .40)	$1,000	$1,500 ($2,500 − $1,000)
2	2,500	$1,000	1,500	600 ($1,500 × .40)	1,600	900
3	2,500	1,600	900	360 ($900 × .40)	1,960	540
4	2,500	1,960	540	40	2,000	500
5	2,500	2,000	500	—	2,000	500
	↑ Original cost of $2,500 does not change. Residual value was not subtracted.	↑ Ending accumulated depreciation of 1 year becomes next year's beginning.	↑ Cost less accumulated depreciation	↑ *Note:* In year 4, only $40 is taken since we cannot depreciate below residual value of $500. In year 5, no depreciation is taken.	↑ Accumulated depreciation balance plus depreciation expense this year.	↑ Book value now equals residual value.

Step 1   Rate is equal to $\dfrac{100\%}{5 \text{ years}} \times 2 = 40\%$.

Or another way to look at it is that the straight-line rate is $\frac{1}{5} \times 2 = \frac{2}{5} = 40\%$.

Step 2

$$\dfrac{\text{Depreciation expense}}{\text{each year}} = \dfrac{\text{Book value of equipment}}{\text{at beginning of year}} \times \dfrac{\text{Depreciation}}{\text{rate}}$$

Step 3   We cannot depreciate the equipment below its residual value ($500). The straight-line method automatically reduced the asset's book value to the residual value. This is not true with the declining-balance method. So you must be careful when you prepare the depreciation schedule.

Now let's check your progress again with another Practice Quiz.

Learning Unit 17–3

Practice Quiz

Complete this Practice Quiz to see how you are doing.

Prepare a depreciation schedule from the following:

Cost of machine: $16,000   Estimated life: 5 years

Rate: 40% (this is twice the straight-line rate)   Residual value: $1,000

✓ **Solutions**

End of year	Cost of machine	Accumulated depreciation at beginning of year	Book value at beginning of year (Cost – Accumulated depreciation)	Depreciation (Book value at beginning of year × Rate)	Accumulated depreciation at end of year	Book value at end of year (Cost – Accumulated depreciation)
1	$16,000	$ –0–	$16,000.00	$6,400.00	$ 6,400.00	$9,600.00
2	16,000	6,400.00	9,600.00	3,840.00	10,240.00	5,760.00
3	16,000	10,240.00	5,760.00	2,304.00	12,544.00	3,456.00
4	16,000	12,544.00	3,456.00	1,382.40	13,926.40	2,073.60
5	16,000	13,926.40	2,073.60	829.44*	14,755.84	1,244.16

* Since we do not reach the residual value of $1,000, another $244.16 could have been taken as depreciation expense to bring it to the estimated residual value of $1,000.

Modified Accelerated Cost Recovery System (MACRS) with Introduction to ACRS (1986, 1989, 2010, 2020)

In Learning Units 17–1 to 17–3, we discussed the depreciation methods used for financial reporting. Since 1981, federal tax laws have been passed that state how depreciation must be taken for income tax purposes. The pandemic has resulted in some assets of businesses being idle. Usually depreciation can be taken on idle assets. (See IRS publication 946 for yearly update.) Assets put in service from 1981 through 1986 fell under the federal **Accelerated Cost Recovery System (ACRS)** tax law enacted in 1981. The Tax Reform Act of 1986 established the **Modified Accelerated Cost Recovery System (MACRS)** for all property placed into service after December 31, 1986. This system, used by businesses to calculate depreciation for tax purposes based on the tax laws of 1986, 1989, and 2020, is also known as the **General Depreciation System (GDS).** Airplanes for commercial use are usually depreciated by MACRS for 7 years or ADS (Alternative Depreciation System) for 12 years. Search "MACRS Depreciation" online to see the latest updates as well as to find a depreciation calculator similar to the one below.

Learn: Depreciation for Tax Purposes Based on the Tax Reform Act of 1986 (MACRS)

Tables 17.4 and 17.5 give the classes of recovery and annual depreciation percentages that MACRS established in 1986. The key points of MACRS are

1. It calculates depreciation for tax purposes.

2. It ignores residual value.

3. Depreciation in the first year (for personal property) is based on the assumption that the asset was purchased halfway through the year. (A new law adds a midquarter convention for all personal property if more than 40% is placed in service during the last 3 months of the taxable year.)

4. Classes 3, 5, 7, and 10 use a 200% declining-balance method for a period of years before switching to straight-line depreciation. You do not have to determine the year in which to switch since Table 17.5 builds this into the calculation.

5. Classes 15 and 20 use a 150% declining-balance method before switching to straight-line depreciation.

6. Classes 27.5 and 31.5 use straight-line depreciation.

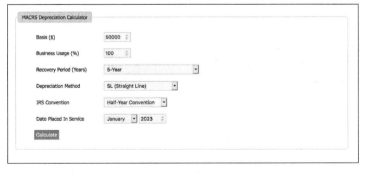

Good Calculator

Racehorses, when put in training, can be depreciated by MACRS.

TABLE 17.4 Modified Accelerated Cost Recovery System (MACRS) for assets placed in service after December 31, 1986

Class recovery period (life)	Asset types
3-year*	Racehorses more than 2 years old or any horse other than a racehorse that is more than 12 years old at the time placed into service; special tools of certain industries.
5-year*	Automobiles (not luxury); taxis; light general-purpose trucks; semiconductor manufacturing equipment; computer-based telephone central-office switching equipment; qualified technological equipment; property used in connection with research and experimentation.
7-year*	Railroad track; single-purpose agricultural (pigpens) or horticultural structures; fixtures; equipment; furniture.
10-year*	New law doesn't add any specific property under this class.
15-year†	Municipal wastewater treatment plants; telephone distribution plants and comparable equipment used for two-way exchange of voice and data communications.
20-year†	Municipal sewers.
27.5-year‡	Only residential rental property.
31.5-year‡	Only nonresidential real property.

*These classes use a 200% declining-balance method before switching to the straight-line method.
†These classes use a 150% declining-balance method before switching to the straight-line method.
‡These classes use a straight-line method.

TABLE 17.5 Annual recovery for MACRS

Recovery year	3-year class (200% D.B.)	5-year class (200% D.B.)	7-year class (200% D.B.)	10-year class (200% D.B.)	15-year class (150% D.B.)	20-year class (150% D.B.)
1	33.00	20.00	14.28	10.00	5.00	3.75
2	45.00	32.00	24.49	18.00	9.50	7.22
3	15.00*	19.20	17.49	14.40	8.55	6.68
4	7.00	11.52*	12.49	11.52	7.69	6.18
5		11.52	8.93*	9.22	6.93	5.71
6		5.76	8.93	7.37	6.23	5.28
7			8.93	6.55*	5.90*	4.89
8			4.46	6.55	5.90	4.52
9				6.55	5.90	4.46*
10				6.55	5.90	4.46
11				3.29	5.90	4.46
12					5.90	4.46
13					5.90	4.46
14					5.90	4.46
15					5.90	4.46
16					3.00	4.46

*Identifies when switch is made to straight line.

Example: Using the same equipment cost of $2,500 for Ajax, prepare a depreciation schedule under MACRS assuming the equipment is a 5-year class and not part of the tax bill of 1989. Use Table 17.5. Note that percent figures from Table 17.5 have been converted to decimals.

End of year	Cost	Depreciation expense	Accumulated depreciation	Book value at end of year
1	$2,500	$500 (.20 × $2,500)	$ 500	$2,000
2	2,500	800 (.32 × $2,500)	1,300	1,200
3	2,500	480 (.1920 × $2,500)	1,780	720
4	2,500	288 (.1152 × $2,500)	2,068	432
5	2,500	288 (.1152 × $2,500)	2,356	144
6	2,500	144 (.0576 × $2,500)	2,500	–0–

Money Tip Consider refinancing your home to obtain a lower fixed interest rate. Determine whether your savings offsets the refinance costs. An interest rate reduction of 1.5% is generally worth the cost in the long run.

Check your understanding of this learning unit with the below Practice Quiz.

Learning Unit 17–4

Practice Quiz

Complete this Practice Quiz to see how you are doing.

1. In 2022, Rancho Corporation bought semiconductor equipment for $80,000. Using MACRS, what is the depreciation expense in year 3?

2. What would depreciation be the first year for a wastewater treatment plant that cost $800,000?

✓ **Solutions**

1. $80,000 \times .1920 = $ $15,360

2. $800,000 \times .05 = $ $40,000

Chapter 17 Review

Topic/Procedure/Formula	Example	You try it*
Straight-line method $$\text{Depreciation expense each year} = \frac{\text{Cost} - \text{Residual value}}{\text{Estimated useful life in years}}$$ For partial years if purchased before 15th of month depreciation is taken.	Truck, $25,000; $5,000 residual value, 4-year life. $$\text{Depreciation expense} = \frac{\$25,000 - \$5,000}{4}$$ = **$5,000** per year	**Calculate depreciation expense** Truck, $50,000; $10,000 residual value; 4-year life.
Units-of-production method $$\frac{\text{Depreciation}}{\text{per unit}} = \frac{\text{Cost} - \text{Residual value}}{\text{Total estimated units produced}}$$ Do not depreciate below residual value even if actual units are greater than estimate.	Machine, $5,000; estimated life in units, 900; residual value, $500. Assume first year produced 175 units. $$\text{Depreciation expense} = \frac{\$5,000 - \$500}{900}$$ $$= \frac{\$4,500}{900}$$ = $5 depreciation per unit 175 units × $5 = **$875** depreciation expense	**Calculate depreciation expense** Machine, $4,000; estimated life in units, 700; residual value, $500. Assume first year produced 150 units.
Declining-balance method An accelerated method. Residual value not subtracted from cost in depreciation schedule. Do not depreciate below residual value. $$\frac{\text{Depreciation expense each year}}{} = \frac{\text{Book value of equipment at beginning of year}}{} \times \frac{\text{Depreciation rate}}{}$$	Truck, $50,000; estimated life, 5 years; residual value, $10,000. $\frac{1}{5} = 20\% \times 2 = 40\%$ (assume double the straight-line rate) Year: 1, Cost: $50,000, Depreciation expense: **$20,000** ($50,000 × .40), Book value at end of year: $30,000 ($50,000 − $20,000) Year: 2, Cost: $50,000, Depreciation expense: **$12,000** ($30,000 × .40), Book value at end of year: $18,000 ($50,000 − $32,000)	**Calculate depreciation expense and book value for 2 years** Truck, $40,000; estimated life, 4 years; residual value, $5,000.
MACRS/Tax Bill of 1989, 2010, 2020 After December 31, 1986, depreciation calculation is modified. Tax Act of 1989, 2010, modifies way to depreciate equipment. The new tax act of 2020 speeds up depreciation.	Auto: $8,000, 5 years. First year, .20 × $8,000 = **$1,600** depreciation expense	Auto: $7,000, 5 years. Second year = ? depreciation expense

Key Terms

Accelerated Cost
  Recovery System
  (ACRS)

Accelerated
  depreciation

Accumulated
  depreciation

Asset cost

Book value

Declining-balance
  method

Depreciation

Depreciation expense

Depreciation schedule

Estimated useful life

General Depreciation
  System (GDS)

Modified Accelerated
  Cost Recovery System
  (MACRS)

Residual value

Salvage value

Straight-line method

Straight-line rate

Trade-in value

Units-of-production
  method

* Worked-out solutions are in Appendix A.

Critical Thinking Discussion Questions with Chapter Concept Check

1. What is the difference between depreciation expense and accumulated depreciation? Why does the book value of an asset never go below the residual value?

2. Compare the straight-line method to the units-of-production method. Should both methods be based on the passage of time?

3. Why is it possible in the declining-balance method for a person to depreciate below the residual value by mistake?

4. Explain the Modified Accelerated Cost Recovery System. Do you think this system will be eliminated in the future?

5. **Chapter Concept Check.** Search the web for a car of your choice and use concepts from this chapter to provide a depreciation schedule for the car.

6. **Chapter Concept Check.** How has the pandemic affected depreciation of assets?

End-of-Chapter Problems

Name _____ Date _____

Check figures for odd-numbered problems in Appendix A.

Drill Problems

From the following facts, complete a depreciation schedule by using the straight-line method:
LU 17–1(2)

Given Cost of Honda Accord Hybrid $40,000

Residual value $10,000

Estimated life   6 years

End of year	Cost of Accord	Depreciation expense for year	Accumulated depreciation at end of year	Book value at end of year
17–1.				
17–2.				
17–3.				
17–4.				
17–5.				
17–6.				

From the following facts, prepare a depreciation schedule using the declining-balance method
(twice the straight-line rate):   *LU 17–3(2)*

Given Chevrolet Colorado $25,000

Residual value $ 5,000

Estimated life   5 years

End of year	Cost of Chevy truck	Accumulated depreciation at beginning of year	Book value at beginning of year	Depreciation expense for year	Accumulated depreciation at end of year	Book value at end of year
17–7.						
17–8.						
17–9.						
17–10.						

End-of-Chapter Problems (Continued)

For the first 2 years, calculate the depreciation expense for a $7,000 car under MACRS. This is a nonluxury car.   *LU 17–4(2)*

	MACRS			**MACRS**
17–11. Year 1		**17–12.** Year 2		

Complete the following table given this information:

Cost of machine	$94,000	Estimated units machine will produce	100,000	
Residual value	$ 4,000	Actual production:	**Year 1**	**Year 2**
Useful life	5 years		60,000	15,000

		Depreciation Expense	
Method		**Year 1**	**Year 2**
17–13. Straight line *LU 17–1(2)*			
17–14. Units of production *LU 17–2(2)*			
17–15. Declining balance *LU 17–3(2)*			
17–16. MACRS (5-year class) *LU 17–4(2)*			

Word Problems

17–17. Shearer's Foods, part of the $374 billion global snack food industry, employs 3,300 people in Brewster, Ohio. If Shearer's purchased a packaging unit for $185,000 with a life expectancy of 695,000 units and a residual value of $46,000, what is the depreciation expense for year 1 if 75,000 units were produced?   *LU 17–2(2)*

excel 17–18. Lena Horn bought a Toyota Tundra on January 1 for $30,000 with an estimated life of 5 years. The residual value of the truck is $5,000. Assume a straight-line method of depreciation. **(a)** What will be the book value of the truck at the end of year 4? **(b)** If the Tundra was bought the first year on April 12, how much depreciation would be taken the first year?   *LU 17–1(2)*

excel 17–19. Jim Company bought a machine for $36,000 with an estimated life of 5 years. The residual value of the machine is $6,000. Calculate **(a)** the annual depreciation and **(b)** the book value at the end of year 3. Assume straight-line depreciation. *LU 17–1(2)*

excel 17–20. Using Problem 17–19, calculate the first 2 years' depreciation, assuming the units-of-production method. This machine is expected to produce 120,000 units. In year 1, it produced 19,000 units, and in year 2, 38,000 units. *LU 17–2(2)*

17–21. CNBC reported that one in five consumers who purchase bitcoin do so using their credit card. Melissa Gamez purchased a used RV with 19,000 miles for $46,900. Originally the RV sold for $70,000 with a residual value of $20,000. After subtracting the residual value, depreciation allowance per mile was $.86. How much was Melissa's purchase price over or below the book value? Does she have any equity that might assist her with purchasing bitcoin? *LU 17–2(1)*

17–22. Volkswagen car sales hit record high sales of $6.23 million. If one of Volkswagen's assembly lines purchased a new quality control computer verifying VIN numbers for $7,985 with a 5-year life and residual value of $1,100, what is the depreciation expense in year 2 for the QC computer? Use the straight-line method. *LU 17–1(2)*

17–23. CNN reported that higher-income shoppers are the biggest bargain shoppers searching Amazon, clicking on their mobile devices and even querying Alexa looking for the best deals. If Rebecca Johnson purchased a state-of-the-art Biologique Recherche facial machine for her home business for $108,000, with a useful life of 3 years and a residual value of $35,000, what would be the book value of the machine after the first year using the straight-line depreciation method? Round your answers to the nearest dollar. *LU 17–1(2)*

17–24. If corporate headquarters for UPS in Atlanta is considering adding to its 96,000+ fleet of delivery vans, what is year 5's depreciation expense using MACRS if one van costs $78,500? *LU 17–4(2)*

End-of-Chapter Problems (Continued)

17–25. You purchased a MacBook Pro computer for your home office for $3,844. Using the straight-line method, how much can you depreciate each year for five years?

17–26. Businesswire.com reported a shortage of inventory for new and used cars during the pandemic mainly due to mass transit riders choosing to purchase a vehicle for safer transportation coupled with the ease of the transformed digital buying process along with supply chain disruptions. New vehicle sales doubled during the pandemic pushing the average price of both new and used vehicles up to an estimated $39,500 in 2021. If a new car loses 20% of its value the first year and 15% annually until year 10 when it is worth 10% of its original cost, what will the estimated value of your $45,680 business vehicle be in year 4? If using the declining-balance method, what is the depreciation expense for year 4 with an estimated life of 5 years and residual value of $19,000? Round answers to the nearest dollar.

17–27. Using the information from Problem 17–26, what is the MACRS depreciation expense for year 4 of your nonluxury business vehicle? Round to the nearest dollar.

Challenge Problems

17–28. A delivered price (including attachments) of a crawler dozer tractor is $135,000 with a residual value of 35%. The useful life of the tractor is 7,700 hours. *LU 17–2(2)*

    a. What is the total amount of depreciation allowed?

    b. What is the amount of depreciation per hour?

    c. If the tractor is operated five days a week for an average of $7\frac{1}{4}$ hours a day, what would be the depreciation for the first year?

    d. If the hours of operation were the same each year, what would be the total number of years of useful life for the tractor? Round years to the nearest whole number.

17–29. Assume a piece of equipment was purchased July 26, 2022, at a cost of $72,000. The estimated residual value is $5,400 with a useful life of 5 years. Assume a production life of 60,000 units. Compute the depreciation for years 2022 and 2023 using **(a)** straight-line and **(b)** units-of-production (in 2022, 5,000 units were produced and in 2023, 18,000 units were produced).   *LU 17–1(2), LU 17–2(2)*

Summary Practice Test

Do you need help? Connect videos have step-by-step worked-out solutions.

1. Leo Lucky, owner of a Pizza Hut franchise, bought a delivery truck for $30,000. The truck has an estimated life of 5 years with a residual value of $10,000. Leo wants to know which depreciation method will be the best for his truck. He asks you to prepare a depreciation schedule using the declining-balance method at twice the straight-line rate.   *LU 17–3(2)*

2. Using MACRS, what is the depreciation for the first year on furniture costing $12,000? *LU 17–4(2)*

excel 3. Abby Matthew bought a new Jeep Commander for $30,000. The Jeep Commander has a life expectancy of 5 years with a residual value of $10,000. Prepare a depreciation schedule for the straight-line method.   *LU 17–1(2)*

excel 4. Car.com bought a Toyota for $28,000. The Toyota has a life expectancy of 10 years with a residual value of $3,000. After 3 years, the Toyota was sold for $19,000. What was the difference between the book value and the amount received from selling the car if Car.com used the straight-line method of depreciation?   *LU 17–1(2)*

5. A machine cost $70,200; it had an estimated residual value of $6,000 and an expected life of 300,000 units. What would be the depreciation in year 3 if 60,000 units were produced? (Round to nearest cent.)   *LU 17–2(2)*

MY MONEY

 ## Work@Home

 What I need to know

The coronavirus pandemic of 2020–2021 placed many in a work from home situation. As a result of that experience, you may now be considering starting a small business or possibly running a business from your home. Consider the pros and cons of being an entrepreneur. The flexibility and freedom that come with doing what you love may outweigh for you the challenges of administrative details, no regular salary, and learning how to be competitive. Having a home office provides one of the best-known tax advantages for a home-based business: a percentage of your household expenses (mortgage, utilities, property taxes, insurance, Internet, vehicle usage, etc.) may be tax deductible. Consult with a tax accountant for details.

 What I need to do

Give thought to your financial goals and whether opening a small/home business is in line with your goals. Part of this consideration should take into account the rationale behind a decision to start a business. Are you seeking to satisfy a particular niche in the market? Are you interested in supplementing your income with this business venture? Is there a clear demand for the type of product or service offering you are considering as part of your business? Should you launch your business through existing online marketplaces such as eBay, Amazon, Facebook Marketplace, or Etsy? These can be complex questions, but determining your answers related to these topics can help you determine if there is an opportunity for success. Consulting your local Small Business Development Center (SBDC) and/or SCORE (Service Corps of Retired Executives) can assist you with this important assessment process.

Determine the amount of time and resources you can allocate toward running your own business. In the beginning there will be a significant amount of time needed to start your business. Determine if you can work another job in addition to the work on your small/home business or if it makes sense to devote 100% of your time toward your business venture. Consider the financial investment needed to launch a business and the sources where you can generate these funds. For instance, you may need to take on partners to help finance the venture or possibly take out loans to cover your expenses. The financing option you choose will impact whether or not you can achieve the business goals you have set for yourself.

 Steps I need to take

1. Identify your business idea and create a business plan and marketing plan to determine the viability of such a business.
2. Determine if there is demand for your product and/or service. Seek consultation from SBDC and/or SCORE.
3. Devote the necessary time and resources to make your business successful.

 Resources I can use

- https://www.thebalance.com/taxes-and-deductions-for-home-based-business-398592—tax considerations for home-based businesses.
- https://www.nerdwallet.com/article/small-business/20-apps-small-business-owners—apps to assist in running your small/home business.
- https://blog.hubspot.com/sales/how-to-write-business-proposal—how to write a business plan.
- https://blog.hubspot.com/marketing/marketing-plan-examples—how to write a marketing plan.

MY MONEY ACTIVITY

- Assume you want to open a small/home business, what would it be?
- Explain the reasons behind your decision to start your own business.
- Make a list of the resources you will need and begin analyzing the market, your capabilities, and financial resources to help determine if you should pursue the business idea.

PERSONAL FINANCE

A KIPLINGER APPROACH

"SAVE MONEY WITH AN ELECTRIC CAR." Kiplinger's. February, 2021.

SAVE MONEY
WITH AN ELECTRIC CAR

Although some are still a pricey luxury, we found EVs that are fun, affordable and eligible for tax incentives. BY DAVID MUHLBAUM

PEOPLE BUY ELECTRIC CARS FOR ALL SORTS OF REASONS—a common one is the I'm-so-green statement they make. In our polarized times, that message has met with backlash, even vandalism. But how about this for a reason to buy electric that plays to a fundamental American (and *Kiplinger's*) value: They can save you money.

That's *can*, not *will*. You'll have to align a number of variables to make the savings work. But the potential is there.

Business Math Issue

Electric cars will depreciate faster than traditional cars.

1. List the key points of the article and information to support your position.

2. Write a group defense of your position using math calculations to support your view. If you are in an online course, post to a discussion board.

Chapter 18
Inventory and Overhead

Learning Unit Objectives

LU 18–1: Assigning Costs to Ending Inventory—Specific Identification; Weighted Average; FIFO; LIFO

1. List the key assumptions of each inventory method.
2. Calculate the cost of ending inventory and cost of goods sold for each inventory method.

LU 18–2: Retail Method; Gross Profit Method; Inventory Turnover; Distribution of Overhead

1. Calculate the cost ratio and ending inventory at cost for the retail method.
2. Calculate the estimated inventory using the gross profit method.
3. Explain and calculate inventory turnover.
4. Explain overhead; allocate overhead according to floor space and sales.

Q Essential Question

How can I use inventory and overhead to understand the business world and make my life easier?

🌐 Math Around the World

The *Wall Street Journal* chapter opener clip, "New Tactics Are Tested To Trim Inventory Glut", illustrates how the pandemic affected inventory control. The pandemic has resulted in many out of stock items as well as a glut of other items. The two methods that a company can use to monitor its inventory are the *perpetual* method and the *periodic* method.

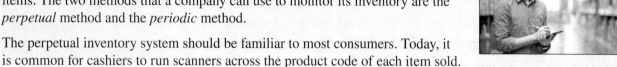

The perpetual inventory system should be familiar to most consumers. Today, it is common for cashiers to run scanners across the product code of each item sold. These scanners read pertinent information into a computer terminal, such as the item's number, department, and price. The computer then uses the **perpetual inventory system** as it subtracts outgoing merchandise from inventory and adds incoming merchandise to inventory. However, as you probably know, the computer cannot be completely relied on to maintain an accurate count of merchandise in stock. Since some products may be stolen or lost, periodically a physical count is necessary to verify the computer count.

With the increased use of computers, many companies are changing to a perpetual inventory system of maintaining inventory records. Some small stores, however, still use the **periodic inventory system.** This system usually does not keep a running account of a store's inventory but relies only on a physical inventory count taken at least once a year. The store then uses various accounting methods to value the cost of its merchandise. In this chapter we discuss the periodic method of inventory.

You may wonder why a company should know the status of its inventory. In Chapter 16 we introduced you to the balance sheet and the income statement. Companies cannot accurately prepare these statements unless they have placed the correct value on their inventory. To do this, a company must know (1) the cost of its ending inventory (found on the balance sheet) and (2) the cost of the goods (merchandise) sold (found on the income statement).

No longer do retailers get a few seasonal deliveries; they now receive new items often, maybe even weekly, to keep their store looking fresh. Frequently, the same type of merchandise flows into a company at different costs. The value assumptions a company makes about the merchandise it sells affect the cost assigned to its ending inventory. Remember that different costs result in different levels of profit on a firm's financial reports.

This chapter begins by using the Blue Company to discuss four common methods (specific identification, weighted average, FIFO, and LIFO) that companies use to calculate the cost of ending inventory and the cost of goods sold. In these methods, the flow of costs does not always match the flow of goods. The chapter continues with a discussion of two methods of estimating ending inventory (retail and gross profit methods), inventory turnover, and the distribution of overhead.

Aha! *A company must declare on its financial statements the inventory method used.*

Steinberg, Julie, and Joe Wallace. "New Tactics Are Tested to Trim Inventory Glut." *The Wall Street Journal*, August 6, 2020.

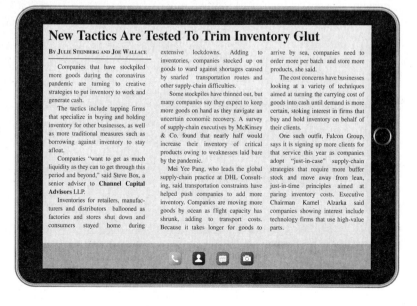

New Tactics Are Tested To Trim Inventory Glut

BY JULIE STEINBERG AND JOE WALLACE

Companies that have stockpiled more goods during the coronavirus pandemic are turning to creative strategies to put inventory to work and generate cash.

The tactics include tapping firms that specialize in buying and holding inventory for other businesses, as well as more traditional measures such as borrowing against inventory to stay afloat.

Companies "want to get as much liquidity as they can to get through this period and beyond," said Steve Box, a senior adviser to **Channel Capital Advisors LLP.**

Inventories for retailers, manufacturers and distributors ballooned as factories and stores shut down and consumers stayed home during extensive lockdowns. Adding to inventories, companies stocked up on goods to ward against shortages caused by snarled transportation routes and other supply-chain difficulties.

Some stockpiles have thinned out, but many companies say they expect to keep more goods on hand as they navigate an uncertain economic recovery. A survey of supply-chain executives by McKinsey & Co. found that nearly half would increase their inventory of critical products owing to weaknesses laid bare by the pandemic.

Mei Yee Pang, who leads the global supply-chain practice at DHL Consulting, said transportation constraints have helped push companies to add more inventory. Companies are moving more goods by ocean as flight capacity has shrunk, adding to transport costs. Because it takes longer for goods to arrive by sea, companies need to order more per batch and store more products, she said.

The cost concerns have businesses looking at a variety of techniques aimed at turning the carrying cost of goods into cash until demand is more certain, stoking interest in firms that buy and hold inventory on behalf of their clients.

One such outfit, Falcon Group, says it is signing up more clients for that service this year as companies adopt "just-in-case" supply-chain strategies that require more buffer stock and move away from lean, just-in-time principles aimed at paring inventory costs. Executive Chairman Kamel Alzarka said companies showing interest include technology firms that use high-value parts.

Learning Unit 18–1:

Assigning Costs to Ending Inventory— Specific Identification; Weighted Average; FIFO; LIFO

Blue Company is a small artist supply store. Its beginning inventory is 40 tubes of art paint that cost $320 (at $8 a tube) to bring into the store. As shown in Figure 18.1, Blue made additional purchases in April, May, October, and December. Note that because of inflation and other competitive factors, the cost of the paint rose from $8 to $13 per tube. At the end of December, Blue had 48 unsold paint tubes. During the year, Blue had 120 paint tubes to sell. Blue wants to calculate (1) the cost of ending inventory (not sold) and (2) the cost of goods sold.

Learn: Specific Identification Method

Companies use the **specific identification method** when they can identify the original purchase cost of an item with the item. For example, Blue Company color codes its paint tubes as they come into the store. Blue can then attach a specific invoice price to each paint tube. This makes the flow of goods and flow of costs the same. Then, when Blue computes its ending inventory and cost of goods sold, it can associate the actual invoice cost with each item sold and in inventory.

To help Blue calculate its inventory with the specific identification method, use the steps that follow.

FIGURE 18.1 Blue Company—a case study

Calculating The Specific Identification Method

Step 1 Calculate the cost of goods (merchandise available for sale).

Step 2 Calculate the cost of the ending inventory.

Step 3 Calculate the cost of goods sold (Step 1 – Step 2).

	Number of units purchased	Cost per unit	Total cost	
Beginning inventory	40	$ 8	$ 320	
First purchase (April 1)	20	9	180	
Second purchase (May 1)	20	10	200	
Third purchase (October 1)	20	12	240	
Fourth purchase (December 1)	20	13	260	
Goods (merchandise) available for sale	120		$1,200	← **Step 1**
Units sold	72			
Units in ending inventory	48			

First, Blue must actually count the tubes of paint on hand. Since Blue coded these paint tubes, it can identify the tubes with their purchase cost and multiply them by this cost to arrive at a total cost of ending inventory. Let's do this now.

	Cost per unit	Total cost	
20 units from April 1	$ 9	$180	
20 units from October 1	12	240	
8 units from December 1	13	104	
Cost of ending inventory		$524	← Step 2

Blue uses the following cost of goods sold formula to determine its cost of goods sold:

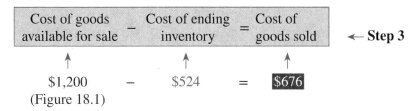

$$\underset{\substack{\uparrow \\ \$1,200}}{\text{Cost of goods available for sale}} - \underset{\substack{\uparrow \\ \$524}}{\text{Cost of ending inventory}} = \underset{\substack{\uparrow \\ \$676}}{\text{Cost of goods sold}} \quad \leftarrow \textbf{Step 3}$$

(Figure 18.1)

Aha! Note that the $1,200 for cost of goods available for sale comes from Figure 18.1. *Remember we are focusing our attention on Blue's purchase costs. Blue's actual selling price does not concern us here.*

Now let's look at how Blue would use the weighted-average method.

Learn: Weighted-Average Method[1]

The **weighted-average method** prices the ending inventory by using an average unit cost. Let's replay Blue Company and use the weighted-average method to find the average unit cost of its ending inventory and its cost of goods sold. Blue would use the steps that follow.

Calculating The Weighted-Average Method

Step 1 Calculate the average unit cost.

Step 2 Calculate the cost of the ending inventory.

Step 3 Calculate the cost of goods sold.

---

[1] Virtually all countries permit the use of the weighted-average method.

In the table that follows, Blue makes the calculation using the above steps.

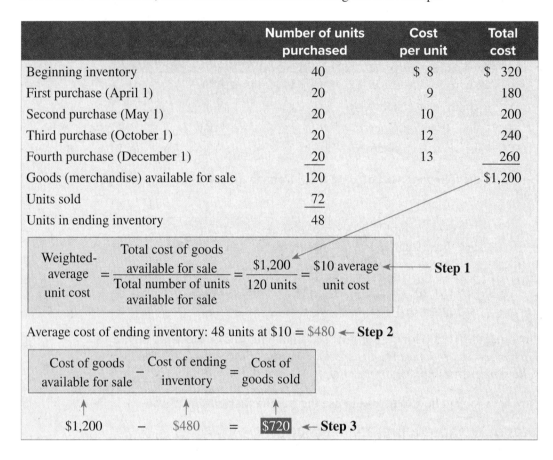

	Number of units purchased	Cost per unit	Total cost
Beginning inventory	40	$ 8	$ 320
First purchase (April 1)	20	9	180
Second purchase (May 1)	20	10	200
Third purchase (October 1)	20	12	240
Fourth purchase (December 1)	20	13	260
Goods (merchandise) available for sale	120		$1,200
Units sold	72		
Units in ending inventory	48		

$$\text{Weighted-average unit cost} = \frac{\text{Total cost of goods available for sale}}{\text{Total number of units available for sale}} = \frac{\$1,200}{120 \text{ units}} = \$10 \text{ average unit cost} \quad \longleftarrow \textbf{Step 1}$$

Average cost of ending inventory: 48 units at $10 = $480 ← **Step 2**

$$\text{Cost of goods available for sale} - \text{Cost of ending inventory} = \text{Cost of goods sold}$$

$$\$1,200 \quad - \quad \$480 \quad = \quad \boxed{\$720} \quad \longleftarrow \textbf{Step 3}$$

Remember that some of the costs we used to determine the average unit cost were higher and others were lower. The weighted-average method, then, calculates an *average unit price* for goods. Companies with similar units of goods, such as rolls of wallpaper, often use the weighted-average method. Also, companies with homogeneous products such as fuels and grains may use the weighted-average method.

Now let's see how Blue Company would value its inventory with the FIFO method.

Learn: FIFO—First-In, First-Out Method

The **first-in, first-out (FIFO)** inventory valuation method assumes that the first goods (paint tubes for Blue) brought into the store are the first goods sold. Thus, FIFO assumes that each sale is from the oldest goods in inventory. FIFO also assumes that the inventory remaining in the store at the end of the period is the most recently acquired goods. This cost flow assumption may or may not hold in the actual physical flow of the goods. An example of a corporation using the FIFO method is Gillette Corporation.

Use the following steps to calculate inventory with the FIFO method.

Calculating The FIFO Inventory

Step 1 List the units to be included in the ending inventory and their costs.

Step 2 Calculate the cost of the ending inventory.

Step 3 Calculate the cost of goods sold.

In the table that follows, we show how to calculate FIFO for Blue using the above steps.

FIFO (bottom up)	Number of units purchased	Cost per unit	Total cost
Beginning inventory	40	$ 8	$ 320
First purchase (April 1)	20	9	180
Second purchase (May 1)	20	10	200
Third purchase (October 1)	20	12	240
Fourth purchase (December 1)	20	13	260
Goods (merchandise) available for sale	120		$1,200
Units sold	72		
Units in ending inventory	48		

20 units from December 1 purchased at $13 $260

20 units from October 1 purchased at $12   ← Step 1 →   240

8 units from May 1 purchased at $10   80

48 units result in an ending inventory cost of $580 ← Step 2

$$\text{Cost of goods available for sale} - \text{Cost of ending inventory} = \text{Cost of goods sold}$$

$1,200 − $580 = $620 ← Step 3

In FIFO, the cost flow of goods tends to follow the physical flow. For example, a fish market could use FIFO because it wants to sell its old inventory first. Note that during inflation, FIFO produces a higher income than other methods. So companies using FIFO during this time must pay more taxes.

We conclude this unit by using the LIFO method to value Blue Company's inventory.

Learn: LIFO—Last-In, First-Out Method

If Blue Company chooses the **last-in, first-out (LIFO)** method of inventory valuation, then the goods sold by Blue will be the last goods brought into the store. The ending inventory would consist of the old goods that Blue bought earlier.

You can calculate inventory with the LIFO method by using the steps that follow.

Calculating The LIFO Inventory

Step 1 List the units to be included in the ending inventory and their costs.

Step 2 Calculate the cost of the ending inventory.

Step 3 Calculate the cost of goods sold.

Now we use the above steps to calculate LIFO for Blue.

LIFO (top down)	Number of units purchased	Cost per unit	Total cost
Beginning inventory	40	$ 8	$ 320
First purchase (April 1)	20	9	180
Second purchase (May 1)	20	10	200
Third purchase (October 1)	20	12	240
Fourth purchase (December 1)	20	13	260
Goods (merchandise) available for sale	120		$1,200
Units sold	72		
Units in ending inventory	48		

40 units of beginning inventory at $8 $320

8 units from April at $9   ← Step 1 →   72

48 units result in an ending inventory cost of $392 ← Step 2

$$\text{Cost of goods available for sale} - \text{Cost of ending inventory} = \text{Cost of goods sold}$$

$1,200   − $392   = $808   ← Step 3

Although LIFO doesn't always match the physical flow of goods, companies do still use it to calculate the flow of costs for products such as DVDs and computers, which have declining replacement costs. Also, during inflation, LIFO produces less income than other methods. This results in lower taxes for companies using LIFO.

Before concluding this unit, we will make a summary for the cost of ending inventory and cost of goods sold under the weighted-average, FIFO, and LIFO methods. From this summary, you can see that in times of rising prices, LIFO gives the highest cost of goods sold ($808). This results in a tax savings for Blue. The weighted-average method tends to smooth out the fluctuations between LIFO and FIFO and falls in the middle.

The key to this discussion of inventory valuation is that different costing methods produce different results. So management, investors, and potential investors should understand the different inventory costing methods and should know which method a particular company uses. For example, Fruit of the Loom, Inc., changed its inventories from LIFO to FIFO due to cost reductions.

Inventory method	Cost of goods available for sale	Cost of ending inventory	Cost of goods sold
Weighted average	$1,200	$480	$1,200 − $480 = $720
		Step 1:	
		Total goods, $1,200	
		Total units, 120	
		$\dfrac{\$1,200}{120} = \$10$	
		Step 2: $10 \times 48 = \$480$	
FIFO	$1,200	Bottom up to inventory level (48)	$1,200 − $580 = $620
		$20 \times \$13 = \$260$	
		$20 \times \$12 = 240$	
		$8 \times \$10 = \underline{80}$	
		$\$580$	
LIFO	$1,200	Top down to inventory level (48)	$1,200 − $392 = $808
		$40 \times \$8 = \$320$	
		$8 \times \$9 = \underline{72}$	
		$\$392$	

Let's check your understanding of this unit with a Practice Quiz.

Money Tip Take inventory of your assets and record all serial numbers. Save receipts as proof of purchase for high dollar items. Take a video of assets and store the file in a fireproof container offsite in the event of a loss to assist with filing an insurance claim.

Learning Unit 18–1

Practice Quiz
Complete this Practice Quiz to see how you are doing.

From the following, calculate (a) the cost of ending inventory and (b) the cost of goods sold under the assumption of (1) weighted-average method, (2) FIFO, and (3) LIFO (ending inventory shows 72 units):

	Number of books purchased for resale	Cost per unit	Total
January 1 inventory	30	$3	$ 90
March 1	50	2	100
April 1	20	4	80
November 1	60	6	360

✓ Solutions

1. a. 72 units of ending inventory × $3.94 = $283.68 cost of ending inventory

($630 ÷ 160)

b.

$$\begin{array}{ccc} \text{Cost of goods} & \text{Cost of ending} & \text{Cost of} \\ \text{available for sale} - \text{inventory} & = & \text{goods sold} \\ \downarrow & \downarrow & \downarrow \\ \$630 \quad - \quad \$283.68 & = & \boxed{\$346.32} \end{array}$$

2. a.

60 units from November 1 purchased at $6	$360
12 units from April 1 purchased at $4	48
72 units Cost of ending inventory	$408

b.

$$\begin{array}{ccc} \text{Cost of goods} & \text{Cost of ending} & \text{Cost of} \\ \text{available for sale} - \text{inventory} & = & \text{goods sold} \\ \downarrow & \downarrow & \downarrow \\ \$630 \quad - \quad \$408 & = & \boxed{\$222} \end{array}$$

3. a.

30 units from January 1 purchased at $3	$ 90
42 units from March 1 purchased at $2	84
72 Cost of ending inventory	$174

b.

$$\begin{array}{ccc} \text{Cost of goods} & \text{Cost of ending} & \text{Cost of} \\ \text{available for sale} - \text{inventory} & = & \text{goods sold} \\ \downarrow & \downarrow & \downarrow \\ \$630 \quad - \quad \$174 & = & \boxed{\$456} \end{array}$$

Retail Method; Gross Profit Method; Inventory Turnover; Distribution of Overhead

Customers want stores to have products available for sale as soon as possible. This has led to outsourced warehouses offshore where tens of thousands of products can be stored ready to be quickly shipped to various stores.

When retailers receive their products, they go into one of their most important assets—their inventory. When the product is sold, it must be removed from inventory so it can be replaced or discontinued. Often these transactions occur electronically at the registers that customers use to pay for products. How is inventory controlled when the register of the store cannot perform the task of adding and subtracting products from inventory?

Convenience stores often try to control their inventory by taking physical inventories. This can be time-consuming and expensive. Some stores draw up monthly financial reports but do not want to spend the time or money to take a monthly physical inventory. The *Wall Street Journal* clip to the left shows how robots are the future of inventory control.

Many stores estimate the amount of inventory on hand. Stores may also have to estimate their inventories when they have a loss of goods due to fire, theft, flood, and the like. This unit begins with two methods of estimating the value of ending inventory—the *retail method* and the *gross profit method*.

BY JENNIFER SMITH

Retailers that have been turning to robots to handle inventory in warehouses are testing whether machines can handle a new task: detecting when store shelves need restocking.

Keeping track of inventory and doing it quickly has become one of the most pressing supply-chain concerns for merchants as they try to put into place new strategies for selling and delivering goods under the quickly changing demands of e-commerce.

Services including rapid home delivery and buy-online-pickup-in-store are pushing retailers to blur the lines between distribution centers and stores—and obscure their view of how many items may be in stock and where the goods are held.

The complicated blending of inventories in stores and warehouses has some retailers testing the use of shelf-scanning robots that roam store aisles and send restocking data back through their networks.

Learn: Retail Method

Many companies use the **retail method** to estimate their inventory. As shown in Figure 18.2, this method does not require that a company calculate an inventory cost for each item. To calculate the $3,500 ending inventory in Figure 18.2, Green Company used the steps that follow:

Smith, Jennifer. "Robots Get New Retail Job." *The Wall Street Journal,* January 28, 2019.

Calculating The Retail Method

Step 1 Calculate the cost of goods available for sale at cost and retail: $6,300; $9,000.

Step 2 Calculate a cost ratio using the following formula:

$$\frac{\text{Cost of goods available for sale at cost}}{\text{Cost of goods available for sale at retail}} = \frac{\$6,300}{\$9,000} = .70$$

Step 3 Deduct net sales from cost of goods available for sale at retail: $9,000 − $4,000.

Step 4 Multiply the cost ratio by the ending inventory at retail: .70 × $5,000.

Now let's look at the gross profit method.

FIGURE 18.2
Estimating inventory
with the retail
method

	Cost	Retail
Beginning inventory	$4,000	$6,000
Net purchases during month	2,300	3,000
Cost of goods available for sale **(Step 1)**	$6,300	$9,000
Less net sales for month		4,000 **(Step 3)**
Ending inventory at retail		$5,000
Cost ratio ($6,300 ÷ $9,000) **(Step 2)**		70%
Ending inventory at cost (.70 × $5,000) **(Step 4)**		$3,500

Learn: Gross Profit Method

To use the **gross profit method** to estimate inventory, the company must keep track of (1) average gross profit rate, (2) net sales at retail, (3) beginning inventory, and (4) net purchases. You can use the following steps to calculate the gross profit method:

Calculating The Gross Profit Method

Step 1 Calculate the cost of goods available for sale (Beginning inventory + Net purchases).

Step 2 Multiply the net sales at retail by the complement of the gross profit rate. This is the estimated cost of goods sold.

Step 3 Calculate the cost of estimated ending inventory (Step 1 – Step 2).

Example: Assume Radar Company has the following information in its records:

Gross profit on sales	30%
Beginning inventory, January 1, 2022	$20,000
Net purchases	$ 8,000
Net sales at retail for January	$12,000

If you use the gross profit method, what is the company's estimated inventory?

The gross profit method calculates Radar's estimated cost of ending inventory at the end of January as follows:

Goods available for sale		
Beginning inventory, January 1, 2022		$20,000
Net purchases		8,000
Cost of goods available for sale		$28,000 ← **Step 1**
Less estimated cost of goods sold:		
Net sales at retail	$12,000	
Cost percentage (100% − 30%) **Step 2 →**	.70	
Estimated cost of goods sold		8,400
Estimated ending inventory, January 31, 2022		$19,600 ← **Step 3**

Note that the cost of goods available for sale less the estimated cost of goods sold gives the estimated cost of ending inventory.

Since this chapter has looked at inventory flow, let's discuss inventory turnover—a key business ratio.

Learn: Inventory Turnover

Inventory turnover is the number of times the company replaces inventory during a specific time. Companies use the following two formulas to calculate inventory turnover:

$$\text{Inventory turnover at retail} = \frac{\text{Net sales}}{\text{Average inventory at retail}}$$

$$\text{Inventory turnover at cost} = \frac{\text{Cost of goods sold}}{\text{Average inventory at cost}}$$

The **average inventory at cost** can be calculated by:

$$\text{Average inventory} = \frac{\text{Beginning inventory} + \text{Ending inventory}}{2}$$

Aha! *Note that inventory turnover at retail is usually lower than inventory turnover at cost. This is due to theft, markdowns, spoilage, and so on.* Also, retail outlets and grocery stores usually have a high turnover, but jewelry and appliance stores have a low turnover.

Now let's use an example to calculate the inventory turnover at retail and at cost.

Example: The following facts are for Abby Company, a local sporting goods store (rounded to the nearest hundredth):

Net sales	$32,000	Cost of goods sold	$22,000
Beginning inventory at retail	$11,000	Beginning inventory at cost	$ 7,500
Ending inventory at retail	$ 8,900	Ending inventory at cost	$ 5,600

With these facts, we can make the following calculations to determine the inventory turnover at retail and at cost.

$$\text{At retail: } \frac{\$32,000}{\left(\frac{\$11,000 + \$8,900}{2}\right)} = \frac{\$32,000}{\$9,950} = \boxed{3.22}$$

$$\text{At cost: } \frac{\$22,000}{\left(\frac{\$7,500 + \$5,600}{2}\right)} = \frac{\$22,000}{\$6,550} = \boxed{3.36}$$

What Turnover Means   Inventory is often a company's most expensive asset. The turnover of inventory can have important implications. Too much inventory results in the use of needed space, extra insurance coverage, and so on. A low inventory turnover could indicate customer dissatisfaction, too much tied-up capital, and possible product obsolescence. A high inventory turnover might mean insufficient amounts of inventory causing stockouts that may lead to future lost sales. If inventory is moving out quickly, perhaps the company's selling price is too low compared to that of its competitors.

In recent years the **just-in-time (JIT) inventory system** from Japan has been introduced in the United States. Under ideal conditions, manufacturers must have suppliers that will provide materials daily as the manufacturing company needs them, thus eliminating inventories. The companies that are using this system, however, have often not been able to completely eliminate the need to maintain some inventory.

Learn: Distribution of Overhead

In Chapter 16 we studied the cost of goods sold and operating expenses shown on the income statement. The operating expenses included **overhead expenses**—expenses that are *not* directly associated with a specific department or product but that contribute indirectly to the running of the business. Examples of such overhead expenses are rent, taxes, and insurance.

Companies must allocate their overhead expenses to the various departments in the company. The two common methods of calculating the **distribution of overhead** are by (1) floor space (square feet) or (2) sales volume.

Calculations by Floor Space   To calculate the distribution of overhead by floor space, use the steps that follow:

Calculating The Distribution of Overhead by Floor Space

Step 1 Calculate the total square feet in all departments.

Step 2 Calculate the ratio for each department based on floor space.

Step 3 Multiply each department's floor space ratio by the total overhead.

Example: Roy Company has three departments with the following floor space:

Department A	6,000 square feet
Department B	3,000 square feet
Department C	1,000 square feet

Money Tip Track your overhead expenses such as electricity, gas, water, sewage, etc., to determine trends. Make adjustments where needed.

The accountant's job is to allocate $90,000 of overhead expenses to the three departments.

To allocate this overhead by floor space:

	Floor space in square feet	**Ratio**	
Department A	6,000	$\dfrac{6,000}{10,000} = 60\%$	
Department B	3,000	$\dfrac{3,000}{10,000} = 30\%$	← **Steps 1 and 2**
Department C	$\dfrac{1,000}{10,000}$ total square feet	$\dfrac{1,000}{10,000} = 10\%$	

Department A	.60 × $90,000 =	$54,000
Department B	.30 × $90,000 =	27,000 ← **Step 3**
Department C	.10 × $90,000 =	9,000
		$90,000

Calculations by Sales   To calculate the distribution of overhead by sales, use the steps that follow:

Calculating The Distribution of Overhead by Sales

Step 1 Calculate the total sales in all departments.

Step 2 Calculate the ratio for each department based on sales.

Step 3 Multiply each department's sales ratio by the total overhead.

Example: Morse Company distributes its overhead expenses based on the sales of its departments. For example, last year Morse's overhead expenses were $60,000. Sales of its two departments were as follows, along with its ratio calculation.

	Sales	**Ratio**	
Department A	$ 80,000	$\dfrac{\$80,000}{\$100,000} = .80$	⎤
Department B	20,000	$\dfrac{\$20,000}{\$100,000} = .20$	⎦ ← **Steps 1 and 2**
Total sales	$100,000		

Since Department A makes 80% of the sales, it is allocated 80% of the overhead expenses.

These ratios are then multiplied by the overhead expense to be allocated.

Department A   .80 × $60,000 = $48,000
Department B   .20 × $60,000 = 12,000 ← **Step 3**
                             $60,000

It's time to try another Practice Quiz.

Learning Unit 18–2

Practice Quiz

Complete this Practice Quiz to see how you are doing.

1. From the following facts, calculate the cost of ending inventory using the retail method (round the cost ratio to the nearest tenth percent):

January 1—inventory at cost	$ 18,000
January 1—inventory at retail	58,000
Net purchases at cost	220,000
Net purchases at retail	376,000
Net sales at retail	364,000

2. Given the following, calculate the estimated cost of ending inventory using the gross profit method:

Gross profit on sales	40%
Beginning inventory, January 1, 2022	$27,000
Net purchases	$ 7,500
Net sales at retail for January	$15,000

3. Calculate the inventory turnover at cost and at retail from the following (round the turnover to the nearest hundredth):

Average inventory at cost	Average inventory at retail	Net sales	Cost of goods sold
$10,590	$19,180	$109,890	$60,990

4. From the following, calculate the distribution of overhead to Departments A and B based on floor space.

Amount of overhead expense to be allocated	Square footage
$70,000	10,000 Department A
	30,000 Department B

✓ Solutions

	Cost	Retail
1. Beginning inventory	$ 18,000	$ 58,000
Net purchases during the month	220,000	376,000
Cost of goods available for sale	$238,000	$434,000
Less net sales for the month		364,000
Ending inventory at retail		$ 70,000
Cost ratio ($238,000 ÷ $434,000)		54.8%
Ending inventory at cost (.548 × $70,000)		$ 38,360

Practice Quiz *Continued*

2. Goods available for sale

Beginning inventory, January 1, 2022		$ 27,000
Net purchases		7,500
Cost of goods available for sale		$ 34,500
Less estimated cost of goods sold:		
Net sales at retail	$ 15,000	
Cost percentage (100% − 40%)	.60	
Estimated cost of goods sold		9,000
Estimated ending inventory, January 31, 2022		$ 25,500

3. $\text{Inventory turnover at cost} = \dfrac{\text{Cost of goods sold}}{\text{Average inventory at cost}} = \dfrac{\$60,900}{\$10,590} = 5.75$

$\text{Inventory turnover at retail} = \dfrac{\text{Net sales}}{\text{Average inventory at retail}} = \dfrac{\$109,890}{\$19,180} = 5.73$

4.

		Ratio		
Department A	10,000	$\dfrac{10,000}{40,000}$ = .25 × $70,000 =		$17,500
Department B	30,000	$\dfrac{30,000}{40,000}$ = .75 × $70,000 =		52,500
				$70,000

Chapter 18 Review

Topic/Procedure/Formula	Examples	You try it*
Specific identification method Identification could be by serial number, physical description, or coding. The flow of goods and flow of costs are the same.	<table><tr><td></td><td>Cost per unit</td><td>Total cost</td></tr><tr><td>April 1, 3 units at</td><td>$7</td><td>$21</td></tr><tr><td>May 5, 4 units at</td><td>8</td><td>32</td></tr><tr><td></td><td></td><td>$53</td></tr></table>If 1 unit from each group is left, ending inventory is: $1 \times \$7 = \$7$ $+1 \times \ 8 = \underline{\ \ 8}$ $\qquad\qquad \$15$ Cost of goods available for sale − Cost of ending inventory = Cost of goods sold $\$53 \ - \ \$15 \ = \ \boxed{\$38}$	**Calculate ending inventory and cost of goods sold** <table><tr><td></td><td>Cost per unit</td><td>Total cost</td></tr><tr><td>May 1, 4 units at</td><td>$9</td><td></td></tr><tr><td>June 6, 3 units at</td><td>10</td><td></td></tr></table>Assume one unit from each group is left.
Weighted-average method Weighted-average unit cost = $\dfrac{\text{Total cost of goods available for sale}}{\text{Total number of units available for sale}}$	<table><tr><td></td><td>Cost per unit</td><td>Total cost</td></tr><tr><td>1/XX, 4 units at</td><td>$4</td><td>$16</td></tr><tr><td>5/XX, 2 units at</td><td>5</td><td>10</td></tr><tr><td>8/XX, 3 units at</td><td>6</td><td>18</td></tr><tr><td></td><td></td><td>$44</td></tr></table>Unit cost $= \dfrac{\$44}{9} = \$4.89$ If 5 units left, cost of ending inventory is 5 units $\times \$4.89 = \boxed{\$24.45}$	**Calculate unit cost and cost of ending inventory** <table><tr><td></td><td>Cost per unit</td><td>Total cost</td></tr><tr><td>1/XX, 6 units at</td><td>$5</td><td>$30</td></tr><tr><td>5/XX, 4 units at</td><td>6</td><td>24</td></tr><tr><td>8/XX, 5 units at</td><td>7</td><td>35</td></tr><tr><td></td><td></td><td>$89</td></tr></table>4 units left
FIFO—first-in, first-out method Sell old inventory first. Ending inventory is made up of last merchandise brought into store.	Using example above: 5 units left: ↓ <table><tr><td>(Last into store)</td><td>3 units at $6</td><td>$18</td></tr><tr><td></td><td>2 units at $5</td><td>10</td></tr><tr><td>Cost of ending inventory</td><td></td><td>$28</td></tr></table>	**Calculate cost of inventory by FIFO** Use weighted-average example.
LIFO—last-in, first-out method Sell last inventory brought into store first. Ending inventory is made up of oldest merchandise in store.	Using weighted-average example: 5 units left: ↓ <table><tr><td>(First into store)</td><td>4 units at $4</td><td>$16</td></tr><tr><td></td><td>1 unit at $5</td><td>5</td></tr><tr><td>Cost of ending inventory</td><td></td><td>$21</td></tr></table>	**Calculate cost of inventory by LIFO** Use weighted-average example.

Chapter 18 Review (Continued)

Topic/Procedure/Formula	Examples	You try it*
Retail method Ending inventory at cost equals: $\dfrac{\text{Cost of goods available at cost}}{\text{Cost of goods available at retail}} \times \begin{array}{c}\text{Ending}\\ \text{inventory}\\ \text{at retail}\end{array}$ (This is cost ratio.)		**Calculate cost of ending inventory at cost and at retail**

Retail method (Examples)

	Cost	Retail
Beginning inventory	$52,000	$ 83,000
Net purchases	28,000	37,000
Cost of goods available for sale	$80,000	$ 120,000
Less net sales for month		80,000
Ending inventory at retail		$ 40,000

$\text{Cost ratio} = \dfrac{\$80,000}{\$120,000} = .67 = 67\%$

Rounded to nearest percent.

Ending inventory at cost, $26,800

(.67 × $40,000)

Retail method (You try it)

	Cost	Retail
Beginning inventory	$60,000	$80,000
Net purchases	28,000	37,000

Assume net sales of $90,000.

(Round ratio to nearest percent.)

Gross profit method

$\begin{array}{c}\text{Beg.}\\ \text{inv.}\end{array} + \begin{array}{c}\text{Net}\\ \text{purchases}\end{array} - \begin{array}{c}\text{Estimated}\\ \text{cost of}\\ \text{goods}\\ \text{sold}\end{array} = \begin{array}{c}\text{Estimated}\\ \text{ending}\\ \text{inventory}\end{array}$

Goods available for sale

Beginning inventory	$30,000
Net purchases	3,000
Cost of goods available for sale	$33,000
Less: Estimated cost of goods sold:	
Net sales at retail	$18,000
Cost percentage (100% − 30%)	.70
Estimated cost of goods sold	12,600
Estimated ending inventory	$ 20,400

Calculate estimated ending inventory

Given: Net sales at retail of $20,000 and a 75% gross profit.

Goods available for sale

Beginning inventory	$40,000
Net purchases	2,000

Inventory turnover at retail and at cost

$\dfrac{\text{Net sales}}{\begin{array}{c}\text{Average inventory}\\ \text{at retail}\end{array}}$ or $\dfrac{\text{Cost of goods sold}}{\begin{array}{c}\text{Average inventory}\\ \text{at cost}\end{array}}$

$\begin{array}{c}\text{Average}\\ \text{inventory}\end{array} = \dfrac{\begin{array}{c}\text{Beginning}\ \ \text{Ending}\\ \text{inventory} + \text{inventory}\end{array}}{2}$

Inventory, January 1 at cost	$20,000
Inventory, December 31 at cost	48,000
Cost of goods sold	62,000

At cost:

$\dfrac{\$62,000}{\left(\dfrac{\$20,000 + \$48,000}{2}\right)} = 1.82$ (inventory turnover at cost)

Calculate inventory turnover at cost

Jan 1 inventory at cost $40,000
Dec 31 inventory at cost $60,000
Cost of goods sold $90,000

Chapter 18 Review (Continued)

Topic/Procedure/Formula	Examples	You try it*
Distribution of overhead Based on floor space or sales volume, calculate: 1. Ratios of department floor space or sales to the total. 2. Multiply ratios by total amount of overhead to be distributed.	Total overhead to be distributed, $10,000 **Floor space** Department A 6,000 sq. ft. Department B 2,000 sq. ft. 8,000 sq. ft. $\text{Ratio A} = \dfrac{6,000}{8,000} = .75$ $\text{Ratio B} = \dfrac{2,000}{8,000} = .25$ Dept. A = .75 × $10,000 = $7,500 Dept. B = .25 × $10,000 = $2,500	**Calculate overhead cost to each department** Total overhead to be distributed, $30,000 **Floor space** Department A 4,000 sq. ft. Department B 6,000 sq. ft.

Key Terms

Average inventory

Distribution of overhead

First-in, first-out (FIFO) method

Gross profit method

Inventory turnover

Just-in-time (JIT) inventory system

Last-in, first-out (LIFO) method

Overhead expenses

Periodic inventory system

Perpetual inventory system

Retail method

Specific identification method

Weighted-average method

* Worked-out solutions are in Appendix A.

Critical Thinking Discussion Questions with Chapter Concept Check

1. Explain how you would calculate the cost of ending inventory and cost of goods sold for specific identification, FIFO, LIFO, and weighted-average methods. Explain why during inflation LIFO results in a tax savings for a business.

2. Explain the cost ratio in the retail method of calculating inventory. What effect will the increased use of computers have on the retail method?

3. What is inventory turnover? Explain the effect of a high inventory turnover during the Christmas shopping season.

4. How is the distribution of overhead calculated by floor space or sales? Give an example of why a store in your area cut back one department to expand another. Did it work?

5. Discuss how levels of inventory have been affected by the economic crises at your local mall.

6. **Chapter Concept Check.** Explain how robots can strengthen the supply chain during a pandemic.

End-of-Chapter Problems

Name _____ Date _____

Check figures for odd-numbered problems in Appendix A.

Drill Problems

18–1. Using the specific identification method, calculate **(a)** the cost of ending inventory and **(b)** the cost of goods sold given the following: *LU 18–1(2)*

Date	Units purchased	Cost per unit	Ending inventory
June 1	15 Echo Show's 360	$275	2 Echo Show's from June
July 1	45 Echo Show's 360	250	15 Echo Show's from July
August 1	60 Echo Show's 360	240	12 Echo Show's from August

From the following, calculate the **(a)** cost of ending inventory (round the average unit cost to the nearest cent) and **(b)** cost of goods sold using the weighted-average method, FIFO, and LIFO (ending inventory shows 61 units).   *LU 18–1(2)*

	Number purchased	Cost per unit	Total
January 1 inventory	40	$4	$160
April 1	60	7	420
June 1	50	8	400
November 1	55	9	495

18–2. Use weighted average:

18–3. Use FIFO:

18–4. Use LIFO:

End-of-Chapter Problems (Continued)

From the following (18–5 to 18–12), calculate the cost of ending inventory and cost of goods sold for the LIFO (18–13), FIFO (18–14), and weighted-average (18–15) methods (make sure to first find total cost to complete the table); ending inventory is 49 units: *LU 18–1(2)*

	Beginning inventory and purchases	Units	Unit cost	Total dollar cost
18–5.	Beginning inventory, January 1	5	$2.00	
18–6.	April 10	10	2.50	
18–7.	May 15	12	3.00	
18–8.	July 22	15	3.25	
18–9.	August 19	18	4.00	
18–10.	September 30	20	4.20	
18–11.	November 10	32	4.40	
18–12.	December 15	16	4.80	_____

18–13. LIFO:

Cost of ending inventory	Cost of goods sold

18–14. FIFO:

Cost of ending inventory	Cost of goods sold

18–15. Weighted average:

Cost of ending inventory	Cost of goods sold

18–16. From the following, calculate the cost ratio (round to the nearest hundredth percent) and the cost of ending inventory to the nearest cent under the retail method.   *LU 18–2(1)*

Net sales at retail for year	$40,000	Purchases—cost	$14,000
Beginning inventory—cost	$27,000	Purchases—retail	$19,000
Beginning inventory—retail	$49,000		

18–17. Complete the following (round answers to the nearest hundredth): *LU 18–2(3)*

a.	b.	c.	d.	e.	f.
Average inventory at cost	Average inventory at retail	Net sales	Cost of goods sold	Inventory turnover at cost	Inventory turnover at retail
$14,000	$21,540	$70,000	$49,800		

Complete the following (assume $90,000 of overhead to be distributed): *LU 18–2(4)*

	Square feet	Ratio	Amount of overhead allocated
18–18. Department A	10,000		
18–19. Department B	30,000		

18–20. Given the following, calculate the estimated cost of ending inventory using the gross profit method. *LU 18–2(2)*

Gross profit on sales	55%	Net purchases	$ 3,900
Beginning inventory	$29,000	Net sales at retail	$17,000

Word Problems

18–21. If Exxon uses FIFO for its inventory valuation, calculate the cost of ending inventory and cost of goods sold if ending inventory is 110 barrels of crude oil. *LU 18–1(2)*

Beginning inventory and purchases	Barrels	Barrel cost	Total cost
Beginning inventory: Jan 1	125	$ 95	$11,875
March 1	50	101	5,050
June 1	65	98	6,370
September 1	75	90	6,750
December 1	50	103	5,150

End-of-Chapter Problems (Continued)

18–22. Marvin Company has a beginning inventory of 12 sets of paints at a cost of $1.50 each. During the year, the store purchased 4 sets at $1.60, 6 sets at $2.20, 6 sets at $2.50, and 10 sets at $3.00. By the end of the year, 25 sets were sold. Calculate **(a)** the number of paint sets in ending inventory and **(b)** the cost of ending inventory under the LIFO, FIFO, and weighted-average methods. Round to nearest cent for the weighted average.   *LU 18–1(2)*

18–23. Better Finance (previously BillFloat), based in San Francisco, California, provides leasing and credit solutions to consumers and small businesses. If Better Finance wants to distribute $45,000 worth of overhead by sales, calculate the overhead expense for each department: *LU 18–2(4)*

New customer sales (NCS)	$ 5,120,000
Current customer new sales (CCNS)	4,480,000
Current customer loan extension sales (CCLES)	3,200,000

18–24. If Comcast is upgrading its cable boxes and has 500 obsolete boxes in ending inventory, what is the cost of ending inventory using FIFO, LIFO, and the weighted-average method?   *LU 18–1(2)*

Beginning inventory and purchases	Boxes	Box cost	Total cost
Beginning inventory: January 1	15,500	$15	$232,500
March 1	6,500	16	104,000
June 1	2,500	20	50,000
September 1	1,500	23	34,500
December 1	1,000	32	32,000

excel **18–25.** May's Dress Shop's inventory at cost on January 1 was $39,000. Its retail value was $59,000. During the year, May purchased additional merchandise at a cost of $195,000 with a retail value of $395,000. The net sales at retail for the year were $348,000. Calculate May's inventory at cost by the retail method. Round the cost ratio to the nearest whole percent. *LU 18–2(1)*

excel **18–26.** A sneaker outlet has made the following wholesale purchases of new running shoes: 12 pairs at $45, 18 pairs at $40, and 20 pairs at $50. An inventory taken last week indicates that 23 pairs are still in stock. Calculate the cost of this inventory by FIFO. *LU 18–1(2)*

excel **18–27.** Over the past 3 years, the gross profit rate for Jini Company was 35%. Last week a fire destroyed all Jini's inventory. Using the gross profit method, estimate the cost of inventory destroyed in the fire, given the following facts that were recorded in a fireproof safe: *LU 18–2(2)*

Beginning inventory	$ 6,000
Net purchases	64,000
Net sales at retail	49,000

End-of-Chapter Problems (Continued)

18–28. Calculate cost of goods sold and ending inventory for Emergicare's bandages orders using FIFO, LIFO and average cost. There are 35 units in ending inventory.   *LU 18–1(2)*

Date	Units purchased	Cost per unit	Total cost
January 1	50	$7.50	$ 375.00
April 1	45	6.75	303.75
June 1	60	6.50	390.00
September 1	55	7.00	385.00
Total	210		$1,453.75

18–29. Your home business uses 350 square feet of your 1,750 square foot home. If household expenses for the year were $17,558, how much was allotted to your business?   *LU 18–2(4)*

18–30. Increased and volatile demand combined with constrained supply challenged the supply chain during the pandemic. A spokesperson for Clorox stated demand for Clorox wipes increased 500% during the pandemic, as reported by CNN Business. If your local grocery store had the following Clorox wipes stock, calculate cost of ending inventory and cost of goods sold using the weighted-average method, FIFO, and LIFO. Ending inventory is 5 units. Round to the nearest cent.   *LU 18–1(1,2)*

Inventory	Number of units	Cost per unit	Total cost
Beginning inventory	60	$4.50	$270.00
Purchase March 1	25	4.75	118.75
Purchase April 1	5	6.25	31.25
Purchase June 1	2	10.25	20.50
Purchase November 1	12	8.90	106.80

18–31. Because the items you sell for your home business are all unique, you use the specific identification method of inventory. Using the below list of items, what is your cost of ending inventory and cost of goods sold?   *LU 18–1(1,2)*

Item	Cost
Prada boots size 8 women	$ 150
Rocking chair wooden	125 sold
Patio set	225 sold
DeWalt tool set	210
Diamond earrings $\frac{1}{2}$ carat	590
Diamond ring 1 carat	3,500 sold
Dog life jacket XL	55
Metal wheelbarrow	100 sold
Vitamix	200

18–32. As the accounting technician for Jones Landscaping, calculate inventory turnover at retail and inventory turnover at cost based on the following (rounded to the nearest hundredth):   *LU 18–2(3)*

Net sales	$57,190
Beginning inventory at retail	19,500
Ending inventory are retail	8,400
Cost of goods sold	42,600
Beginning inventory at cost	13,295
Ending inventory at cost	9,927

End-of-Chapter Problems (Continued)

Challenge Problems

18–33. Monroe Company had a beginning inventory of 350 cans of paint at $12 each on January 1 at a cost of $4,200. During the year, the following purchases were made:

February 15	280 cans at $14.00
April 30	110 cans at $14.50
July 1	100 cans at $15.00

Monroe marks up its goods at 40% on cost. At the end of the year, ending inventory showed 105 units remaining. Calculate the amount of sales assuming a FIFO flow of inventory.   *LU 18–1(2)*

18–34. Logan Company uses a perpetual inventory system on a FIFO basis. Assuming inventory on January 1 was 800 units at $8 each, what is the cost of ending inventory at the end of October 5?   *LU 18–1(2)*

Received				Sold	
Date	**Quantity**	**Cost per unit**		**Date**	**Quantity**
Apr. 15	220	$5		Mar. 8	500
Nov. 12	1,900	9		Oct. 5	200

Summary Practice Test

Do you need help? Connect videos have step-by-step worked-out solutions.

1. Writing.com has a beginning inventory of 16 sets of pens at a cost of $2.12 each. During the year, Writing.com purchased 8 sets at $2.15, 9 sets at $2.25, 14 sets at $3.05, and 13 sets at $3.20. By the end of the year, 29 sets were sold. Calculate **(a)** the number of pen sets in stock and **(b)** the cost of ending inventory under LIFO, FIFO, and weighted-average methods. *LU 18–1(2)*

2. Lee Company allocates overhead expenses to all departments on the basis of floor space (square feet) occupied by each department. The total overhead expenses for a recent year were $200,000. Department A occupied 8,000 square feet; Department B, 20,000 square feet; and Department C, 7,000 square feet. What is the overhead allocated to Department C? In your calculations, round to the nearest whole percent. *LU 18–2(4)*

3. A local college bookstore has a beginning inventory costing $80,000 and an ending inventory costing $84,000. Sales for the year were $300,000. Assume the bookstore markup rate on selling price is 70%. Based on the selling price, what is the inventory turnover at cost? Round to the nearest hundredth. *LU 18–2(3)*

4. Dollar Dress Shop's inventory at cost on January 1 was $82,800. Its retail value was $87,500. During the year, Dollar purchased additional merchandise at a cost of $300,000 with a retail value of $325,000. The net sales at retail for the year were $295,000. Calculate Dollar's inventory at cost by the retail method. Round the cost ratio to the nearest whole percent. *LU 18–2(1)*

Summary Practice Test (Continued)

5. On January 1, Randy Company had an inventory costing $95,000. During January, Randy had net purchases of $118,900. Over recent years, Randy's gross profit in January has averaged 45% on sales. The company's net sales in January were $210,800. Calculate the estimated cost of ending inventory using the gross profit method.   *LU 18–2(2)*

🔍 Pandemic Effect

 What I need to know

COVID-19 created a social and economic disruption worldwide. Words and phrases such as social distancing, masks, herd immunity, isolation, shelter-in-place, SARS, stay-at-home order, and quarantine became commonplace. To date COVID-19, which began in Wuhan, China, and was reported to the World Health Organization (WHO) on December 31, 2019, has infected over 240 million worldwide and killed over 4.9 million, approximately .062% of the population.

No one could have predicted COVID-19 and how it has affected humanity. We found ourselves in constantly evolving situations across all aspects of our lives. We dealt with tremendous uncertainty which contributed to high levels of stress and anxiety. Births were unwitnessed, people died alone, weddings and funerals were canceled, businesses had to close or reduce hours due to labor shortages, elective procedures and surgeries were delayed due to no hospital space available or no staffing, bidding wars on residential real estate became commonplace, supply chain backups created pent-up demand for many industries including automotive, housing prices soared, travel virtually stopped, borders were closed. Social activities were placed on hold and our academic pursuits were altered into many different formats. Our workplaces underwent incredible changes to incorporate many protocols for dealing with our new reality. Even tasks such as shopping were changed as we dealt with scarcity and panic buying of common household items. We may never view toilet paper the same again.

Now that we have been through such a time, we need to look at how this event impacted us. This includes the way in which we view our finances and the decisions we make in our financial futures. We have all most likely experienced some changes in our personal lives related to how we view our health, priorities, goals, and quality of life. We have gained some valuable insights and it can be helpful to reflect on our experiences during the pandemic and the practical applications we can use in our lives going forward.

 What I need to do

As with many events in our lifetimes, there are lessons to be learned since experience is the greatest teacher. COVID-19 is no exception. We all learned to adjust to the new situation we faced. Reflecting deeply on the lessons learned through our experiences during the pandemic can be a good way to create practical applications from the events we encountered. Assess your priorities in life and how they may have changed as a result of the pandemic. These priorities may include both personal and financial goals and perhaps they shifted. Identify some positive things resulting from your experiences during the pandemic. You may feel a deeper sense of closeness to family and friends having been through a time in which in-person social encounters were drastically limited. Your current financial situation and future financial goals may have been impacted. Your future career pursuits may have altered as you began to see the impact on the business community and workplace environments.

 Steps I need to take

1. Assess what is important to you both personally and financially.
2. Consider your life priorities and goals.
3. Create plans for your personal and financial future.

 Resources I can use

- Success Coach (mobile app)—life management goal setting and tracking.

MY MONEY ACTIVITY

- Reflect on what your life was like during the pandemic.
- How has the way you view your education, career aspirations, finances, relationships, and the future changed?

A KIPLINGER APPROACH

"A New Chapter for This Bookstore." Kiplinger's. March 2021.

MAKING IT WORK

A New Chapter for This Bookstore

Katrina and COVID haven't deterred the owner of a used bookstore from catering to his clientele.

PROFILE

WHO: John Gilman, age 73

OCCUPATION: Owner of Kaboom Books

WHERE: Houston, Texas

How did you become a used bookstore owner in Houston?
It all started about 43 years ago, when I lived in New Orleans. I had too many books, and it behooved me to get out from under them. So I sold books at the flea market in the French Quarter for about a year and a half, at which point I recognized that the customer base was interesting and intelligent and that I was good at selling stuff. So I opened my first shop in the French Quarter, which my wife and I ran for more than 30 years. Before Hurricane Katrina in 2005, I had just finished remodeling our store. After Katrina, I realized business would go into the toilet for about 10 years, and so it seemed to me that the only logical thing was to come to Houston.

How has the pandemic impacted your store during the past year?
The first quarter was excellent, but the second quarter, we immediately shut down because Houston became a hot spot. We stayed shut down for the second and the third quarter. But if you're in small business for a certain amount of time, eventually you understand that some years there are untoward events that just aren't predictable, like Katrina. We had some money put aside, and we folded that money back into the business to keep it going. And we spent our time just tidying up, making sure we had the maximum use of space, and hoping that we could open back up.

How has business been since you reopened? When we reopened in October, I told myself, if we do half of what we would normally do this time of year, I would be accepting of that. We changed our hours to just three days a week, six hours a day, and we've done as much business in those shortened days as we would with our usual eight-hour days. It makes it a bit hectic, but I've been very pleased. Thankfully, we have a very loyal clientele. And it seems people are reading a lot more this year.

Bookstores have long struggled to compete with Amazon. How do you do that during a pandemic? I don't try to compete. What I do is handle what I know, and I don't handle what I know other people are long and strong in. As a used bookstore owner,

the service is having stuff that people aren't finding elsewhere, in very good condition and for a reasonable price. My customers are delighted by this and only with the internet do they express their disgust.

So, you don't sell online? We did try selling online for a while, but I never did like that type of selling. We make enough money out of the shop that we don't have to. We have been using much more social media. I'm a pretty private person; I'm not really nuts about being on social media. But my wife handles it, and it has become a much larger part of our sphere.

What's it like to shop at the store? My basic layout for a bookshop has confirmedly been, from the beginning, labyrinthian. We've got 84 sections, from Earth Sciences to Astronomy, Physics, Social Sciences and so on. But if you're less than 4 feet tall, it really does feel like a labyrinth. Children love it, and I think adults find it cheerful as well. **EMMA PATCH**

Business Math Issue

A small business cannot compete with Amazon's high inventory.

1. List the key points of the article and information to support your position.

2. Write a group defense of your position using math calculations to support your view. If you are in an online course, post to a discussion board.

Chapter 19

Sales, Excise, and Property Taxes

Learning Unit Objectives

LU 19–1: Sales and Excise Taxes

1. Compute sales tax on goods sold involving trade and cash discounts and shipping charges.
2. Explain and calculate excise tax.

LU 19–2: Property Tax

1. Calculate the tax rate in decimal.
2. Convert tax rate in decimal to percent, per $100 of assessed value, per $1,000 of assessed value, and in mills.
3. Compute property tax due.

Essential Question

How can I use sales, excise, and property taxes to understand the business world and make my life easier?

🌐 Math Around the World

The *Wall Street Journal* chapter opener, "A Tax Too Far", reveals the tax advantage of moving to Florida. The article was written two months before the pandemic. One should consider all factors, including taxes, before making a lifetime move.

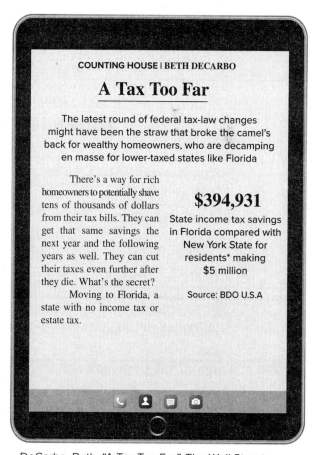

In Learning Unit 19–1 you will learn how sales taxes are calculated. Learning Unit 19–1 discusses the excise tax that is collected in addition to the sales tax. Learning Unit 19–2 explains the use of property tax.

COUNTING HOUSE | BETH DECARBO

A Tax Too Far

The latest round of federal tax-law changes might have been the straw that broke the camel's back for wealthy homeowners, who are decamping en masse for lower-taxed states like Florida

There's a way for rich homeowners to potentially shave tens of thousands of dollars from their tax bills. They can get that same savings the next year and the following years as well. They can cut their taxes even further after they die. What's the secret?

Moving to Florida, a state with no income tax or estate tax.

$394,931

State income tax savings in Florida compared with New York State for residents* making $5 million

Source: BDO U.S.A

DeCarbo, Beth. "A Tax Too Far." *The Wall Street Journal* (October 1, 2020).

Learning Unit 19–1:

Sales and Excise Taxes

Today, many states have been raising their sales tax and excise tax.

Learn: Sales Tax

If you recently bought an item on eBay you may have noticed that sales tax is collected. Before 2018 online sales were exempt.

In many cities, counties, and states, the sellers of certain goods and services collect **sales tax** and forward it to the appropriate government agency. Forty-five states have a sales tax. Of the 45 states, 28 states and the District of Columbia exempt food; 44 states and the District of Columbia exempt prescription drugs. The following Tax Foundation map shows sales tax rates by state.

Sales tax is usually computed electronically by the new cash register systems and scanners. However, it is important to know how sellers calculate sales tax manually. The example of a car battery will show you how to manually calculate sales tax.

Amount of
sales tax

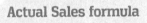

P
($1.08)

B × R
($21.50) (.05)

$21.50 + $1.08 = $22.58
(sale)   (tax
          amount)

Example:

Selling price of a Sears battery	$32.00	Shipping charge	$3.50
Trade discount to local garage	$10.50	Sales tax	5%

Manual calculation

$32.00 − $10.50 = $21.50 taxable
                ×   .05
                  $ 1.08 tax
                + 21.50 taxable
                + 3.50 shipping
                $26.08 total price with tax and shipping

Check

100% is base + 5% is tax = 105%
1.05 × $21.50 = $22.58
                + 3.50 shipping
                $26.08

In this example, note how the trade discount is subtracted from the selling price before any cash discounts are taken. If the buyer is entitled to a 6% cash discount, it is calculated as follows:

Aha!

.06 × $21.50 = $1.29

Remember cash discounts are not taken on sales tax or shipping charges.

Calculating Actual Sales   Managers often use the cash register to get a summary of their total sales for the day. The total sales figure includes the sales tax. So the sales tax must be deducted from the total sales. To illustrate this, let's assume the total sales for the day were $40,000, which included a 7% sales tax. What were the actual sales?

Actual Sales formula

$$\text{Actual sales} = \frac{\text{Total sales}}{1 + \text{Tax rate}}$$

Hint: $40,000 is 107% of actual sales.

$$\text{Actual sales} = \frac{\$40,000}{1.07} = \boxed{\$37,383.18}$$

Total sales
100% sales
+ 7% tax
107% ⟶ 1.07

Copyright © McGraw Hill   (puzzle)frender/Getty Images

How High are Sales Taxes in Your State?

Combined State & Average Local Sales Tax Rates, January 2021

WA 9.23% #4

VT 6.24% #36

NH

ME 5.50% #42

MT

ND 6.96% #27

MN 7.46% #17

NY 8.52% #10

OR

ID 6.03% #37

SD 6.40% #32

WI 5.43% #43

MI 6.00% #38

PA 6.34% #34

WY 5.33% #44

IA 6.94% #28

NV 8.23% #13

UT 7.19% #21

NE 6.94% #29

IL 8.82% #7

IN 7.00% #24

OH 7.23% #20

WV 6.50% #31

VA 5.73% #41

MA 6.25% #35

CA 8.68% #9

CO 7.72% #16

KS 8.69% #8

MO 8.25% #12

KY 6.00% #38

RI 7.00% #24

AZ 8.40% #11

NM 7.83% #15

OK 8.95% #6

AR 9.51% #3

TN 9.55% #1

NC 6.98% #26

CT 6.35% #33

SC 7.46% #18

NJ 6.60% #30

MS 7.07% #23

AL 9.22% #5

GA 7.32% #19

DE

TX 8.19% #14

LA 9.52% #2

MD 6.00% #38

AK 1.76% #46

FL 7.08% #22

DC 6.00% (#38)

HI 4.44% #45

City, county and municipal rates vary. These rates are weighted by population to compute an average local tax rate. The sales taxes in Hawaii, New Mexico, and South Dakota have broad bases that include many business-to-business services. D.C's rank does not affect states' ranks, but the figure in parentheses indicates where it would rank if included.

Sources: Sales Tax Clearinghouse; Tax Foundation calculations; State Revenue Department websites

Combined State & Average Local Sales Tax Rates

Lower — Higher

1 Special taxes in Montana's resort areas are not included in our analysis.

Thus, the store's actual sales were $37,383.18. The actual sales plus the tax equals $40,000.

Check

$37,383.18 × .07 = $ 2,616.82   sales tax

+ 37,383.18   actual sales

$ 40,000.00   total sales including sales tax

Learn: Excise Tax

Governments (local, federal, and state) levy **excise tax** on particular products and services. This can be a sizable source of revenue for these governments.

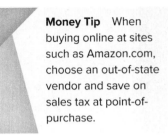

Money Tip When buying online at sites such as Amazon.com, choose an out-of-state vendor and save on sales tax at point-of-purchase.

Consumers pay the excise tax in addition to the sales tax. The excise tax is based on a percent of the *retail* price of a product or service. This tax, which varies in different states, is imposed on luxury items or nonessentials. Examples of products or services subject to the excise tax include airline travel, telephone service, alcoholic beverages, jewelry, furs, fishing rods, tobacco products, and motor vehicles. Although excise tax is often calculated as a percent of the selling price, the tax can be stated as a fixed amount per item sold. The following example calculates excise tax as a percent of the selling price.[1]

> **Example:** On June 1, Angel Rowe bought a fur coat for a retail price of $5,000. Sales tax is 7% with an excise tax of 8%. Her total cost is as follows:
>
> $5,000
> + 350 sales tax (.07 × $5,000)
> + 400 excise tax (.08 × $5,000)
> $5,750

Let's check your progress with a Practice Quiz.

---

[1] If excise tax were a stated fixed amount per item, it would have to be added to the cost of goods or services before any sales tax was taken. For example, a $100 truck tire with a $4 excise tax would be $104 before the sales tax was calculated.

Practice Quiz

Complete this Practice Quiz to see how you are doing.

From the following shopping list, calculate the total sales tax. Food items are excluded from sales tax, which is 8%.

Chicken	$6.10	Orange juice	$1.29	Shampoo	$4.10
Lettuce	$ .75	Laundry detergent	$3.65		

✓ Solutions

Shampoo $4.10
Laundry detergent $\underline{+\ 3.65}$
$$\$7.75 \times .08 = \boxed{\$.62}$$

Learning Unit 19–2:
Property Tax

When you own property, you must pay property tax. In this unit we listen in on a conversation between a property owner and a tax assessor.

Learn: Defining Assessed Value

Bill Adams was concerned when he read in the local paper that the property tax rate had been increased. Bill knows that the revenue the town receives from the tax helps pay for fire and police protection, schools, and other public services. However, Bill wants to know how the town set the new rate and the amount of the new property tax.

Bill went to the town assessor's office to get specific details. The assessor is a local official who estimates the fair market value of a house. Before you read the summary of Bill's discussion, note the following formula:

Assessed Value formula

$$\text{Assessed value} = \text{Assessment rate} \times \text{Market value}$$

Bill: What does *assessed value* mean?

Assessor: **Assessed value** is the value of the property for purposes of computing property taxes. We estimated the market value of your home at $210,000. In our town, we assess property at 30% of the market value. Thus, your home has an assessed value of $63,000 ($210,000 × .30). Usually, assessed value is rounded to the nearest dollar.

Bill: I know that the **tax rate** multiplied by my assessed value ($63,000) determines the amount of my property tax. What I would like to know is how did you set the new tax rate?

Learn: Determining the Tax Rate

Assessor: In our town first we estimate the total amount of revenue needed to meet our budget. Then we divide the total of all assessed property into this figure to get the *tax rate.* The formula looks like this:[2]

Tax Rate formula

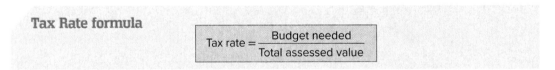

$$\text{Tax rate} = \frac{\text{Budget needed}}{\text{Total assessed value}}$$

Our town budget is $125,000, and we have a total assessed property value of $1,930,000. Using the formula, we have the following:

$$\frac{\$125,000}{\$1,930,000} = \$.0647668 = \boxed{.0648} \text{ tax rate per dollar}$$

Note that the rate should be rounded up to the indicated digit, *even if the digit is less than 5.* Here we rounded to the nearest ten thousandth.

---

[2] Remember that exemptions to total assessed value include land and buildings used for educational and religious purposes and the like.

Learn: How the Tax Rate Is Expressed

Assessor: We can express the .0648 tax rate per dollar in the following forms:

By percent	Per $100 of assessed value	Per $1,000 of assessed value	In mills
6.48%	$6.48	$64.80	64.80
(Move decimal two places to right.)	(.0648 × 100)	(.0648 × 1,000)	$\left(\dfrac{.0648}{.001}\right)$

A **mill** is $\frac{1}{10}$ of a cent or $\frac{1}{1,000}$ of a dollar (.001). To represent the number of mills as a tax rate per dollar, we divide the tax rate in decimal by .001. Rounding practices vary from state to state. Colorado tax bills are now rounded to the thousandth mill. An alternative to finding the rate in mills is to multiply the rate per dollar by 1,000, since a dollar has 1,000 mills. In the problems in this text, we round the mills per dollar to the nearest hundredth.

Learn: How to Calculate Property Tax Due[3]

Assessor: The following formula will show you how we arrive at your **property tax:**

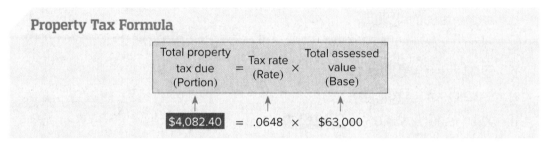

Property Tax Formula

$$\underset{\text{(Portion)}}{\text{Total property tax due}} = \underset{\text{(Rate)}}{\text{Tax rate}} \times \underset{\text{(Base)}}{\text{Total assessed value}}$$

$$\$4{,}082.40 = .0648 \times \$63{,}000$$

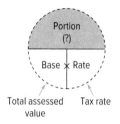

We can use the other forms of the decimal tax rate to show you how the property tax will not change even when expressed in various forms:

By percent	Per $100	Per $1,000	Mills
6.48% × $63,000	$\dfrac{\$63{,}000}{\$100} = 630$	$\dfrac{\$63{,}000}{\$1{,}000} = 63$	Property tax due
= $4,082.40	630 × $6.48	63 × $64.80	= Mills × .001 × Assessed value
	= $4,082.40	= $4,082.40	= 64.80 × .001 × $63,000
			= $4,082.40

Aha! *Keep in mind you always round up when calculating the tax rate—even if the digit being rounded is less than 5.*

Now it's time to try the Practice Quiz.

---

[3] Some states have credits available to reduce what the homeowner actually pays. For example, 42 out of 50 states give tax breaks to people over age 65. In Alaska, the state's homestead exemption reduces the property tax of a $168,000 house from $1,512 to $253.

Learning Unit 19–2

Practice Quiz

Complete this Practice Quiz to see how you are doing.

From the following facts: (1) calculate the assessed value of Bill's home; (2) calculate the tax rate for the community in decimal (to the nearest ten thousandth); (3) convert the decimal to **(a)** %, **(b)** per $100 of assessed value, **(c)** per $1,000 of assessed value, and **(d)** mills (to the nearest hundredth); and (4) calculate the property tax due on Bill's home **(a)** in decimal, **(b)** per $100, **(c)** per $1,000, and **(d)** in mills.

Given

Assessed market value	40%	Total budget needed	$ 176,000
Market value of Bill's home	$210,000	Total assessed value	$1,910,000

✓ Solutions

1. $.40 \times \$210,000 = $ $\boxed{\$84,000}$

2. $\dfrac{\$176,000}{\$1,910,000} = \boxed{.0922 \text{ per dollar}}$

3. **a.** $.0922 = \boxed{9.22\%}$

b. $.0922 \times 100 = \boxed{\$9.22}$

   c. $.0922 \times 1,000 = \boxed{\$92.20}$

d. $\dfrac{.0922}{.001} = \boxed{92.2 \text{ mills}}$ (or $.0922 \times 1,000$)

4. **a.** $.0922 \times \$84,000 = \boxed{\$7,744.80}$

   b. $\$9.22 \times 840 = \boxed{\$7,744.80}$

   c. $\$92.20 \times 84 = \boxed{\$7,744.80}$

   d. $92.20 \times .001 \times \$84,000 = \boxed{\$7,744.80}$

Chapter 19 Review

Topic/Procedure/Formula	Examples	You try it*
Sales tax Sales tax is not calculated on trade discounts. Shipping charges, etc., also are not subject to sales tax. $\text{Actual sales} = \dfrac{\text{Total sales}}{1 + \text{Tax rate}}$ Cash discounts are calculated on sale price before sales tax is added on.	**Calculate sales tax** Purchased 12 bags of mulch at $59.40; 10% trade discount; 5% sales tax. $59.40 − $5.94 = $53.46 $53.46 × .05 $2.67 sales tax Any cash discount would be calculated on $53.46.	**Calculate sales tax** 14 bags of mulch at $62.80; 8% trade discount; 6% sales tax
Excise tax Excise tax is calculated separately from sales tax and is an additional tax. It is based as a percent of the selling price. It could be stated as a fixed amount per item sold. In that case, the excise tax would be added to the cost of the item before any sales tax calculations. Rates for excise tax vary.	Jewelry $4,000 retail price Sales tax 7% Excise tax 10% $4,000 + 280 sales tax + 400 excise tax $4,680	**Calculate cost of jewelry** $6,000 retail price Sales tax 5% Excise tax 10%
Assessed value Assessment rate × Market value	$100,000 house; rate, 30%; $30,000 assessed value.	**Calculate assessed value** $200,000 house; rate, 40%.
Tax rate $\dfrac{\text{Budget needed}}{\text{Total assessed value}} = \text{Tax rate}$ (Round rate up to indicated digit even if less than 5.)	$\dfrac{\$800,000}{\$9,200,000} = .08695 = .0870$ tax rate per $1	**Calculate tax rate** Budget needed, $700,000; Total assessed value, $8,400,000. (Round up to 4 digits.)
Expressing tax rate in other forms 1. Percent: Move decimal two places to right. Add % sign. 2. Per $100: Multiply by 100. 3. Per $1,000: Multiply by 1,000. 4. Mills: Divide by .001.	1. .0870 = 8.7% 2. .0870 × 100 = $8.70 3. .0870 × 1,000 = $87 4. $\dfrac{.0870}{.001} = 87$ mills	**Using the above tax rate, calculate tax rate in:** 1. Percent 2. Per $100 3. Per $1,000 4. Mills
Calculating property tax $\dfrac{\text{Total property}}{\text{tax due}} = \text{Tax rate} \times \dfrac{\text{Total assessed}}{\text{value}}$ Various forms: 1. Percent × Assessed value 2. Per $100: $\dfrac{\text{Assessed value}}{\$100} \times \text{Rate}$ 3. Per $1,000: $\dfrac{\text{Assessed value}}{\$1,000} \times \text{Rate}$ 4. Mills: Mills × .001 × Assessed value	*Example:* Rate, .0870 per $1; $30,000 assessed value 1. (.087)8.7% × $30,000 = $2,610 2. $\dfrac{\$30,000}{\$100} = 300 \times \$8.70 = \$2,610$ 3. $\dfrac{\$30,000}{\$1,000} = 30 \times \$87 = \$2,610$ 4. $\dfrac{.0870}{.001} = 87$ mills 87 mills × .001 × $30,000 = $2,610	**Calculate property tax for various forms given:** $.0950 per $1; $40,000 assessed value

Chapter 19 Review (Continued)

Key Terms

Assessed value	Personal property	Sales tax
Excise tax	Property tax	Tax rate
Mill	Real property	

* Worked-out solutions are in Appendix A.

Critical Thinking Discussion Questions with Chapter Concept Check

1. Explain sales and excise taxes. Should all states have the same tax rate for sales tax?

2. Explain how to calculate actual sales when the sales tax was included in the sales figure. Is a sales tax necessary?

3. How is assessed value calculated? If you think your value is unfair, what could you do?

4. What is a mill? When we calculate property tax in mills, why do we use .001 in the calculation?

5. **Chapter Concept Check.** Search the web to find the latest information on taxing online sales. Do you think it is fair? Defend your position using concepts learned in this chapter.

6. How did the pandemic affect online shopping sites such as eBay?

End-of-Chapter Problems

Name _____   Date _____

Check figures for odd-numbered problems in Appendix A.

Drill Problems

Calculate the following: *LU 19–1(1, 2)*

Retail selling price	Sales tax (6%)	Excise tax (10%)	Total price including taxes
19–1. $800			
19–2. $1,200			

Calculate the actual sales since the sales and sales tax were rung up together. Assume a 6% sales tax and round your answer to the nearest cent.   *LU 19–1(1)*

19–3. $\dfrac{\$88,000}{}$

19–4. $\dfrac{\$26,000}{}$

Calculate the assessed value of the following pieces of property: *LU 19–2(2)*

Assessment rate	Market value	Assessed value
19–5. 30%	$130,000	
19–6. 80%	$210,000	

Calculate the tax rate in decimal form to the nearest ten thousandth: *LU 19–2(2)*

Required budget	Total assessed value	Tax rate per dollar
19–7. $920,000	$39,500,000	

Complete the following:

Tax rate per dollar	In percent	Per $100	Per $1,000	Mills
19–8. .0956				
19–9. .0699				

Complete the amount of property tax due to the nearest cent for each situation: *LU 19–2(3)*

Tax rate	Assessed value	Amount of property tax due
19–10. 40 mills	$ 65,000	
19–11. $42.50 per $1,000	$105,000	
19–12. $8.75 per $100	$125,000	
19–13. $94.10 per $1,000	$180,500	

End-of-Chapter Problems (Continued)

Word Problems

excel **19–14.** Be careful when signing a work-for-hire agreement if you are a songwriter. You may lose all rights to your song in the copyright law world. If your song sells thousands on iTunes, you do not get to share in any of the publishing income. If you live in New Jersey and iTunes sold five of your songs for a total of $65,000, what is the tax owed at 7.0%?   *LU 19–1(1)*

excel **19–15.** Don Chather bought a new HP computer for $1,995. This included a 6% sales tax. What is the amount of sales tax and the selling price before the tax?   *LU 19–1(1)*

19–16. Homeowners enjoy many benefits, including a federal tax deduction for state and local property taxes paid. Fishers, Indiana, was voted one of the top 100 best places to live by *Money* magazine. With a population of 86,357, a median home price of $236,167, and estimated property taxes at 10.6 mills, how much does the average homeowner pay in property taxes?   *LU 19–2(3)*

19–17. The median home price in Arlington, Virginia, is $634,000. If the assessment rate is 100%, what is the assessed value?   *LU 19–2(2)*

excel **19–18.** Bemidji, Minnesota, needed $3,850,000 for its budget. If total assessed value of property in Bemidji was $353,211,009, what was the tax rate expressed as a percent, per $100, per $1,000, and in mills?   *LU 19–2(2)*

19–19. Lois Clark bought a ring for $6,000. She must still pay a 5% sales tax and a 10% excise tax. The jeweler is shipping the ring, so Lois must also pay a $40 shipping charge. What is the total purchase price of Lois's ring?   *LU 19–1(1, 2)*

19–20. Blunt County needs $700,000 from property tax to meet its budget. The total value of assessed property in Blunt is $110,000,000. What is the tax rate of Blunt? Round to the nearest ten thousandth. Express the rate in mills.   *LU 19–2(1, 2)*

19–21. Bill Shass pays a property tax of $3,200. In his community, the tax rate is 50 mills. What is Bill's assessed value?   *LU 19–2(2)*

e)cel 19–22. The home of Bill Burton is assessed at $80,000. The tax rate is 18.50 mills. What is the tax on Bill's home?   *LU 19–2(3)*

19–23. New Hampshire ranked as the #1 most expensive state for property taxes. The median property tax was $4,636. If the rate was $1.25 per $100 of assessed value, how much did a homeowner owe for a property assessed at $378,150?   *LU 19–2(3)*

19–24. Bill Blake pays a property tax of $2,500. In his community, the tax rate is 55 mills. What is Bill's assessed value? Round to the nearest dollar.   *LU 19–2(2)*

19–25. Assume the property tax rate for Minneapolis is $8.73 per square foot, and the Denver rate is $2.14 a square foot. If 3,500 square feet is occupied at each location, what is the difference paid in property taxes?   *LU 19–2(3)*

19–26. Marlina Fields thought she was financing $3,000 for the purchase of a puppy at Pet City in Colorado Springs, Colorado. Turns out she didn't read the contract and she was only leasing the puppy. After the 5-year lease ended for a total cost to Marlina of $7,000, she would need to pay a balloon payment of $3,000 to actually own her pet. She had a newfound understanding of how important it is to read and understand any contract before signing. How much sales tax did Marlina pay for her puppy if the state, city, and county combined sales tax was 8.20%?

19–27. *Money* magazine rated Reston, Virginia, as the #1 place to live in the United States in 2021 if you work from home. With over 55 miles of paved trails connecting 73 parks, Reston boasts two golf courses and four man-made lakes. The median home price is $434,000. If you purchase a home for $475,000 and the assessed value is 50%, what is your property tax at $12.65 per $1,000? Round to the dollar.

19–28. If you qualified for the third stimulus check of $1,400 issued during the first or second quarter of 2021 and spent all of it on taxable items in your hometown with a sales tax rate of 9.23%, how much did you spend on your items and how much did you spend on taxes? Round to the nearest cent.

End-of-Chapter Problems (Continued)

Challenge Problems

19–29. Ginny Fieg expanded her beauty salon by increasing her space by 20%. Ginny paid property taxes of $2,800 at 22 mills. The new rate is now 24 mills. As Ginny's accountant, estimate what she may have to pay for property taxes this year. Round the final answer to the nearest dollar. In the calculation, round assessed value to the nearest dollar. *LU 19–2(2)*

e)cel 19–30. Art Neuner, an investor in real estate, bought an office condominium. The market value of the condo was $250,000 with a 70% assessment rate. Art feels that his return should be 12% per month on his investment after all expenses. The tax rate is $31.50 per $1,000. Art estimates it will cost $275 per month to cover general repairs, insurance, and so on. He pays a $140 condo fee per month. All utilities and heat are the responsibility of the tenant. Calculate the monthly rent for Art. Round your answer to the nearest dollar (at intermediate stages). *LU 19–2(2)*

Summary Practice Test

Do you need help? Connect videos have step-by-step worked-out solutions.

1. Carol Shan bought a new Apple iPad at Best Buy for $299. The price included a 5% sales tax. What are the sales tax and the selling price before the tax?   *LU 19–1(1)*

2. Jeff Jones bought a ring for $4,000 from Zales. He must pay a 7% sales tax and 10% excise tax. Since the jeweler is shipping the ring, Jeff must also pay a $30 shipping charge. What is the total purchase price of Jeff's ring?   *LU 19–1(1, 2)*

3. The market value of a home in Boston, Massachusetts, is $365,000. The assessment rate is 40%. What is the assessed value?   *LU 19–2(1)*

4. Jan County needs $910,000 from its property tax to meet the budget. The total value of assessed property in Jan is $180,000,000. What is Jan's tax rate? Round to the nearest ten thousandth. Express the rate in mills (to the nearest tenth).   *LU 19–2(2)*

5. The home of Nancy Billows is assessed at $250,000. The tax rate is 4.95 mills. What is the tax on Nancy's home?   *LU 19–2(3)*

6. V's Warehouse has a market value of $880,000. The property in V's area is assessed at 35% of the market value. The tax rate is $58.90 per $1,000 of assessed value. What is V's property tax? *LU 19–2(3)*

MY MONEY

Q Up Close and Personal with Your Taxes

 What I need to know

There is a saying that the only certain things in life are death and taxes. However, that is not to say we cannot attempt to alter these certainties to our advantage, well, at least the taxes. Taxes are a part of nearly every aspect of our financial lives, and we must operate within the confines of these tax structures. Arming ourselves with the knowledge of how taxes work and how we can best position ourselves financially with a personal tax plan is to our advantage. Taking the time to plan for the taxes you pay on a yearly basis will help to ensure you are using every opportunity available to lower the amount of tax you pay.

 What I need to do

It is natural to primarily think about your taxes every spring when you are completing your income tax returns for the prior year. However, it is in your best interest to consider tax implications throughout the year to take full advantage of any tax credits or deductions for which you may qualify. For instance, when you donate a box of clothes, create a receipt of the fair value of the clothes. This receipt will be valuable in the event you are able to use this donation as part of your deductions come tax time.

Many employers offer the benefit of a flexible spending account (FSA) which is funded tax-free by the employee. The portion of your salary you place into an FSA is not taxed as income and can be used for out-of-pocket expenses such as medical, dental, and child care. To show the value of this benefit let's look at an example. Assume you make a salary of $50,000 per year and decide to place $2,500 into an FSA. If you have a tax rate of 20%, you will receive a benefit of $500 (20% × $2,500) from using the flexible spending account since the amount of your FSA is funded with pre-tax dollars. It is important to note many of these FSA plans require you to use all the money in the account or risk losing any amount not spent on qualified expenses during the year. So, if you decide to take advantage of this employer benefit be sure to use realistic estimates of the expenses you plan to incur during the FSA plan year.

 Steps I need to take

1. Estimate the tax rate you pay based on your annual income.
2. Spend time researching tax credits and deductions for which you may qualify.
3. Seek out employer benefits which may reduce your taxes owed.

 Resources I can use

- https://www.nerdwallet.com/article/taxes/tax-planning—learning the basics of personal tax planning and helpful advice for creating your plan.
- https://thecollegeinvestor.com/21804/federal-tax-brackets/—federal tax brackets based on your yearly income for 2021.

MY MONEY ACTIVITY

- Research employer-provided flexible spending accounts (FSA).
- Identify 5 expenses which are covered under this FSA.
- Identify 5 expenses which are not covered under this FSA.
- Would you take advantage of an FSA at your employer? Why or why not?

PERSONAL FINANCE

A KIPLINGER APPROACH

"Moving to Lower-Tax States." Kiplinger's. May, 2021.

On The Go

Moving to Lower-Tax States

United Van Lines, a national moving company, says the most commonly cited reasons for moving in 2020 were to take a new job and to be closer to family. But for high earners and retirees, low taxes may also be a factor. Most of the states with the highest percentage of outbound moves in 2020 had above-average tax rates, according to Kiplinger's state-by-state tax guide, while those with the highest percentage of inbound moves tended to have lower tax rates.

New York City and Weehawken, N.J.

Boise, Idaho

States with the largest percentage of outbound moves in 2020:

1. New Jersey
2. New York
3. Illinois
4. Connecticut
5. California

SOURCE: UNITED VAN LINES

States with the largest percentage of inbound moves in 2020:

1. Idaho
2. South Carolina
3. Oregon
4. South Dakota
5. Arizona

Business Math Issue

New employees will always move to states with lower taxes.

1. List the key points of the article and information to support your position.

2. Write a group defense of your position using math calculations to support your view. If you are in an online course, post to a discussion board.

Chapter 20

Life, Fire, and Auto Insurance

Learning Unit Objectives

LU 20–1: Life Insurance

1. Explain the types of life insurance; calculate life insurance premiums.
2. Explain and calculate cash value and other nonforfeiture options.

LU 20–2: Fire Insurance

1. Explain and calculate premiums for fire insurance of buildings and their contents.
2. Calculate refunds when the insured and the insurance company cancel fire insurance.
3. Explain and calculate insurance loss when coinsurance is not met.

LU 20–3: Auto Insurance

1. Explain and calculate the cost of auto insurance.
2. Determine the amount paid by the insurance carrier and the insured after an auto accident.

Firms Hit by Covid Want Insurers to Pay. They Won't.

At issue is 'business interruption' coverage and what meets the physical-damage test

By Leslie Scism

One of the biggest legal fights in the history of insurance has begun.

A cavalcade of restaurateurs, retailers and others hurt by pandemic shutdowns have sued to force their insurers to cover billions in business losses. A video berating the industry ran for most of June on a giant screen in New York's Times Square, four times each hour around the clock.

"Insurance companies: Do the right thing," was the chorus at the end of the video. Repeating the words were a musician, a dancer, a chef, a rabbi, comedian Whoopi Goldberg–and a New Orleans plaintiffs' lawyer, John Houghtaling II, who paid for the video.

Millions of businesses across the U.S. have "business interruption"

insurance. The pandemic, no question, interrupted their businesses.

But insurance companies have largely refused to pay claims under this coverage, citing a standard requirement for physical damage. That is a legacy of its origins in the early 1900s as part of property insurance protecting manufacturers from broken boilers or other failing equipment that closed factories. The insurance is also known as "business income" coverage.

More than half of property policies in force today specifically exclude viruses. The firms filing the lawsuits mostly hold policies without that exclusion. Their argument for getting around the physical-damage requirement is that the coronavirus sticks to surfaces and renders workplaces unsafe.

Scism, Leslie. "Firms Hit By Covid Want Insurers to Pay. They Won't." *The Wall Street Journal*, July 1, 2020.

Essential Question

How can I use life, fire, and auto insurance to understand the business world and make my life easier?

🌐 Math Around the World

The *Wall Street Journal* chapter opener clip, "Firms Hit by Covid Want Insurers to Pay. They won't.", shows that businesses with interruption insurance expected their claims due to the pandemic to be covered and paid. Insurance companies are not paying.

Regardless of the type of insurance you buy—life, auto, nursing home, property, or fire—be sure to read and understand the policy before you buy the insurance. It has been reported that half of the people in the United States who have property insurance have not read their policy and 60% do not understand their policy. If you do not understand your life, fire, or auto insurance policies, this chapter should answer many of your questions. Today you can download your insurance company's app and manage your policy. We begin by studying life insurance.

Life Insurers Halt Sales As Hopes for Profit Dim

By Leslie Scism

U.S. insurers are doing the once unthinkable, turning away business from some Americans who want a life-insurance policy.

The driving force behind the action: a collapse in interest rates tied to the spread of the coronavirus and an expectation from insurers that rates won't rebound significantly soon.

Life insurers earn much of their profit by investing customers' premiums in bonds until claims come due. In simplest terms, when they price policies, they make assumptions about how much interest income they will earn investing these premiums. The less they earn, the more they

may need to collect in premium or fees to turn a profit.

A wave of stopgap measures is hitting potential buyers even as some companies say more consumers are seeking out life insurance during the pandemic. In addition to suspending sales of some popular products and raising prices, insurers are scaling back policy sizes and reducing benefits.

"In 33 years, I have never seen more changes come more quickly to the life-insurance products we sell,"

Scism, Leslie. "Life Insurers Halt Sales as Hopes for Profits Dim." *The Wall Street Journal*, May 11, 2020.

Learning Unit 20–1:
Life Insurance

The *Wall Street Journal* clip, "Life Insurers Halt Sales As Hopes for Profit Dim" shows insurance companies are cutting sales of life insurance due to the pandemic.

Bob Brady owns Bob's Deli. He is 40 years of age, married, and has three children. Bob wants to know what type of life insurance protection will best meet his needs. Following is a discussion between an insurance agent, Rick Jones, and Bob.

Bob: I would like to buy a life insurance policy that will pay my partner $200,000 in the event of my death. My problem is that I do not have much cash. You know, bills, bills, bills. Can you explain some types of life insurance and their costs?

Rick: Let's begin by explaining some life insurance terminology. The **insured** is you— the **policyholder** receiving coverage. The **insurer** is the company selling the insurance policy. Your partner is the **beneficiary.** As the beneficiary, your partner is named in the policy to receive the insurance proceeds at the death of the insured (that's you, Bob). The amount stated in the policy, say, $200,000, is the **face amount** of the policy. The **premium** (determined by **statisticians** called *actuaries*) is the periodic payments you agree to make for the cost of the insurance policy. You can pay premiums annually, semiannually, quarterly, or monthly. The more frequent the payment, the higher the total cost due to increased paperwork, billing, and so on. Now we look at the different types of insurance.

Learn: Types of Insurance

In this section Rick explains term insurance, straight life (ordinary life), 20-payment life, 20-year endowment, and universal life insurance.

Term Insurance[1]

Rick: The cheapest type of life insurance is **term insurance,** but it only provides *temporary* protection. Term insurance pays the face amount to your partner (beneficiary) only if you die within the period of the insurance (1, 5, 10 years, and so on).

For example, let's say you take out a 5-year term policy. The insurance company automatically allows you to renew the policy at increased rates until age 70. A new policy called **level premium term** may be less expensive than an annual term policy since each year for, say, 50 years, the premium will be fixed.

The policy of my company lets you convert to other insurance types without a medical examination. To determine your rates under 5-year term insurance, check this table (Table 20.1). The annual premium at 40 years per $1,000 of insurance is $3.52. We use the following steps to calculate the total yearly premium.

Calculating Annual Life Insurance Premiums

Step 1 Look up the age of the insured (for females, subtract 3 years) and the type of insurance in Table 20.1. This gives the premium cost per $1,000. For the time being, if you identify as nonbinary, you will need to apply as either male or female.

Step 2 Divide the amount of coverage by $1,000 and multiply the answer by the premium cost per $1,000.

---

[1] A new term policy is available that covers policyholders until their expected retirement age.

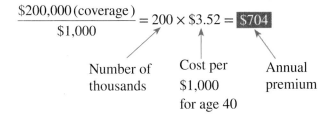

$$\frac{\$200,000 \, (\text{coverage})}{\$1,000} = 200 \times \$3.52 = \boxed{\$704}$$

Number of thousands   Cost per $1,000 for age 40   Annual premium

From this formula you can see that for $704 per year for the next 5 years, we, your insurance company, offer to pay your partner $200,000 in the event of your death. At the end of the 5th year, you are not entitled to any cash from your paid premiums. If you do not renew your policy (at a higher rate) and die in the 6th year, we will not pay your partner anything. Term insurance provides protection for only a specific period of time.

Airport flight insurance is a type of term insurance.

Bob: Are you telling me that my premium does not build up any cash savings that you call **cash value?**

Age	Five-year term	Age	Straight Age	Age	Twenty-payment life	Age	Twenty-year endowment
20	1.85	20	5.90	20	8.28	20	13.85
21	1.85	21	6.13	21	8.61	21	14.35
22	1.85	22	6.35	22	8.91	22	14.92
23	1.85	23	6.60	23	9.23	23	15.54
24	1.85	24	6.85	24	9.56	24	16.05
25	1.85	25	7.13	25	9.91	25	17.55
26	1.85	26	7.43	26	10.29	26	17.66
27	1.86	27	7.75	27	10.70	27	18.33
28	1.86	28	8.08	28	11.12	28	19.12
29	1.87	29	8.46	29	11.58	29	20.00
30	1.87	30	8.85	30	12.05	30	20.90
31	1.87	31	9.27	31	12.57	31	21.88
32	1.88	32	9.71	32	13.10	32	22.89
33	1.95	33	10.20	33	13.67	33	23.98
34	2.08	34	10.71	34	14.28	34	25.13
35	2.23	35	11.26	35	14.92	35	26.35
36	2.44	36	11.84	36	15.60	36	27.64
37	2.67	37	12.46	37	16.30	37	28.97
38	2.95	38	13.12	38	17.04	38	30.38
39	3.24	39	13.81	39	17.81	39	31.84
40	3.52	40	14.54	40	18.61	40	33.36
41	3.79	41	15.30	41	19.44	41	34.94
42	4.04	42	16.11	42	20.31	42	36.59
43	4.26	43	16.96	43	21.21	43	38.29
44	4.50	44	17.86	44	22.15	44	40.09

TABLE 20.1 Life insurance rates for males (for females, subtract 3 years from the age. Non-binary individuals currently need to apply as male or female.)[2]

[2] The life insurance tables in this chapter show premiums for a sampling of age groups, options, and coverage available to those under 45 years of age.

Rick: The term insurance policy does not build up cash savings. Let me show you a policy that does build up cash value. This policy is straight life.

Straight Life (Ordinary Life)

Rick: Straight-life insurance provides *permanent* protection rather than the temporary protection provided by term insurance. The insured pays the same premium each year or until death.[3] The premium for straight life is higher than that for term insurance because straight life provides both protection and a built-in cash savings feature. According to our table (Table 20.1), your annual premium, Bob, would be

$$\frac{\$200,000}{\$1,000} = 200 \times \$14.54 = \boxed{\$2,908} \text{ annual premium}$$

Face value is usually the amount paid to the beneficiary at the time of the insured's death.

Bob: Compared to term, straight life is quite expensive.

Rick: Remember that term insurance has no cash value accumulating, as straight life does. Let me show you another type of insurance—20-payment life—that builds up cash value.

Twenty-Payment Life

Rick: A **20-payment life** policy is similar to straight life in that 20-payment life provides permanent protection and cash value, but you (the insured) pay premiums for only the first 20 years. After 20 years you own **paid-up insurance.** According to my table (Table 20.1), your annual premium would be

$$\frac{\$200,000}{\$1,000} = 200 \times \$18.61 = \boxed{\$3,722} \text{ annual premium}$$

Bob: The 20-payment life policy is more expensive than straight life.

Rick: This is because you are only paying for 20 years. The shorter period of time does result in increased yearly costs. Remember that in straight life you pay premiums over your entire life. Let me show you another alternative that we call 20-year endowment.

Twenty-Year Endowment

Rick: The **20-year endowment** insurance policy is the most expensive. It is a combination of term insurance and cash value. For example, from age 40 to 60, you receive term insurance protection in that your partner would receive $200,000 should you die. At age 60, your protection *ends* and you receive the face value of the policy that equals the $200,000 cash value. Let's use my table again (Table 20.1) to see how expensive the 20-year endowment is:

$$\frac{\$200,000}{\$1,000} = 200 \times \$33.36 = \boxed{\$6,672} \text{ annual premium}$$

In summary, Bob, following is a review of the costs for the various types of insurance we have talked about:

	5-year term	Straight life	20-payment life	20-year endowment
Premium cost per year	$704	$2,908	$3,722	$6,672

Before we proceed, I have another policy that may interest you—universal life.

---

[3] In the following section on nonforfeiture values, we show how a policyholder in later years can stop making payments and still be covered by using the accumulated cash value built up.

Universal Life Insurance

Rick: **Universal life** is basically a **whole-life** insurance plan with flexible premium schedules and death benefits. Under whole life, the premiums and death benefits are fixed. Universal has limited guarantees with greater risk on the holder of the policy. For example, if interest rates fall, the policyholder must pay higher premiums, increase the number of payments, or switch to smaller death benefits in the future.

Bob: That policy is not for me—too much risk. I'd prefer fixed premiums and death benefits.

Rick: OK, let's look at how straight life, 20-payment life, and 20-year endowment can build up cash value and provide an opportunity for insurance coverage without requiring additional premiums. We call these options **nonforfeiture values.**

Nonforfeiture Values

Rick: Except for term insurance, the other types of life insurance build up cash value as you pay premiums. These policies provide three options should you, the policyholder, ever want to cancel your policy, stop paying premiums, or collect the cash value. As shown in Figure 20.1, these options are cash value; **reduced paid-up insurance;** and **extended term insurance.**

Option 1: Cash value (cash surrender value)
a. Receive cash value of policy.
b. Policy is terminated.
The longer the policy has been in effect, the higher the cash value because more premiums have been paid in.
Option 2: Reduced paid-up insurance
a. Cash value buys protection without paying new premiums.
b. Face amount of policy is related to cash value buildup and age of insured. The **face amount is less than original policy.**
c. Policy continues for life (at a reduced face amount).
Option 3: Extended term insurance
a. Original face amount of policy continues for a certain period of time.
b. Length of policy depends on cash value built up and on insured's age.
c. This option results automatically if policyholder doesn't pay premiums and fails to elect another option.

FIGURE 20.1
Nonforfeiture options

For example, Bob, let's assume that at age 40 we sell you a $200,000 straight-life policy. Assume that at age 55, after the policy has been in force for 15 years, you want to stop paying premiums. From this table (Table 20.2), I can show you the options that are available.

TABLE 20.2 Nonforfeiture options based on $1,000 face value

Years insurance policy in force	Straight life				20-payment life				20-year endowment			
			EXTENDED TERM				EXTENDED TERM				EXTENDED TERM	
	Cash value	Amount of paid-up insurance	Years	Days	Cash value	Amount of paid-up insurance	Years	Days	Cash value	Amount of paid-up insurance	Years	Days
5	29	86	9	91	71	220	19	190	92	229	23	140
10	96	259	18	76	186	521	28	195	319	520	30	160
15	148	371	20	165	317	781	32	176	610	790	35	300
20	265	550	21	300	475	1,000		Life	1,000	1,000		Life

Option 1: Cash value

$$\frac{\$200,000}{\$1,000} = 200 \times \$148$$
$$= \$29,600$$

Option 2: Reduced paid-up insurance

$$\frac{\$200,000}{\$1,000} = 200 \times \$371$$
$$= \$74,200$$

Option 3: Extended term insurance

Bob could continue this $200,000 policy for 20 years and 165 days.

Insight into Health and Business Insurance Often people who interview for a new job are more concerned with the salary offered than the whole health care package such as eye care, dental care, hospital and doctor care, and so on. Be sure you know exactly what the new job offers in health insurance. For employees, company health insurance and life insurance benefits can be an important job consideration.

Some of the key types of business insurance that you may need as a business owner include fire insurance, business interruption insurance (business loss until physical damages are fixed), casualty insurance (insurance against a customer's suing your business due to an accident on company property), workers' compensation (insurance against injuries or sickness from being on the job), and group insurance (life, health, and accident).

Although group health insurance costs have soared recently, many companies still pay the major portion of the cost. Some companies also provide health insurance benefits for retirees. As health costs continue to rise, we can expect to see some changes in this employee benefit.

Companies vary in the type of life insurance benefits they provide to their employees. This insurance can be a percent of the employee's salary with the employee naming the beneficiary; or in the case of key employees, the company can be the beneficiary.

If as an employer you need any of the types of insurance mentioned in this section, be sure to shop around for the best price. If you are in the job market, consider the benefits offered by a company as part of your salary and make your decisions accordingly.

 Aha! *Don't forget to subtract 3 years from a female's (and those applying as female) age when using the life insurance tables.* In the next unit, we look specifically at fire insurance. Now let's check your understanding of this unit with a Practice Quiz.

Practice Quiz

Complete this Practice Quiz to see how you are doing.

1. Bill Boot, age 39, purchased a $60,000, 5-year term life insurance policy. Calculate his annual premium from Table 20.1. After 4 years, what is his cash value?

2. Ginny Katz, age 32, purchased a $78,000, straight-life policy. Calculate their annual premium if they applied as a female. If after 10 years they want to surrender their policy, what options and what amounts are available to them?

✓ Solutions

1. $\dfrac{\$60,000}{\$1,000} = 60 \times \$3.24 = \boxed{\$194.40}$ No cash value in term insurance.

2. $\dfrac{\$78,000}{\$1,000} = 78 \times \$8.46^* = \boxed{\$659.88}$

   Option 1: Cash value $78 \times \$96 = \boxed{\$7,488}$

   Option 2: Paid up $78 \times \$259 = \boxed{\$20,202}$

   Option 3: Extended term $\boxed{\text{18 years and 76 days}}$

   *For females, and those applying as females, we subtract 3 years.

Fire Insurance

Periodically, some areas of the United States, especially California, have experienced drought followed by devastating fires. These fires spread quickly and destroy wooded areas and homes. When the fires occur, the first thought of the owners is the adequacy of their **fire insurance.** Homeowners are made more aware of the importance of fire insurance that provides for the replacement value of their home. Out-of-date fire insurance policies can result in great financial loss.

In this unit, Alice Swan meets with her insurance agent, Bob Jones, to discuss fire insurance needs for her new dress shop at 4 Park Plaza. (Alice owns the building.)

Alice: What is *extended coverage?*

Bob: Your basic fire insurance policy provides financial protection if fire or lightning damages your property. However, the extended coverage protects you from smoke, chemicals, water, or other damages that firefighters may cause to control the fire. We have many options available.

Alice: What is the cost of a fire insurance policy?

Bob: Years ago, if you bought a policy for 2, 3, 5, or more years, reduced rates were available. Today, with rising costs of reimbursing losses from fires, most insurance companies write policies for 1 to 3 years. The cost of a 3-year policy premium is 3 times the annual premium. Because of rising insurance premiums, your total costs are cheaper if you buy one 3-year policy than three 1-year policies.

Alice: For my purpose, I will need coverage for 1 year. Before you give me the premium rates, what factors affect the cost of my premium?

Bob: In your case, you have several factors in your favor that will result in a lower premium. For example, (1) your building is brick, (2) the roof is fire-resistant, (3) the building is located next to a fire hydrant, (4) the building is in a good location (not next to a gas station) with easy access for the fire department, and (5) the goods within your store are not as flammable as, say, those of a paint store. I have a table here (Table 20.3) that gives an example of typical fire insurance rates for buildings and contents (furniture, fixtures, etc.).

TABLE 20.3 Fire insurance rates per $100 of coverage for buildings and contents

| | CLASSIFICATION OF BUILDING | | | |
| | CLASS A | | CLASS B | |
Rating of area	Building	Contents	Building	Contents
1	.28	.35	.41	.54
2	.33	.47	.50	.60
3	.41	.50	.61	.65

Let's assume your building has an insured value of $190,000, is rated Class B, and has an area rating of 2. We insure your contents for $80,000. Using the rates shown in Table 20.3, we would calculate your total annual premium for building and contents as follows:

Fire insurance premium equals premium for building and premium for contents.

$$\text{Premium} = \frac{\text{Insured value}}{\$100} \times \text{Rate}$$

Building

$$\frac{\$190,000}{\$100} = 1,900 \times \$.50 = \$950$$

Contents

$$\frac{\$80,000}{\$100} = 800 \times \$.60 = \$480$$

Total premium = $950 + $480 = $1,430

For our purpose, we round all premiums to the nearest cent. In practice, the premium is rounded to the nearest dollar.

Learn: Canceling Fire Insurance

Alice: What if my business fails in 7 months? Do I get back any portion of my premium when I cancel?

Bob: If the insured—that's you, Alice—cancels or wants a policy for less than 1 year, we use this **short-rate table** (Table 20.4). These rates are higher because it is more expensive to process a policy for a short time. For example, if you cancel at the end of 7 months, the premium cost is 67% of the annual premium. We would calculate your refund as follows:

Short-rate premium = Annual premium × Short rate

$958.10　=　$1,430　×　.67

Refund = Annual premium − Short-rate premium

$471.90 =　$1,430　−　$958.10

Alice: Let's say that I don't pay my premium or follow the fire codes. What happens if your insurance company cancels me?

Bob: If the insurance company cancels you, the company is *not* allowed to use the short-rate table. To calculate what part of the premium the company may keep,[4] you can prorate the premium based on the actual days that have elapsed. We can illustrate the amount of your refund by assuming you are canceled after 7 months:

For insurance company:

$$\text{Charge} = \$1,430 \text{ annual premium} \times \frac{7 \text{ months elapsed}}{12}$$

Charge = $834.17

For insured:

Refund = $1,430 annual premium − $834.17 charge

Refund = $595.83

Money Tip Make a video of the contents of your home, garage, sheds, etc., to inventory your belongings. Keep the video in a safety deposit box or fireproof box offsite. In the event of a loss, you will have a recording of your belongings. You may be amazed how much can be forgotten without documentation.

Note that when the insurance company cancels the policy, the refund ($595.83) is greater than if the insured cancels ($471.90).

---

[4] Many companies use $\frac{\text{Days}}{365}$.

TABLE 20.4 Fire insurance short-rate and cancellation table

Time policy is in force		Percent of annual rate to be charged	Time policy is in force		Percent of annual rate to be charged
Days:	5	8%	Months:	5	52%
	10	10		6	61
	20	15		7	67
	25	17		8	74
Months:	1	19		9	81
	2	27		10	87
	3	35		11	96
	4	44		12	100

Learn: Coinsurance

Alice: My friend tells me that I should meet the coinsurance clause. What is coinsurance?

Bob: Usually, fire does not destroy the entire property. **Coinsurance** means that you and the insurance company *share* the risk. The reason for this coinsurance clause[5] is to encourage property owners to purchase adequate coverage.

Alice: What is adequate coverage?

Bob: In the fire insurance industry, the usual rate for coinsurance is 80% of the current replacement cost. This cost equals the value to replace what was destroyed. If your insurance coverage is 80% of the current value, the insurance company will pay all damages up to the face value of the policy.

Alice: Hold it, Bob! Will you please show me how this coinsurance is figured?

Bob: Yes, Alice, I'll be happy to show you how we figure coinsurance. Let's begin by looking at the following steps so you can see what amount of the insurance the company will pay.

Money Tip Request quotes from your insurance provider using different deductibles for each policy. Choose the one that meets your financial needs. Keep the amount of each deductible in an interest-earning account for easy access if and when it is needed.

[5] In some states (including Wisconsin), the clause is not in effect for losses under $1,000.

Calculating What Insurance Company Pays with Coinsurance Clause

Step 1 Set up a fraction. The numerator is the actual amount of the insurance carried on the property. The denominator is the amount of insurance you should be carrying on the property to meet coinsurance (80% times the replacement value).

Step 2 Multiply the fraction by the amount of loss (up to the face value of the policy).

Let's assume for this example that you carry $60,000 fire insurance on property that will cost $100,000 to replace. If the coinsurance clause in your policy is 80% and you suffer a loss of $20,000, your insurance company will pay the following:

Although there are many types of property and homeowner's insurance policies, they usually include fire protection.

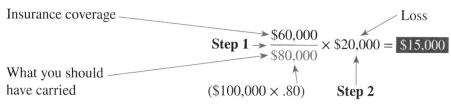

This kind of limited insurance payment for a loss is often called an **indemnity.** If you had had actual insurance coverage of $80,000, then the insurance company would have paid $20,000. Remember that if the coinsurance clause is met, the most an insurance company will pay is the face value of the policy.

You are now ready for the following Practice Quiz.

Learning Unit 20–2

Practice Quiz

Complete this Practice Quiz to see how you are doing.

1. Calculate the total annual premium of a warehouse that has an area rating of 2 with a building classification of B. The value of the warehouse is $90,000 with contents valued at $30,000.

2. If the insured in problem 1 cancels at the end of month 9, what are the costs of the premium and the refund?

3. Jones insures a building for $120,000 with an 80% coinsurance clause. The replacement value is $200,000. Assume a loss of $60,000 from fire. What will the insurance company pay? If the loss was $160,000 and coinsurance *was* met, what would the insurance company pay?

✓ Solutions

1. $\dfrac{\$90,000}{\$100} = 900 \times \$.50 = \$450$

    $\dfrac{\$30,000}{\$100} = 300 \times \$.60 = \dfrac{180}{\boxed{\$630}} \leftarrow$ total premium

2. $\$630 \times .81 = \boxed{\$510.30}$ $\qquad$ $\$630 - \$510.30 = \boxed{\$119.70}$

3. $\dfrac{\$120,000}{\$160,000} = \dfrac{3}{4} \times \$60,000 = \boxed{\$45,000}$

    $\underset{\uparrow}{}$
    $(.80 \times \$200,000)$ $\qquad$ $\boxed{\$160,000}$ never more than face value

Auto insurance is important to have if you drive. And it can be expensive. Many insurance companies provide mobile apps that track your driving and can save you as much as 50% if you qualify as a good driver. Apps, like Metromile pictured below, can even help you find your lost or stolen car. Due to the pandemic, insurance companies issued policyholders a 15% rebate on auto insurance for three months since cars were being driven infrequently. Insurance rates often increase when a driver is involved in an accident. Some insurance companies give reduced rates to accident-free drivers—a practice that has encouraged drivers to be more safety conscious. For example, State Farm Insurance offers a discount to drivers who maintain a safety record. An important factor in safe driving is the use of a seat belt. Make it a habit to always put on your seat belt.

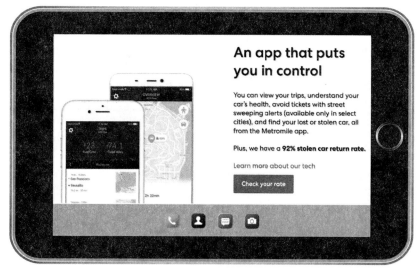

Metromile

Money Tip Although insurance company mobile apps can save you money by tracking your driving (if you qualify as a good driver), the cost is reduced privacy. Because of this negative, several companies are offering incentives beyond discounts such as GPS car tracking, mileage tracking, parking reminders, etc., to help offset the privacy loss and encourage the use of their apps.

In this unit we follow Shirley as she learns about auto insurance. Shirley, who just bought a new auto, has never purchased auto insurance. So she called her insurance agent, Bob Long, who agreed to meet her for lunch. We will listen in on their conversation.

Shirley: Bob, where do I start?

Bob: Our state has two kinds of **liability insurance,** or **compulsory insurance,** that by law you must buy (regulations and requirements vary among states). Liability insurance covers any physical damages that you inflict on others or their property. You must buy liability insurance for the following:

1. **Bodily injury** to others: 10/20. This means that the insurance company will pay damages to people injured or killed by your auto up to $10,000 for injury to one person per accident or a total of $20,000 for injuries to two or more people per accident.

2. **Property damage** to someone else's property: 5. The insurance company will pay up to $5,000 for damages that you have caused to the property of others.

Now we leave Shirley and Bob for a few moments as we calculate Shirley's premium for compulsory insurance.

Liability insurance includes
1. **Bodily injury**—injury or death to people in passenger car or other cars, etc.
2. **Property damage**—injury to other people's autos, trees, buildings, hydrants, etc.

TABLE 20.5
Compulsory
insurance (based on
class of driver)

BODILY INJURY TO OTHERS		DAMAGE TO SOMEONE ELSE'S PROPERTY	
Class	10/20	Class	5M*
10	$ 55	10	$129
17	98	17	160
18	80	18	160
20	116	20	186

Explanation of 10/20 and 5

10	20	5
Maximum paid to one person per accident for bodily injury	Maximum paid for total bodily injury per accident	Maximum paid for property damage per accident

* M means thousands.

Learn: Calculating Premium for Compulsory Insurance[6]

Insurance companies base auto insurance rates on the territory you live in, the class of driver (class 10 is experienced driver with driver training), whether the auto is for business use, how much you drive the car, the age of the car, and the make of the car (symbol). Shirley lives in Territory 5 (suburbia). She is classified as 17 because she is an inexperienced operator licensed for less than 6 years. Her car is age 3 and symbol 4 (make of car). We use Table 20.5 to calculate Shirley's compulsory insurance. Note that the table rates in this unit are not representative of all areas of the country. In case of lawsuits, the minimum coverage may not be adequate. Some states add surcharges to the premium if the person has a poor driving record. The tables are designed to show how rates are calculated. From Table 20.5, we can determine Shirley's premium for compulsory insurance as follows:

The tables we use in this unit are for Territory 5. Other tables are available for different territories.

$$
\begin{array}{ll}
\text{Bodily injury} & \$\ 98 \\
+ \text{ Property damage} & \underline{\ 160} \\
& \$258
\end{array}
$$

Remember that the $258 premium represents minimum coverage. Assume Shirley hits two people and the courts award them $13,000 and $5,000, respectively. Shirley would be responsible for $3,000 because the insurance company would pay only up to $10,000 per person and a total of $20,000 per accident.

Although total damages of $18,000 are less than $20,000, the insurance company pays only $15,000.

	(1)	(2)	
	$13,000 +	$5,000 =	$18,000
Paid by insurance company →	− 10,000 −	5,000 =	− 15,000
Paid by Shirley →	$ 3,000 + $	0 = $	3,000

We return to Shirley and Bob. Bob now shows Shirley how to calculate her optional insurance coverage. Remember that optional insurance coverages (Tables 20.6 to 20.10) are added to the costs in Table 20.5.

TABLE 20.6 Bodily
injury

Class	15/30	20/40	20/50	25/50	25/60	50/100	100/300	250/500	500/1,000
10	27	37	40	44	47	69	94	144	187
17	37	52	58	63	69	104	146	228	298
18	33	46	50	55	60	89	124	193	251
20	41	59	65	72	78	119	168	263	344

[6] Some states may offer medical payment insurance (a supplement to policyholders' health and accident insurance) as well as personal injury protection against uninsured or underinsured motorists.

Class	10M	25M	50M	100M
10	132	134	135	136
17	164	166	168	169
18	164	166	168	169
20	191	193	195	197

TABLE 20.7 Damage to someone else's property

Learn: Calculating Optional Insurance Coverage

Bob: In our state, you can add optional bodily injury to the compulsory amount. If you finance your car, the lender may require specific amounts of optional insurance to protect its investment. I have two tables (Tables 20.6 and 20.7) here that we use to calculate the option of 250/500/50. This means that in an accident the insurance company will pay $250,000 per person, up to $500,000 per accident, and up to $50,000 for property damage.

Bob then explains the tables to Shirley. By studying the tables, you can see how insurance companies figure bodily injury and damage to someone else's property. Shirley is Class 17:

Bodily injury

250/500 = $228

Property damage

50M = +168

$396   premium for optional bodily injury and property damage

Note: These are additional amounts to compulsory.

Shirley: Is that all I need?

Bob: No, I would recommend two more types of optional coverage: **collision** and **comprehensive.** Collision provides protection against damages to your car caused by a moving vehicle. It covers the cost of repairs less **deductibles** (amount of repair you cover first before the insurance company pays the rest) and depreciation.[7] In collision, insurance companies pay the resale or book value. So as the car gets older, after 5 or more years, it might make sense to drop the collision. The decision depends on how much risk you are willing to assume. Comprehensive covers damages resulting from theft, fire, falling objects, and so on. Now let's calculate the cost of these two types of coverage—assuming a $100 deductible for collision and a $200 deductible for comprehensive—with some more of my tables (Tables 20.8 and 20.9).

Collision and comprehensive are optional insurance types that pay only the insured. Note that Tables 20.8 and 20.9 are based on territory, age, and car symbol. The higher the symbol, the more expensive the car.

	Class	Age	Symbol	Premium	
Collision	17	3	4	$191 ($148 + $43)	Cost to reduce deductibles
Comprehensive	17	3	4	+ 56 ($52 + $4)	
				$247	

Total premium for collision and comprehensive

Shirley: Anything else?

Bob: I would also recommend that you buy towing and substitute transportation coverage. The insurance company will pay up to $25 for each tow. Under substitute transportation, the insurance company will pay you $12 a day for renting a car, up to $300 total. Again, from another table (Table 20.10), we find the additional premium for towing and substitute transportation is $20 ($16 + $4).

---

[7] In some states, repair to glass has no deductible and many insurance companies now use a $500 deductible instead of $300.

Life, Fire, and Auto Insurance • Chapter 20   605

Copyright © McGraw Hill

TABLE 20.8 Collision

Classes	Age group	Symbols 1–3 $300 ded.	Symbol 4 $300 ded.	Symbol 5 $300 ded.	Symbol 6 $300 ded.	Symbol 7 $300 ded.	Symbol 8 $300 ded.	Symbol 10 $300 ded.
10–20	1	180	180	187	194	214	264	279
	2	160	160	166	172	190	233	246
	3	148	148	154	166	183	221	233
	4	136	136	142	160	176	208	221
	5	124	124	130	154	169	196	208

These classes would use all this information.

To find the premium, use the age and symbol only.

Additional cost to reduce deductible

Class	From $300 to $200	From $300 to $100
10	13	27
17	20	43
18	16	33
20	26	55

TABLE 20.9 Comprehensive

Classes	Age group	Symbols 1–3 $300 ded.	Symbol 4 $300 ded.	Symbol 5 $300 ded.	Symbol 6 $300 ded.	Symbol 7 $300 ded.	Symbol 8 $300 ded.	Symbol 10 $300 ded.
10–25	1	61	61	65	85	123	157	211
	2	55	55	58	75	108	138	185
	3	52	52	55	73	104	131	178
	4	49	49	52	70	99	124	170
	5	47	47	49	67	94	116	163

Additional cost to reduce deductible: From $300 to $200 add $4

TABLE 20.10 Transportation and towing

Substitute transportation	$16
Towing and labor	4

We leave Shirley and Bob now as we make a summary of Shirley's total auto premium in Table 20.11.

No-Fault Insurance Some states have **no-fault insurance,** a type of auto insurance that was intended to reduce premium costs on bodily injury. With no fault, one forfeits the right to sue for *small* claims involving medical expense, loss of wages, and so on. Each person collects the bodily injury from his or her insurance company no matter who is at fault. In reality, no-fault insurance has not reduced premium costs, due to large lawsuits, fraud, and operating costs of insurance companies. Many states that were once considering no fault are no longer pursuing its adoption. Note that states with no-fault insurance require the purchase of *personal-injury protection (PIP)*. The most successful no-fault law seems to be in Michigan, since it has tough restrictions on the right to sue along with unlimited medical and rehabilitation benefits.

Premiums for collision, property damage, and comprehensive are not reduced by no fault.

TABLE 20.11
Worksheet for calculating Shirley's auto premium

Compulsory insurance	Limits	Deductible	Premium
Bodily injury to others	$10,000 per person $20,000 per accident	None	$ 98 (Table 20.5)
Damage to someone else's property	$5,000 per accident	None	$160 (Table 20.5)
Options			
Optional bodily injury to others	$250,000 per person $500,000 per accident	None	$228 (Table 20.6)
Optional property damage	$50,000 per accident	None	$168 (Table 20.7)
Collision	Actual cash value	$100	$191 (Table 20.8) ($148 + $43)
Comprehensive	Actual cash value	$200	$ 56 (Table 20.9) ($52 + $4)
Substitute transportation	Up to $12 per day or $300 total	None	$ 16 (Table 20.10)
Towing and labor	$25 per tow	None	$ 4 (Table 20.10)
			$921 Total premium

Learn: Calculating What the Insurance Company and the Insured Pay after an Auto Accident

When an automobile accident occurs, the insurance company pays up to the maximum of insurance coverage. The insured pays whatever is left. Because you will be financially responsible for damages not covered by your insurance, you should always get quotes from various companies for different coverage limits as well as deductibles. The cost to change to the next level of coverage may not be significant but lack of better coverage can be very costly.

Let's look at a typical example.

Example: Mario Andreety was at fault in an auto accident. He destroyed a fence and tree in a yard, causing damages of $2,500. He injured three passengers in the car he hit, which resulted in the following medical expenses: passenger 1, $25,250, passenger 2, $17,589, passenger 3, $12,567. The damage to the BMW he hit amounted to $45,888. Finally, the damage to his own vehicle came to $9,772.

a. If Mario has 15/30/10 coverage with a $500 deductible for collision and $100 deductible for comprehensive, how much will the insurance company pay? How much will Mario pay?

Money Tip Drive carefully. Tickets and accidents significantly affect the cost of your automobile insurance premiums. When paying your premiums, consider how important it is to have insurance when you need it and what a gift it is to have insurance and not need it.

Insurance company pays	Mario pays
$2,500	$0 Property damage (He has $10,000 property damage.)
$15,000	$25,250 − $15,000 = $10,250 Passenger 1 (His coverage is $15,000 per person bodily injury with an accident maximum of $30,000.)
$15,000	$17,589 − $15,000 = $2,589 Passenger 2 (The $30,000 [$15,000 + $15,000] maximum bodily injury coverage has been met.)
$0 max	$12,567 − $0 = $12,567 Passenger 3 ($30,000 per accident has been reached.)
$7,500	$45,888 − $7,500 = $38,388 (His coverage is $10,000 property damage with $2,500 used.)
$9,272	$500 ($9,772 personal vehicle − $500 collision deductible)
$49,272	$64,294

b. If Mario has 50/100/50 with a $500 deductible for collision and $100 deductible for comprehensive, how much will the insurance company pay? How much will Mario pay?

Insurance company pays	Mario pays
$2,500	$0 Property damage
$25,250	$0 Passenger 1
$17,589	$0 Passenger 2 ($25,250 + $17,589 = $42,839)
$12,567	$0 Passenger 3 ($42,839 + $12,567 = $55,406 < $100,000)
$45,888	$0 ($2,500 + $45,888 = $48,388 < $50,000)
$9,272	$500 ($9,772 − $500)
$113,066	$500

Clearly, Mario would be better off paying higher premiums for better coverage. His premiums won't amount to anywhere near $64,000 a year but, after only one accident with insufficient coverage, his out-of-pocket costs could easily exceed that amount.

It's time to take your final Practice Quiz in this chapter.

Practice Quiz

Complete this Practice Quiz to see how you are doing.

1. Calculate the annual auto premium for Mel Jones who lives in Territory 5, is a driver classified 18, and has a car with age 4 and symbol 7. His state has compulsory insurance, and Mel wants to add the following options:

   a. Bodily injury, 100/300.

   b. Damage to someone else's property, 10M.

   c. Collision, $200 deductible.

   d. Comprehensive, $200 deductible.

   e. Towing.

2. Calculate how much the insurance company and Carl Burns, the insured, pay if Carl carries 10/20/5 with $500 deductible for collision and $100 deductible for comprehensive and is at fault in an auto accident causing the following damage: $7,981, personal property; $6,454, injury to passenger 1; $4,239, injury to passenger 2; $25,250, injury to passenger 3; and $12,120 damage to Carl's car.

✓ Solutions

1. **Compulsory**

Bodily	$ 80		(Table 20.5)
Property	160		(Table 20.5)
Options			
Bodily	124		(Table 20.6)
Property	164		(Table 20.7)
Collision	192	($176 + $16)	(Table 20.8)
Comprehensive	103	($99 + $4)	(Table 20.9)
Towing	4		(Table 20.10)
Total annual premium	$827		

2. Carl's coverage is $10,000 per person bodily injury with an accident maximum of $20,000 bodily injury and $5,000 property damage.

Insurance company pays		**Carl pays**
$ 5,000		$ 2,981 Property damage
6,454		0 Passenger 1
4,239		0 Passenger 2 ($6,454 + $4,239 = $10,693)
9,307	($20,000 − $10,693)	15,943 Passenger 3 ($25,250 − $9,307)
11,620		500
$36,620		$19,424

Chapter 20 Review

Topic/Procedure/Formula	Examples	You try it*
Life insurance Using Table 20.1, per $1,000: $$\frac{\text{Coverage desired}}{\$1,000} \times \text{Rate}$$ For females, subtract 3 years.	**Given** $80,000 of insurance desired; age 34; male. **1.** 5-year term: $$\frac{\$80,000}{\$1,000} = 80 \times \$2.08 = \boxed{\$166.40}$$ **2.** Straight life: $$\frac{\$80,000}{\$1,000} = 80 \times \$10.71 = \boxed{\$856.80}$$ **3.** 20-payment life: $$\frac{\$80,000}{\$1,000} = 80 \times \$14.28 = \boxed{\$1,142.40}$$ **4.** 20-year endowment: $$\frac{\$80,000}{\$1,000} = 80 \times \$25.13 = \boxed{\$2,010.40}$$	**Given** $90,000 of insurance desired; age 36; male. **Calculate these premiums:** 1. 5-year term 2. Straight life 3. 20-payment life 4. 20-year endowment
Nonforfeiture values **By Table 20.2** Option 1: Cash surrender value. Option 2: Reduced paid-up insurance policy continues for life at reduced face amount. Option 3: Extended term—original face policy continued for a certain period of time.	A $50,000 straight-life policy was issued to Jim Rose at age 28. At age 48 Jim wants to stop paying premiums. What are his nonforfeiture options? Option 1: $\dfrac{\$50,000}{\$1,000} = 50 \times \$265$ $\qquad\qquad = \boxed{\$13,250}$ Option 2: $50 \times \$550 = \boxed{\$27,500}$ Option 3: $\boxed{\text{21 years 300 days}}$	**Given** $60,000 straight-life policy issued to Ron Lee at age 30. At age 50, Ron wants to stop paying premium. **Calculate his nonforfeiture options**
Fire insurance Per $100 $$\text{Premium} = \frac{\text{Insurance value}}{\$100} \times \text{Rate}$$ Rate can be for buildings or contents.	**Given** Area 3; Class B; building insured for $90,000; contents, $30,000. Building: $\dfrac{\$90,000}{\$100} = 900 \times \$.61$ $\qquad\qquad = \boxed{\$549}$ Contents: $\dfrac{\$30,000}{\$100} = 300 \times \$.65$ $\qquad\qquad = \boxed{\$195}$ Total: $549 + $195 = $\boxed{\$744}$	**Calculate fire insurance premium** Area 3; Class B; insurance for $80,000; contents, $20,000.
Canceling fire insurance—short-rate **Table 20.4 (canceling by policyholder)** $$\frac{\text{Short-rate}}{\text{premium}} = \frac{\text{Annual}}{\text{premium}} \times \frac{\text{Short}}{\text{rate}}$$ $$\text{Refund} = \frac{\text{Annual}}{\text{premium}} - \frac{\text{Short-rate}}{\text{premium}}$$ If insurance company cancels, do not use Table 20.4.	Annual premium is $400. Short rate is .35 (cancel end of 3 months). $400 \times .35 = $140 Refund = $400 − $140 = $\boxed{\$260}$	**Calculate refund** Annual premium is $600; insurance cancels after 4 months.

Chapter 20 Review (Continued)

Topic/Procedure/Formula	Examples	You try it*
Canceling by insurance company $\text{Annual premium} \times \dfrac{\text{Months elapsed}}{12}$ (Refund is higher since company cancels.)	Using example above, assume the insurance company cancels at end of 3 months. $\$400 \times \frac{1}{4} = \$100$ Refund $= \$400 - \$100 = \boxed{\$300}$	**Calculate refund from example above if insurance company cancels**
Coinsurance Amount insurance company pays: $\dfrac{\text{Actual} \longrightarrow \text{Insurance carried}}{\text{What} \quad \text{(Face value)}} \times \text{Loss}$ insurance $\dfrac{\text{(Face value)}}{\text{Insurance required}} \times \text{Loss}$ coverage $\longrightarrow$ to meet coinsurance should (Rate $\times$ Replacement value) have been Insurance company never pays more than the face value.	**Given** Face value, \$30,000; replacement value, \$50,000; coinsurance rate, 80%; loss, \$10,000; insurance to meet required coinsurance, \$40,000. $\dfrac{\$30,000}{\$40,000} \times \$10,000 = \boxed{\$7,500}$ paid by insurance company ($\$50,000 \times .80$)	**Calculate coinsurance Given** Face value, \$40,000; replacement, \$60,000; rate, 80%; loss, \$9,000.

Auto insurance
Compulsory Required insurance.
Optional Added to cost of compulsory.
Bodily injury—pays for injury to person caused by insured.
Property damage—pays for property damage (not for insured auto).
Collision—pays for damages to insured auto.
Comprehensive—pays for damage to insured auto for fire, theft, etc.
Towing.
Substitute transportation.
	Calculate the annual premium. Driver class 10; compulsory 10/20/5.
	Optional
	Bodily—100/300
	Property—10M
	Collision—age 3, symbol 10, \$100 deductible
	Comprehensive—\$300 deductible
	
	                   ($\$55 + \$129$)
	10/20/5 \$184 Table 20.5
	Bodily 94 Table 20.6
	Property 132 Table 20.7
	                   ($\$233 + \$27$)
	Collision 260 Table 20.8
	Comprehensive 178 Table 20.9
	Total premium $\boxed{\$848}$
		Calculate annual premium
		Driver class 10; compulsory 10/20/5.
		Optional
		Bodily—100/300
		Property—10M
		Collision—age 5, symbol 8, \$100 deductible
		Comprehensive—\$300 deductible

Chapter 20 Review (Continued)

Key Terms

Beneficiary	Face value	Property damage
Bodily injury	Fire insurance	Reduced paid-up insurance
Cash value	Indemnity	
Coinsurance	Insured	Short-rate table
Collision insurance	Insurer	Statisticians
Comprehensive insurance	Level premium term	Straight-life insurance
	Liability insurance	Term insurance
Compulsory insurance	No-fault insurance	20-payment life
Deductibles	Nonforfeiture values	20-year endowment
Extended term insurance	Paid-up insurance	Universal life
	Policyholder	Whole life
Face amount	Premium	

* Worked-out solutions are in Appendix A.

Critical Thinking Discussion Questions with Chapter Concept Check

1. Compare and contrast term insurance versus whole-life insurance. At what age do you think people should take out life insurance?

2. What is meant by *nonforfeiture values*? If you take the cash value option, should it be paid in a lump sum or over a number of years?

3. How do you use a short-rate table? Explain why an insurance company gets less in premiums if it cancels a policy than if the insured cancels.

4. What is coinsurance? Do you feel that an insurance company should pay more than the face value of a policy in the event of a catastrophe?

5. Explain compulsory auto insurance, collision, and comprehensive. If your car is stolen, explain the steps you might take with your insurance company.

6. "Health insurance is not that important. It would not be worth the premiums." Please take a stand.

7. **Chapter Concept Check.** Based on concepts in the chapter, what would it cost you to set up a life insurance policy that would fit your needs?

8. **Chapter Concept Check.** How much auto insurance do you need? Base your response on how much you pay versus how much your insurance company pays in the event of an accident.

End-of-Chapter Problems

Name _____ Date _____

Check figures for odd-numbered problems in Appendix A.

Drill Problems

Calculate the annual premium for the following policies using Table 20.1 (for females subtract 3 years from the table).   *LU 20–1(1)*

	Amount of coverage (face value of policy)	Age and sex of insured	Type of insurance policy	Annual premium
20–1.	$200,000	42 F	Straight life	
20–2.	$200,000	42 M	20-payment life	
20–3.	$75,000	29 F	5-year term	
20–4.	$50,000	27 F	20-year endowment	

Calculate the following nonforfeiture options for Lee Chin, age 42, who purchased a $200,000 straight-life policy. At the end of year 20, Lee stopped paying premiums.   *LU 20–1(2)*

20–5. Option 1: Cash surrender value

20–6. Option 2: Reduced paid-up insurance

20–7. Option 3: Extended term insurance

Calculate the total cost of a fire insurance premium (rounded to nearest cent) for a building and its contents given the following: *LU 20–2(1)*

		Rating of area	Class	Building	Contents	Total premium cost
e)cel	20–8.	3	B	$90,000	$40,000	

Calculate the short-rate premium and refund of the following: *LU 20–2(2)*

		Annual premium	Canceled after	Short-rate premium	Refund
e)cel	20–9.	$700	8 months by insured		
e)cel	20–10.	$360	4 months by insurance company		

End-of-Chapter Problems (Continued)

Complete the following: *LU 20–2(3)*

	Replacement value of property	Amount of insurance	Kind of policy	Actual fire loss	Amount insurance company will pay
e)cel 20–11.	$100,000	$60,000	80% coinsurance	$22,000	
20–12.	$60,000	$40,000	80% coinsurance	$42,000	

Calculate the annual auto insurance premium for the following: *LU 20–3(1)*

20–13. Britney Sper, Territory 5
Class 17 operator
Compulsory, 10/20/5   _____

Optional

a. Bodily injury, 500/1,000   _____

b. Property damage, 25M   _____

c. Collision, $100 deductible   _____
Age of car is 2; symbol of car is 7

d. Comprehensive, $200 deductible   _____
Total annual premium   _____

Word Problems

20–14. The average Roman's lifespan 2,000 years ago was 22 years. In 1900, a person was expected to live 47.3 years. In 2021, life expectancy was 78.99. If you are a 47-year-old female, what annual premium would you pay for a $200,000, 5-year term life insurance policy? What will be the cash value after 3 years?   *LU 20–1(1, 2)*

20–15. CBS News reported four ways to cut down on the cost of life insurance: (1) Shop around for the best rates from reputable companies, (2) improve your life expectancy by quitting smoking, etc., (3) buy life insurance when you are young, and (4) negotiate for lower premiums. Warren Kawano, age 34, was quoted $20 per month for a $100,000 term life insurance policy. Compare this to the rates in Table 20.1.   *LU 20–1(1)*

e)cel 20–16. Kathleen Osness, a 38-year-old massage therapist, decided to take out a limited-payment life policy. She chose this since she expects her income to decline in future years. Kathleen decided to take out a 20-year payment life policy with a coverage amount of $90,000. Could you advise Kathleen about what her annual premium will be? If she decides to stop paying premiums after 15 years, what will be her cash value?   *LU 20–1(1)*

20–17. Life insurance for a single parent is critical. Protect your children. The good news is life insurance for most single parents is very inexpensive. Buy term life insurance with a 20-year term, multiply your annual income by 7 to 10 times to determine how much to buy but get what you can afford now. Compare premiums. Choose a beneficiary. Janette Raffa, a single mom, has two young children and wants to take out an additional $300,000 of 5-year term insurance. Janette is 40 years old. What will be her additional annual premium? In 3 years, what cash value will have been built up? *LU 20–1(1)*

20–18. Roger's office building has a $320,000 value, a 2 rating, and a B building classification. The contents in the building are valued at $105,000. Could you help Roger calculate his total annual premium? *LU 20–2(1)*

e)cel **20–19.** Abby Ellen's toy store is worth $400,000 and is insured for $200,000. Assume an 80% coinsurance clause and that a fire caused $190,000 damage. What is the liability of the insurance company? *LU 20–2(3)*

20–20. To an insurer, you are a statistic. Your premiums are based on your risk factors, including your credit rating. Bad credit increases the amount you pay for your premiums. Make certain to check your credit report annually for accuracy. Calculate the premium for someone in class 20 for 10/20/5. Then determine how much the premium will be for 50/100/50. What is the difference between the two? *LU 20–3(1)*

20–21. As given via the Internet, auto insurance quotes gathered online could vary from $947 to $1,558. A class 18 operator carries compulsory 10/20/5 insurance. He has the following optional coverage: bodily injury, 500/1,000; property damage, 50M; and collision, $200 deductible. His car is 1 year old, and the symbol of the car is 8. He has comprehensive insurance with a $200 deductible. Using your text, what is the total annual premium? *LU 20–3(1)*

e)cel **20–22.** Earl Miller insured his pizza shop for $100,000 for fire insurance at an annual rate per $100 of $.66. At the end of 11 months, Earl canceled the policy since his pizza shop went out of business. What was the cost of Earl's premium and his refund? *LU 20–2(2)*

End-of-Chapter Problems (Continued)

e⟨cel **20–23.** Warren Ford insured his real estate office with a fire insurance policy for $95,000 at a cost of $.59 per $100. Eight months later the insurance company canceled his policy because of a failure to correct a fire hazard. What did Warren have to pay for the 8 months of coverage? Round to the nearest cent. *LU 20–2(2)*

20–24. If you had 10/20/5 coverage and were in a car accident causing injury to three people with injuries totaling $15,000, $9,000, and $5,000, how much would you have to pay out of pocket? If you had 50/100/50 coverage for the same scenario, how much would you have to pay out of pocket? What is the difference? *LU 20–3(1)*

20–25. Tina Grey bought a new Honda Civic and insured it with only 10/20/5 compulsory insurance. Driving up to her ski chalet one snowy evening, Tina hit a parked van and injured the couple inside. Tina's car had damage of $4,200, and the van she struck had damage of $5,500. After a lengthy court suit, the injured persons were awarded personal injury judgments of $16,000 and $7,900, respectively. What will the insurance company pay for this accident, and what is Tina's responsibility? *LU 20–3(1)*

20–26. Rusty Reft, who lives in Territory 5, carries 10/20/5 compulsory liability insurance along with optional collision that has a $300 deductible. Rusty was at fault in an accident that caused $3,600 damage to the other auto and $900 damage to his own. Also, the courts awarded $15,000 and $7,000, respectively, to the two passengers in the other car for personal injuries. How much will the insurance company pay, and what is Rusty's share of the responsibility?   *LU 20–3(1)*

20–27. Marika Katz bought a new Blazer and insured it with only compulsory insurance 10/20/5. Driving up to her summer home one evening, Marika hit a parked car and injured the couple inside. Marika's car had damage of $7,500, and the car she struck had damage of $5,800. After a lengthy court suit, the couple struck were awarded personal injury judgments of $18,000 and $9,000, respectively. What will the insurance company pay for this accident, and what is Marika's responsibility?   *LU 20–3(1)*

20–28. In Problem 20–27, what will the insurance company pay and what is Marika's responsibility if Marika has 25/50/25 coverage with $200 deductible for collision instead of 10/20/5?   *LU 20–3(2)*

20–29. Dave Ramsey (daveramsey.com) recommends only term life insurance and suggests having 10 to 12 times your annual income in life insurance. If you earn $85,000 per year, are 35 years old, and are nonbinary applying as a male, how much term life insurance do you need and what will it cost?   *LU 20–1(1)*

End-of-Chapter Problems (Continued)

20–30. Louis Hamilton is conducting an annual review of his insurance policies. He wants to make certain he has enough fire insurance coverage for the building he calls "his garage" valued at $175,000 and its contents valued at $475,000. Calculate his total annual premium with a rating of 3 and a B building and content classification. *LU 20–2(1)*

20–31. Selling items you find at garage sales, thrift stores, and online has provided you with a budding business. Your goods available for sale are stored in a building valued at $150,000 and insured for $75,000. A fire causes $65,000 in damage. Assume an 80% coinsurance clause. How much will the insurance company cover? *LU 20–2(3)*

Challenge Problems

20–32. Money.cnn.com states, "The single most important reason to own life insurance is to provide support for your dependents." Insurance4usa.com states, "Professionals suggest you have 8 to 12 times your income in life insurance." Pat and Bonnie Marsh are calculating how much life insurance they need. They have two young children with no college fund set up. Bonnie is a 35-year-old stay-at-home mom. Pat, 39, earns $68,000 per year. How much life insurance do you recommend each person have? (Note that a spouse who stays at home to raise the family generates the equivalent of a salary that needs to be taken into account. Assume Bonnie's salary is $25,000.) What will be the cost of straight-life insurance for both policies if the lowest recommended amount is used? What is the monthly premium owed? *LU 20–1(1)*

20–33. Lou Ralls insured a building and contents (area 2, class B) for $150,000. After 1 month, he canceled the policy. The next day he received a cancellation notice by the company. It stated that he was being canceled due to his previous record. How does Lou save by this insurance cancellation versus his planned cancellation? *LU 20–2(1, 2)*

Summary Practice Test

Do you need help? Connect videos have step-by-step worked-out solutions.

1. Howard Slater, age 44, an actor, expects his income to decline in future years. He decided to take out a 20-year payment life policy with a $90,000 coverage. What will be Howard's annual premium? If he decides to stop paying premiums after 15 years, what will be his cash value?   *LU 20–1(1, 2)*

2. J.C. Monahan, age 40, bought a straight-life insurance policy for $210,000. Calculate her annual premium. If after 20 years J.C. no longer pays her premiums, what nonforfeiture options will be available to her?   *LU 20–1(1, 2)*

3. The property of Pote's Garage is worth $900,000. Pote has a $375,000 fire insurance policy that contains an 80% coinsurance clause. What will the insurance company pay on a fire that causes $450,000 damage? If Pote meets the coinsurance, how much will the insurance company pay?   *LU 20–2(3)*

4. Lee Collins insured her pizza shop with a $90,000 fire insurance policy at a $1.10 annual rate per $100. At the end of 7 months, Lee's pizza shop went out of business so she canceled the policy. What is the cost of Lee's premium and her refund?   *LU 20–2(2)*

5. Charles Prose insured his real estate office with a $300,000 fire insurance policy at $.78 annual rate per $100. Nine months later the insurance company canceled his policy because Charles failed to correct a fire hazard. What was Charles's cost for the 9-month coverage? Round to the nearest cent.   *LU 20–2(2)*

6. Roger Laut, who lives in Territory 5, carries 10/20/5 compulsory liability insurance along with optional collision that has a $1,000 deductible. Roger was at fault in an accident that caused $4,800 damage to the other car and $8,800 damage to his own car. Also, the courts awarded $19,000 and $9,000, respectively, to the two passengers in the other car for personal injuries. How much does the insurance company pay, and what is Roger's share of the responsibility?   *LU 20–3(1)*

MY MONEY

Q Insuring Your Future

 What I need to know

Insurance planning, also known as risk management, is a critical part of healthy financial planning. Knowing which policies you require and being properly insured can ease the financial impact and help to reduce stress in the event of a loss (accident, theft, fire, death, etc.). Determining an accurate assessment of your potential financial risks will assist you in locating the type and coverage amounts of insurance you need. Determining the value of your personal property will assist you in determining the amount of insurance coverage appropriate to your unique situation. Remember that insurance is what protects you from the risks associated with loss events. Different types of insurance to consider are health, life, auto, long-term disability, property, homeowners or renters, long-term care, identity theft, umbrella policies, among others.

 What I need to do

Research a variety of insurance plans and options to gain an understanding of the types of insurance coverages available. Assess your personal situation to determine the insurance policies best suited to your needs. When selecting your insurance coverages be sure to note the items that will be covered by the policy and those that are not. In addition, familiarize yourself with the types of loss events that will be covered to avoid any surprises should you need to file a claim on a loss. Understanding the coverage limits of your policy is also helpful in determining the level of coverage you need to adequately protect your assets.

Selecting the appropriate coverage amounts you need will be based on the value of your insured assets. For your personal property (car, home, possessions, etc.) you will want to assess the fair market value of these items and what they would cost you to replace. Documenting your purchases with receipts is a great way to account for the value of your personal property. Record serial numbers and VIN's. Make a list of your assets (car, house, furnishings, etc.) with their associated values to make sure you have the correct amount of coverage included in your insurance policies. Consider options from multiple insurance providers and learn about the types and premiums of insurance policies available from each to assist you in making the appropriate coverage decisions for your unique situation. Conduct research on each company you are considering to ensure they are financially sound and reputable.

 Steps I need to take

1. List the types of insurance policies required to protect health, life, auto, long-term disability and property.
2. Review your insurance policies annually to ensure what you want covered is covered and to understand the items and amounts covered.
3. Stay organized and document large purchases (television, car, furniture).
4. Research other insurance coverages available based on your personal situation.

 Resources I can use

- https://www.daveramsey.com/blog/types-insurance-cant-go-without—lists a variety of insurance options you may want to consider.
- https://www.daveramsey.com/blog/how-much-car-insurance—details the basics about car insurance and helps you determine how much car insurance you need.

MY MONEY ACTIVITY

- Request quotes from three different reputable car insurance providers for your vehicle.
- Note the difference in insurance premiums from each.
- Repeat this activity for homeowners/renters/life/health/long-term disability insurance.
- What is the value and/or savings in combining your car and home/renters policies?

PERSONAL FINANCE

A KIPLINGER APPROACH

"Time for an Insurance Review." Kiplinger's. September 2020.

FAMILY FINANCES

Time for an Insurance Review

You may need to update your policies in light of COVID-19.
BY DANIEL BORTZ

THE HENDERSONS SLASHED THEIR AUTO INSURANCE PREMIUM BY $60 A MONTH WHEN THEY STOPPED DRIVING THEIR SECOND CAR.

THE CORONAVIRUS HAS MADE one thing abundantly clear: We all need to be prepared for an emergency. Which prompts the question: When did you last review your insurance policies?

You may want to make some changes to your auto, homeowners and life insurance policies in light of COVID-19. "You could be paying for coverage that you don't need anymore, or you could be lacking coverage in some areas because of the pandemic," says Carmen Balber, executive director at Consumer Watchdog, a non-profit consumer-advocacy organization.

Auto insurance. Chances are you received a discount from your auto insurer automatically. In April, a number of large insurance carriers offered their customers discounts of, typically, 15%.

Many drivers realized they were barely using their cars and asked their insurers for even bigger premium adjustments. Scott and McKenzie Henderson of Lehi, Utah, started driving a lot less when they began working from home in March. The couple, who own two cars, decided to stop using their Dodge Stratus and share McKenzie's Kia Optima. When they informed their insurance company, USAA, it lowered their insurance premiums to $25 a month, down from $85. "It made sense for us to start using only one car, since neither of us was commuting to work," says Scott.

Dino Selita of Staten Island, N.Y., also saved money by asking Geico, his insurer, to adjust his auto premium. Selita says he hardly drove his car in March, April or May. "I was basically leaving my house once a week to buy groceries," he says. Selita's insurer gave him a $500 credit.

"It only took a 10-minute phone call [to my insurance agent]," he says.

"If your car is basically parked in your driveway, you may not need the same level of collision coverage that you currently have," says Balber. Moreover, drivers who change their car's status from "business" to "pleasure" save an average of $172 a year, according to the State of Auto Insurance 2020 report from The Zebra, an insurance brokerage.

Shopping around for a lower rate now is also a good idea, Balber says. Given the stiff competition among carriers, you may have even more negotiating power when threatening to leave your insurer.

Business Math Issue

Post pandemic means insurance companies will not negotiate with you.

1. List the key points of the article and information to support your position.

2. Write a group defense of your position using math calculations to support your view. If you are in an online course, post to a discussion board.

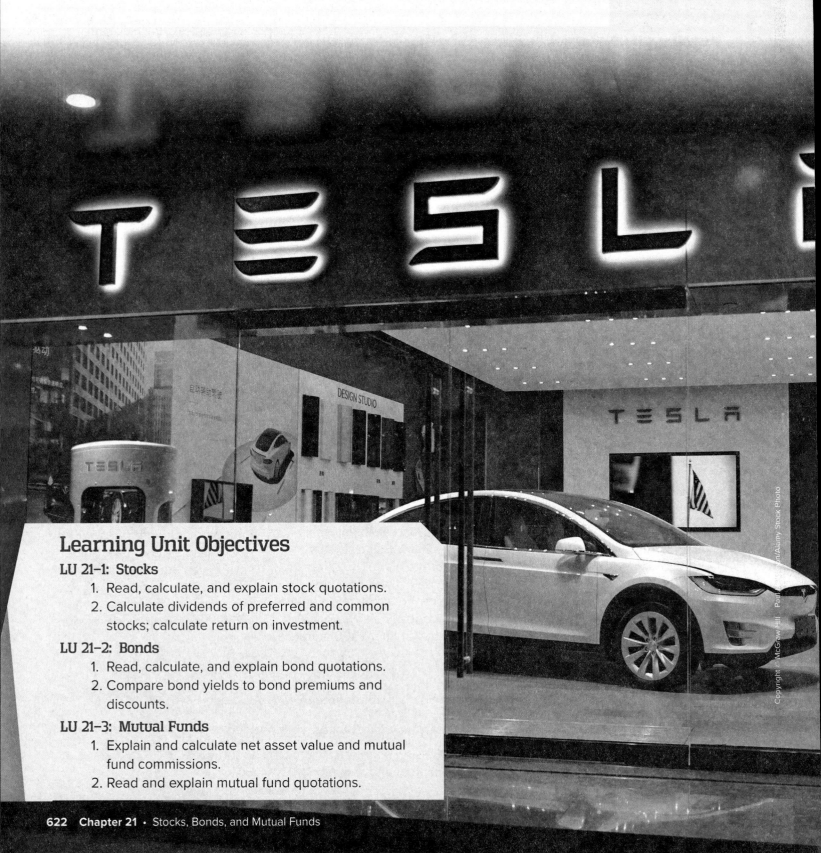

Chapter 21
Stocks, Bonds, and Mutual Funds

Learning Unit Objectives

LU 21-1: Stocks

1. Read, calculate, and explain stock quotations.
2. Calculate dividends of preferred and common stocks; calculate return on investment.

LU 21-2: Bonds

1. Read, calculate, and explain bond quotations.
2. Compare bond yields to bond premiums and discounts.

LU 21-3: Mutual Funds

1. Explain and calculate net asset value and mutual fund commissions.
2. Read and explain mutual fund quotations.

Tesla to Split Stock 5 for 1 After Price Surge

BY MARIA ARMENTAL

Tesla Inc. said it would enact a 5-for-1 stock split after a share-price surge over recent months vaulted the electric-vehicle maker to the status of most valuable car company as Chief Executive Elon Musk navigated the pandemic.

Enthusiasm in Tesla's stock has been fueled by four consecutive quarters of profit and Wall Street's expectation the company will post its first-ever full-year profit. Shares rose more than 6% in after-hours trading following the Tuesday announcement after closing lower at $1,374.39.

The split wouldn't change Tesla's market value–which as of Tuesday stood at more than Toyota Motor Corp. and Ford Co. combined–but it would potentially make the stock more appealing to a broader base of investors. Essentially, an investor who owned 1,000 Tesla shares before the split would own 5,000 shares after the split but the share price would be divided by 5.

Tesla's move follows one from Apple Inc., which last month said its board had approved a 4-for-1 stock split, aiming to make the stock more accessible to more investors.

Tesla, which started trading publicly in 2010, has seen its stock more than triple in value this year–and nearly sixfold over the past 12 months.

Mr. Musk, who has criticized the company's stock price as too high, said in January on Twitter that a stock split was worth discussing at the company's annual shareholder meeting. That meeting has been pushed back to September amid the pandemic. Several investors had urged Mr. Musk to pursue a stock split.

Armental, Maria. "Tesla to Split Stock 5 for 1 after Price Surge." *The Wall Street Journal* (August 12, 2020).

Essential Question

How can I use stocks, bonds, and mutual funds to understand the business world and make my life easier?

Math Around the World

When you make financial investments there is always some degree of risk. Should you invest in Tesla stock, or just keep your money in cash? The *Wall Street Journal* chapter opener clip, "Tesla to Split Stock 5 for 1 After Price Surge", discusses Tesla's strategy for a stock split.

Before we explain the concept of stock, consider the following general investor principles: (1) know your risk tolerance and the risk of the investments you are considering—determine whether you are a low-risk conservative investor or a high-risk speculative investor; (2) know your time frame—how soon you need your money; (3) know the liquidity of the investments you are considering—how easy it is to get your money; (4) know the return you can expect on your money—how much your money should earn; and (5) do not put "all your eggs in one basket"—diversify with a mixture of stocks, bonds, and cash equivalents. It is most important that before you seek financial advice from others, you go to the library and/or the Internet for information. When you do your own research first, you can judge the advice you receive from others.

This chapter introduces you to the major types of investments—stocks, bonds, and mutual funds. These investments indicate the performance of the companies they represent and the economy of the country at home and abroad.

Stocks

We begin this unit with an introduction to the basic stock terms. Then we explain the reason why people buy stocks, newspaper stock quotations, dividends on preferred and common stocks, and return on investment.

Learn: Introduction to Basic Stock Terms

Companies sell shares of ownership in their company to raise money to finance operations, plan expansion, and so on. These ownership shares are called **stocks.** The buyers of the stock (**stockholders**) receive **stock certificates** verifying the number of shares of stock they own.

The two basic types of stock are **common stock** and **preferred stock.** Common stockholders have voting rights. Preferred stockholders do not have voting rights, but they receive preference over common stockholders in **dividends** (payments from profit) and in the company's assets if the company goes bankrupt. **Cumulative preferred stock** entitles its owners to a specific amount of dividends in 1 year. Should the company fail to pay these dividends, the **dividends in arrears** accumulate. The company pays no dividends to common stockholders until the company brings the preferred dividend payments up to date.

> If you own 50 shares of common stock, you are entitled to 50 votes in company elections. Preferred stockholders do not have this right.

Learn: Why Buy Stocks?

Some investors own stock because they think the stock will become more valuable, for example, if the company makes more profit, new discoveries, and the like. Other investors own stock to share in the profit distributed by the company in dividends (cash or stock).

For various reasons, investors at different times want to sell their stock or buy more stock. Strikes, inflation, or technological changes may cause some investors to think their stock will decline in value. These investors may decide to sell. Then the law of supply and demand takes over. As more people want to sell, the stock price goes down. Should more people want to buy, the stock price would go up. During the pandemic, stock prices tumbled. Six months after the start of the pandemic, stocks reached record highs. It will be interesting to see what the long-term effect, if any, the pandemic has on the stock market.

How Are Stocks Traded?   Stock exchanges provide an orderly trading place for stock. You can think of these exchanges as an auction place. Only **stockbrokers** and their representatives are allowed to trade on the floor of the exchange. Stockbrokers charge commissions for stock trading—buying and selling stock for investors. As you might expect, in this age of the Internet, stock trades can also be made on the Internet. Electronic trading is growing each day. Let's look at how to read stock quotations.

Learn: How to Read Stock Quotations[1]

We will use Hershey Company stock to learn how to read the stock quotations. Note the following listing of Hershey Company stock:

52 WEEKS			YLD			NET
HI	LO	STOCK (SYM)	%	PE	LAST	CHG
150.24	147.33	Hershey (HSY)	2.14	29.45	148.33	2.26

[1] For centuries, stocks were traded and reported in fraction form. In 2001 the New York Stock Exchange and NASDAQ began the conversion to decimals, which is how stocks are reported today.

The highest price at which Hershey stock traded during the past 52 weeks was $150.24 per share. This means that during the year someone was willing to pay $150.24 for one share of stock.

The lowest price at which Hershey stock traded during the year was $147.33 per share. The symbol that Hershey uses for trading is HSY.

The **stock yield** percent tells stockholders that the dividend per share is returning a rate of 2.14% to investors. This 2.14% is based on the closing price. Hershey declared a dividend of $3.17. The calculation is

Stock Yield Formula

$$\frac{\text{Stock}}{\text{yield}} = \frac{\text{Annual dividend per share}}{\text{Today's last price per share}} = \frac{\$3.17}{\$148.33} = 2.14\% \quad \text{(rounded to nearest hundredth percent)}$$

The 2.14% return may seem low to people who could earn a better return on their money elsewhere. Remember that if the stock price rises and you sell, your investment may result in a high rate of return.

The Hershey stock is selling at $148.33; it is selling at 29.45 times its **earnings per share (EPS).** Earnings per share are not listed on the stock quote.

$$\text{Earnings per share} = \text{Last price} \div \text{Price-earnings ratio}$$
$$(\$5.04) \quad = (\$148.33) \div \quad (29.45)$$

The **price-earnings ratio,** or **PE ratio,** measures the relationship between the closing price per share of stock and the annual earnings per share. For Hershey we calculate the following price-earnings ratio.

Round PE to the nearest hundredth.

$$\text{PE ratio} = \frac{\text{Last price per share of stock}}{\text{Annual earnings per share}} = \frac{\$148.33}{\$5.04} = 29.43^*$$

*off due to rounding

If the PE ratio column shows ". . . ," this means the company has no earnings. The PE ratio will often vary depending on quality of stock, future expectations, economic conditions, and so on.

The last trade of the day, called the closing price, was at $148.33 per share.

On the *previous day,* the closing price was $146.07 (not given). The result is that the last price is up $2.26 from the *previous day.*

Learn: Dividends on Preferred and Common Stocks

If you own stock in a company, the company may pay out dividends. (Not all companies pay dividends.) The amount of the dividend is determined by the net earnings of the company listed in its financial report.

Earlier we stated that cumulative preferred stockholders must be paid all past and present dividends before common stockholders can receive any dividends. Following is an example to illustrate the calculation of dividends on preferred and common stocks for 2023 and 2024.

Example: The stock records of Jason Corporation show the following:

> Preferred stock issued: 20,000 shares.   In 2023, Jason paid no dividends.
>
> Preferred stock cumulative at $.80 per share.   In 2024, Jason paid $512,000 in dividends.
>
> Common stock issued: 400,000 shares.

Since Jason declared no dividends in 2023, the company has $16,000 (20,000 shares × $.80 = $16,000) dividends in arrears to preferred stockholders. The dividend of $512,000 in 2024 is divided between preferred and common stockholders as follows:

	2023	2024	
Dividends paid 0		$512,000	
Preferred stockholders*	Paid: 0 Owe: Preferred, $16,000 (20,000 shares × $.80)	Paid for 2023 (20,000 shares × $.80)	$ 16,000
		Paid for 2024	16,000
Common stockholders	0		$ 32,000
		Total dividend	$512,000
		Paid preferred for 2023 and 2024	– 32,000
		To common	$480,000
		$\dfrac{\$480,000}{400,000 \text{ shares}} = \$1.20$ per share	

*For a discussion of par value (arbitrary value placed on stock for accounting purposes) and cash and stock dividend distribution, check your accounting text.

Aha!   *Shares are typically traded in groups of 100 called **round lots**. Purchases of fewer than 100 shares of stock are called **odd lots**.*

Learn: Calculating Return on Investment

Now let's learn how to calculate a return on your investment if you bought a different stock than Hershey. Let's assume you decided to buy stock of the Kraft Heinz Company given the following:

Bought 200 shares at $39.09.

Sold at end of 1 year 200 shares at $41.10.

1% commission rate on buying and selling stock.

Current $1.21 dividend per share in effect.

Bought		**Sold**	
200 shares at $39.09	$7,818.00	200 shares at $41.10	$8,220.00
+ Broker's commission		− Broker's commission	
(.01 × $7,818)	+ 78.18	(.01 × $8,220.00)	− 82.20
Total cost	$7,896.18	Total receipt	$8,137.80

Note: A commission is charged on both the buying and selling of stock.

Total receipt	$8,137.80
Total cost	− 7,896.18
Net gain	$ 241.62
Dividends	+ 242.00 (200 shares × $1.21)
Total gain	$ 483.62

Portion → $\dfrac{\$483.62}{\$7,896.18}$ = **6.12%** rate of return (to nearest hundredth percent)

↑ Base

It's time for another Practice Quiz.

Learning Unit 21–1

Practice Quiz

Complete this Practice Quiz to see how you are doing.

1. From the following Pinterest, Inc. stock quotation **(a)** explain the letters, **(b)** estimate the company's earnings per share, and **(c)** show how "YLD %" was calculated.

52 WEEKS			YLD			NET
HI	**LO**	**STOCK (SYM)**	**%**	**PE**	**LAST**	**CHG**
73.90	48.25	Pinterest, Inc. (PINS)	2.5	14	72.25	+0.46
(A)	(B)	(C)	(D)	(E)	(F)	(G)

2. **Given:** 30,000 shares of preferred cumulative stock at $.70 per share; 200,000 shares of common; 2020, no dividend; 2021, $109,000. How much is paid to each class of stock in 2021?

✓ Solutions

1. **a.** (A) Highest price traded in last 52 weeks.

   (B) Lowest price traded in past 52 weeks.

   (C) Name of corporation is Pinterest, Inc. (symbol PINS).

   (D) Yield for year is 2.5%.

   (E) Pinterest, Inc. stock sells at 14 times its earnings.

   (F) The last price (closing price for the day) is $72.25.

   (G) Stock is up $.46 from closing price yesterday.

   b. $\text{EPS} = \dfrac{\$72.25}{14} = \$5.16 \text{ per share}$

   c. $\dfrac{?}{\$72.25} = 2.5\%$ $\$72.25 \times 2.5\% = \$1.80^*$

2. **Preferred:** 30,000 × $.70 = $21,000   Arrears 2020

                                              + 21,000   2021

                                              $42,000

   Common: $67,000 ($109,000 − $42,000)

Bonds

Have you heard of the Rule of 115? This rule is used as a rough measure to show how quickly an investment will triple in value. To use the rule, divide 115 by the rate of return your money earns. For example, if a bond earns 5% interest, divide 115 by 5. This measure estimates that your money in the bond will triple in 23 years.

This unit begins by explaining the difference between bonds and stocks. Then you will learn how to read bond quotations and calculate bond yields.

Aha! *When you own stock, you own a share of a company. When you own a bond, you are lending the company money, similar to how banks lend money.*

Learn: Reading Bond Quotations

Sometimes companies raise money by selling bonds instead of stock. When you buy stock, you become a part owner in the company. To raise money, companies may not want to sell more stock and thus dilute the ownership of their current stock owners, so they sell bonds. **Bonds** represent a promise from the company to pay the face amount to the bond owner at a future date, along with interest payments at a stated rate. See the *Wall Street Journal* clip, "Tesla Marshals Its Cash," showing how Tesla made a bond payment.

Once a company issues bonds, they are traded as stock is. If a company goes bankrupt, bondholders have the first claim to the assets of the corporation—before stockholders. As with stock, changes in bond prices vary according to supply and demand. Brokers also charge commissions on bond trading. These commissions vary.

Learn: How to Read the Bond Section of the Newspaper

The bond section of the newspaper shows the bonds that are traded that day. The information given on bonds differs from the information given on stocks. The newspaper states bond prices in *percents of face amount, not in dollar amounts* as stock prices are stated. Also, bonds are usually in denominations of $1,000 (the face amount).

When a bond sells at a price below its face value, the bond is sold at a discount. Why? The interest that the bond pays may not be as high as the current market rate. When this happens, the bond is not as attractive to investors, and it sells for a **discount.** The opposite could, of course, also occur. The bond may sell at a **premium,** which means that the bond sells for more than its face value or the bond interest is higher than the current market rate.

Let's look at this newspaper information given for Aflac bonds:

Bond quotes are stated in percents of the face value of the bond and not in dollars as stock is. Interest is paid semiannually.

Tesla Marshals Its Cash

Auto maker makes $920 million bond payment

By Akane Otani and Sam Goldfarb

Tesla Inc. delivered its largest-ever bond payment Friday, a move that likely used up nearly a quarter of its cash at a time when the company faces increasing scrutiny from regulators and investors.

The electric-car maker had issued $920 million in convertible, senior notes five years ago, a time when rapid growth and optimism around its Model S electric car drove a furious rally in Tesla shares.

Reflecting that, the strike price for the notes–or the level at which the notes would be fulfilled by a conversion into Tesla stock–was $359.87 a share, or 42.5% above the level at that time.

Yet because shares traded below that level recently, Tesla had to make good on its obligation using cash. That likely wiped out a substantial portion of the $3.69 billion

Please turn to page B2

Otani, Akane and Sam Goldfarb. "Tesla Marshals its Cash." *The Wall Street Journal* (March 1, 2019).

Bonds	Current yield	Vol.	Close	Net change
Aflac 428	4.02%	214,587	99.50	−1

Note: Bond prices are stated as a percent of face amount.

The name of the company is Aflac. It produces a wide range of insurance coverage. The interest on the bond is 4%. The company pays the interest semiannually. The bond matures (comes due) in 2028. The total interest for the year is $40 (.04 × $1,000). Remember that the face value of the bond is $1,000. Now let's show this with the following formula:

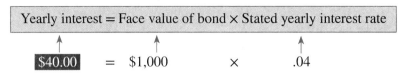

$$\text{Yearly interest} = \text{Face value of bond} \times \text{Stated yearly interest rate}$$

$$\boxed{\$40.00} \quad = \quad \$1,000 \quad \times \quad .04$$

Note this bond is selling for more than $1,000 since its interest is very attractive compared to other new offerings.

We calculate the 4.02% yield by dividing the total annual interest of the bond by the total cost of the bond. (For our purposes, we will omit the commission cost.) We will calculate more bond yields in a moment.

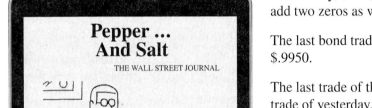

$$\frac{\text{Yearly interest}}{\text{Cost of bond at closing}} = \frac{\$40 \ (.04 \times \$1,000)}{\$995 \ (.9950 \times \$1,000)}$$

$$= \boxed{4.02\%} \quad \text{This is the same as 99.50\%.}$$

On this day, $214,587 worth of bonds were traded. Note that we do *not* add two zeros as we did to the sales volume of stock.

The last bond traded on this day was 99.50% of face value, or in dollars, $.9950.

The last trade of the day was down 1% of the face value from the last trade of yesterday. In dollars this is 1% = $10.

$$1\% = .01 \times \$1,000 = \$10$$

Thus, the closing price on this day, 99.50% + 1%, equals yesterday's close of 100.50% ($1,005). Note that *yesterday's close is not listed in today's quotations.*

 Remember: Bond prices are quoted as a percent of $1,000 but without the percent sign. A bond quote of 99 means 99% of $1,000, or $990.

Learn: Calculating Bond Yields

The Aflac bond (selling at a discount) pays 4% interest when it is yielding investors 4.02%.

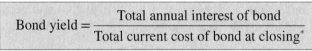

$$\text{Bond yield} = \frac{\text{Total annual interest of bond}}{\text{Total current cost of bond at closing}^*}$$

*We assume this to be the buyer's purchase price.

The following example will show us how to calculate **bond yields.**

Example: Jim Smith bought 5 bonds of Aflac at the closing price of 99.50 (remember that in dollars 99.50% is $995). Jim's total cost excluding commission is:

$$5 \times \$995 = \$4,975$$

What is Jim's interest?

No matter what Jim pays for the bonds, he will still receive interest of $40 per bond (.04 × $1,000). Jim bought the bonds at $995 each, resulting in a bond yield of 4.02%. Let's calculate Jim's yield to the nearest tenth percent:

$$\frac{\$200.00}{\$4,975} = 4.02\% \quad \text{(5 bonds} \times \$40 \text{ interest per bond per year)}$$

Now let's try another Practice Quiz.

Money Tip Spend your money wisely. Cut back on frivolous spending to provide more discretionary income for an emergency fund and retirement savings. Spend less money than you earn. Make your money make more money through investing and save for the future.

Practice Quiz

Complete this Practice Quiz to see how you are doing.

Bonds	Yield	Sales	Close	Net change
Aetna 6.375 33	6.4	20	100.375	.875

From the above bond quotation, **(1)** calculate the cost of 5 bonds at closing (disregard commissions) and **(2)** check the current yield of 6.4%.

✓ **Solutions**

1. $100.375\% = 1.00375 \times \$1,000 = \$1,003.75 \times 5 = \boxed{\$5,018.75}$

2. $6.375\% = .06375 \times \$1,000 = \$63.75$ annual interest

    $\dfrac{\$63.75}{\$1,003.75} = 6.35\% = \boxed{6.4\%}$

Learning Unit 21–3:
Mutual Funds

In recent years, mutual funds have increased dramatically and people in the United States have invested billions in mutual funds. Investors can choose from several fund types—stock funds, bond funds, international funds, balanced funds (stocks and bonds), and so on. This learning unit tells you why investors choose mutual funds and discusses the net asset value of mutual funds, mutual fund commissions, and how to read a mutual fund quotation.

Learn: Why Investors Choose Mutual Funds

The main reasons investors choose mutual funds are the following:

1. **Diversification.** When you invest in a mutual fund, you own a small portion of many different companies. This protects you against the poor performance of a single company but not against a sell-off in the market (stock and bond exchanges) or fluctuations in the interest rate.

2. **Professional management.** You are hiring a professional manager to look after your money when you own shares in mutual funds. The success of a particular fund is often due to the person(s) managing the fund.

   Aha! *Some investors invest in a mutual fund simply because of who the mutual fund manager is.*

3. **Liquidity.** Most funds will buy back your fund shares whenever you decide to sell.

4. **Low fund expenses.** Competition forces funds to keep their expenses low to maximize their performance. Because stocks and bonds in a mutual fund represent thousands of shareholders, funds can trade in large blocks, reducing transaction costs.

5. **Access to foreign markets.** Through mutual funds, investors can conveniently and inexpensively invest in foreign markets.

Learn: Net Asset Value

Investing in a **mutual fund** means that you buy shares in the fund's portfolio (group of stocks and/or bonds). The value of your mutual fund share is expressed in the share's **net asset value (NAV),** which is the dollar value of one mutual fund share. You calculate the NAV by subtracting the fund's current liabilities from the current market value of the fund's investments and dividing this by the number of shares outstanding.

$$NAV = \frac{\text{Current market value of fund's investments} - \text{Current liabilities}}{\text{Number of shares outstanding}}$$

The NAV helps investors track the value of their fund investment. After the market closes on each business day, the fund uses the closing prices of the investments it owns to find the dollar value of one fund share, or NAV. This is the price investors receive if they sell fund shares on that day or pay if they buy fund shares on that day.

Learn: Commissions When Buying Mutual Funds

The following table is a quick reference for the cost of buying mutual fund shares. Commissions vary from 0% to $8\frac{1}{2}$% depending on how the mutual fund is classified.

Classification	Commission charge*	Offer price to buy
No-load (NL) fund	No sales charge	NAV (buy directly from investment company)
Low-load (LL) fund	3% or less	NAV + commission % (buy directly from investment company or from a broker)
Load fund	$8\frac{1}{2}$% or less	NAV + commission % (buy from a broker)

*On a front-end load, you pay a commission when you purchase the fund shares, while on a back-end load, you pay when you redeem or sell. In general, if you hold the shares for more than 5 years, you pay no commission charge.

The offer price to buy a share for a low-load or load fund is the NAV plus the commission. Now let's look at how to read a mutual fund quotation.

Learn: How to Read a Mutual Fund Quotation

We will be studying the Franklin Templeton Funds. Cindy Joelson has invested in the Growth A Fund with the hope that over the years this will provide her with financial security when she retires. Cindy turns to the *Wall Street Journal* and looks up the Franklin Templeton Growth A quotation.

The name of the fund is Growth A, which has the investment objective of growth A securities as set forth in the fund's prospectus (document giving information about the fund). Note that this is only one fund in the Franklin Templeton Funds family of funds.

- The $128.76 figure is the NAV plus the sales commission.
- The fund has increased $0.93 from the NAV quotation of the previous day.
- The fund has a 14.7% return this year (January through December). This assumes reinvestments of all distributions. Sales charges are not reflected.

Financial analysts recommend that individual retirement accounts contain some mixture of stocks and bonds. Retirement accounts should be heavily invested in stocks while the investor is young and gradually shift holdings to bonds as retirement approaches. Mutual funds can invest in a variety of securities such as stocks, bonds, money market instruments, real estate, and similar assets. Investors may use a bond fund for income. Consider the following example:

Example: Bonnie and Pat Meyer are in their retirement years. They just received $250,000 after taxes from the sale of their vacation home and decided to invest the money in a bond mutual fund. They chose a no-load mutual fund that yields 4.5%. How much will they receive each year? How much would they need to invest if they want to earn $15,000 per year?

Step 1. $I = PRT = \$250,000 \times .045 \times 1 =$ $\boxed{\$11,250}$

Step 2. $P = \dfrac{I}{RT} = \dfrac{\$15,000}{.045 \times 1} =$ $\boxed{\$333,333.33}$

If Bonnie and Pat invest $250,000, they will receive $11,250 in interest each year. If they need to earn $15,000 in interest each year, they must invest an additional $83,333.33: $333,333.33 − $250,000 = $83,333.33.

Money Tip A will provides peace of mind to those surviving. Review your will annually to ensure it is up to date. If you do not have a will, write one. There are many free resources online. Put your will in a safe deposit box or fireproof safe and ensure survivors can locate it.

"Mutual Funds." *The Wall Street Journal* (August 18, 2020).

Now let's check your understanding of this unit with a Practice Quiz.

Practice Quiz

Complete this Practice Quiz to see how you are doing.

From the following mutual fund quotation of the Fidelity Invest GrowCoK complete the following:

1. NAV

   _____

2. NAV change

   _____

3. Total return, YTD

   _____

4. You are interested in earning $6,500 each year on a no-load 5% yield mutual fund. How much must you invest?

✓ Solutions

1. **30.15**

2. **+0.53**

3. **41.0**

4. $P = \dfrac{1}{RT} = \dfrac{\$6,500}{.05 \times 1} = \boxed{\$130,000}$

Fund	NAV	Net Chg	YTD %Ret
Fidelity Freedom			
FF2020	16.36	+0.06	4.0
FF2025	14.51	+0.05	3.9
FF2030	17.94	+0.07	3.6
Freedom2020 K	16.34	+0.05	4.0
Freedom2025 K	14.49	+0.05	3.9
Freedom2030 K	17.93	+0.07	3.7
Freedom2035 K	15.06	+0.07	3.0
Freedom2040 K	10.54	+0.05	2.7
Fidelity Invest			
Balanc	26.70	+0.06	9.5
BluCh	144.43	+2.08	34.1
Contra	16.43	+0.14	20.6
ContraK	16.45	+0.13	20.6
CpInc r	9.99	+0.01	0.2
GroCo	30.08	+0.52	40.8
GrowCoK	30.15	+0.53	41.0
InvGrBd	12.27	+0.01	7.9
LowP r	47.06	+0.07	−6.0
Magin	12.05	+0.11	18.1
OTC	16.23	+0.17	26.9
Puritn	25.13	+0.16	11.2
SrsEmrgMkt	21.18	+0.18	2.7
SrsGlobal	12.77	+0.08	−3.0
SrsGroCoRetail	25.26	+0.44	42.3
SrsIntlGrw	18.61	+0.15	6.2
SrsIntlVal	8.98	+0.03	−9.3
TotalBond	11.51	...	7.2
Fidelity SAI			
TotalBd	11.00	+0.01	6.6
Fidelity Selects			
Softwr r	24.13	+0.21	25.5
First Eagle Funds			
GlbA	57.67	+0.36	−0.5
FPA Funds			
FPACres	31.56	−0.06	−3.8
Franklin A1			
CA TF A1 p	7.81	−0.01	4.0
IncomeA1 p	2.12	...	−6.3
FrankTemp/Frank Adv			
IncomeAdv	2.10	−0.01	−6.3
FrankTemp/Franklin A			
Growth A p	128.76	+0.93	14.7
RisDv A p	72.58	+0.24	4.8
FrankTemp/Franklin C			
Income C t	2.15	−0.01	−6.6
FrankTemp/Temp Adv			
GlBondAdv p	9.83	−0.02	−5.3
Guggenheim Funds Tru			
TotRtnBdFdClInst	29.71	+0.03	11.3
Harbor Funds			
CapApInst	100.67	+1.85	32.9
Harding Loevner			
IntlEq	24.70	+0.18	3.8
Invesco Funds Y			
DevMktY	45.10	+0.24	−1.1
JPMorgan I Class			
CoreBond	12.45	+0.01	6.9
EqInc	17.90	...	−6.8
JPMorgan R Class			

"Mutual Funds." *The Wall Street Journal* (August 18, 2020).

Chapter 21 Review

Topic/Procedure/Formula	Examples	You try it*
Stock yield $\dfrac{\text{Annual dividend per share}}{\text{Today's last price per share}}$ (Round yield to nearest hundredth percent.)	Annual dividend, $.72 Today's last price, $42.375 $\dfrac{\$.72}{\$42.375} = $ 1.70%	**Calculate stock yield to nearest hundredth percent** Annual dividend, $.88 Today's closing price, $53.88
Price-earnings ratio $PE = \dfrac{\text{Last price per share of stock}}{\text{Annual earnings per share}}$ (Round answer to nearest whole number.)	From previous example: Last price, $42.375 Annual earnings per share, $4.24 $\dfrac{\$42.375}{\$4.24} = 9.99 = $ 10	**Calculate PE ratio** From previous example: Closing price, $53.88 Annual earnings per share, $3.70
Dividends with cumulative preferred stock Cumulative preferred stock is entitled to all dividends in arrears before common stock receives dividend	2022 dividend omitted; in 2023, $400,000 in dividends paid out. Preferred is cumulative at $.90 per share; 20,000 shares of preferred issued and 100,000 shares of common issued. To preferred: 20,000 shares × $.90 = $18,000 In arrears 2022: 20,000 shares × .90 = 18,000 Dividend to preferred $36,000 To common: $364,000 ($400,000 − $36,000) $\dfrac{\$364,000}{100,000 \text{ shares}} = \$3.64$ dividend to common per share	**Calculate dividends to preferred and common stock** 2022, no dividend; 2023, $300,000 Preferred—$.80 cumulative, 30,000 shares issued Common—60,000 shares issued
Cost of a bond Bond prices are stated as a percent of the face value. Bonds selling for less than face value result in bond discounts. Bonds selling for more than face value result in bond premiums.	Bill purchases 5 $1,000, 12% bonds at closing price of $103\frac{1}{4}$. What is his cost (omitting commissions)? $103\frac{1}{4}\% = 103.25\% = 1.0325$ in decimal $1.0325 \times \$1,000$ bond = $1,032.50 per bond 5 bonds × $1,032.50 = $5,162.50	**Calculate cost of bonds** 6 $1,000, 3% bonds at 102.25

636 **Chapter 21** · Stocks, Bonds, and Mutual Funds

Chapter 21 Review (Continued)

Topic/Procedure/Formula	Examples	You try it*
Bond yield $$\frac{\text{Total annual interest of bond}}{\text{Total current cost of bond at closing}}$$ (Round to nearest tenth percent.)	Calculate bond yield from last example on one bond. $$\frac{\overset{(\$1,000 \times .12)}{\$120}}{\$1,032.50} = \boxed{11.6\%}$$	**Calculate bond yield** 4% bond selling for $1,011.20
Mutual fund $$NAV = \frac{\begin{array}{c}\text{Current market value}\\\text{of fund's investment}\end{array} - \begin{array}{c}\text{Current}\\\text{liabilities}\end{array}}{\text{Number of shares outstanding}}$$	The NAV of the Scudder Income Bond Fund was $12.84. The NAV change was 0.01. What was the NAV yesterday? $\boxed{\$12.83}$	**Calculate yesterday's NAV** Today—$12.44 Change—.05

Key Terms

Bond yield

Bonds

Common stocks

Cumulative preferred stock

Discount

Dividends

Dividends in arrears

Earnings per share (EPS)

Mutual fund

Net asset value (NAV)

Odd lot

PE ratio

Preferred stock

Premium

Price-earnings ratio

Round lot

Stock certificate

Stock yield

Stockbrokers

Stockholders

Stocks

* Worked-out solutions are in Appendix A.

Critical Thinking Discussion Questions with Chapter Concept Check

1. Explain how to read a stock quotation. What are some of the red flags of buying stock?

2. What is the difference between odd and round lots? Explain why the commission on odd lots could be quite expensive.

3. Explain how to read a bond quote. What could be a drawback of investing in bonds?

4. Compare and contrast stock yields and bond yields. As a conservative investor, which option might be better? Defend your answer.

5. Explain what NAV means. What is the difference between a load and a no-load fund? How safe are mutual funds?

6. **Chapter Concept Check.** Track the stock price of Delta Airlines. What has happened to it since the pandemic? How has passenger travel changed?

End-of-Chapter Problems

Name _____ Date _____

Check figures for odd-numbered problems in Appendix A.

Drill Problems

Calculate the cost (omit commission) of buying the following shares of stock: *LU 21–1(1)*

21–1. 500 shares of Delta Air Lines, Inc. (DAL) at $50.61

21–2. 1,200 shares of Apple, Inc. (AAPL) at $125.50

Calculate the yield of each of the following stocks (rounded to the nearest tenth percent):   *LU 21–1(1)*

Company	Yearly dividend	Closing price per share	Yield
21–3. Facebook, Inc. (FB)	$10.09	$306.18	____
21–4. Best Buy Co., Inc. (BBY)	$6.84	$119.87	____

Calculate the price-earnings ratio (21–5) and closing price per share (21–6) (to nearest whole number) or stock price as needed: *LU 21–1(1)*

Company	Earnings per share	Closing price per share	Price-earnings ratio
21–5. Morgan Stanley (MS)	$7.64	$78.59	—
21–6. American Express	$3.85	_____	26

21–7. Calculate the total cost of buying 400 shares of CVS at $102.90. Assume a 2% commission.   *LU 21–1(1)*

21–8. If in Problem 21–1 the 500 shares of Delta Air Lines, Inc. stock were sold at $50, what would be the loss? Commission is omitted.   *LU 21–1(1)*

21–9. Given: 20,000 shares cumulative preferred stock ($2.25 dividend per share); 40,000 shares common stock. Dividends paid: 2021, $8,000; 2022, 0; and 2023, $160,000. How much will preferred and common stockholders receive each year?   *LU 21–1(2)*

For each of these bonds, calculate the total dollar amount you would pay at the quoted price (disregard commission or any interest that may have accrued):   *LU 21–2(1)*

	Company	Bond price	Number of bonds purchased	Dollar amount of purchase price
21–10.	Petro	87.75	3	_____
21–11.	Wang	114	2	_____

For the following bonds, calculate the total annual interest, total cost, and current yield (to the nearest tenth percent):   *LU 21–2(2)*

	Bond	Number of bonds purchased	Selling price	Total annual interest	Total cost	Current yield
21–12.	Sharn $11\frac{3}{4}$ 33	2	115	_____	_____	_____
21–13.	Wang $6\frac{1}{2}$ 29	4	68.125	_____	_____	_____

21–14. From the following calculate the net asset values. Round to the nearest cent.   *LU 21–3(1)*

	Current market value of fund investment	Current liabilities	Number of shares outstanding	NAV
a.	$5,550,000	$770,000	600,000	_____
b.	$13,560,000	$780,000	840,000	_____

21–15. From the following mutual fund quotation, complete the blanks: *LU 21–3(2)*

					TOTAL RETURN		
	Inv. obj.	NAV	NAV chg.	YTD	4 wks.	1 yr.	
EuGr	ITL	12.04	−0.06	+8.2	+0.9	+9.6	

NAV _____   NAV change _____

Total return, 1 year _____

Copyright © McGraw Hill

End-of-Chapter Problems (Continued)

Word Problems

excel 21–16. Ryan Neal bought 1,200 shares of Ford (F) at $15.98 per share. Assume a commission of 2% of the purchase price. What is the total cost to Ryan?   *LU 21–1(1)*

excel 21–17. Assume in Problem 21–16 that Ryan sells the stock for $22.25 with the same 2% commission rate. What is the bottom line for Ryan?   *LU 21–1(1)*

excel 21–18. Jim Corporation pays its cumulative preferred stockholders $1.60 per share. Jim has 30,000 shares of preferred and 75,000 shares of common. In 2021, 2022, and 2023, due to slowdowns in the economy, Jim paid no dividends. Now in 2024, the board of directors decided to pay out $500,000 in dividends. How much of the $500,000 does each class of stock receive as dividends?   *LU 21–1(2)*

excel 21–19. Whirlpool Corporation (WHR) earns $4.80 per share. Today the stock is trading at $234.81. The company pays an annual dividend of $17.07. Calculate **(a)** the price-earnings ratio (rounded to the nearest whole number) and **(b)** the yield on the stock (to the nearest tenth percent).   *LU 21–1(1)*

excel 21–20. Jimmy Comfort was interested in pursuing a second career after retiring from the military. He signed up with Twitter to help network with individuals in his field. Within 1 week, he received an offer from a colleague to join her start-up business in Atlanta, Georgia. Along with his salary, he receives 100 shares of stock each month. If the stock is worth $4.50 a share, what is the value of the 100 shares he receives each month?   *LU 21–1(1, 2)*

21–21. The following bond was quoted in *The Wall Street Journal:*   *LU 21–2(1)*

Bonds	Curr. yld.	Vol.	Close	Net chg.
NJ 4.125 35	3.5	5	96.875	$+1\frac{1}{2}$

Five bonds were purchased yesterday, and 5 bonds were purchased today. How much more did the 5 bonds cost today (in dollars)?

21–22. DailyFinance.com reported one $40 share of Coca-Cola's (KO) stock bought in 1919, with dividends reinvested, would be worth $9.8 million today. If the price-earnings ratio was 28.42 at that time, what were the annual earnings per share? Round to the nearest cent.   *LU 21–1(1)*

21-23. Dairy Queen, as part of Warren Buffet's Berkshire Hathaway (BRKA) with 6,400 locations in the USA, gave away free ice cream cones to celebrate its 75th anniversary. If Warren Buffet has a bond bought at 105.25 at $4\frac{3}{4}$ 35, what is the current yield to the nearest percent?   *LU 21–2(2)*

21–24. Abby Sane decided to buy corporate bonds instead of stock. She desired to have the fixed-interest payments. She purchased 5 bonds of Meg Corporation $11\frac{3}{4}$ 34 at 88.25. As the stockbroker for Abby (assume you charge her a $5 commission per bond), please provide her with the following: **(a)** the total cost of the purchase, **(b)** total annual interest to be received, and **(c)** current yield (to nearest tenth percent).   *LU 21–2(1)*

excel 21–25. Mary Blake is considering whether to buy stocks or bonds. She has a good understanding of the pros and cons of both. The stock she is looking at is trading at $59.25, with an annual dividend of $3.99. Meanwhile, the bond is trading at 96.25, with an annual interest rate of $11\frac{1}{2}\%$. Calculate for Mary her yield (to the nearest tenth percent) for the stock and the bond.   *LU 21–1(1), LU 21–2(1)*

21–26. Wall Street performs a sort of "financial alchemy" enabling individuals to benefit from institutions lending money to them, according to Adam Davidson, cofounder of NPR's "Planet Money." Individuals can invest small amounts of their money in a 401(k), pooling their capital and spreading the risk. If you invested in Fidelity New Millennium, FMILX, how much would you pay for 80 shares if the 52-week high is $32.26, the 52-week low is $26.38, and the NAV is $31.88?   *LU 21–3(1)*

End-of-Chapter Problems (Continued)

21–27. Louis Hall read in the paper that Fidelity Growth Fund has an NAV of $16.02. He called Fidelity and asked how the NAV was calculated. Fidelity gave him the following information:

Current market value of fund investment	$8,550,000
Current liabilities	$ 860,000
Number of shares outstanding	480,000

Did Fidelity provide Louis with the correct information? *LU 21–3(1)*

21–28. Lee Ray bought 130 shares of a mutual fund with an NAV of $13.10. This fund also has a load charge of $8\frac{1}{2}\%$. **(a)** What is the offer price and **(b)** what did Lee pay for his investment? *LU 21–3(1)*

21–29. Ron and Madeleine Couple received their 2023 Form 1099-DIV (dividends received) in the amount of $1,585. Ron and Madeleine are in the 28% bracket. What would be their tax liability on the dividends received? *LU 21–1(2)*

21–30. Bob Eberhart wants to retire in 10 years. He's heard that he needs either $1 million or 10 to 12 times his current income of $60,000 saved. He doesn't have either. His goal now is to save, save, save. Saving can double his nest egg if the stock market continues to deliver 7% annually in this decade. He sold his second car and is saving $500 per month. How many mutual fund shares can he purchase monthly with an NAV of $22.74?

21–31. When should you start investing? The Motley Fool recommends as soon as reasonably possible. The younger you start investing, the less you need to invest; and the richer you can get due to the time value of money. Starting young also provides more recovery time from market downturns. Just make certain your high-interest debt is paid off or under control and you have saved at least a 3-month emergency fund. Wanting to invest in companies that support the environment you decided to purchase NextEra Energy (NEE). If you have $3,500 to invest and NEE is selling at $80.94 per share, how many shares can you purchase? Round to the nearest tenth. *LU 21–1(1)*

21–32. Interestingengineering.com discusses what they refer to as the battle between Wall Street fund managers and online investors. Reddit, a social news platform, coordinated a "short squeeze" in early 2021 of GameStop shares resulting in a hedge fund losing billions of dollars on its short position and ultimately having to declare bankruptcy. If GameStop earns $3.78 per share, is trading at $154.69, and pays an annual dividend of $1.24, calculate (a) the price earnings ratio (rounded to the nearest whole number) and (b) the stock yield (rounded to the nearest tenth percent). *LU 21–1(1)*

21–33. Three growth stock mutual funds had returns over the past 10 years of 7.9%, 9.3%, and 9.3%, respectively, reported *Kiplinger* magazine. If you purchased the mutual fund with an average annual 9.3% return 10 years ago and invested $12,750, how much would it be worth today?   *LU 21–3(1)*

Challenge Problems

21–34. Here's an example of how breakpoint discounts on sales commissions for mutual fund investors work:

Sales charge

> Less than $25,000, 5.75%
>
> $25,000 to $49,999, 5.50%
>
> $50,000 to 99,999, 4.75%
>
> $100,000 to $249,999, 3.75%

Nancy Dolan is interested in the T Rowe Price Mid Cap Fund. Assume the NAV is 19.43. **(a)** What minimum amount of shares must Nancy purchase to have a sales charge of 5.50%? **(b)** What are the minimum shares Nancy must purchase to have a sales charge of 4.75%? **(c)** What are the minimum shares Nancy must purchase to have a sales charge of 3.75%? **(d)** What would be the total purchase price for **(a)**, **(b)**, or **(c)**? Round up to the nearest share even if it is less than 5.   *LU 21–3(1)*

21–35. On September 6, Irene Westing purchased one bond of Mick Corporation at 98.50. The bond pays $8\frac{3}{4}$ interest on June 1 and December 1. The stockbroker told Irene that she would have to pay the accrued interest and the market price of the bond and a $6 brokerage fee. What was the total purchase price for Irene? Assume a 360-day year (each month is 30 days) in calculating the accrued interest. (*Hint:* Final cost = Cost of bond + Accrued interest + Brokerage fee. Calculate time for accrued interest.)   *LU 21–2(2)*

Summary Practice Test

Do you need help? Connect videos have step-by-step worked-out solutions.

1. Russell Slater bought 700 shares of Disney stock at $106.50 per share. Assume a commission of 4% of the purchase price. What is the total cost to Russell?   *LU 21–1(1)*

2. HM Company earns $2.50 per share. Today, the stock is trading at $18.99. The company pays an annual dividend of $.25. Calculate **(a)** the price-earnings ratio (to the nearest whole number) and **(b)** the yield on the stock (to the nearest tenth percent).   *LU 21–1(1)*

3. The stock of Aware is trading at $4.90. The price-earnings ratio is 4 times earnings. Calculate the earnings per share (to the nearest cent) for Aware.   *LU 21–1(1)*

4. Tom Fox bought 8 bonds of UXY Company $3\frac{1}{2}$ 35 at 84 and 4 bonds of Foot Company $4\frac{1}{8}$ 36 at 93. Assume the commission on the bonds is $3 per bond. What was the total cost of all the purchases?   *LU 21–2(1)*

5. Leah Long bought one bond of Vick Company for 147. The original bond was 8.25 30. Leah wants to know the current yield to the nearest tenth percent. Help Leah with the calculation.   *LU 21–2(2)*

6. Cumulative preferred stockholders of Rale Company receive $.80 per share. The company has 70,000 shares outstanding. For the last 9 years, Rale paid no dividends. This year, Rale paid $400,000 in dividends. What is the amount of dividends in arrears that is still owed to preferred stockholders?   *LU 21–1(2)*

7. Bill Roundy bought 800 shares of a mutual fund with an NAV of $14.10. This fund has a load charge of 3%. **(a)** What is the offer price and **(b)** what did Bill pay for the investment? *LU 21–3(1)*

Plan Your Estate Today

 What I need to know

Estate planning can be an intimidating task to complete as you consider your mortality. However, if you want to avoid a family feud, keep your minor children out of foster care, distribute your assets the way you choose, and share your wishes for your funeral: plan your estate TODAY. To maintain control over your estate, you need to legally inform the state of your wishes. You can do this through completing five documents: a will, a revocable living trust, a financial power of attorney, a durable power of attorney and a letter of instruction. When the time comes to carry out your wishes you will be glad you took the time to plan, execute, and discuss your wishes with your family. The time in which your estate decisions are carried out will undoubtedly be emotional for those you leave behind. You want to be sure you have expressed and documented your wishes in detail to make the process easier for your family and friends.

 What I need to do

Share your wishes with your family and friends both orally and in writing so they are aware of your wishes. As part of this process you will need to decide who is best suited to make decisions on your behalf and care for minor children. The person(s) you designate as Financial Power of Attorney and Durable Power of Attorney for Health Care should be considered carefully to ensure they possess the capacity to make these difficult decisions. Financial Power of Attorney allows your designated person to make legal and financial decisions on your behalf if you are unable to do so whereas the Durable Power of Attorney for Health Care will concentrate on decisions related to your health care should you become too ill or incapacitated to make decisions for yourself.

Creating a will is an important part of estate planning. Your will documents and provides direction for carrying out your wishes after you die. In addition to your will, you should also establish a living revocable trust used to avoid probate, protect privacy, and minimize estate taxes. A letter of instruction is a document that provides specific information to your loved ones concerning important personal matters such as your preferences for a funeral or not. A common practice, as well, is to write your loved ones letters storing these with your estate documents. As with any effective planning it is important to conduct an annual review of the documents you are creating as part of your estate planning to account for any changes you may want to make. For instance, you may have a new grandchild you want to ensure is named. A good practice as part of this annual review is to remind your family members where and how to access your estate plans.

 Steps I need to take

1. Make your estate plans TODAY to prepare effectively for your passing and the healthy financial future of your dependents.
2. Create a will, living revocable trust, Financial Power of Attorney, and Durable Power of Attorney for Health Care for each adult as part of your comprehensive estate plan.
3. Review your plans on an annual basis to account for life changes.

 Resources I can use

- Tomorrow—Insurance & Wills (mobile app)—estate planning assistance with items such as wills, living trusts, and power of attorney.
- https://www.suzeorman.com/blog/tag/estate-planning—discussion of a variety of topics to consider in effective estate planning.

MY MONEY ACTIVITY

- Research where you can obtain legal forms recognized by your state for a will, living revocable trust, Financial Power of Attorney, and Durable Power of Attorney for Health Care.
- Complete the forms and have them notarized. Store in a safety deposit box or fireproof file. Inform your family and friends where this information is being stored.
- Consider having an estate attorney review your documents to ensure they conform to the legal requirements in your area.

PERSONAL FINANCE

A KIPLINGER APPROACH

"Briefing Information about the Markets and Your Money GAMESTOP:
WHAT YOU NEED TO KNOW." Kiplinger's. April, 2021.

BRIEFING

INFORMATION ABOUT THE MARKETS AND YOUR MONEY.

GAMESTOP: WHAT YOU NEED TO KNOW

The recent gyrations in the shares of GameStop, a struggling brick-and-mortar retailer of video games and consoles, has captivated the nation and brought business news to prime time. The epic battle between "smart money" hedge funds and an aggressive group of social media day traders shook the stock market briefly, claimed a short-seller pelt or two, and garnered the attention of regulators and Congress. Expect some changes on the regulatory front, but long-term investors will do best to ignore the noise and focus on the fundamentals that drive stock prices over time, including sales and profits.

GameStop's wild ride was fueled by hordes of traders who congregate at r/WallStreetBets, a community on Reddit that has grown in recent weeks from some 2 million followers to more than 8 million. The video-game retailer at first would seem an unlikely object of fascination. At the beginning of 2020, the company hadn't made a profit in two years, and in a world where gamers buy and download games online, its prospects were dim. Then came the pandemic, causing foot traffic at mall-based stores to collapse. GameStop became an obvious target for short sellers, who profit when a company's share price declines.

David versus Goliath?

That's when the mob at WallStreetBets decided to wage war, openly motivated by a desire to profit at the expense of the rich hedge funds and other institutional investors on the short side of the trade. The Redditors realized that if they bought en masse, with many using options strategies to control a significant number of shares at a fraction of their cost, they could deal a blow to the shorts. (For more on short selling and the "short squeeze," see "Street Smart," on page 43.)

And that's just what happened. The tug-of-war sent GameStop shares from $17 at the start of the year to as high as $483 intraday on January 29 (the chart below shows closing prices). The battle was waged on several online trading platforms and spread to other heavily shorted stocks. Some brokers restricted trading in the stocks for a time, with Robinhood in particular drawing criticism for the move. Robinhood said the decision was forced by requests for higher deposits, on account of volatility, from the firms that clear its stock trades.

Accusations of market manipulation flew from both sides, and questions arose about the internet and market integrity, the suitability of high-risk securities for certain sets of customers, options pricing and more. Regulators are scrutinizing the debacle. Congress is holding hearings, and class-action lawyers are on the case. (For more on potential regulation, see "Ahead," on page 9.)

This story's not over. But bear in mind: Whether they've come from irreverent posts on Reddit, the message boards and unsolicited faxes of the '90s, or the proverbial shoeshine boy, dubious hot-stock tips have always attracted plenty of followers. The GameStop episode "is herd behavior in a modern era—familiar, but leveraging today's tools," says Katie Nixon, chief investment officer at Northern Trust Wealth Management. Serious investors who rely on fundamentals, valuations and diversification to allocate capital wisely have little new to learn from this spectacle. **DAN BURROUGHS**

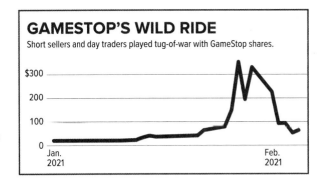

GAMESTOP'S WILD RIDE
Short sellers and day traders played tug-of-war with GameStop shares.

Business Math Issue

Social media cannot affect stock prices.

1. List the key points of the article and information to support your position.

2. Write a group defense of your position using math calculations to support your view. If you are in an online course, post to a discussion board.

Chapter 22

Business Statistics

Learning Unit Objectives

LU 22–1: Mean, Median, and Mode

1. Define and calculate the mean.
2. Explain and calculate a weighted mean.
3. Define and calculate the median.
4. Define and identify the mode.

LU 22–2: Frequency Distributions and Graphs

1. Prepare a frequency distribution.
2. Prepare bar, line, and circle graphs.
3. Calculate price relatives and cost comparisons.

LU 22–3: Measures of Dispersion (Optional)

1. Explain and calculate the range.
2. Define and calculate the standard deviation.
3. Estimate percentage of data by using standard deviations.

Weekly Covid-19 cases in U.S. nursing homes

15 thousand

Note: Analysis excludes nursing homes that failed federal data-quality checks in one or more weeks or that reported Covid-related deaths that exceeded total deaths.
Source: Centers for Medicare & Medicaid Services

"Weekly Covid-19 cases in U.S. Nursing Homes" Infographic. *The Wall Street Journal* (August 22, 2020).

℮ Essential Question

How can I use business statistics to understand the business world and make my life easier?

🌐 Math Around the World

In this chapter we look at various techniques that analyze and graphically represent business statistics. For example, in the *Wall Street Journal* chapter opener clip, "Weekly Covid-19 cases in U.S. nursing homes", we see how COVID-19 has affected patients in nursing homes. At the time of writing, nursing home residents have been vaccinated and vaccines are being made available to most age groups. Statistics have played an important role in mitigating the pandemic. Learning Unit 22–1 discusses the mean, median, and mode. Learning Unit 22–2 explains how to gather data by using frequency distributions and express these data visually in graphs. Emphasis is placed on whether graphs are indeed giving accurate information. The chapter concludes with an introduction to index numbers—an application of statistics—and an optional learning unit on measures of dispersion.

Pepper ...
And Salt
THE WALL STREET JOURNAL

"*I'm going to bombard you with graphs until you agree.*"

Mean, Median, and Mode

Companies frequently use averages and measurements to guide their business decisions. The mean and median are the two most common averages used to indicate a single value that represents an entire group of numbers. The mode can also be used to describe a set of data.

Learn: Mean

The accountant of Bill's Sport Shop told Bill, the owner, that the average daily sales for the week were $150.14. The accountant stressed that $150.14 was an average and did not represent specific daily sales. Bill wanted to know how the accountant arrived at $150.14.

Pepper ...
And Salt

THE WALL STREET JOURNAL

"It's my new social distancing desk."

The accountant went on to explain that he used an arithmetic average, or **mean** (a measurement), to arrive at $150.14 (rounded to the nearest hundredth). He showed Bill the following formula:

Mean Formula

$$\text{Mean} = \frac{\text{Sum of all values}}{\text{Number of values}}$$

The accountant used the following data:

	Sun.	Mon.	Tues.	Wed.	Thur.	Fri.	Sat.
Sport Shop sales	$400	$100	$68	$115	$120	$68	$180

To compute the mean, the accountant used these data:

$$\text{Mean} = \frac{\$400 + \$100 + \$68 + \$115 + \$120 + \$68 + \$180}{7} = \boxed{\$150.14}$$

When values appear more than once, businesses often look for a **weighted mean.** The format for the weighted mean is slightly different from that for the mean. The concept, however, is the same except that you weight each value by how often it occurs (its frequency). Thus, considering the frequency of the occurrence of each value allows a weighting of each day's sales in proper importance. To calculate the weighted mean, use the following formula:

Weighted Mean Formula

$$\text{Weighted mean} = \frac{\text{Sum of products}}{\text{Sum of frequencies}}$$

Let's change the sales data for Bill's Sport Shop and see how to calculate a weighted mean:

	Sun.	Mon.	Tues.	Wed.	Thur.	Fri.	Sat.
Sport Shop sales	$400	$100	$100	$80	$80	$100	$400

Value	Frequency	Product
$400	2	$ 800
100	3	300
80	2	160
		$1,260

The weighted mean is $\dfrac{\$1,260}{7} =$ $180

Note how we multiply each value by its frequency of occurrence to arrive at the product. Then we divide the sum of the products by the sum of the frequencies.

When you calculate your grade point average (GPA), you are using a weighted average. The following formula is used to calculate GPA:

GPA Formula

$$GPA = \frac{\text{Total points}}{\text{Total credits}}$$

Now let's show how Jill Rivers calculated her GPA to the nearest tenth.

Given A = 4; B = 3; C = 2; D = 1; F = 0

Courses	Credits attempted	Grade received	Points (Credits × Grade)
Introduction to Computers	4	A	16 (4 × 4)
Psychology	3	B	9 (3 × 3)
English Composition	3	B	9 (3 × 3)
Business Law	3	C	6 (2 × 3)
Business Math	3	B	9 (3 × 3)
	16		49

$\dfrac{49}{16} =$ 3.1

When high or low numbers do not significantly affect a list of numbers, the mean is a good indicator of the center of the data. If high or low numbers do have an effect, the median may be a better indicator to use.

Learn: Median

The **median** is another measurement that indicates the center of the data. An average that has one or more extreme values is not distorted by the median. For example, let's look at the following yearly salaries of the employees of Rusty's Clothing Shop.

Alice Knight	$95,000	Jane Wang	$67,000
Jane Hess	27,000	Bill Joy	40,000
Joel Floyd	32,000		

Note how Alice's salary of $95,000 will distort an average calculated by the mean.

$$\frac{\$95,000 + \$27,000 + \$32,000 + \$67,000 + \$40,000}{5} = \$52,200$$

The $52,200 average salary is considerably more than the salary of three of the employees. So it is not a good representation of the store's average salary. The following *Wall Street Journal* clip, "Filling Open Slots," shows how educators are trying to better match their programs with new job opportunities. We use the following steps to find the median of a group of numbers.

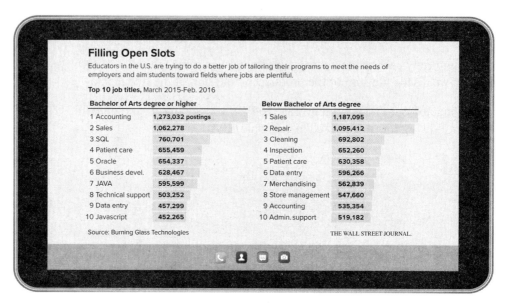

Filling Open Slots

Educators in the U.S. are trying to do a better job of tailoring their programs to meet the needs of employers and aim students toward fields where jobs are plentiful.

Top 10 job titles, March 2015-Feb. 2016

Bachelor of Arts degree or higher		Below Bachelor of Arts degree	
1 Accounting	1,273,032 postings	1 Sales	1,187,095
2 Sales	1,062,278	2 Repair.	1,095,412
3 SQL	760,701	3 Cleaning	692,802
4 Patient care	655,459	4 Inspection	652,260
5 Oracle	654,337	5 Patient care	630,358
6 Business devel.	628,467	6 Data entry	596,266
7 JAVA	595,599	7 Merchandising	562,839
8 Technical support	503,252	8 Store management	547,660
9 Data entry	457,299	9 Accounting	535,354
10 Javascript	452,265	10 Admin. support	519,182

Source: Burning Glass Technologies

THE WALL STREET JOURNAL.

Reprinted with permission of *The Wall Street Journal,* Copyright © 2022 Dow Jones & Company, Inc. All Rights Reserved.

Finding The Median of a Group of Values

Step 1 Orderly arrange values from the smallest to the largest.

Step 2 Find the middle value.

    a. *Odd number of values:* Median is the middle value. You find this by first dividing the total number of numbers by 2. The next-higher number is the median.

    b. *Even number of values:* Median is the average of the two middle values.

For Rusty's Clothing Shop, we find the median as follows:

1. Arrange values from smallest to largest:
    $27,000; $32,000; $40,000; $67,000; $95,000

2. Since we have a total number of five values and 5 is an odd number, we divide 5 by 2 to get $2\frac{1}{2}$. The next-higher number is 3, so our median is the third-listed number, $40,000. If Jane Hess ($27,000) were not on the payroll, we would find the median as follows:

1. Arrange values from smallest to largest:
    $32,000; $40,000; $67,000; $95,000

2. Average the two middle values:
    $$\frac{\$40,000 + \$67,000}{2} = \$53,500$$

Note that the median results in two salaries below and two salaries above the average.

Now we'll look at another measurement tool—the mode.

Learn: Mode

The **mode** is a measurement that also records values. In a series of numbers, the value that occurs most often is the mode. If all the values are different, there is no mode. If two or more numbers appear most often, you may have two or more modes. Note that we do not have to arrange the numbers in the lowest-to-highest order, although this could make it easier to find the mode.

Example: 3, 4, 5, 6, 3, 8, 9, 3, 5, 3

3 is the mode since it is listed 4 times.

Aha!

Use a bar graph to find the mode if you do not have a list of the data set.

Now let's check your progress with a Practice Quiz.

Learning Unit 22–1

Practice Quiz

Complete this Practice Quiz to see how you are doing.

Barton Company's sales reps sold the following last month:

Sales rep	Sales volume	Sales rep	Sales volume
A	$16,500	C	$12,000
B	15,000	D	48,900

Calculate the mean and the median. Which is the better indicator of the center of the data? Is there a mode?

✓ Solutions

$$\text{Mean} = \frac{\$16,500 + \$15,000 + \$12,000 + \$48,900}{4} = \boxed{\$23,100}$$

$$\text{Median} = \frac{\$15,000 + \$16,500}{2} = \boxed{\$15,750}$$

$12,000, \boxed{\$15,000, \$16,500,}$ $48,900. Note how we arrange numbers from smallest to highest to calculate median.

Median is the better indicator since in calculating the mean, the $48,900 puts the average of $23,100 much too high. There is no mode.

Frequency Distributions and Graphs

In this unit you will learn how to gather data and illustrate these data. Today, computer software programs can make beautiful color graphics. But how accurate are these graphics? The *Wall Street Journal* clip below, "What's Wrong With this Picture?" gives an example of graphics that do not agree with the numbers beneath them. The clip reminds all readers to check the numbers illustrated by the graphics. This is an old clip that is still relevant today.

What's Wrong With this Picture? Utility's Glasses Are Never Empty

By Kathleen Deveny
Staff Reporter of The Wall Street Journal

When Les Waas, an investor in Philadelphia Suburban Corp., paged through the company's 1994 annual report, he was impressed by what he saw.

The water utility had used a series of charts to represent its revenues, net income and book value per share, among other results. Each figure was represented by the level of water in a glass. Each chart showed strong growth.

Then Mr. Waas looked a little more carefully. The bars in the chart seemed to indicate far more impressive growth than the numbers beneath them. A chart showing the growth in the number of Philadelphia Suburban's water customers, for ex-

ample, seemed to indicate the company's customer base had more than tripled since 1990. But the numbers actually increased only 6.4%.

The reason for the disparity: The charts don't begin at zero. Even an empty glass in the accompanying chart would represent a customer base of 230,000.

Number of Metered Water Customers (thousands)

1990	1991	1992	1993	1994
235	237	245	247	250

Reprinted with permission of *The Wall Street Journal,* Copyright © 2022 Dow Jones & Company, Inc. All Rights Reserved.

Collecting raw data and organizing the data is a prerequisite to presenting statistics graphically. Let's illustrate this by looking at the following example.

A computer industry consultant wants to know how much college freshmen are willing to spend to set up a computer in their dormitory rooms. After visiting a local college dorm, the consultant gathered the following data on the amount of money 20 students spent on computers:

Price of computer	Tally	Frequency
$ 1,000	ⵘ	5
2,000	I	1
3,000	ⵘ	5
4,000	I	1
5,000	II	2
6,000	II	2
7,000	I	1
8,000	I	1
9,000	I	1
10,000	I	1

$1,000 $7,000 $4,000 $1,000 $ 5,000 $1,000 $3,000
5,000 2,000 3,000 3,000 3,000 8,000 9,000
3,000 6,000 6,000 1,000 10,000 1,000

Note that these raw data are not arranged in any order. To make the data more meaningful, the consultant made the **frequency distribution** table shown on the left. Think of this distribution table as a way to organize a list of numbers to show the patterns that may exist.

As you can see, 25% ($\frac{5}{20} = \frac{1}{4} = 25\%$) of the students spent $1,000 and another 25% spent $3,000. Only four students spent $7,000 or more.

 Aha!

Typically between 5 and 20 classes are used in a frequency distribution for ease in analyzing the data.

Now let's see how we can use bar graphs.

Learn: Bar Graphs

Bar graphs help readers see the changes that have occurred over a period of time. This is especially true when the same type of data is repeatedly studied.

Let's return to our computer consultant example and make a bar graph of the computer purchases data collected by the consultant. Note that the height of the bar represents the frequency of each purchase. Bar graphs can be vertical or horizontal.

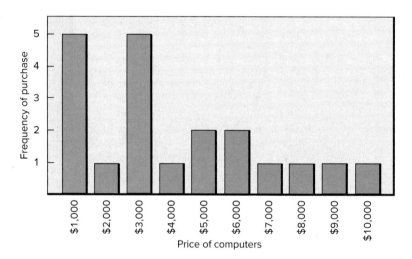

The following *Wall Street Journal* clip, "Nest Egg," uses bar graphs to show how long $1 million in savings may last during retirement.

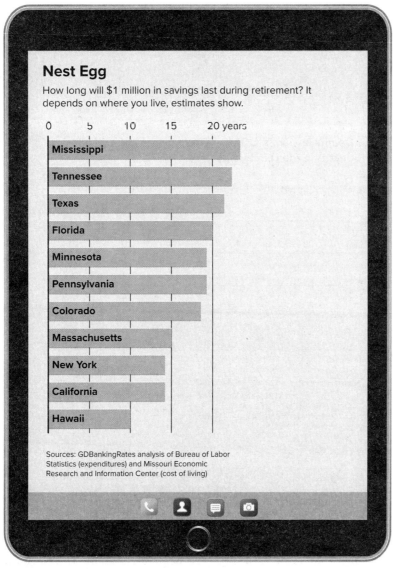

Nest Egg

How long will $1 million in savings last during retirement? It depends on where you live, estimates show.

Sources: GDBankingRates analysis of Bureau of Labor Statistics (expenditures) and Missouri Economic Research and Information Center (cost of living)

"'Nest egg' infographic." *The Wall Street Journal* (August 25, 2020).

In the following *Wall Street Journal* clip, "Deaths involving COVID-19," the bar chart shows how COVID-19 has affected different age groups.

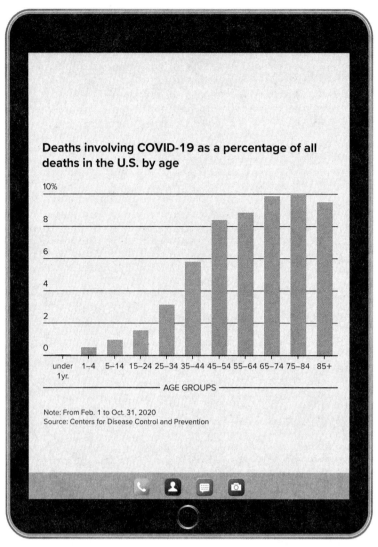

"'Deaths involving Covid-19 as a Percentage of All Deaths in the U.S. by Ages' Infographic." *The Wall Street Journal* (November 16, 2020).

We can simplify this bar graph by grouping the prices of the computers. The grouping, or *intervals,* should be of equal sizes.

A bar graph for the grouped data follows.

Class	Frequency
$1,000– $3,000.99	11
3,001– 5,000.99	3
5,001– 7,000.99	3
7,001– 9,000.99	2
9,001– 11,000.99	1

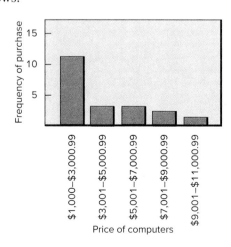

In the following *Wall Street Journal* clip, "3-D Appeal," the bar chart shows how much is spent on 3-D printing worldwide.

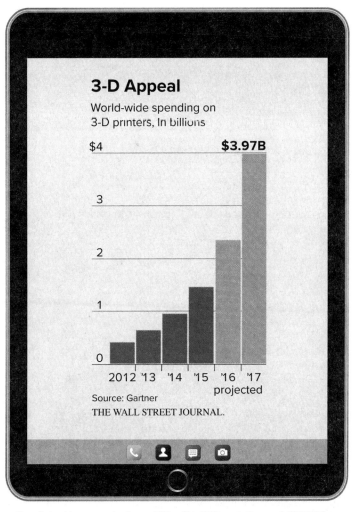

Next, let's see how we can use line graphs.

Learn: Line Graphs

A **line graph** shows trends over a period of time. Often separate lines are drawn to show the comparison between two or more trends.

The *Wall Street Journal* clip, "Aging America," shows the growth in the percentage of Americans 65 and older.

We conclude our discussion of graphics with the use of the circle graph.

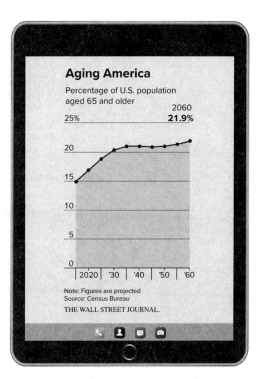

.15 × 360° =   54.0
.11 × 360° =   39.6
.36 × 360° =   129.6
.38 × 360° =   136.8
360.0

Learn: Circle Graphs

Circle graphs, often called *pie charts,* are especially helpful for showing the relationship of parts to a whole. The entire circle represents 100%, or 360°; the pie-shaped pieces represent the subcategories. Note how the circle graph in the *Wall Street Journal* clip "Threats From the Net" uses pie charts to show attitudes on cybersecurity.

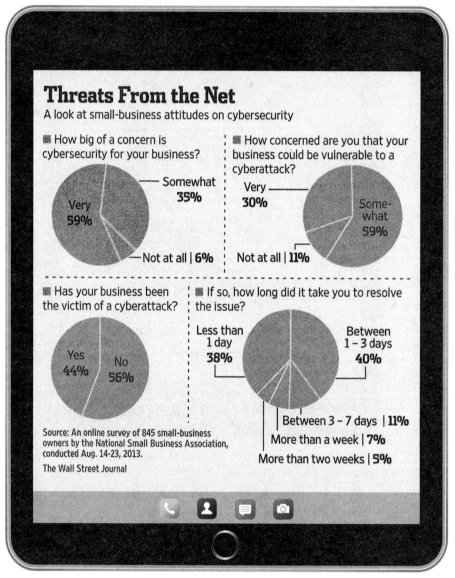

To draw a circle graph (or pie chart), begin by drawing a circle. Then take the percentages and convert each percentage to a decimal. Next multiply each decimal by 360° to get the degrees represented by the percentage. Circle graphs must total 360°. Note the following *Wall Street Journal* clip "An Enduring Challenge" shows the use of a pie chart.

Aha!

You can use Excel, and many other software programs, to create graphs easily. Try it!

We conclude this unit with a brief discussion of index numbers.

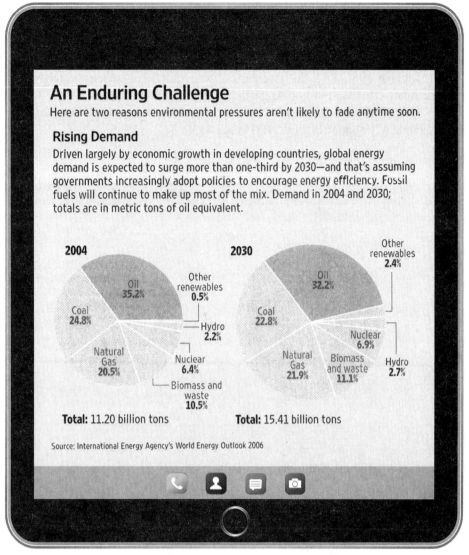

Learn: An Application of Statistics: Index Numbers

The financial section of a newspaper often gives different index numbers describing the changes in business. These **index numbers** express the relative changes in a variable compared with some base, which is taken as 100. The changes may be measured from time to time or from place to place. Index numbers function as percents and are calculated like percents.

Frequently, a business will use index numbers to make comparisons of a current price relative to a given year. For example, a calculator may cost $9 today relative to a cost of $75 some 30 years ago. The **price relative** of the calculator is $\frac{\$9}{\$75} \times 100 = 12\%$. The calculator now costs 12% of what it cost some 30 years ago. A price relative, then, is the current price divided by some previous year's price—the base year—multiplied by 100.

Price Relative Formula

$$\text{Price relative} = \frac{\text{Current price}}{\text{Base year's price}} \times 100$$

Money Tip All too often, the incorrect beneficiary is listed on life insurance policies, retirement accounts, and bank accounts. Create a list and review your accounts annually to verify the correct beneficiary is listed. Ensure a survivor knows where to locate the list and your estate documents, in the event of death.

Index numbers can also be used to estimate current prices at various geographic locations. The frequently quoted Consumer Price Index (CPI), calculated and published monthly by the U.S. Bureau of Labor Statistics, records the price relative percentage cost of many goods and services nationwide compared to a base period. Table 22.1 gives a portion of the CPI that uses 1982–84 as its base period. Note that the table shows, for example, that the price relative for food in Los Angeles is 130.9% of what it cost in 1982–84. Thus, Los Angeles food costs amounting to $100.00 in 1982–84 now cost $130.90. (Convert 130.9% to the decimal 1.309; multiply by $100 to get $130.90.)

Once again, we complete the unit with a Practice Quiz.

TABLE 22.1
Consumer Price
Index (in percent)

Expense	Atlanta	Chicago	New York	Los Angeles
Food	131.9	130.3	139.6	130.9
Clothing	133.8	124.3	121.8	126.4
Medical care	177.6	163.0	172.4	163.3

Practice Quiz

Complete this Practice Quiz to see how you are doing.

1. The following is the number of sales made by 20 salespeople on a given day. Prepare a frequency distribution and a bar graph. Do not use intervals for this example.

5	8	9	1	4	4	0	3	2	8
8	9	5	1	9	6	7	5	9	10

2. Assuming the following market shares for diapers 5 years ago, prepare a circle graph:

Pampers	32%	Huggies	24%
Luvs	20%	Others	24%

3. Today a new Explorer costs $35,000. In 1991 the Explorer cost $19,000. What is the price relative? Round to the nearest tenth percent.

✓ Solutions

1.

Number of sales	Tally	Frequency
0	|	1
1	||	2
2	|	1
3	|	1
4	||	2
5	|||	3
6	|	1
7	|	1
8	|||	3
9	||||	4
10	|	1

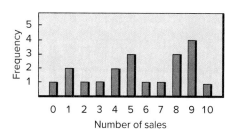

2.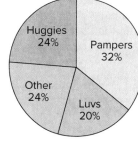

$$.32 \times 360° = 115.20°$$
$$.20 \times 360° = 72.00°$$
$$.24 \times 360° = 86.40°$$
$$.24 \times 360° = 86.40°$$

3. $\dfrac{\$35,000}{\$19,000} \times 100 = 184.2\%$

Learning Unit 22–3:
Measures of Dispersion (Optional)

In Learning Unit 22–1 you learned how companies use the mean, median, and mode to indicate a single value, or number, that represents an entire group of numbers, or data. Often it is valuable to know how the information is scattered (spread or dispersed) within a data set. A **measure of dispersion** is a number that describes how the numbers of a set of data are spread out or dispersed.

This learning unit discusses three measures of dispersion—range, standard deviation, and normal distribution. We begin with the range—the simplest measure of dispersion.

Learn: Range

The **range** is the difference between the two extreme values (highest and lowest) in a group of values or a set of data. For example, often the actual extreme values of hourly temperature readings during the past 24 hours are given but not the range or difference between the high and low readings. To find the range in a group of data, subtract the lowest value from the highest value.

$$\boxed{\text{Range} = \text{Highest value} - \text{Lowest value}}$$

Thus, if the high temperature reading during the past 24 hours was 90° and the low temperature reading was 60° the range is 90° − 60°, or 30°. The range is limited in its application because it gives only a general idea of the spread of values in a data set.

Example: Find the range of the following values: 83.6, 77.3, 69.2, 93.1, 85.4, 71.6.
Range = 93.1 − 69.2 = 23.9

Learn: Standard Deviation

Since the **standard deviation** is intended to measure the spread of data around the mean, you must first determine the mean of a set of data. The following diagram shows two sets of data—A and B. In the diagram, the means of A and B are equal. Now look at how the data in these two sets are spread or dispersed.

Data set A	Data set B
X X X X X	X X X X X X
0 1 2 3 4 5 6 7 8 9 10 11 12 13	0 1 2 3 4 5 6 7 8 9 10 11 12 13
Mean = (1 + 2 + 5 + 10 + 12) ÷ 5 = 6	Mean = (4 + 4 + 5 + 8 + 9) ÷ 5 = 6

Note that although the means of data sets A and B are equal, A is more widely dispersed, which means B will have a smaller standard deviation than A.

Aha! *A statistics calculator will allow you to calculate the mean, variance, standard deviation, and many more basic and advanced statistics calculations easily.*

To find the standard deviation of an ungrouped set of data, use the following steps:

Finding The Standard Deviation

Step 1 Find the mean of the set of data.

Step 2 Subtract the mean from each piece of data to find each deviation.

Step 3 Square each deviation (multiply the deviation by itself).

Step 4 Sum all squared deviations.

Step 5 Divide the sum of the squared deviations by $n - 1$, where n equals the number of pieces of data. This is the variance of the data set.

Step 6 Find the square root ($\sqrt{\ }$) of the number obtained in Step 5 (use a calculator). This is the standard deviation. (The square root is a number that when multiplied by itself equals the amount shown inside the square root symbol.)

Two additional points should be made. First, Step 2 sometimes results in negative numbers. Since the sum of the deviations obtained in Step 2 should always be zero, we would not be able to find the average deviation. This is why we square each deviation—to generate positive quantities only. Second, the standard deviation we refer to is used with *sample* sets of data, that is, a collection of data from a population. The population is the *entire* collection of data. When the standard deviation for a population is calculated, the sum of the squared deviations is divided by n instead of by $n - 1$. In all problems that follow, sample sets of data are being examined.

Example: Calculate the standard deviations for the sample data sets A and B given in the previous diagram. Round the final answer to the nearest tenth. Note that Step 1—find the mean—is given in the diagram.

Standard deviation of data sets A and B: The table on the left uses Steps 2 through 6 to find the standard deviation of data set A, and the table on the right uses Steps 2 through 6 to find the standard deviation of data set B.

Data	Step 2 Data − Mean	Step 3 (Data − Mean)²		Data	Step 2 Data − Mean	Step 3 (Data − Mean)²
1	$1 - 6 = -5$	25		4	$4 - 6 = -2$	4
2	$2 - 6 = -4$	16		4	$4 - 6 = -2$	4
5	$5 - 6 = -1$	1		5	$5 - 6 = -1$	1
10	$10 - 6 = \ \ 4$	16		8	$8 - 6 = \ \ 2$	4
12	$12 - 6 = \ \ 6$	36		9	$9 - 6 = \ \ 3$	9
	Total 0	94 **(Step 4)**			Total 0	22 **(Step 4)**

Step 5: Divide by $n - 1$: $\dfrac{94}{5 - 1} = \dfrac{94}{4} = 23.5$

Step 6: The square root of $\sqrt{23.5}$ is 4.8 (rounded).

The standard deviation of data set A is **4.8.**

Step 5: Divide by $n - 1$: $\dfrac{22}{5 - 1} = \dfrac{22}{4} = 5.5$

Step 6: The square root of $\sqrt{5.5}$ is 2.3.

The standard deviation of data set B is **2.3.**

As suspected, the standard deviation of data set B is less than that of set A. The standard deviation value reinforces what we see in the diagram.

Learn: Normal Distribution

One of the most important distributions of data is the **normal distribution.** In a normal distribution, data are spread *symmetrically* about the mean. A graph of such a distribution looks like the bell-shaped curve in Figure 22.1. Many data sets are normally distributed. Examples are the life span of automobile engines, women's heights, and intelligence quotients.

FIGURE 22.1
Standard deviation
and the normal
distribution

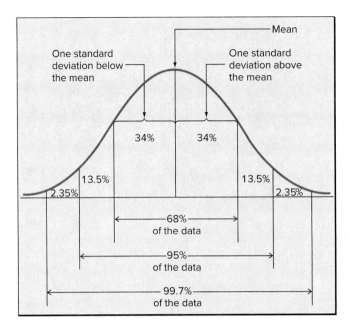

In a normal distribution, the mean, median, and mode are all equal. Additionally, when the data are normally distributed, the **Empirical Rule** (Three Sigma Rule) applies, stating that for a normal distribution, also known as a bell curve, approximately

- 68% of the observations will fall within ± 1 standard deviation of the mean,
- 95% will fall within ± 2 standard deviations of the mean, and
- 99.7% will fall within ± 3 standard deviations of the mean.

Figure 22.1 above illustrates these facts.

Example: Assume that the mean useful life of a particular lightbulb is 2,000 hours and is normally distributed with a standard deviation of 300 hours. Calculate the useful life of the lightbulb with **(a)** one standard deviation of the mean and **(b)** two standard deviations of the mean; also **(c)** calculate the percent of lightbulbs that will last 2,300 hours or longer.

a. The useful life of the lightbulb one standard deviation from the mean is one standard deviation above *and* below the mean.

$2,000 \pm 300 = 1,700$ and $2,300$ hours

The useful life is somewhere between 1,700 and 2,300 hours.

b. The useful life of the lightbulb within two standard deviations of the mean is within two standard deviations above *and* below the mean.

$2,000 \pm 2(300) = 1,400$ and $2,600$ hours

c. Since 50% of the data in a normal distribution lie below the mean and 34% represent the amount of data one standard deviation above the mean, we must calculate the percent of data that lie beyond one standard deviation above the mean.

$100\% - (50\% + 34\%) = \boxed{16\%}$

So 16% of the bulbs should last 2,300 hours or longer.

It's time for another Practice Quiz.

Money Tip Start achieving financial security today! Reduce spending, pay off debt, and start saving. Set small milestones, increasing them as you reach each goal. It's worth the effort and will make tomorrow financially brighter.

Practice Quiz

Complete this Practice Quiz to see how you are doing.

1. Calculate the range for the following data: 58, 13, 17, 26, 5, 41.

2. Calculate the variance and the standard deviation for the following sample set of data: 113, 92, 77, 125, 110, 93, 111. Round answers to the nearest tenth.

3. If the mean tax refund for the year is $3,000 with a $300 standard deviation, what is the refund range within three standard deviations above and below the mean?

✓ Solutions

1. $58 - 5 =$ **53 range**

2. $113 + 92 + 77 + 125 + 110 + 93 + 111)/7 = 103.$

3.

Data	Data − Mean	(Data − Mean)²
113	$113 - 103 = \quad 10$	100
92	$92 - 103 = -11$	121
77	$77 - 103 = -26$	676
125	$125 - 103 = \quad 22$	484
110	$110 - 103 = \quad 7$	49
93	$93 - 103 = -10$	100
111	$111 - 103 = \quad 8$	64
	Total	1,594

$1,594 \div (7 - 1) =$ **265.7** variance

$\sqrt{265.6666667} =$ **16.3** standard deviation

4. $\$3,000 +/- (\$300 \times 3) =$ Between $2,100 and $3,900

Chapter 22 Review

Topic/Procedure/Formula	Examples	You try it*
Mean $\dfrac{\text{Sum of all values}}{\text{Number of values}}$	Age of team players: 22, 28, 31, 19, 15 $\text{Mean} = \dfrac{22 + 28 + 31 + 19 + 15}{5}$ $= \boxed{23}$	**Calculate mean** 41, 29, 16, 15, 18
Weighted mean $\dfrac{\text{Sum of products}}{\text{Sum of frequencies}}$	S. M. T. W. Th. F. S. Sales $90 $75 $80 $75 $80 $90 $90 <table><tr><td>**Value**</td><td>**Frequency**</td><td>**Product**</td></tr><tr><td>$90</td><td>3</td><td>$270</td></tr><tr><td>75</td><td>2</td><td>150</td></tr><tr><td>80</td><td>2</td><td>160</td></tr><tr><td></td><td>7</td><td>$580</td></tr></table> $\text{Mean} = \dfrac{\$580}{7} = \boxed{\$82.86}$	**Calculate weighted mean** S. M. T. W. Th. Fr. S. Sales 80 90 100 80 80 90 90

Median
 	
1. Arrange values from smallest to largest.
2. Find the middle value.
 	
  a. Odd number of values: median is middle value.
 	
  $\left(\dfrac{\text{Total number of numbers}}{2}\right)$
  Next-higher number is median.
 	
  b. Even number of values: average of two middle values.
	12, 15, 8, 6, 3
	
	1. 3 6 8 12 15
	2. $\dfrac{5}{2} = 2.5$
	
	Median is third number, $\boxed{8.}$
		Calculate median
		
		14, 16, 9, 7, 4

Mode Value that occurs most often in a set of numbers	6, 6, 8, 5, 6 Mode is 6	**Find mode** 7, 7, 4, 3, 2, 7
Frequency distribution Method of listing numbers or amounts not arranged in any particular way by columns for numbers (amounts), tally, and frequency	Number of smoothies consumed in one day: 1, 5, 4, 3, 4, 2, 2, 3, 2, 0 <table><tr><td>**Number of sodas**</td><td>**Tally**</td><td>**Frequency**</td></tr><tr><td>0</td><td>I</td><td>1</td></tr><tr><td>1</td><td>I</td><td>1</td></tr><tr><td>2</td><td>III</td><td>3</td></tr><tr><td>3</td><td>II</td><td>2</td></tr><tr><td>4</td><td>II</td><td>2</td></tr><tr><td>5</td><td>I</td><td>1</td></tr></table>	**Prepare frequency distribution** Number of coffees consumed in one day: 1, 4, 5, 8, 2, 2, 3, 0

Chapter 22 Review (Continued)

Topic/Procedure/Formula	Examples	You try it*
Bar graphs Height of bar represents frequency. Bar graph used for grouped data. Bar graphs can be vertical or horizontal.	From smoothies example above: Frequency vs Number of sodas bar graph	**From coffee example above, prepare bar graph**
Line graphs Shows trend. Helps to put numbers in order.	**Sales** 2021 $1,000 2022 2,000 2023 3,000 Line graph 2021–2023	**Prepare line graph** **Sales** 2021 $5,000 2022 3,000 2023 2,000
Circle graphs Circle = 360° % × 360° = Degrees of pie to represent percent Total should = 360°	60% favor diet soda 40% favor sugared soda Pie chart: Sugared 40%, Diet 60% $.60 \times 360° = 216°$ $.40 \times 360° = \underline{144°}$ $\boxed{360°}$	**Create circle graph** 70% coffee drinkers 30% non-coffee-drinkers
Price relative Price relative = $\dfrac{\text{Current price}}{\text{Base year's price}} \times 100$	A compact sedan's sticker price was $8,799 in 1982. Today it is $14,900. Price relative = $\dfrac{\$14,900}{\$8,799} \times 100 = \boxed{169.3}$ (rounded to nearest tenth percent)	**Calculate price relative** Old price, $ 9,000 Today's price, 12,000
Range (optional) Range = Highest value − Lowest value	Calculate range of the data set consisting of 5, 9, 13, 2, 8 Range = 13 − 2 = 11	**Calculate range** 6, 8, 14, 2, 9

Chapter 22 Review *(Continued)*

Topic/Procedure/Formula	Examples	You try it*
Standard deviation (optional) 1. Calculate mean. 2. Subtract mean from each piece of data. 3. Square each deviation. 4. Sum squares. 5. Divide sum of squares by $n - 1$, where n = number of pieces of data. 6. Take square root of number obtained in Step 5, to find the standard deviation.	Calculate the standard deviation of this set of data: 7, 2, 5, 3, 3. 1. Mean $= \dfrac{20}{5} = 4$ 2. $7 - 4 = 3$ $\quad 2 - 4 = -2$ $\quad 5 - 4 = 1$ $\quad 3 - 4 = -1$ $\quad 3 - 4 = -1$ 3. $\quad (3)^2 = 9$ $\quad (-2)^2 = 4$ $\quad\ (1)^2 = 1$ $\quad (-1)^2 = 1$ $\quad \underline{(-1)^2 = 1}$ 4. 16 5. $16 \div 4 = 4$ 6. Standard deviation $= \boxed{2}$	**Calculate standard deviation** 8, 1, 6, 2, 2

Key Terms

Bar graph	Line graph	Normal distribution
Circle graph	Mean	Price relative
Empirical Rule	Measure of dispersion	Range
Frequency distribution	Median	Standard deviation
Index numbers	Mode	Weighted mean

* Worked-out solutions are in Appendix A.

Critical Thinking Discussion Questions with Chapter Concept Check

1. Explain the mean, median, and mode. Give an example that shows you must be careful when you read statistics in an article.

2. Explain frequency distributions and the types of graphs. Locate a company annual report and explain how the company shows graphs to highlight its performance. Does the company need more or fewer of these visuals? Could price relatives be used?

3. Explain the statement, "Standard deviations are not accurate."

4. **Chapter Concept Check.** Visit the Apple website. Gather new statistics on the iPad, Apple Watch, and/or iPhone. Use concepts in this chapter to create a presentation.

End-of-Chapter Problems

Name _____   Date _____

Check figures for odd-numbered problems in Appendix A.

Drill Problems (*Note:* Problems for optional Learning Unit 22–3 follow the Challenge Problem 22–24)

Calculate the mean (to the nearest hundredth):   *LU 22–1(1)*

22–1. 9, 3, 2, 11

22–2. 5, 4, 8, 12, 15

22–3. $55.83, $66.92, $108.93

22–4. $1,001, $68.50, $33.82, $581.95

e)cel   **22–5.** Calculate the grade point average: A = 4, B = 3, C = 2, D = 1, F = 0 (to nearest tenth). *LU 22–1(2)*

Courses	Credits	Grade
Computer Principles	3	B
Business Law	3	C
Logic	3	D
Biology	4	A
Marketing	3	B

22–6. Find the weighted mean (to the nearest tenth):   *LU 22–1(2)*

Value	Frequency	Product
4	7	
8	3	
2	9	
4	2	

Find the median: *LU 22–1(3)*

22–7. 55, 10, 19, 38, 100, 25

22–8. 95, 103, 98, 62, 31, 15, 82

Find the mode: *LU 22–1(4)*
22–9. 8, 9, 3, 4, 12, 8, 8, 9
22–10. 22, 19, 15, 16, 18, 18, 5, 18

End-of-Chapter Problems (Continued)

22–11. Given: Truck cost   2012 $30,000

             Truck cost   2008 $21,000

Calculate the price relative (rounded to the nearest tenth percent). *LU 22–2(3)*

excel 22–12. Given the following sales of Finn Corporation, prepare a line graph (run sales from $5,000 to $20,000). *LU 22–2(2)*

2021 $ 8,000

2022   11,000

2023   13,000

2024   18,000

22–13. Prepare a frequency distribution from the following weekly salaries of teachers at Pikes Peak Community College. Use the following intervals: *LU 22–2(1)*

$200–$299.99

$300–$399.99

$400–$499.99

$500–$599.99

$210	$505	$310	$380	$275
290	480	550	490	200
286	410	305	444	368

22–14. Prepare a bar graph from the frequency distribution in Problem 22–13. *LU 22–2(2)*

22–15. How many degrees on a circle graph would be given to each of the following? *LU 22–2(2)*

Wear digital watch   42%

Wear traditional watch   51

Wear no watch   7

Word Problems

excel 22–16. The first Super Bowl on January 15, 1967, charged $42,000 for a 30-second commercial. Create a line graph for the following Super Bowl 30-second commercial costs: 2011, $3,100,000; 2012, $3,500,000; 2013 and 2014, $4,000,000; 2015, $4,500,000; 2016, $5,000,000; 2017 and 2018, $5,020,000; 2019, $5,250,000; and 2020 and 2021, $5,600,000.   *LU 22–2(2)*

excel 22–17. The American Kennel Club announced the "Most Popular Dogs in the U.S." Labrador retrievers remained number one for the 28th consecutive year. French bulldogs came in second followed by German shepherds, golden retrievers, and bulldogs. Create a circle graph for Dogs for Life Kennel Club with the following members: 52 Labrador retrievers, 33 French bulldogs, 22 German shepherds, 15 golden retrievers, and 10 bulldogs. *LU 22–2(2)*

22–18. Despite tuition skyrocketing, a college education is still valuable. Recent calculations by the Federal Reserve Bank in San Francisco demonstrate a college degree is worth $830,000 in lifetime earnings compared to the average high school education. If graduates earn $40,632, $35,554, $42,192, $33,432, $69,479 and $43,589, what is the standard deviation for this sample? Round to a whole number for each calculation. *LU 22–3(2)*

End-of-Chapter Problems (Continued)

22–19. Costcotravel.com provided a member with the following information regarding her upcoming travel. Construct a circle graph for the member.   *LU 22–2(2)*

Transportation	35%
Hotel	28
Food and entertainment	20
Miscellaneous	17

22–20. Jim Smith, a marketing student, observed how much each customer spent in a local convenience store. Based on the following results, prepare **(a)** a frequency distribution and **(b)** a bar graph. Use intervals of $0–$5.99, $6.00–$11.99, $12.00–$17.99, and $18.00–$23.99.   *LU 22–2(2)*

$18.50	$18.24	$ 6.88	$9.95
16.10	3.55	14.10	6.80
12.11	3.82	2.10	
15.88	3.95	5.50	

22–21. Angie's Bakery bakes bagels. Find the weighted mean (to the nearest whole bagel) given the following daily production for June: *LU 22–1(2)*

200	150	200	150	200
150	190	360	360	150
190	190	190	200	150
360	400	400	150	200
400	360	150	400	360
400	400	200	150	150

22–22. The United Nations states the gender pay gap will not close for 70 years. Women across the world earn $0.76 for every $1.00 of what men earn. Construct a bar graph reflecting the following Harvard University study on pay for women based on $1.00 for men: financial specialists, $0.66; physicians, $0.71; aircraft pilots, $0.71; accountants, $0.76; lawyers, $0.82; and nurses, $0.89.   *LU 22–2(2)*

Challenge Problems

22–23. Listed below are annual revenues for a few travel agencies prior to the pandemic:

AAA Travel Agency	$86,700,000
Riser Group	63,200,000
Casto Travel	62,900,000
Balboa Travel	36,200,000
Hunter Travel Managers	36,000,000

(a) What would be the mean and the median? **(b)** What is the total revenue percent of each agency? **(c)** Prepare a circle graph depicting the percents.   *LU 22–1(1, 2), LU 22–2(2)*

End-of-Chapter Problems (Continued)

22–24. Review the two circle graphs for recommendations on how to budget. Then look at the circle graph budget for Ron Rye and his family for a month. Ron would like you to calculate the percent (to the hundredth) for each part of the circle graph along with the appropriate number of degrees. Should Ron adjust his spending? If yes, how?
LU 22–2(2)

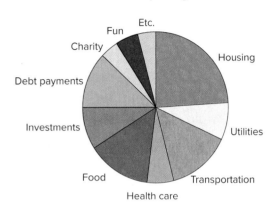

Recommended percentage of overall spending: A

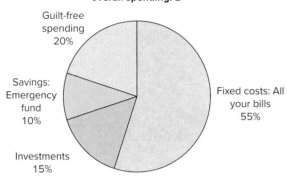

Recommended percentage of overall spending: B

Ron Rye Family Budget

Learning Unit 22–3 Optional Assignments

Name _____   Date _____

Check figures for odd-numbered problems in Appendix A.

Drill Problems

1. Calculate the range for the following set of data: 117, 98, 133, 52, 114, 35.   *LU 22–3(1)*

Calculate the standard deviation for the following sample sets of data. Round the final answers to the nearest tenth.   *LU 22–3(2)*

2. 83.6, 92.3, 56.5, 43.8, 77.1, 66.7 (mean = 70)

3. 7, 3, 12, 17, 5, 8, 9, 9, 13, 15, 6, 6, 4, 5

4. 41, 41, 38, 27, 53, 56, 28, 45, 47, 49, 55, 60

Word Problems

excel 5. The mean useful life of car batteries is 48 months. They have a standard deviation of 3. If the useful life of batteries is normally distributed, calculate (a) the percent of batteries with a useful life of less than 45 months and (b) the percent of batteries that will last longer than 54 months.   *LU 22–3(2)*

6. The average weight of a particular box of crackers is 24.5 ounces with a standard deviation of 0.8 ounce. The weights of the boxes are normally distributed. What percent of the boxes (a) weigh more than 22.9 ounces and (b) weigh less than 23.7 ounces?   *LU 22–3(2)*

Learning Unit 22–3 Optional Assignments (Continued)

7. An examination is normally distributed with a mean score of 77 and a standard deviation of 6. Find the percent of individuals scoring as indicated below.   *LU 22–3(2)*

   a. Between 71 and 83

   b. Between 83 and 65

   c. Above 89

   d. Less than 65

   e. Between 77 and 65

8. Listed below are the sales figures in thousands of dollars for a group of insurance salespeople. Calculate the mean sales figure and the standard deviation.   *LU 22–3(1, 2)*

$117	$350	$400	$245	$420
223	275	516	265	135
486	320	285	374	190

9. The time in seconds it takes for 20 individual sewing machines to stitch a border onto a particular garment is listed below. Calculate the mean stitching time and the standard deviation to the nearest hundredth.   *LU 22–3(1, 2)*

67	69	64	71	73
58	71	64	62	67
62	57	67	60	65
60	63	72	56	64

Summary Practice Test

Do you need help? Connect videos have step-by-step worked-out solutions.

excel 1. In July, Lee Realty sold 10 condominiums at the following prices: $140,000; $166,000; $80,000; $98,000; $185,000; $150,000; $108,000; $114,000; $142,000; and $250,000. Calculate the mean and median. *LU 22–1(1, 3)*

2. Lowes counted the number of customers entering the store for a week. The results were 1,100; 950; 1,100; 1,700; 880; 920; and 1,100. What is the mode? *LU 22–1(4)*

3. This semester Hung Lee took four 3-credit courses at Riverside Community College. She received an A in accounting and C's in history, psychology, and algebra. What is her cumulative grade point average (assume A = 4 and C = 2) to the nearest hundredth? *LU 22–1(2)*

4. Pete's Variety Shop reported the following sales for the first 20 days of May. Prepare a frequency distribution for Pete's. *LU 22–2(1)*

$100	$400	$600	$400	$600
100	600	300	500	700
200	600	700	500	200
100	600	100	700	700

excel 5. Leeds Company produced the following number of maps during the first 5 weeks of last year. Prepare a bar graph. *LU 22–2(2)*

Week	Maps
1	800
2	600
3	400
4	700
5	300

Summary Practice Test (Continued)

6. Laser Corporation reported record profits of 30%. It stated in the report that the cost of sales was 40% with expenses of 30%. Prepare a circle graph for Laser.   *LU 22–2(2)*

7. Today a new Explorer costs $39,900. In 1990, Explorers cost $24,000. What is the price relative to the nearest tenth percent?   *LU 22–2(3)*

*8. Calculate the standard deviation for the following set of data: 7, 2, 5, 3, 3, 10. Round the final answer to the nearest tenth.   *LU 22–3(2)*

* Optional problem.

🔍 My Health

 What I need to know

Throughout the My Money segments in this book, you have no doubt learned that managing your finances is an important step to achieving your financial goals. The same can be said about managing your health. The hard work you put into reaching your financial goals won't matter much if you are not around to reap the rewards of your diligence. A healthy lifestyle is just as important, if not more so, than a healthy financial position. Creating good health habits early on in life will make these changes more of a lifestyle than just a fad or temporary activity. As with the habits you are learning about managing your finances, managing your health will demand a concerted effort. Just as effective financial planning leads to achieving your goals, healthy living will yield positive results in your overall health.

 What I need to do

Take an active approach to life. Engage in a regular form of exercise as simple as walking or maybe joining a competitive sports league. Find ways to exercise throughout your day such as taking the steps versus the escalator/elevator or maybe walk with a co-worker over your lunch break. Over time you will find the act of exercising becomes easier and more enjoyable as you establish good health habits. A fitness tracker will help you stay on track with your exercising goals and track your progress. There are many easy-to-use fitness trackers available for your smartphone or computer.

A balanced diet is another part of creating and maintaining a healthy lifestyle. Changes to your eating habits do not have to be drastic. Finding ways to cut total calories you consume daily is a great way to make positive strides, for instance, reducing your sugar intake by switching from soda to water or opting for a handful of almonds versus a cookie. Also, selecting fresh fruits and vegetables versus prepackaged processed foods will leave you feeling fuller and provide you with key nutrients your body needs. If you consume alcohol, do so in moderation.

Stay on track by maintaining your healthy lifestyle. Regular checkups with your health care professionals will keep you on the right path to a healthy life. These professionals will also provide you with helpful ways in which you can continue to maintain your healthy lifestyle based on your unique health situation. If you are experiencing stress, seek out ways in which you can reduce the stressful situations in your life. Getting the right amount of restful sleep will not only help with your stress management but also allow your body to reenergize itself. Earning your education and working at your career are important toward helping you reach your educational, professional and financial goals. However, you should strive for a good balance between the amount of time you spend educating, working and the time you spend on other activities in your life. Finding the appropriate work-life balance for yourself is of vital importance to a healthy lifestyle.

 Steps I need to take

1. Find practical ways in which you can lead an active lifestyle daily.
2. Be aware of how your diet impacts your health and make smart food and drink choices.
3. Maintain good health habits and achieve a positive work-life balance.

 Resources I can use

- MyFitnessPal (mobile app)—manage your diet and track your progress on healthy activities.
- https://health.clevelandclinic.org/11-simple-health-habits-worth-adopting-into-your-life/—easy healthy habits you can incorporate into your daily life.

MY MONEY ACTIVITY

- For 3 weeks, commit to exercising 30 minutes per day, 3 days per week.
- After 3 weeks evaluate this exercise. You may be surprised at the outcome.

PERSONAL FINANCE

A KIPLINGER APPROACH

PRICES SURGE FOR VACATION HOMES

The median sales price for homes in "seasonal towns" rose 19% in December from a year earlier, to $408,000, according to a new report from real estate firm Redfin. Demand for second and vacation homes at popular vacation destinations rose 84% year over year—more than double the demand for a primary home, the report adds. Redfin economist Taylor Marr says the popularity of vacation homes follows the rise in remote work due to the coronavirus pandemic. More families are spending time outdoors, and those who can afford it are opting to move to less-crowded parts of the country. Others, Marr says, are choosing to spend as much time as possible during the pandemic at vacation destinations, including for work.

"PRICES SURGE FOR VACATION HOMES." Kiplinger's. May, 2021.

YOURS, MINE & OURS

Ally recently surveyed consumers about how they approach finances with their partners. Here's what the survey found:

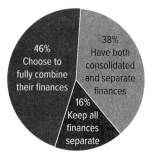

For those who have combined finances, most have joint checking (73%) or savings (74%). More than half of couples who fully consolidate their finances are more likely to have a monthly budget they try to stick to, and nearly half are more likely to accomplish a financial goal they set. About one-fourth share responsibility for managing the finances, while another fourth say one partner manages most of the finances, but the other partner is very informed.

"YOURS, MINE & OURS." Kiplinger's. May, 2021.

U.S. GETS AN *F* FOR WORK BENEFITS

The U.S. ranks worst among the world's developed countries for work benefits, according to a study by Zenefits, a human resources software provider. The research looked at paid leave, sick leave, parental leave, health benefits, retirement, unemployment and work-life balance. The U.S. ranked 30th, behind Mexico, South Korea, Latvia and the Czech Republic. Denmark was the top-ranked country, followed by the Netherlands, Finland, Sweden and Switzerland.

"U.S. GETS AN F FOR WORK BENEFITS." Kiplinger's. May, 2021.

From *The Kiplinger Letter*

STUDENTS FACE LONG-TERM CHALLENGES

The negative health outcomes of not having in-person school are piling up, such as damage to social and emotional development and mental health issues. America's K-12 students face a dire level of lost learning in coming years, caused by worse learning outcomes for most students in prolonged virtual settings. Issues range from young kids not learning to read to falling a year behind in math. Affected kids could take a serious economic hit, according to an estimate from consulting firm McKinsey. A student not returning to full classes by January 2021 could lose $60,000 to $80,000 in lifetime earnings. Some have now been out longer. The damage is worse for Black and Hispanic students compared with white students. The problem will be at the forefront of education policy debates for years.

"STUDENTS FACE LONG-TERM CHALLENGES." Kiplinger's. May, 2021.

Business Math Issue

Statistics can never be misleading.

1. List the key points of the article and information to support your position.

2. Write a group defense of your position using math calculations to support your view. If you are in an online course, post to a discussion board.

Appendix A

Check Figures

Worked-Out Solutions to You Try It Problems

Odd-Numbered Drill and Word Problems for End-of-Chapter Problems.

Challenge Problems (all).

Cumulative Reviews (all).

Odd-Numbered Additional Assignments by Learning Unit from Appendix A.

Worked-Out Solutions to You Try It Problems, Check Figures to Drill and Word Problems (Odds), Challenge Problems, and Cumulative Reviews

Chapter 1

You Try It

1. $571 \rightarrow$ Five hundred seventy-one

 $7,943 \rightarrow$ Seven thousand, nine hundred forty-three

2. $691 = 691 = 690$

 Identify   Less
 digit   than 5

3. $429,685 \rightarrow 429,685 \rightarrow 400,000$

 Identify   Less
 digit   than 5

4. $\begin{array}{r} \overset{1}{7}6 \\ +38 \\ \hline 114 \end{array}$

5. $\begin{array}{r} \overset{512}{6\cancel{2}9} \\ -134 \\ \hline 495 \end{array}$

6. $\begin{array}{r} 491 \\ \times 28 \\ \hline 3928 \\ 982 \\ \hline 13,748 \end{array}$

 $13 \times 10 = 130$ (attach 1 zero)
 $13 \times 1,000 = 13,000$ (attach 3 zeros)

7. $\begin{array}{r} 5\,R15 \\ 16\overline{)95} \\ 80 \\ \hline 5 \end{array}$

 $4,000 \div 100 = 40$ (drop 2 zeros)
 $4,000 \div 1,000 = 4$ (drop 3 zeros)

End-of-Chapter Problems

1–1. 105

1–3. 154

1–5. 13,580

1–7. 113,690

1–9. 38

1–11. 3,600

1–13. 1,074

1–15. 31,110

1–17. 340,531

1–19. 126,000

1–21. 90

1–23. 86 R4

1–25. 405

1–27. 1,616

1–29. 24,876

1–31. 17,989; 18,000

1–33. 80

1–35. 133

1–37. 216

1–39. 19 R21

1–41. 7,690; 6,990

1–43. 70,470; 72,000

1–45. 700

1–47. $27,738

1–49. $240; $200; $1,200; $1,080

1–51. $2,436; $3,056; $620 more

1–53. 905,600

1–55. 1,080

1–57. a. $4,569

     b. $4,600

     c. $31

1–59. $879 decrease

1–61. $1,872,000

Appendix A (Continued)

1–63. $4,815; $250,380

1–65. $64,180

1–67. 200,000; 10,400,000

1–69. $1,486

1–71. Average $33; no concern

1–73. $40 per sq yard

1–75. $7,680 difference between drugstore and bakery

1–76. $12,000 difference

Chapter 2

You Try It

1. $\dfrac{3}{10}$ proper, $\dfrac{9}{8}$ improper, $1\dfrac{4}{5}$ mixed

2. $\dfrac{18}{7} = 2\dfrac{4}{7}$ $\qquad 5\dfrac{1}{7} = \dfrac{35+1}{7} = \dfrac{36}{7}$

3. $\dfrac{16 \div 8}{24 \div 8} = \dfrac{2}{3}$

4. $\dfrac{20}{50} = 20\overline{)50}$
$\dfrac{40}{10}$

$\qquad\qquad 10\overline{)20}$
$\qquad\qquad\dfrac{20}{0}$

10 is greatest common denominator

5. $\dfrac{16}{31} = \dfrac{}{310}$
$310 \div 31 = 10 \qquad 10 \times 16 = 160$

6. $\dfrac{3}{7} + \dfrac{2}{7} = \dfrac{5}{7}$ $\quad \dfrac{5}{7} - \dfrac{2}{7} = \dfrac{3}{7}$ $\qquad \dfrac{5}{8} = \dfrac{25}{40}$
$\qquad\qquad\qquad\qquad\qquad\qquad\qquad +\dfrac{3}{40} = \dfrac{3}{40}$
$\qquad\qquad\qquad\qquad\qquad\qquad\qquad\qquad \dfrac{28}{40} = \dfrac{7}{10}$

7. Prime numbers 2, 3, 5, 7, 11, 13, 17

8. $\dfrac{1}{2} + \dfrac{1}{4} + \dfrac{1}{5} = \begin{array}{c} 2 \\ 1 \end{array}\Big/ \begin{array}{ccc} 2 & 4 & 5 \\ 2 & 5 \end{array}$
$\qquad\qquad\qquad 2 \times 1 \times 2 \times 5 = 20\,\text{LCD}$

9. $2\dfrac{1}{4}$
$+3\dfrac{3}{4}$
$\overline{5\dfrac{4}{4}} = 6$

10. $11\dfrac{1}{3}$ $\quad 10\dfrac{4}{3}$
$-2\dfrac{2}{3}$ $\quad -2\dfrac{2}{3}$
$\qquad\qquad \overline{8\dfrac{2}{3}}$

11. $\dfrac{4}{5} \times \dfrac{25}{26} = \dfrac{\overset{2}{\cancel{4}}}{\cancel{5}_{1}} \times \dfrac{\overset{5}{\cancel{25}}}{\cancel{26}_{13}} = \dfrac{10}{13}$

12. $2\dfrac{1}{4} \times 3\dfrac{1}{4} = \dfrac{9}{4} \times \dfrac{13}{4} = \dfrac{117}{16} = 7\dfrac{5}{16}$

13. $\dfrac{1}{8} \div \dfrac{1}{4} = \dfrac{1}{\cancel{8}_{2}} \times \cancel{4} = \dfrac{1}{2}$

14. $3\dfrac{1}{4} \div 1\dfrac{4}{5} = \dfrac{13}{4} \div \dfrac{9}{5} = \dfrac{13}{4} \times \dfrac{5}{9} = \dfrac{65}{36}$

End-of-Chapter Problems

2–1. Proper

2–3. Improper

2–5. $61\dfrac{2}{5}$

2–7. $\dfrac{59}{3}$

2–9. $\dfrac{11}{13}$

2–11. 60 $(2 \times 2 \times 3 \times 5)$

2–13. 96 $(2 \times 2 \times 2 \times 2 \times 2 \times 3)$

2–15. $\dfrac{13}{21}$

2–17. $15\dfrac{5}{12}$

2–19. $\dfrac{5}{6}$

2–21. $7\dfrac{4}{9}$

2–23. $\dfrac{5}{16}$

2–25. $\dfrac{3}{25}$

2–27. $\dfrac{1}{3}$

2–29. $\dfrac{7}{18}$

2–31. $215,658

2–33. $500,000 infected

2–35. $35\dfrac{1}{4}$ hours

2–37. $10\dfrac{3}{4}$ hours

2–39. $6\dfrac{1}{2}$ gallons

2–41. $875

2–43. $\dfrac{23}{36}$

2–45. $3,667

2–47. $3\frac{3}{4}$ lbs apple; $8\frac{1}{8}$ cups flour; $\frac{5}{8}$ cup marg; $5\frac{15}{16}$ cups sugar; 5 teaspoons cin.

2–49. 400 people

2–51. 275 gloves

2–53. $450

2–55. $45\frac{3}{16}$

2–57. $62,500,000; $37,500,000

2–59. $\frac{3}{8}$

2–61. $2\frac{3}{5}$ hours

2–63. $8\frac{31}{48}$ feet; Yes

2–64. a. 400 homes **b.** $320,000

    c. 3,000 people **d.** 2,500 people

    e. $112.50 **f.** $8,800,000

Chapter 3

You Try It

1. .8256 → Ten thousandths place

2. .841 = .8
    Less than 5

3. $\frac{9}{1,000} = .009$

    $\frac{3}{10,000} = .0003$

4. $\frac{1}{7} = .142 = .1$

5. $5\frac{4}{5} = \frac{4}{5} = .80 + 5 = 5.80$

6. .865 $\frac{865}{\ }$ $\frac{865}{1}$ $\frac{865}{1,000}$ (attach 3 zeros)

7.
```
 1.7      5 10
 3.0     6.00
  .8    -4.10
 ----    1.90
 5.5
```

8.
```
  3.49 (2 places)
  .015 (3 places)
  ----
  1745
  349
  .05235
```

9.
```
      1.5
33)49.5
   33
   ---
   165
   165
   ---
     0
```

10.
```
        .46 = .5
3.2)1.4 80
    128
    ---
    200
    192
```

11. $6.92 \times 100 = 692$ (move 2 places to right)
    $6.92 \div 100 = .0692$ (move 2 places to left)

End-of-Chapter Problems

3–1. Hundredths

3–3. .7; .74; .739

3–5. 5.8; 5.83; 5.831

3–7. 6.6; 6.56; 6.556

3–9. $4,822.78

3–11. .08

3–13. .06

3–15. .91

3–17. 16.61

3–19. $\frac{71}{100}$

3–21. $\frac{125}{10,000}$

3–23. $\frac{825}{1,000}$

3–25. $\frac{7,065}{10,000}$

3–27. $28\frac{48}{100}$

3–29. .005

3–31. .0085

3–33. 818.1279

3–35. 3.4

3–37. 2.32

3–39. 1.2; 1.26791

3–41. 4; 4.0425

3–43. 24,526.67

3–45. 161.29

3–47. 6.82

3–49. .04

3–51. .63

3–53. 2.585

3–55. .0086

3–57. 486

3–59. 3.950

3–61. 7,913.2

3–63. .583

3–65. $19.57

3–67. $0.75

3–69. $119.47

Appendix A (Continued)

3–71. $29.00

3–73. 91 million

3–75. $423.16

3–77. $105.08

3–79. $24,996.78

3–81. 6.7 million more claims

3–83. 0.9% difference

3–85. $6,465.60

3–86. Yes, $16,200

3–87. $560.45

Cumulative Review 1, 2, 3

1. $216

2. $200,000

3. $50,560,000

4. $25.50

5. $225,000

6. $750

7. $369.56

8. $130,000,000

9. $63.64

Chapter 4

You Try It

Sample

1. Pete Co. 24-111-9

   Pay to the order of Reel Bank Pete Co. 24-111-9

   Pay to the order of Reel Bank for deposit only Pete Co. 24-111-9

Checkbook

Beg. balance		$300
2. *Less:* NSF	$50	
ATM service charge	20	70
Ending balance		$230

End-of-Chapter Problems

4–1. $4,720.33

4–3. $4,705.33

4–5. $753

4–7. $540.82

4–9. $577.95

4–11. $998.86

4–13. $1,530

4–15 $136.45

4–16. $1,862.13

4–17. $3,061.67

Chapter 5

You Try It

1. $E + 15 = 14$
$$\frac{-15 \quad -15}{E \quad = \quad -1}$$

2. $B - 40 = 80$
$$\frac{+40 \quad +40}{B \quad = \quad 120}$$

3. $\dfrac{\cancel{5}C}{\cancel{5}} = 75$
$C = 15$

4. $\dfrac{A}{6} = 60 \quad (\cancel{6})\dfrac{A}{\cancel{6}} = 6(60)$
$A = 360$

5. $\dfrac{C}{4} + 10 = 17$
$$\frac{-10 \quad -10}{\dfrac{C}{4} \quad = \quad 7}$$
$(\cancel{4})\dfrac{C}{\cancel{4}} = 7(4)$
$C = 28$

6. $7(B - 10) = 35$
$7B - 70 = 35$
$$\frac{+70 \quad +70}{\dfrac{\cancel{7}B}{\cancel{7}} = \dfrac{105}{7}}$$
$B = 15$

7. $5B + 3B = 16$
$\dfrac{\cancel{8}B}{\cancel{8}} = \dfrac{16}{8}$
$B = 2$

Sit. 1. $P - \$53 = \$110$
$$\frac{+53 \quad +53}{P \quad = \$163}$$

Sit. 2. $\dfrac{1}{7}B = \$6,000$

$7\left(\dfrac{B}{\cancel{7}}\right) = 6,000(7)$

$B = \$42,000$

Sit. 3.
$$9S - S = 640$$
$$\frac{8S}{8} = \frac{640}{8}$$
$$S = 80 \qquad 9S = 720$$

Sit. 4.
$$9S + S = 640$$
$$\frac{10S}{10} = \frac{640}{10}$$
$$S = 64 \qquad 9S = 576$$

Sit. 5.
$$400(3N) + 300N = 15,000$$
$$\frac{1,500N}{1,500} = \frac{15,000}{1,500}$$
$$N = 10$$
$$3N = 30$$

Sit. 6.
$$400S + 300(40 - S) = 15,000$$
$$400S + 12,000 - 300S = 15,000$$
$$100S + 12,000 = 15,000$$
$$-12,000 \quad -12,000$$
$$\frac{100S}{100} = \frac{3,000}{100}$$
$$S = 30$$
$$40 - S = 10$$

End-of-Chapter Problems

5–1. $X = 440$

5–3. $Q = 300$

5–5. $Y = 15$

5–7. $Y = 12$

5–9. $P = 25$

5–11. Fred 25; Lee 35

5–13. Josh, 16; Jessica, 240

5–15. 50 shorts; 200 T-shirts

5–17. $B = 70$

5–19. $N = 63$

5–21. $Y = 7$

5–23. $P = \$610.99$

5–25. Pete = 90; Bill = 450

5–27. 48 TP rolls; 240 wet wipes

5–29. $A = 135$

5–31. $M = 60$

5–33. 211 Boston; 253 Colorado Springs

5–35. $W = 129$

5–37. Shift 1: 3,360; shift 2: 2,240

5–39. 22 boxes of hammers 18 boxes of wrenches

5–41. 135,797 lenders

5–42. a. 2.5

   b. 15 miles

   c. 6 hours

5–43. $B = 4$

Chapter 6

You Try It

1. $.92 = 92\%$

   $.009 = .9\%$

   $5.46 = 546\%$

2. $\frac{2}{9} = 22.222\% = 22.22\%$

3. $78\% = .0078$ (2 places to left)
   $96\% = .96$ (2 places to left)
   $246\% = 2.46$ (2 places to left)

   $7\frac{3}{4}\% = 7.75\% = .0775$

   $\frac{3}{4}\% = .75\% = .0075$

   $\frac{1}{2}\% = .50\% = .0050$

4. $\frac{3}{5} = .60 = 60\%$

5. $74\% \rightarrow 74 \times \frac{1}{100} = \frac{74}{100} = \frac{37}{50}$

   $\frac{1}{5}\% \rightarrow \frac{1}{5} \times \frac{1}{100} = \frac{1}{500}$

   $121\% \rightarrow 121 \times \frac{1}{100} = \frac{121}{100} = 1\frac{21}{100}$

   $17\frac{1}{5}\% \rightarrow \frac{86}{5} \times \frac{1}{100} = \frac{86}{500} = \frac{43}{250}$

   $17.75\% \rightarrow 17\frac{3}{4}\% = \frac{71}{4} \times \frac{1}{100} = \frac{71}{400}$

6. Portion ($1,600$) = Base ($2,000$) × Rate (.80)

7. Rate (25%) = $\dfrac{\text{Portion (\$500)}}{\text{Base (\$2,000)}}$

8. Base ($1,000$) = $\dfrac{\text{Portion (\$200)}}{\text{Rate (.20)}}$

9. $\dfrac{\text{Difference in price (\$100)}}{\text{Base (orig. \$500)}} = 20\%$

End-of-Chapter Problems

6–1. 88%

6–3. 40%

6–5. 356.1%

6–7. .04

Appendix A (Continued)

6–9. .643

6–11. 1.19

6–13. 8.3%

6–15. 87.5%

6–17. $\dfrac{1}{25}$

6–19. $\dfrac{19}{60}$

6–21. $\dfrac{27}{400}$

6–23. 10.5

6–25. 102.5

6–27. 156.6

6–29. 114.88

6–31. 16.2

6–33. 141.67

6–35. 10,000

6–37. 17,777.78

6–39. 108.2%

6–41. 110%

6–43. 400%

6–45. 59.40

6–47. 1,100

6–49. 40%

6–51. +20%

6–53. 80%

6–55. $10,000

6–57. $640 per month

6–59. 677.78%

6–61. 6%

6–63. $2,434.50

6–65. 21,068,800 more

6–67. 900

6–69. $742,500

6–71. $220,000

6–73. 33.3%

6–75. 21,068,800

6–77. $39,063.83

6–79. $138.89

6–81. $1,900

6–83. $102.50

6–85. 3.7%

6–87. $2,571

6–89. $41,176

6–91. 40%

6–93. 585,000

6–94. a. 68%

    b. 125%

    c. $749,028

    d. $20

    e. 7 people

6–95. $55,429

Chapter 7

You Try It

1. $700 \times .20 = \$140$

2.
$$\begin{array}{r} 1.00 \\ -\ .20 \\ \hline .80 \end{array} \qquad \$700 \times .80 = \$560$$

3. Seller will pay the freight

4. $\dfrac{\$240}{.40} = \$600$

    $(100\% - 60\%)$

5.
$$\begin{array}{cc} \$200 & \$188 \\ \times\ .06 & \times\ .08 \\ \hline \$12.00 & \$15.04 \end{array}$$

$$\begin{array}{rl} \$188.00 & \qquad\qquad .8648\ \text{NPER} \\ -\ 15.04 & \ .94 \times .92 = \times\ \$200 \\ \hline \$172.96 & \qquad\qquad \$172.96 \end{array}$$

6. $.94 \times .92 \times \$2,000 = \$1,729.60$

7.
$$\begin{array}{l} 1.0000 \\ -\ .8648 \quad (.94 \times .92) \\ \hline \ .1352 \times \$2,000 = \$270.40 \end{array}$$

8.
$$\begin{array}{l} \$2,000 \\ -\quad 80 \quad \text{(Freight and returns)} \\ \hline \$1,920 \times .02 = \$38.40 \end{array}$$

9. April 12, May 2

10.
$$\begin{array}{l} \$700 \\ -100 \\ \hline \$600 \quad \times .98 = \quad \$588 \\ \qquad\qquad\qquad\quad +100 \\ \qquad\qquad\qquad\ \ \hline \qquad\qquad\qquad\quad \$688 \end{array}$$

11. No discount; pay full $700

12. November 10; November 30

13. $\$300/.98 = \$306.12$

End-of-Chapter Problems

7–1. 9504; .0496; $59.52; $1,140.48

7–3. 893079; .106921; $28.76; $240.24

7–5. $369.70; $80.30

7–7. $1,392.59; $457.41

7–9. June 28; July 18

7–11. June 15; July 5

7–13. July 10; July 30

7–15. $138; $6,862

7–17. $2; $198

7–19. $408.16; $291.84

7–21. $219.80 TD, $879.20 NP

7–23. .648; .352; $54.56; $100.44

7–25. $15,668.73 NP; $1,331.27 TD

7–27. $5,100; $5,250

7–29. $5,850

7–31. $1,357.03

7–33. $8,173.20

7–35. $8,333.33; $11,666.67

7–37. $99.99

7–39. $489.90; $711.10

7–41. $4,658.97

7–43. $1,083.46; $116.54

7–45. $5,008.45

7–47. $781.80 paid

7–48. **a.** $1,500

    b. 8.34%

    c. $164.95

    d. $16,330.05

    e. $1,664.95

7–49. $4,794.99

Chapter 8

You Try It

1. $S = C + M$
$S = \$400 + \$200$
$S = \$600$

2. $\dfrac{\$50}{\$200} = 25\%$

$\dfrac{\$50}{.25} = \$200$

3. $S = C + M$
$S = \$8 + .10(\$8)$
$S = \$8 + \$.80$
$S = \$8.80$

4. $S = C + M$
$\$200 = C + .60\,C$
$\dfrac{\$200}{1.60} = \dfrac{1.60C}{1.60}$
$\$125 = C$

5. $M = S - C$
$(\$2,500) = (\$4,500) - (\$2,000)$

6. $\dfrac{\$700}{\$2,800} = 25\%$

$\dfrac{\$700}{.50} = \$1,400$

7. $S = C + M$
$S = \$800 + .40(S)$
$\dfrac{-.40 \qquad\qquad -.40}{}$
$\dfrac{.60S}{.60} = \dfrac{\$800}{.60}$
$S = \$1333.33$

8. $S = C + M$
$\$2,000 = C + .70(\$2,000)$
$\$2,000 = C + \$1,400$
$\dfrac{-1,400}{\$600} = \dfrac{-1,400}{C}$

9. $\dfrac{.47}{1 + .47} = \dfrac{.47}{1.47} = 32\%$ rounded

10. $\begin{array}{l}\$50 \\ \times\,.20 \\ \hline \$10\end{array}$ $\dfrac{\$10}{\$50} = 20\%$

11. $TS = TC + TM$
$TS = \$9 + .30(\$9)$
$TS = \$9 + \$2.7$
$TS = \$11.70$
$\dfrac{\$11.70}{45} = \$.26$

12. $\dfrac{\$70,000}{\$20} = 3,500$ units

End-of-Chapter Problems

8–1. $600; $2,600

8–3. $4,285.71

8–5. $6.90; 45.70%

8–7. $450; $550

8–9. $110.83

8–11. $34.20; 69.8%

8–13. 11%

8–15. $3,830.40; $1,169.60; 23.39%

8–17. $16,250; $4.00

8–19. $166.67

8–21. $14.29

Appendix A (Continued)

8–23. $600; $262.50

8–25. $84

8–27. 42.86%

8–29. $320

8–31. 20,000

8–33. $44; 56%

8–35. $195

8–37. $.59

8–39. $2.31

8–41. 12,000

8–43. $7.65

8–45. $266

8–46. $94.98; $20.36; loss

Cumulative Review 6, 7, 8

1. 650,000

2. $296.35

3. $133

4. $2,562.14

5. $48.75

6. $259.26

7. $1.96; $1.89

Chapter 9

You Try It

1. 38 hrs × $9.25 = $351.50

2. Reg $ Overtime $
(40 × $7) + (3 × $10.50)
$280   + $31.50 = $311.50
           gross pay

3. 2,250 × $.79 = $1,777.50

4. 600 × $.79 = $474
300 × $.88 = +264
$738

5. $175,000 × .07 = $12,286.96

6. $6,000 × .05 = $300
$2,000 × .09 =   180
$4,000 × .12 =   480
$960

7. $600 + ($6,000 × .04)
$600 + 240 = $840

8. Gross $490.00 $ 490.00
Less: FIT $0.70   − 79.80
SS   30.38   483.00
Med.   7.11   7
$451.81 $7.00 × .10 = $0.70

9. Social Security = $142,800 × .062 = $8,853.60

Medicare = $160,000 × .0145 = $2,320

10. $1,400
− 865
$ 535
$38.20 + ($535 × .12) = $102.40

11. FUTA $200 × .006 = $1.20
SUTA $200 × .054 = $10.80

End-of-Chapter Problems

9–1. 37; $331.15

9–3. $12.00; $452

9–5. $1,680

9–7. $60

9–9. $13,000

9–11. $4,500

9–13. $11,900; $6,900; $138; $388

9–15. $49.60; $130.50

9–17. $174.40; $124.00; $29.00; $1,672.60

9–19. $752.60; $85.20

9–21. $1,315.28

9–23. $2,130.72

9–25. $825

9–27. $1,128.75

9–29. $357; $8,853.60

9–31. $1,084.70

9–32. a. $420.19

b. $422.33

c. $2.14

9–33. $1,653.60, $193.13 understated; $52

Chapter 10

You Try It

1. $4,000 × .03 × $\frac{18}{12}$ = $180

2. $3,000 × .04 × $\frac{45}{365}$ = $14.79

Feb 22   53
Jan 8   − 8
45

3. $3,000 × .04 × $\frac{45}{360}$ = $15.00

4. $\$2,000 \times .04 \times \dfrac{90}{360} = \$20$

5. $\dfrac{\$20}{.04 \times \dfrac{90}{360}} = \$2,000$

6. $\dfrac{\$20}{\$2,000 \times \dfrac{90}{360}} = 4\%$

7. $\dfrac{20}{\$2,000 \times .04} = .25 \times 360 = 90 \text{ days}$

8. $\$4,000 \times .04 \times \dfrac{30}{360} = \$13.33$

$\begin{array}{r} \$400.00 \\ -\ \ 13.33 \\ \hline \$386.67 \end{array}$

$\$4,000 - 386.67 = \$3,613.33$

$\$3,613.33 \times .04 \times \dfrac{40}{360} = \$16.06$

$\$300 - \$16.06 = \$283.94$
$\$3,613.33 - \$283.94 = \$3,329.39$

$\$3,329.39 \times .04 \times \dfrac{20}{360} = \$7.40$

$\$3,329.39 + \$7.40 = \$3,336.79$
Total interest $= \$13.33 + \$16.06 + \$7.40 = \$36.79$

End-of-Chapter Problems

10–1. $303.75; $9,303.75

10–3. $1,012.50; $21,012.50

10–5. $28.23; $613.23

10–7. $20.38; $1,020.38

10–9. $73.78; $1,273.78

10–11. $1,904.76

10–13. $4,390.61 balance due

10–15. $618.75; $15,618.75

10–17. $2,377.70; Save $1.08

10–19. 4.7 years

10–21. $21,596.11

10–23. $714.87; $44.87

10–25. 266 days

10–27. $2,608.65

10–29. $18,666.85

10–31. 12.37%

10–33. 15 days

10–35. 5.6%

10–37. 179 days

10–39. **a.** $1,000

       b. 8%

       c. $280; $1,400

10–40. $7.82; $275.33

Chapter 11

You Try It

1. $\$4,000 \times .02 \times \dfrac{30}{360} = \$6.67$

$\begin{array}{r} \$4,000.00 \\ -\ \ \ \ \ \ 6.67 \\ \hline \$3,993.33 \text{ Proceeds} \end{array}$

2. $\$15,000 \times .04 \times \dfrac{40}{360} = \$66.67$

$\begin{array}{r} \$15,000.00 \\ -\ \ \ \ \ 66.67 \end{array}$ $\quad \dfrac{\$66.67}{\$14,933.33 \times \dfrac{40}{360}} = 4.02\%$

3. $\begin{array}{r} \text{Dec } 15 \quad 349 \\ \text{Nov } 5 \quad -309 \\ \hline 40 \text{ days} \end{array}$

$\$2,000 \times .03 \times \dfrac{60}{360} = \$10$

$MV = \$2,010 \qquad \text{(Left to go)}$

$\$2,010 \times .05 \times \dfrac{20}{360} = \$5.58$

$\$2,010 - \$5.58 = \$2,004.42 \text{ Proceeds}$

End-of-Chapter Problems

11–1. $93.33; $5,906.67

11–3. 25 days

11–5. $51,451.39; 57; $733.18; $50,718.21

11–7. 1.003%

11–9. $7,566.67; 6.9%

11–11. $8,937

11–13. $5,309.80

11–15. $5,133.33; 56; $71.87; $5,061.46

11–17. $4,836.44

11–19. ¥20,188

11–21. $13,294.85

11–23. **a.** $90.13

       b. $177.50

       c. 3.64%

       d. 3.61%

11–24. $2,127.66; 9.57%

Chapter 12

You Try It

1. $\begin{array}{rr} \$200 & \$\ \ \ 208 \\ \times\ 1.04 & \times\ \ 1.04 \\ \hline \$208 & \$216.32 \end{array}$

2. $\$4,000 \times 1.4258 \ (3\% \ 12 \text{ periods})$
$= \$5,703.20$

Appendix A (Continued)

3. Table 1.0609 (3% 2 periods)

$$\underrightarrow{\quad}$$
6.09%

$4,000 \times 1.0609 = \$4,243.60$
$ -4,000.00$
$$\overline{ \$\ \ 243.60}$$

$\dfrac{\$243.60}{\$4,000.00} = 6.09\%$

4. Table .7880 (1.5% 16 periods)
$ \times \$6,000$
$$\overline{ \$4,728}$$

End-of-Chapter Problems

12–1. 4; 1%; $598.35; $23.35

12–3. $15,450; $450

12–5. 6.14%

12–7. 16; $1\frac{1}{2}\%$; .7880; $4,728

12–9. 28; $\frac{1}{2}\%$; .8697; $15,306.72

12–11. $17,600.72

12–13. $64,188

12–15. Mystic $4,775, Four Rivers $3,728

12–17. $25,734.40

12–19. $12,698

12–21. $12,900.87

12–23. $51,210

12–25. No. $13,439 (compounding) or $11,161.50 (p. v.)

12–27. $3,739.20

12–29. $27,757.40

12–31. $5,217.68

12–33. $20,016

12–34. $105,878.50

12–35. $689,125; $34,125 Bank B

Chapter 13

You Try It

1. 4.4399 (7% 4 periods)
$\times 6,000$
$$\overline{\$26,639.40}$$

2. $6,000 \times 5.7507 = \$34,504.20$ (7% 5 periods)
$ -6000.00$
$$\overline{\$28,504.20}$$

3. 5.2421 (4% 6 periods)
$\times \$\ 20,000$
$$\overline{ \$104,842}$$

4. .0302 (5% 20 periods)
$\times \$400,000$
$$\overline{\$\ 12,080}$$

End-of-Chapter Problems

13–1. $67,431.50

13–3. $53,135.10

13–5. $3,118.59

13–7. End of first year $2,405.71

13–9. $1,410

13–11. $3,397.20

13–13. $137,286; $1,721,313

13–15. $38,841.30

13–17. $900,655

13–19. $421,885.11

13–21. $13,838.25

13–23. Annuity $12,219.18 or $12,219.93

13–25. $3,625.60

13–27. $111,013.29

13–29. $404,313.97

13–31. $1,043,565.60

13–33. $199.29

13–34. $120,747.09

Cumulative Review 10, 11, 12, 13

1. Annuity $2,058.62 or $2,058.59

2. $3,355.56

3. $116,963.02

4. $3,113.92

5. $5,797.92

6. $18,465.20

7. $17,518.05

8. $55,251

Chapter 14

You Try It

1. $5,400 amount financed
$ - 100$
$$\overline{\$5,300}$$

2. $129.99 \times 60 =$
$\$7,799.40$
$-\ 5,300.00$
$$\overline{\$2,499.40\,\text{FC}}$$

3. $7,799.40 + $100 = $7,899.40

4. $\dfrac{\$2,499.40}{\$5,300.00} \times \$100 = 47.16$ (between 16.25% and 16.50%)

5. $\dfrac{\$2,499.40 + \$5,300}{60} = \$129.99$

$\dfrac{\$5,300}{1,000} = 5.3 \times 24.32 = \$128.9$

(off due to using 16% instead of using between 16.25% and 16.50%)

6. $5,000 × .035 = $175
$275 − $175 = $100
$5,000 − $100 = $4,900

7. 12 days × $200 = $2,400

4 days × $120 [$200 − $80] = $480

14 days × $180 [$120 + $60] = $2,520

Total = $5,400

$5,400/30 = $180 daily balance

Finance charge = $180 × 2.5% = $4.50

End-of-Chapter Problems

14–1. Finance charge $5,240

14–3. Finance charge $1,279.76; 12.75%–13%

14–5. $119.39; $119.37

14–7. $2,741; $41.12

14–9. $1,354.08

14–11. **a.** $4,050

b. $1,656

c. $5,756

d. $40.89, falls between 14.25% and 14.50%

e. $95.10

14–13. $1,245; $18.68; $1,318.68

14–15. $940.36

14–17. $298.12

14–19. 5% APR

14–21. Peg is correct

14–22. 15.48%

Chapter 15

You Try It

1. $\dfrac{\$70,000}{\$1,000} = 70 \times \$4.49045 = \$314.33$

2. 30 years $= \times \$ \quad\begin{array}{r} 360 \text{ payments} \\ 314.33 \end{array}$

$\overline{\$113,158.80} − \$70,000 = \$43,158.80$ interest

3.

Payment	Interest	Principal reduction	Balance
1	$204.17	$110.16	$69,889.84

$\left(\$70,000 \times .035 \times \dfrac{1}{12} = \$204.17\right)$ ($314.33 − $204.17) ($70,000 − $110.16)

Payment	Interest	Principal reduction	Balance
2	$203.85	$110.48	$69,779.36

$\left(\$69,889.84 \times .035 \times \dfrac{1}{12}\right)$ ($314.33 − $203.85) ($69,889.84 − $110.48)

End-of-Chapter Problems

15–1. $628.66

15–3. $989.49

15–5. $88,743.20

15–7. $1,423.43; $179,326.57

15–9. $70,789.60

15–11. $635.92; $116,931.20

15–13. Payment 3, $119,271.85

15–15. $57,873.60

15–17. $87.43

15–19. No. They would spend $57,676.71 more on interest.

15–21. $589.90

15–22. **a.** $92,495.50

b. $1,128.83

c. $213,878.80

Chapter 16

You Try It

1.

$ 400	40%
+ 600	60%
$1,000	100%

2.

	2023	2022	Change	%	
Cash	$8,000	$2,000	$6,000	300%	$\dfrac{\$6,000}{\$2,000}$

3. $400 − $20 − $5 = $375 net sales

4. $50 + $200 − $20 = $230

5. $400 − $250 = $150 gross profit

6. $210 − $180 = $30 net income

7.

2024	2023	2022
1,200	800	1,000
120%	80%	100%
$\left(\dfrac{1,200}{1,000}\right)$	$\dfrac{200}{1,000}$	

8. $\dfrac{\$40,000}{\$160,000} = .25$

Appendix A (Continued)

9. $\dfrac{\$40,000 - \$2,000 - \$3,000}{\$160,000} = \dfrac{\$35,000}{\$160,000} = .22$

10. $\dfrac{\$4,000}{\left(\dfrac{\$60,000}{360}\right)} = 24$ days

11. $\dfrac{\$180,000}{\$70,000} = 257.14$

12. $\dfrac{\$16,000}{-\$110,000} = -14.55\%$

13. $\dfrac{\$60,000}{70,000} = .86$

14. $\dfrac{\$16,000}{\$60,000} = .27$

End-of-Chapter Problems

16–1. Total assets $84,000

16–3. Inventory −16.67%;
mortgage note +13.79%

16–5. Net sales 13.62%;
Net earnings 2020 47.92%

16–7. Depreciation $100; + 16.67%

16–9. 1.43; 1.79

16–11. .20; .23

16–13. .06; .08

16–15. 13.57%

16–17. 87.74%; 34.43%; .13; 55.47%

16–19. 2024 68% sales

16–21. $9,447 net worth

16–23. Net income 615.14% decrease

16–25. $3,470; 6.8%; $431; .8%

16–26. 3.5; 2.3

Chapter 17

You Try It

1. $\dfrac{\$50,000 - \$10,000}{4} = \dfrac{\$40,000}{4} = \$10,000$ per year

2. $\dfrac{\$4,000 - \$500}{700} = \dfrac{\$3,500}{700} = \$5$ depreciation per unit
$150 \times \$5 = \$750$

3.

Year	Cost	Depreciation expense	Book value at end of year
1	$40,000	$20,000	$20,000
		($40,000 × .50)	
2	$20,000	$10,000 =	$10,000
		($20,000 × .50)	

4. .20 × $7,000 = $1,400 depreciation expense

End-of-Chapter Problems

17–1. Book value (end of year) $35,000

17–3. Book value (end of year) $25,000

17–5. Book value (end of year) $15,000

17–7. Book value (end of year) $15,000

17–9. Book value (end of year) $5,400

17–11. $2,240

17–13. $18,000

17–15. $22,560

17–17. $15,000

17–19. $6,000; $18,000

17–21. $6,760 below

17–23. $83,667

17–25. $768.80

17–27. $5,262

17–28. a. $87,750
    b. $11.40
    c. $21,489
    d. 4 years

17–29. $13,320; $1.11

Chapter 18

You Try It

1. $4 \times 9 = 36$
$3 \times 10 = \underline{30}$
       66 total cost
$1 \times \$9 = \$9$
$1 \times \$10 = \dfrac{\$10}{\$19}$
$\$66 - 19 = \$47$ Cost of goods sold

2. $\dfrac{89}{15} = \$5.93$ unit cost
$4 \times \$5.93 = \$23.72$

3. FIFO $\quad 4 \times \$7 = \$28$

4. LIFO $\quad 4 \times \$5 = \$20$

5.

	Cost	Retail
Cost of goods available for sale	$88,000	$117,000
		− 90,000
Net sales		$27,000

Cost ratio: $\dfrac{\$88,000}{\$117,000} = 75\%$
$.75 \times \$27,000 = \$20,250$

6. Cost of goods available for sale $42,000

     Net sales at retail $20,000

                                $\times .25$

     COGS at retail   5,000

     Ending inventory $37,000

7. $\dfrac{\$90,000}{\left(\dfrac{\$40,000 + \$60,000}{2}\right)} = \dfrac{\$90,000}{\$50,000} = 1.8$

8. Total sq. ft. for dept. 10,000

    .40 to Dept A $\$30,000 \times .40 = \$12,000$
    .60 to Dept B $30,000 \times .60 = 18,000$

End-of-Chapter Problems

18–1. $7,180; $22,635

18–3. $543; $932

18–5. $10

18–7. $36

18–9. $72

18–11. $140.80

18–13. $147.75; $345.60

18–15. $188.65; $304.70

18–17. 3.56; 3.25

18–19. .75; $67,500

18–21. $10,550; $24,645

18–23. $45,000

18–25. $55,120

18–27. $38,150

18–29. $3,511.16

18–31. $1,205 EI; $3,950 COGS

18–33. $13,499.50

18–34. $1,900

Chapter 19

You Try It

1. $62.80 − $5.02 = $57.78

                    $\times .06$

                 3.47 sales tax

2. $6,000 + $300 + $600 = $6,900

3. $200,000 × .40 = $80,000 assessed value

4. $\dfrac{\$700,000}{\$8,400,000} = .0833$

5. 1. 8.33%   2. $8.33

    3. $83.3   4. $\dfrac{.0833}{.001} = 83.3 = 83$ mills

6. 1. 9.5% × $40,000 = $3,800

    2. $\dfrac{\$40,000}{\$100} = 400 \times \$9.50 = \$3,800$

    3. $\dfrac{\$40,000}{\$1,000} = 40 \times \$95 = \$3,800$

    4. $\dfrac{\$.0950}{.001} = 95 \times .001 \times \$40,000 = \$3,800$

End-of-Chapter Problems

19–1. $928

19–3. $83,018.87

19–5. $39,000

19–7. $.0233

19–9. 6.99%; $6.99; $69.90; 69.90

19–11. $4,462.50

19–13. $16,985.05

19–15. $112.92

19–17. $634,000

19–19. $6,940

19–21. $64,000

19–23. $4,726.88

19–25. $23,065 more in Minn.

19–27. $3,005

19–29. $3,665

19–30. $979

Chapter 20

You Try It

1. 1. $\dfrac{\$90,000}{\$1,000} = 90 \times \$2.44 = \$219.60$

    2. $\dfrac{\$90,000}{\$1,000} = 90 \times \$11.84 = \$1,065.60$

    3. $\dfrac{\$90,000}{\$1,000} = 90 \times \$15.60 = \$1,404.00$

    4. $\dfrac{\$90,000}{\$1,000} = 90 \times \$27.64 = \$2,487.60$

2. Option 1 : $\dfrac{\$60,000}{\$1,000} = 60 \times \$265 = \$15,900$

    Option 2: 60 × $550 = $33,000

    Option 3: 21 yr 300 days

3. $\dfrac{\$80,000}{\$100} = 800 \times \$.61 = \$488$

    $\dfrac{\$20,000}{\$100} = 200 \times \$.65 = \underline{\$130}$

                    Total $\underline{\$618}$

4. 600 × $.44 = $264

    Refund $600 − $264 = $336

Appendix A (Continued)

5. $\$600 \times \dfrac{1}{3} = \$200$

   $\$600 - \$200 = \$400$

6. $\dfrac{\$40,000}{\$60,000} \times \$9,000 = \$6,000$

7. 10/20/5 $184 ($55 + $129)

Bodily	94
Property	132
Collision	196
Comprehensive	178
Total premium	$784

End-of-Chapter Problems

20–1. $2,762

20–3. $138.75

20–5. $53,000

20–7. 21 years, 300 days

20–9. $518; $182

20–11. $16,500

20–13. $1,067

20–15. $208 vs $240

20–17. $801 No cash value

20–19. $118,750

20–21. $1,100

20–23. $373.67

20–25. $22,900; $10,700

20–27. $24,000; $16,300

20–29. Annual premium between $1,895.50 and $2,274.60

20–31. $40,625

20–32. $7,512.64; $1,942.00; $787.89

20–33. $176.00

Chapter 21

You Try It

1. $\dfrac{\$.88}{\$53.88} = 1.63\%$

2. $\dfrac{\$53.88}{\$3.70} = 14.56 = 15$

3. $30,000 \times \$.80 = 24,000$
   $30,000 \times \$.80 = \underline{24,000}$
   $\overline{48,000}$ to preferred

   $\$300,000$
   $\underline{-\ \ 48,000}$
   $\overline{\$252,000} \div 60,000 = \$4.20$ to common

4. $\$1,022.25 \times 6 = \$6,133.50$

5. $\dfrac{\$40}{1,011.20} = 3.96\%$

6. $\$12.44 + \$.05 = \$12.49$

End-of-Chapter Problems

21–1. $25,305

21–3. 3.3%

21–5. 10

21–7. $41,983.20

21–9. 2021   preferred $8,000

   2022   0

   2023   preferred $127,000

   common $33,000

21–11. $2,280

21–13. $260; $2,725; 9.5%

21–15. $12.04; −$.06; 9.6%

21–17. Gain $6,606.48

21–19. 49, 7.3%

21–21. $4,843.75; $75

21–23. 4.5%

21–25. Stock 6.7%; bond 11.9%

21–27. Yes, $16.02

21–29. $443.80

21–31. 43.2 shares

21–33. $31,025

21–34. a. 1,287 shares

   b. 2,574 shares

   c. 5,147 shares

   d. $26,381.76 for (a); $52,388.43 for (b); $103,756.44 for (c)

21–35. $1,014.33

Chapter 22

You Try It

1. $\dfrac{41 + 29 + 16 + 15 + 18}{5} = 23.8$

2.

Value	Frequency	Product
80	2	160
90	3	270
100	$\dfrac{1}{6}$	$\dfrac{100}{690}$

$\text{Mean} = \dfrac{690}{6} = 115$

3. 4 7 ⑨ 14 16

4. 7

5.

Coffees consumed	Tally	Frequency
0			1
1			1
2				2
3			1
4			1
5			1
6		0
7		0
8			1

6.

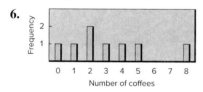

7.

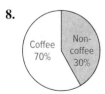

8.

Coffee 70%
Non-coffee 30%

$.70 \times 360° = 252°$
$.30 \times 360° = 108°$

9. $\dfrac{\$12,000}{\$9,000} \times 100 = 133.3$

10. Range = 14 − 2 = 12

11. 1. $\text{Mean} = \dfrac{19}{5} = 3.8$

2. $8 - 3.8 = 4.2$
$1 - 3.8 = -2.8$
$6 - 3.8 = 2.2$
$2 - 3.8 = -1.8$
$2 - 3.8 = -1.8$

3. $(4.2)^2 = 17.64$
$(-2.8)^2 = 7.84$
$(2.2)^2 = 4.84$
$(-1.8)^2 = 3.24$
$(-1.8)^2 = 3.24$

4. 36.8

5. $36.8 \div 4 = 9.2$

6. Standard deviation = 3.03

End-of-Chapter Problems

22–1. 6.25

22–3. $77.23

22–5. 2.7

22–7. 31.5

22–9. 8

22–11. 142.9

22–13. $200–$299.99 卌

22–15. Traditional watch 183.6°

22–17.

22–19. Transportation 126°
Hotel 100.8°
Food 72°
Miscellaneous 61.2°

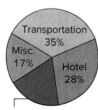

22–21. 250

22–23. **a.** 57,000,000 mean

62,900,000 median

b. AAA = 30.42%

Riser = 22.18%

Casto = 22.07%

Appendix A (Continued)

Balbon = 12.70%

Hunter = 12.63%

   c. 109.51°, 79.85°, 79.45°,
45.72°, 45.47°

22–24. 24.94%; 15.42%; 10.88%; 13.15%; 18.59%; 17.01%

     89.78°, 55.51°, 39.17°, 47.34°, 66.92°, 61.24°

Optional Assignment

1. 98

3. 4.3

5. 16%; 2.5%

7. 68%; 81.5%; 2.5%; 2.5%; 47.5%

9. 5.02

Check Figures (Odds) to Additional Assignments by Learning Unit from Appendix A

LU 1–1

1. a. Eight thousand, eight hundred twenty-one

  d. Fifty-eight thousand, three

3. a. 80; 380; 860; 5,980; 210

  c. 21,000; 1,000; 4,000; 10,000

5. a. Hundreds place

  c. Ten thousands place

  e. Billions place

7. Five hundred sixty-five

9. $375,985

11. Two thousand, nineteen

LU 1–2

1. a. 1,006

  c. 1,319

  d. 179

3. a. Estimated 50; 52

  c. Estimated 10; 9

5. $71,577

7. $19,973

9. 12,797 lbs

11. Estimated $9,400; $9,422

13. $746 discount

LU 1–3

1. a. Estimated 4,000; actual 2,400

  c. Estimated 15,000,000; actual 16,184,184

3. a. Estimated 1,000; actual 963 R5

  c. Estimated 20; actual 25 R8

5. 5,040

7. $78

9. 27

11. $43,200

13. 40 stacks and 23 "extra" bricks

LU 2–1

1. a. Improper

  b. Proper

  c. Improper

  d. Mixed number

  e. Improper

  f. Mixed number

3. a. $\frac{46}{5}$ **c.** $\frac{31}{7}$ **f.** $\frac{53}{3}$

5. a. $6;\frac{6}{7}$ **b.** $15;\frac{2}{5}$ **e.** $12;\frac{8}{11}$

7. $\frac{13}{4}$

9. $\frac{17}{25}$

11. $\frac{60}{100}$

13. $\frac{7}{12}$

LU 2–2

1. a. 32 **b.** 180 **c.** 480 **d.** 252

3. a. $\frac{1}{3}$ **b.** $\frac{2}{3}$ **e.** $6\frac{1}{8}$ **h.** $56\frac{5}{6}$

5. $3\frac{1}{40}$ yards

7. $17\frac{5}{12}$ miles

9. $4\frac{8}{9}$ hours

LU 2–3

1. a. 1 **b.** $\frac{1}{4}$ **g.** 35 **i.** $1\frac{17}{18}$

3. a. $1\frac{1}{4}$ **b.** 3 **g.** 24 **l.** $\frac{4}{7}$

5. $39,000

7. 714

9. $20\frac{2}{3}$ miles

11. $412,000

LU 3–1

1. a. .62 **b.** .6 **c.** .953
  d. .401 **e.** .06

3. a. Hundredths place

  d. Thousandths place

5. a. $\frac{2}{5}$ **b.** $\frac{11}{25}$

  g. $\frac{5}{16}$ **l.** $9\frac{1}{25}$

7. .286

9. $\frac{566}{1,000}$

11. .333

13. .0020507

LU 3–2

1. a. 31.608 **b.** 5.2281 **d.** 3.7736

3. a. .3 **b.** .1 **c.** 1,480.0 **d.** .1

5. a. 6,870 **c.** .0272

  e. 34,700 **i.** 8,329.8

7. $4.53

9. $111.25

11. 15

LU 4–1

1. a. $430.64 **b.** 3 **c.** $867.51

3. a. Neuner Realty Co.

  b. Kevin Jones

  h. $2,756.80

LU 4–2

1. $1,435.42
3. Add $3,000; deduct $22.25
5. $2,989.92
7. $1,315.20

LU 5–1

1. a. $4N = 180$ e. $14 + \dfrac{N}{3} = 18$
   h. $2N + 3N + 8 = 68$

LU 5–2

1. $80
3. $45 telephone; $135 utility
5. 51 tickets—Sherry;
   408 tickets—Linda
7. 12 necklaces ($48);
   36 watches ($252)
9. $157.14

LU 6–1

1. a. 8%   b. 72.9%
   i. 503.8%   l. 80%
3. a. 70%   c. 162.5%
   h. 50%   n. 1.5%
5. a. $\dfrac{1}{4}$ b. .375   c. 50%
   d. $.66\overline{6}$ n. $1\dfrac{1}{8}$
7. 2.9%
9. $\dfrac{39}{100}$
11. $\dfrac{9}{10,000}$

LU 6–2

1. a. $20,000; 30%; $4,000
   c. $7.00; 12%; $.84
3. a. 33.3%   b. 3%   c. 27.5%
5. a. −1,584; −26.6%
   d. −20,000; −16.7%
7. $9,000
9. $3,196
11. 329.5%

LU 7–1

1. a. $120   b. $360   c. $50
   d. $100   e. $380

3. a. $75   b. $21.50; $40.75
5. a. .7125; .2875   b. .7225; .2775
7. $3.51
9. $81.25
11. $315
13. 45%

LU 7–2

1. a. February 18; March 10
   d. May 20; June 9
   e. October 10; October 30
3. a. .97; $1,358   c. .98; $367.99
5. a. $16.79; $835.21
7. $16,170
9. a. $439.29   b. $491.21
11. $209.45
13. a. $765.31   b. $386.99

LU 8–1

1. a. $19.90   b. $2.72
   c. $4.35   d. $90   e. $116.31
3. a. $2; 80%   b. $6.50; 52%
   c. $.28; 28.9%
5. a. $1.52   b. $225
   c. $372.92   d. $625
7. a. $199.60   b. $698.60
9. a. $258.52   b. $90.48
11. a. $212.50   b. $297.50
13. $8.17

LU 8–2

1. a. $10.00   b. $57.50
   c. $34.43   d. $27.33   e. $.15
3. a. $6.94   b. $882.35   c. $30
   d. $171.43   e. $0.36
5. a. 28.57%   b. 33.33%
   d. 53.85%
7. $346.15
9. 39.39%
11. $2.29
13. 63.33%

LU 8–3

1. a. $80; $120   b. $525; $1,574.98
3. a. $410   b. $18.65
5. a. $216; $324; $5.14
   b. $45; $63.90; $1.52
7. 17%
9. $21.15
11. $273.78
13. $.79

LU 8–4

1. a. $6.00   b. $11.11
3. a. 16,667   b. 7,500
5. 5,070
7. 22,222

LU 9–1

1. a. $427.50; 0; $427.50
   b. $360; $40.50; $400.50
3. a. $438.85   b. $615.13
5. a. $5,200   b. $3,960
   c. $3,740   d. $4,750
7. $723.00
9. $3,846.25
11. $2,032.48

LU 9–2

1. a. $2,300; $2,300
3. $0; $2,000
5. $236.30 FIT
7. $143.75
9. $613.51
11. $690.01

LU 10–1

1. a. $240   b. $1,080   c. $1,275
3. a. $131.25   b. $4.08   c. $98.51
5. a. $515.63   b. $6,015.63
7. a. $5,459.66
9. $659.36
11. $360

Appendix A (Continued)

LU 10–2

1. **a.** $4,371.44 **b.** $4,545.45
   c. $3,433.33
3. **a.** 60; .17 **b.** 120; .33
   c. 270; .75 **d.** 145; .40
5. 5%
7. $250
9. $3,000
11. 119 days

LU 10–3

1. **a.** $2,568.75; $1,885.47; $920.04;$0
3. $4,267.59
5. $4,715.30; $115.30

LU 11–1

1. I; B; D; I; D; I; B; D
3. **a.** 2% **c.** 13%
5. $15,963.75
7. $848.75; $8,851.25
9. $14,300
11. $7,855

LU 11–2

1. **a.** $5,075.00
   b. $16,480.80
   c. $994.44
3. **a.** $14.76
   b. $223.25
   c. $3.49
5. $4,031.67
7. $8,262.74
9. $5,088.16
11. $721.45

LU 12–1

1. **a.** $573.25 year 2
   b. $3,115.57 year 4
3. **a.** $25,306; $5,306
   b. $16,084; $6,084

5. $7,430.50
7. $8,881.20
9. $2,129.40
11. $3,207.09; $207.09
13. $3,000; $3,469; $3,498

LU 12–2

1. **a.** .9804 **b.** .3936
   c. .5513
3. **a.** $1,575.50; $924.50
   b. $2,547.02; $2,052.98
5. $14,509.50
7. $13,356.98
9. $16,826.40
11. $652.32
13. $18,014.22

LU 13–1

1. **a.** $1,000; $2,080; $3,246.40
3. **a.** $6,888.60 **b.** $6,273.36
5. $325,525
7. $13,412
9. $30,200.85
11. $33,650.94

LU 13–2

1. **a.** $2,638.65
   b. $6,375.24; $7,217.10
3. $2,715.54
5. $24,251.85
7. $47,608
9. $456,425
11. Accept Jason $265,010

LU 13–3

1. **a.** $4,087.50 **b.** $21,607
   c. $1,395 **d.** $201.45
   e. $842.24
3. $16,200

5. $24,030
7. $16,345
9. $8,742

LU 14–1

1. **a.** $1,200; $192
   b. $9,000; $1,200
3. **a.** 14.75% **b.** 10% **c.** 11.25%
5. **a.** $3,528 **b.** $696 **c.** $4,616
7. **a.** $22,500 **b.** $4,932 **c.** $29,932
9. **a.** $20,576 **b.** 12.75%

LU 14–2

1. **a.** $465; $8,535
   b. $915.62; $4,709.38
3. **a.** $332.03 **b.** $584.83
   c. $384.28
5. Final payment $784.39
7. $51.34
9. $35
11. $922.48
13. 7.50% to 7.75%

LU 15–1

1. **a.** $1,095.31 **b.** $965.24
   c. $3,150.22 **d.** $2,694.47
3. **a.** $92.86 6.7% **b.** $151.86 7.8%
5. $733.16
7. **a.** $1,560.70 **b.** $1,443.04
9. **a.** $117.68 **b.** $51,284.40
11. $294,589

LU 15–2

1. **a.** $1,691.03; $513.54; $1,177.49
3. #4 balance outstanding $190,128.26
5. $112,897.60
7. **a.** $696.04 **b.** $81,812.80
9. $44,271.43
11. $30,022.20

LU 16–1

1. Total assets $224,725
3. Merch. inventory 13.90%; 15.12%

LU 16–2

1. Net income $57,765
3. Purchases 73.59%; 71.43%

LU 16–3

1. Sales 2023, 93.5%; 2022, 93.2%
3. .22
5. 59.29%
7. .83
9. COGS 119.33%; 111.76%; 105.04%
11. .90
13. 5.51%
15. 11.01%

LU 17–1

1. a. 4% b. 25%
   c. 10% d. 20%
3. a. $2,033; $4,667
   b. $1,850; $9,750
5. $8,625 depreciation per year
7. $2,800 depreciation per year
9. $95
11. a. $12,000 b. $6,000
    c. $18,000 d. $45,000

LU 17–2

1. a. $.300 b. $.192 c. $.176
3. a. $.300, $2,600
   b. $.192, $300,824
5. $5,300 book value end of year 5
7. a. $.155 b. $20,001.61

LU 17–3

1. a. 8% b. 20% c. 25%
3. a. $4,467; $2,233
   b. $3,867; $7,733
5. $121, year 6
7. a. 28.57% b. $248 c. $619

9. a. 16.67% b. $2,500
   c. $10,814 d. $2,907

LU 17–4

1. a. 33%; $825; $1,675
3. Depreciation year 8, $346
5. $125
7. a. $15,000 b. $39,000
   c. $21,600 d. 2001
9. $68,440

LU 18–1

1. a. $5,120; $3,020
   b. $323,246; $273,546
3. $35,903; $165,262
5. $10,510.20; $16,345
7. $37.62; $639.54
9. $628.40
11. $3,069; $952; $2,117

LU 18–2

1. a. $85,700; $143,500; .597;
   $64,500; $38,507
3. $85,000
5. $342,000; $242,500; 5.85; 6.29
7. $60,000; $100,000; $40,000
9. $70,150
11. $5,970
13. 3.24; 3.05
15. $32,340; $35,280;
    $49,980; $29,400

LU 19–1

1. a. $26.80; $562.80
   b. $718.80; $12,698.80
3. a. $20.75; $43.89; $463.64
5. Total is (a) $1,023; (b) $58.55
7. $5.23; $115.23
9. $2,623.93
11. $26.20
13. $685.50

LU 19–2

1. a. $68,250 b. $775,450
3. a. $7.45; $74.50; 74.50
5. $9.10
7. $8,368.94
9. $42,112
11. $32,547.50

LU 20–1

1. a. $9.27; 25; $231.75
3. a. $93.00; $387.50; $535.00; $916.50
5. $1,242.90
7. $14,265
9. $47.50 more
11. $68,750

LU 20–2

1. a. $488 b. $2,912
3. a. $68,000; $60,000
   b. $41,600; $45,000
5. $1,463
7. $117,187.50
9. $336,000
11. a. $131,250 b. $147,000

LU 20–3

1. a. $98; $160; $258
3. a. $312 b. $233
   c. $181 d. $59; $20
5. a. $647 b. $706
7. $601
9. $781
11. $10,000; $8,000
13. $60,000; $20,000
15. $19.50; $110.50

LU 21–1

1. a. $43.88 f. 49
3. $27.06
5. $1,358.52 gain
7. $18,825.15
9. $7.70

Appendix A (Continued)

LU 21–2

1. a. IBM   b. $10\frac{1}{4}$ c. 2032
   d. $102.50   e. 102.375
3. a. $1,025   b. $1,023.75
5. a. $3,075
   b. $307.50
7. a. $30 discount
   b. $16.25 premium
   c. $42.50 premium
9. a. $625   b. $375 discount
   c. $105   d. 16.8%
11. 7.8%; 7.2%
13. 8.98%

LU 21–3

1. $11.90
3. $15.20
5. +$.14
7. 7.6%
9. $1.45; $18.45
11. $.56; $14.66
13. $1,573.50
15. $123 loss
17. a. 2032; 2034
    b. 9.3% Comp USA; 6.9% GMA
    c. $1,023.75 Comp USA
       $1,016.25 GMA
    d. Both at premium
    e. $1,025 Comp USA;
       $1,028.75 GMA

LU 22–1

1. a. 20.4   b. 83.75   c. 10.07
3. a. 59.5
   b. 50
5. a. 63.7; 62; 62
7. $1,500,388.50
9. $10.75
11. $9.98

LU 22–2

1. 18: JHT II 7
3. 25–30: JHT III 8
5. 7.2°
7. 145–154: IIII 4
9. 98.4°; 9.9°; 70.5°; 169.2°; 11.9°

Appendix A (Continued)

Appendix B

Metric System

John Sullivan: Angie, I drove into the gas station last night to fill the tank up. Did I get upset! The pumps were not in gallons but in liters. This country (U.S.) going to metric is sure making it confusing.

Angie Smith: Don't get upset. Let me first explain the key units of measure in metric, and then I'll show you a convenient table I keep in my purse to convert metric to U.S. (also called customary system), and U.S. to metric. Let's go on.

The metric system is really a decimal system in which each unit of measure is exactly 10 times as large as the previous unit. In a moment, we will see how this aids in conversions. First, look at the middle column (Units) of this to see the basic units of measure:

U.S.	Thousands	Hundreds	Tens	Units	Tenths	Hundredths	Thousandths
Metric	Kilo-	Hecto-	Deka-	Gram	Deci-	Centi-	Milli-
	1,000	100	10	Meter	.1	.01	.001
				Liter			
				1			

- Weight: Gram (think of it as $\frac{1}{30}$ of an ounce).
- Length: Meter (think of it for now as a little more than a yard).
- Volume: Liter (a little more than a quart).

To aid you in looking at this, think of a decimeter, a centimeter, or a millimeter as being "shorter" (smaller) than a meter, whereas a dekameter, hectometer, and kilometer are "larger" than a meter. For example:

1 centimeter $= \frac{1}{100}$ of a meter; or 100 centimeters equals 1 meter.

1 millimeter $= \frac{1}{1,000}$ meter; or 1,000 millimeters equals 1 meter.

1 hectometer $= 100$ meters.

1 kilometer $= 1,000$ meters.

Remember we could have used the same setup for grams or liters. Note the summary here.

Length	Volume	Mass
1 meter:	1 liter:	1 gram:
= 10 decimeters	= 10 deciliters	= 10 decigrams
= 100 centimeters	= 100 centiliters	= 100 centigrams
= 1,000 millimeters	= 1,000 milliliters	= 1,000 milligrams
= .1 dekameter	= .1 dekaliter	= .1 dekagram
= .01 hectometer	= .01 hectoliter	= .01 hectogram
= .001 kilometer	= .001 kiloliter	= .001 kilogram

Practice these conversions and check solutions.

Appendix B (*Continued*)

Practice Quiz

Convert the following:

1. 7.2 meters to centimeters

2. .89 meter to millimeters

3. 64 centimeters to meters

4. 350 grams to kilograms

5. 7.4 liters to centiliters

6. 2,500 milligrams to grams

✓ Solutions

1. 7.2 meters = 7.2 × 100 = 720 centimeters (remember, 1 meter = 100 centimeters)

2. .89 meter = .89 × 1,000 = 890 millimeters (remember, 1 meter = 1,000 millimeters)

3. 64 centimeters = 64/100 = .64 meters (remember, 1 meter = 100 centimeters)

4. 350 grams = $\dfrac{350}{1,000}$ = .35 kilogram (remember 1 kilogram = 1,000 grams)

5. 7.4 liters = 7.4 × 100 = 740 centiliters (remember, 1 liter = 100 centiliters)

6. 2,500 milligrams = $\dfrac{2,500}{1,000}$ = 2.5 grams (remember, 1 gram = 1,000 milligrams)

Angie: Look at the table of conversions and I'll show you how easy it is. Note how we can convert liters to gallons. Using the conversion from metric to U.S. (liters to gallons), we see that you multiply numbers of liters by .26, so for 37.95 liters we get 37.95 × .26 = 9.84 gallons.

Common conversion factors for U.S./metric					
A. To convert from U.S. to	**Metric**	**Multiply by**	**B. To convert from metric to**	**U.S.**	**Multiply by**
Length:			*Length:*		
Inches (in)	Meters (m)	.025	Meters (m)	Inches (in)	39.37
Feet (ft)	Meters (m)	.31	Meters (m)	Feet (ft)	3.28
Yards (yd)	Meters (m)	.91	Meters (m)	Yards (yd)	1.1
Miles	Kilometers (km)	1.6	Kilometers (km)	Miles	.62
Weight:			*Weight:*		
Ounces (oz)	Grams (g)	28	Grams (g)	Ounces (oz)	.035
Pounds (lb)	Grams (g)	454	Grams (g)	Pounds (lb)	.0022
Pounds (lb)	Kilograms (kg)	.45	Kilograms (kg)	Pounds (lb)	2.2
Volume or capacity:			*Volume or capacity:*		
Pints	Liters (L)	.47	Liters (L)	Pints	2.1
Quarts	Liters (L)	.95	Liters (L)	Quarts	1.06
Gallons (gal)	Liters (L)	3.8	Liters (L)	Gallons	.26

John: How would I convert 6 miles to kilometers?

Angie: Take the number of miles times 1.6; thus 6 miles × 1.6 = 9.6 kilometers.

John: If I weigh 120 pounds, what is my weight in kilograms?

Angie: 120 times .45 (use the conversion table) equals 54 kilograms.

John: OK. Last night, when I bought 16.6 liters of gas, I really bought 4.3 gallons (16.6 liters times .26).

Practice Quiz

Convert the following:

1. 10 meters to yards
2. 110 quarts to liters
3. 78 kilometers to miles
4. 52 yards to meters
5. 82 meters to inches
6. 292 miles to kilometers

✓ Solutions

1. 10 meters × 1.1 = 11 yards
2. 110 quarts × .95 = 104.5 liters
3. 78 kilometers × .62 = 48.36 miles
4. 52 yards × .91 = 47.32 meters
5. 82 meters = 39.37 = 3,228.34 inches
6. 292 miles × 1.6 = 467.20 kilometers

Appendix B: Problems

Drill Problems

Convert:

1. 65 centimeters to meters
2. 7.85 meters to centimeters
3. 44 centiliters to liters
4. 1,500 grams to kilograms
5. 842 millimeters to meters
6. 9.4 kilograms to grams
7. .854 kilogram to grams
8. 5.9 meters to millimeters
9. 8.91 kilograms to grams
10. 2.3 meters to millimeters

Appendix B (Continued)

Convert, rounding to the nearest tenth:

11. 50.9 kilograms to pounds

12. 8.9 pounds to grams

13. 395 kilometers to miles

14. 33 yards to meters

15. 13.9 pounds to grams

16. 594 miles to kilometers

17. 4.9 feet to meters

18. 9.9 feet to meters

19. 100 yards to meters

20. 40.9 kilograms to pounds

21. 895 miles to kilometers

22. 1,000 grams to pounds

23. 79.1 meters to yards

24. 12 liters to quarts

25. 2.92 meters to feet

26. 5 liters to gallons

27. 8.7 meters to feet

28. 8 gallons to liters

29. 1,600 grams to pounds

30. 310 meters to yards

Word Problems

31. A metric ton is 39.4 bushels of corn. China bought 450,000 metric tons of U.S. corn, valued at $105 million, for delivery after September 30. Convert the number of bushels purchased from metric tons to bushels of corn.

Glossary/Index

Note: Page numbers followed by n indicate material found in footnotes.

Specific identification method *This method calculates the cost of ending inventory by identifying each item remaining to invoice price,* 542–547.

Standard deviation *Measures the spread of data around the mean,* 664–665.

State income tax (SIT) *Taxation rate imposed by individual states. State rates vary. Some states do not have a state income tax,* 299.

Statement of cash flows, 479n

Statements. *see* **Bank statement**

State Unemployment Tax Act (SUTA) *Tax paid by employer. Rate varies depending on amount of unemployment the company experiences,* 300.

Statistician *A person who is skilled at compiling statistics,* 592.

Statistics
  frequency distribution, 655.
  graphs, 655–663.
  index numbers, 661–663.
  mean, 650–651.
  measures of dispersion, 664–667.
  median, 651–652.
  mode, 653.
  normal distribution, 665–667.

Step approach to finding greatest common divisor, 43.

Stinson, Rivan, 338.

Stockbrokers *People who with their representatives do the trading on the floor of the stock exchange,* 624, 626.

Stock certificate *Evidence of ownership in a corporation,* 624.

Stockholder *One who owns stock in a company,* 481, 624.

Stockholders' equity *Assets less liabilities,* 481, 483.
  return on, 495.

Stocks *Ownership shares in the company sold to buyers, who receive stock certificates,* 624.
  common, 624, 625.
  dividends, 624, 625.
  preferred, 624, 625.
  quotations, 624–625.
  return on investment, 626.
  terminology, 624.
  trading, 624, 626.

Stock yield *Dividend per share divided by the closing price per share,* 625.

Straight commission *Wages calculated as a percent of the value of goods sold,* 293.

Straight-life insurance *Protection (full value of policy) results from continual payment of premiums by insured. Until death or retirement, nonforfeiture values exist for straight life,* 594.

Straight-line method *Method of depreciation that spreads an equal amount of depreciation each year over the life of the assets,* 517–518.

Straight-line rate (rate of depreciation) *One divided by number of years of expected life,* 522.

Stress, 417.

Subprime loan *A loan with a rate higher than prime due to uncertainty of payment,* 456.

Subtraction
  of decimals, 86.
  of fractions, 50–51.
  of mixed numbers, 50–51.
  for solving equations, 145, 146.
  of whole numbers, 11.

Subtrahend *In a subtraction problem, smaller number that is being subtracted from another,* 11. *Example: 30 in* 150–30 = 120.

Sum *Total in the adding process,* 10.

SUTA. *see* **State Unemployment Tax Act**

Taxes, 312, 588.
  depreciation and, 525–527.
  excise, 576.
  FICA, 297.
  incentives, on electric vehicles, 538.
  income, 298, 573.
  property, 578–580.
  rates, 589.
  sales, 574–575.
  state, 299–300.
  unemployment, 299–300.

Tax Reform Act of 1986, 525.

Technology
  apps, 111, 117, 121.
  in banks, 111, 117, 121–123.
  product obsolescence, 517.

Term
  of promissory notes, 343.

Term insurance *Inexpensive life insurance that provides protection for a specific period of time. No nonforfeiture values exist for term,* 592.

Term of the annuity *Time period from the first to last payment of a stream of payments,* 392.

Term policy *Period of time that the policy is in effect,* 592, 592n

Terms of the sale *Criteria on invoice showing when cash discounts are available, such as rate and time period,* 219.

Tesla Inc., 623, 629.

Time *Expressed as years or fractional years, used to calculate simple interest,* 317, 321.

T-Mobile US Inc., 3.

Tootsie Roll Industries, 7.

Total assets *Total of current assets and plant and equipment,* 482.

Total current assets *Total of all assets that the company will consume or convert to cash within 1 year,* 482.

Total current liabilities *Total obligations that the company must pay within 1 year,* 482.

Total debt to total assets ratio *Amount of debt as a percent of total assets,* 495.

Total liabilities and stockholders' equity *Total current liabilities, long-term liabilities, stock, and retained earnings. This total represents all the claims on assets-prior and present claims of creditors, owners' residual claims, and any other claims,* 483.

Total liabilities *Total of current and long-term liabilities,* 482.

Total operating expenses *Total of all the individual expenses,* 489.

Total plant and equipment *Total of building and land, including machinery and equipment,* 482.

Total stockholders' equity *Total of stock plus retained earnings,* 483.

Trade discount amount *List price less net price,* 209.

Trade discount rate *Trade discount amount given in percent,* 210.
  complement of, 212–213.

Trade discount *Reduction off original selling price (list price) not related to early payment,* 209.
  chain, 214–216.
  complement method, 212–213.
  discount sheets, 210.
  formula, 210.
  net price equivalent rate, 214.
  single, 212.
  single equivalent discount rate, 215.
  word problems, 212–214, 226–227.

Trade-in value *Estimated value of a plant asset after depreciation is taken (or end of useful life),* 516.

Traditional IRAs, 387.

TransUnion, 360.

Treasury bill *Loan to the federal government for 91 days (13 weeks), 182 days (26 weeks), or 1 year,* 344.

Trend analysis *Analyzing each number as a percentage of a base year,* 493.

Truth in Lending Act *Federal law that requires sellers to inform buyers, in writing, of (1) the finance charge and (2) the annual percentage rate. The law doesn't dictate what can be charged,* 430, 431.

Truth in Savings law, 369.

Turnover. *see* **Asset turnover; Inventory turnover**

20-payment life *Provides permanent protection and cash value, but insured pays premiums for first 20 years,* 594–596.

20-year endowment *Most expensive life insurance policy. It is a combination of term insurance and cash value,* 594, 595.

Unemployment tax *Tax paid by the employer that is used to aid unemployed persons,* 299–300.

United Parcel Service (UPS), 208, 209.

United Van Lines, 589.

Units-of-production method *Depreciation method that estimates amount of depreciation based on usage,* 520–521.

Universal life *Whole life insurance plan with flexible premium and death benefits. This life plan has limited guarantees,* 595–596.

Unknown *The variable we are solving for,* 143.
  solving equations for, 142–147.
  solving word problems for, 149–152.

Unlike fractions *Proper fractions with different denominators,* 46, 47, 50.